Innovative Process Development in Metallurgical Industry

Vaikuntam Iyer Lakshmanan
Raja Roy • V. Ramachandran
Editors

Innovative Process Development in Metallurgical Industry

Concept to Commission

Springer

Editors
Vaikuntam Iyer Lakshmanan
Process Research Ortech, Inc.
Mississauga, ON, Canada

Raja Roy
Process Research Ortech, Inc.
Mississauga, ON, Canada

V. Ramachandran
Consulting Engineer
Scottsdale, AZ, USA

ISBN 978-3-319-21598-3 ISBN 978-3-319-21599-0 (eBook)
DOI 10.1007/978-3-319-21599-0

Library of Congress Control Number: 2015947279

Springer Cham Heidelberg New York Dordrecht London

Printed on acid-free paper

Springer International Publishing AG Switzerland is part of Springer Science+Business Media (www.springer.com)

This book is dedicated to
Sarada, Manju and Vasanti
for their
infinite patience and
selfless lifelong support.

Foreword

When I think about innovation in mining today, a recent passage in the *MIT Technology Review* comes to mind: "Innovative companies evolve; those that do not, wither."

The mining industry should pay heed. Take the gold mining sector. The 12-year run-up in the gold price led many companies to pursue production at all costs; the good times, after all, seemed like they would never end. But end they did, and gold miners, if you will pardon the pun, have been digging themselves out ever since. Many base metal producers face similar headwinds. In response, companies, quite rightly, are becoming far more disciplined about capital allocation. Questionable projects are being delayed or shelved and operational excellence and efficiency is paramount.

But one often overlooked consequence of the unrelenting focus on growth is that technology has not kept pace with operational needs. Innovation, simply put, has been allowed to wither. With this backdrop in mind, I could not have been more thrilled when Lucky Lakshmanan informed me recently about the publication of this book, and honoured that he asked me to write the foreword. It could not have come at a better time. The scientists, metallurgists and mining engineers that produced the works that follow are among the finest minds working in mining today. They see the industry's current predicament as an opportunity—an opportunity to address mining's challenges through innovation.

Those challenges are formidable: high operating costs, low grade, complex ore bodies in environmentally sensitive locations are just some of the obstacles that mining faces. Barrick Gold, the company that I work for, is by no means immune from these challenges, but its collaborative approach to innovation is generating practical solutions to the complex problems that the company and many of its peers face today.

One great example of this is the TCM (total carbonaceous matter), project at Barrick's Goldstrike mine in Nevada. This large project involved the construction of a new leaching circuit that uses calcium thiosulfate instead of cyanide to recover gold from a complex, double-refractory ore. The new circuit began operating in late 2014 and is the first commercial use of thiosulfate for gold processing in the world. It preserved jobs and key infrastructure at Goldstrike, and accelerated the processing of millions of ounces of gold.

But most importantly, the TCM project underscores the kind of technological advancements that are possible in mining when companies take a collaborative approach to innovation. When mining experts work hand in hand with frontline operators to tackle complex operational problems, the odds of success are greatly enhanced and the step changes needed to sustain our industry become attainable. This may seem like the most natural and obvious thing in the world, but I cannot tell you how many times in my career that I have seen a silo mentality, or lack of trust, or simple bureaucratic inefficiency stifle innovation.

While I am proud of Barrick's partnership approach to innovation, there are numerous ways to develop collaborative processes to facilitate innovation. This book will serve as a roadmap. It is not just important reading for mining companies, along with their suppliers and partners, it is must reading. The mining industry cannot afford to wither any longer.

Toronto, ON, Canada Jim Gowans

Preface

This book can truly be considered as a labour of love. It is a result of an undying passion for the betterment of society through research and development in the field of Metallurgy for several decades. We live in a fast changing world with innovative products hitting the market everyday. Behind the glamour and glitz lies the dedication of a number of researchers who toil day and night to develop new products and processes that make it all possible. Over the years, the editors have felt the need for a book that formalizes the steps to be taken to develop an innovative product or process starting from the very first innovative idea. We feel that unlike invention, innovation is a journey and not an incident. It is a continuous process, not a discrete event. When we were contacted by Springer to write a book, we suggested this topic to them. After receiving the go ahead from the publisher, we conceptualized the framework of the book and contacted innovators from the field of Metallurgy, not alone from university, but also from industry with hands-on experience and proven track record in taking innovation to commercialization to contribute to this book. The contributed articles were then sent to independent reviewers for peer review process. This book is targeted not only to researchers in the field of metallurgy, but also to senior managers, CEOs and government officials, who can help in creating an atmosphere to bring innovation to commercialization, thereby generating employment and creating wealth in a competitive environment in a sustainable fashion.

We would like to express our sincere gratitude to all the contributors to this book, which include Dr. Barun Gorain, Dr. Nathan Stubina, Professor Alexander McLean, Dr. T. Emi, Dr. Md Abdul Halim, Dr. Shiv Vijayan, Dr. Corale Brierley, Dr. Peter Kondos, Mr. Alan Taylor, Dr. Ananth Seshan, Mr. Mark Vancas, Dr. Krishna Parameswaran, Dr. Dave King, Mr. Michael Dehn, Dr. Donald R Fosnacht, Dr. Iwao Iwasaki, Dr. Richard F. Kiesel, Dr. David J. Englund, Dr. Rodney L. Bleifuss, Dr. M. E. Mlinar, Dr. David W. Hendrickson and Mr. Tim Robinson. We are indebted to all the reviewers including Dr. Michael King, Dr. Jacques McMullen, Dr. Phil Mackey, Prof. Ravi Ravindran, Mr. Tim Robinson, Dr. David Dew, Prof. Mansoor Barati, Prof. Mike Moats, Mr. Warren Galloway, Mr. Peter Mathews, Dr. Jim Lommen, Dr. Bruce Conard, Mr. Peter Dahlberg, Mr. Lance Engelbrecht, Ms. Luisa Marino, Prof. Vladimiros G. Papangelakis, Mr. Harold R. Kokal, Dr. Jim Brierley and Dr. Funsho Ojebuoboh for their review and comments and meeting the tight schedule despite their other commitments. We deeply appreciate the writing of the Foreword to this book by Jim Gowans,

Co-President, Barrick Gold Corporation. Thanks are due to Ms. Lorraine Sequeira for administrative support. Special gratitude is expressed to Ms. Ania Levinson and Ms. Brinda Megasyamalan from our publisher Springer for their constant support and encouragement.

Mississauga, ON, Canada — Vaikuntam Iyer Lakshmanan
Mississauga, ON, Canada — Raja Roy
Scottsdale, AZ, USA — V. Ramachandran

Contents

About the Editors

V.I. Lakshmanan, Ph.D. is an internationally renowned teacher, scientist and innovator in the areas of sustainable development and clean technologies. He has more than 40 years of hands-on experience in technology commercialization and skills development initiatives with both private and public sectors including the United Nations. He has successfully guided process technologies from concept to commercialization for resource, energy and chemical industries. Born and educated in India, Dr. Lakshmanan moved to Canada in 1974 after serving as a Lecturer in the University of Birmingham, UK. His corporate experience includes roles in Noranda, Eldorado Nuclear, ORTECH Corporation and Process Research ORTECH, the last of which he founded and guided to become a global leader in sustainable process technologies development through its laboratory and pilot plant.

Dr. Lakshmanan is a Fellow of the Canadian Academy of Engineering and an Adjunct Professor at the Department of Materials Science and Engineering with the University of Toronto. He is an active mentor of new entrepreneurs and Chairman of the Canada India Foundation. Dr. Lakshmanan received the prestigious Queen Elizabeth II Diamond Jubilee Medal in 2013 for his services to civil society.

V. Ram Ramachandran obtained his Ph.D. in Metallurgical Engineering from Colorado School of Mines. Ram has over 39 years of Research and Development experience in non-ferrous metal industry with emphasis on process development and process improvements including water conservation and treatment. He is an author/co-author of 30 technical papers and Joint editor of Proceedings of technical symposia by TMS. Ram is Co-author of the book "Extractive Metallurgy of Nickel, Cobalt and Platinum Group Metals" published by Elsevier in 2011. He received Milton. E. Wadsworth Hydrometallurgy Award from SME in 2001 and Distinguished Services Award from TMS in 2008.

For the last 15 years, Ram has been working as a Consulting Engineer for the non-ferrous metallurgical industry.

Raja Ram Mohan Roy, Ph.D. obtained his undergraduate degree in Metallurgical Engineering from Indian Institute of Technology, Kanpur and Ph.D. in Materials Science and Engineering from the Ohio State University. Raja has conducted research and development work for a number of years at the University of Toronto and Process Research Ortech, Inc. Raja has over

20 years of experience in extractive metallurgy and materials science with emphasis on process flow sheet development. He has supervised and managed a large number of projects involving bench scale and pilot scale test work. He received Light Metals Recycling Award from TMS in 1998. He has co-authored 40 technical papers and was co-editor of Chloride 2011 symposium organized by TMS.

About the Authors

Corale L. Brierley Dr. Corale L. Brierley, Principal, Brierley Consultancy LLC, provides technical and business consultation to the mining and chemical industries and government agencies, offering clients expertise and experience in minerals bioleaching, management of metal-bearing wastes and business development related to these areas. She has over 80 technical publications and 5 patents and is internationally recognized from a long career, comprising 23 years as an international consultant; 2 years with Newmont Mining Corporation; 8 years of managing Advanced Minerals Technology, Inc., a metal's biotechnology company; and 10 years of applied R&D at New Mexico Institute of Mining and Technology. She is a member of the Society for Mining, Metallurgy and Exploration and the Mining Foundation of the Southwest and has received several awards from these organizations. Dr. Brierley is an elected member and Vice President of the U.S. National Academy of Engineering, which honors those who have demonstrated unusual accomplishment in the pioneering of new and developing fields of technology.

Michael Dehn Mr. Dehn's principal occupation is a Partner of Avanti Management and Consulting Limited. Mr. Dehn is also the President, CEO, and/or Director of several small publically listed junior mining companies. Mr. Dehn is a graduate of the University of Waterloo with a B.Sc. in Earth Science. Mr. Dehn started his career in the mining industry with Goldcorp, Inc. spending 12 years based in the Toronto Head Office, managing many property acquisitions and exploration programs for Goldcorp. Mr. Dehn joined Goldcorp 6 months before the High Grade Zone Discovery and was one of the managers of the Goldcorp Challenge that introduced open-source exploration targeting to the world.

Toshihiko Emi Dr. Emi holds Honorary Professorships at the University of Science and Technology, Beijing, and at North Eastern University, China. He is Honorary Director of the Research Institute for Iron and Steel, Sha Steel, China. He formerly served as Professor at Tohoku University in Sendai, Japan. He served as Research Director and was a Board member for many years at Kawasaki Steel (now JFE Steel) and concurrently President, Rheotechnology Co., Japan. Dr. Emi is the recipient of a number of international honours and awards including the Nishiyama Medal from the Iron and Steel Institute of Japan, Howe Memorial Lecturer and Distinguished Member of the Iron and Steel Society of AIME, now Association of Iron and Steel

Technology, USA, and the Brinell Medal from the Royal Academy of Engineering in Sweden. He has published over 250 papers in archival journals and conference proceedings together with book, book chapters, and reviews.

Donald R. Fosnacht Dr. Donald R. Fosnacht, is a practicing metallurgist, who currently leads the Center for Applied Research and Technology Development for the Natural Resources Research Institute, University of MN Duluth as Departmental Director. He had been an executive with the former Inland Steel Company which is now part of Arcelor Mittal, Inc. He also was President of Steel Profitability Consulting. He has vast experience in steel manufacturing from iron ore development to the design and manufacturing of advanced steel products. He currently is working on strategies for advanced technologies for iron ore conversion, renewal energy development, recovery of TiO_2 from ilmenite ores, and new methods for processing non-ferrous metals.

B.K. Gorain Dr. Barun Gorain has more than 25 years experience in mineral processing plant operations, capital projects, and technology management with a focus on improving operations profitability and project economics. His industry experience includes working for Barrick Gold (presently in their Corporate office in Toronto since 2005), Teck (Cominco), Mt. Isa Mines (now Glencore), Coal India and Hindustan Zinc (now Vedanta). Dr. Gorain has a Ph.D. in Metallurgical Engineering from JKMRC (University of Queensland) in Australia and a B.Tech. degree in Mineral Engineering from Indian School of Mines, Dhanbad.

M.A. Halim Dr. M.A. Halim is a Senior Process Metallurgist at Process Research ORTECH, Inc., Mississauga, Ontario, Canada. He received his Ph.D. in Environmental Engineering from Kyushu University, Japan in 2006. Dr. Halim has over 10 years of experience in hydrometallurgical process development, minerals processing, and wastewater treatment. He has written more than 50 scientific papers, several patents, and a chapter on Titanium Dioxide: Production, Properties, and Applications. Dr. Halim received several awards including JSPS Fellowship, Research Fellowship from National University of Singapore, and IRD-NSERC Fellowship of Canada.

David King Dr. David King has worked for over 45 years for a major engineering/procurement/construction management company and its predecessors. During this time, he held a number of senior technical positions. His career included working on studies of all levels, basic and detailed engineering, and startup/commissioning, and he also managed numerous studies and small projects. David has lived and worked in many places around the world. Most of his career has been spent working in the non-ferrous metallurgical industry, specializing in the field of hydrometallurgy, on projects involving copper, nickel, uranium, gold, zinc, and rare earths among other metals and minerals, quite a few of which used innovative technology.

Peter D. Kondos Peter Kondos was born in Athens, Greece, where he completed his Diploma of Engineering at the National Technical University of

Athens, followed by M. Eng. on Computers in Mining and a Ph.D. in Hydrometallurgy, both at McGill University.

Peter has worked at:

- The Mineral Sciences Laboratories at CANMET in Ottawa.
- The Noranda Technology Centre in Montreal, where he ended up leading the Hydrometallurgy group.
- An electronic powders manufacturer for a year as R&D Manager of New Products.
- Inco Technical Services in Mississauga as a Mini-plant Operations Manager for the Voisey's Bay hydrometallurgical process.

Since 2004, he is with Barrick Gold Corporation, where he currently holds the position of Sr. Director, Strategic Technology Solutions.

Alexander McLean With degrees in Applied Chemistry and Metallurgy from the Royal College of Science and Technology and the University of Glasgow, Dr. McLean spent 5 years at McMaster University before moving to Jones and Laughlin Steel Corporation in Pittsburgh. He joined the University of Toronto in 1970 and in 1984 the Ferrous Metallurgy Research Group was formed with Professor McLean as Director. He served as Department Chair from 1992 to 1997 and was appointed Professor Emeritus in 2002. He is an Honorary Member and elected Fellow of several organizations including AIME, the Iron and Steel Institute of Japan, the Royal Society of Canada, and the Canadian Academy of Engineering. He is a recipient of the Queen Elizabeth II Diamond Jubilee Medal and received Honorary Doctorates from the University of Miskolc and the University of Strathclyde. He has published extensively in the areas of iron and steelmaking and materials processing.

Krishna Parameswaran Krishna Parameswaran received a Bachelor of Technology degree in metallurgical engineering from the Indian Institute of Technology (Mumbai, India) in 1968, a Master of Science degree in metallurgical engineering from the University of Missouri-Rolla (now Missouri Institute of Science and Technology) in 1971 and a Doctor of Philosophy in metallurgy from the Pennsylvania State University in 1974.

Dr. Parameswaran is currently President of tfgMM Strategic Consulting in Scottsdale, Arizona. He retired from ASARCO LLC after 34 years of service on March 31, 2015. His position prior to retirement was Director of Environmental Services and Compliance Assurance in Asarco's Environmental Affairs Department. In this position, he was responsible for providing oversight for the company's environmental compliance and permitting activities and for coordinating its sustainable development and compliance auditing programs. In June 2005, he co-edited a book entitled "Sustainable Mining Practices: A Global Perspective." He is the author/co-author of chapters in forthcoming books relating to sustainability considerations in mining, metallurgical operations and management. In September 2014, he was appointed by the U.S. Secretary of Interior to the Bureau of Land Management (BLM) Arizona Resource Advisory Council (RAC) for a 3-year term representing mineral and energy interests.

Tim Robinson Tim Robinson graduated from University of Melbourne with B.E. (Chem. Eng.) honors. He has worked in copper, zinc, nickel, and lead industries. He has worked in Australia, South America, and North America in the fields of mineral processing, smelting, refining, hydrometallurgy, and metal fabrication. He has started up over a dozen copper tankhouses around the world.

Ananth Seshan Dr. Ananth Seshan is the CEO and Managing Director of 5G Automatika Ltd., a high technology software product company in Canada. The company is headquartered in Ottawa and has operations in the UK, Mexico, and India. Ananth has been a consultant to several large manufacturing organizations and utilities worldwide for the last three decades in the areas of robotics, automation, and of late, digital manufacturing. He was an invited member of a Special Task Force set up by Industry Canada during the turn of the millennium to explore the viability of applying advanced manufacturing technologies as a strategy to build competitive advantage in the manufacturing segment in Canada. He has been an invited speaker in many international forums on manufacturing automation and has won several awards for innovation and entrepreneurship. He is presently a member of the Board of Directors of MESA International, a global community of manufacturers, producers and industry leaders, Chairman of the Asset Performance Management Working Group of MESA, and Senior Member of the International Society of Automation. Ananth has more than 30 publications in journals and international conference proceedings and owns two patents.

Nathan M. Stubina Nathan M. Stubina is McEwen Mining's Managing Director. He previously worked for over 25 years at various major mining companies, including Barrick Gold Corp., Noranda, Inc. and Falconbridge Limited. Nathan holds a Ph.D. in Metallurgy and Materials from the University of Toronto and a B.E. in Mining and Metallurgical Engineering from McGill University. He is a member of the Association of Professional Engineers of Ontario, is the current V.P. International of the CIM (Canadian Institute of Mining, Metallurgy and Petroleum), and is a Past-President of MetSoc (Metallurgy and Materials Society).

Alan Taylor Alan is Principal of ALTA Metallurgical Services (ALTA) in Melbourne Australia with over 40 years experience in the global mining and metallurgical industry. He has worked in metallurgical consulting, process and project development, engineering/construction, presentation of short courses and convening of international conferences, focussing mainly on the treatment of copper, nickel, cobalt, uranium, and gold ores.

Mark F. Vancas Mark F. Vancas obtained Bachelor of Science degree in Chemical Engineering at the University of Arizona in 1970. He is currently SX Process Specialist with MFV Engineering, Inc. He has over 45 years of professional experience including 17 years of operating experience at Magma Copper Co. and 28 years of engineering design and construction experience with BATEMAN and M3 Engineering in the non-ferrous minerals industry with emphasis on extractive hydrometallurgical and electrometallurgical unit operations, including leaching, solution purification, solvent extraction, electrowinning, anode casting, and electrorefining. Leaching has included in situ and heap/pad leaching with and without crushing and agglomerating in a variety of locations, cultures, and climates as well as agitated leaching of ores and concentrates in acid media with bacterial enhancement. Solvent extraction and electrowinning experience has resulted in the development of many industrial innovations; some of which are now recognized as industrial norms (i.e., permanent cathodes used in electrowinning); and some of which have been patented. Experience also includes metallurgical plant management, plant construction, process design, and project engineering.

Shiv Vijayan After completing bachelor and postgraduate degrees in chemical engineering from the Indian Institute of technology in Madras, India, Sivaraman (Shiv) Vijayan pursued doctoral research in emulsion separation at the Swiss Federal Institute of Technology in Lausanne, Switzerland, and was awarded D.Sc. (Tech) in 1974. Dr. Vijayan gained experience in research and development, and teaching as a faculty member at McMaster University, Canada, and later at the University of Florida in Gainesville and the University of Western Ontario in London, Canada before heading chemical technology, process simulation, fuel reprocessing, and waste processing areas with Atomic Energy of Canada in Pinawa and Chalk River, Canada. After over 30 years of pursuing development of various separation processes and their implementation, Dr. Vijayan has been serving as a specialist consultant since 2013 with Process Research Ortech in Mississauga, Canada in the environmental area advancing technologies and educational training in separation sciences and waste management. Dr. Vijayan is an author or co-author of over 80 papers, 2 patents, 70 proprietary technology reports, and he is a registered professional engineer in the province of Ontario, Canada.

The Need for Process Innovation 1

V.I. Lakshmanan, Raja Roy, and Ram Ramachandran

The Mining and Metallurgical Industry is highly capital intensive and creates environmental challenges. In order to be sustainable, it needs to be innovative in developing processes that are efficient and minimize environmental damage. To meet these goals, we need to be innovative and understand in detail the process steps that will have to be brought together for process development. This will require knowledge of chemistry, chemical engineering, metallurgy, and environmental science.

Innovation is defined as the introduction of something "new," viz., a new idea, method, or device. In the area of science, *innovation* means technologies or practices that are new to a given society. Unless the technology or practice is disseminated and used in society, it is not considered as an innovation.

Science, technology, and innovation are the keys for continued progress in the industrial world. Technical innovation is the key to economic and social development and environmental progress. In-depth scientific and technical research plays a vital role in the development of new processes through innovation.

Why Innovate? The world does not remain stagnant; As Heraclitus (535 BCE) said, "One cannot step twice in the same stream" (Perisco 2011). Innovative thinking begins with great questions, not answers. Demographics, cultural changes, and new technologies outpace a company's business model and force it to innovate to stay in business. The needs to innovate are: (a) Economic Challenge, (b) Global Competition, and (c) Customers' Choice for Products and Services. Innovation involves: (a) Profitable Implementation of Strategic Creativity, (b) Willingness to challenge status quo, (c) Pay off for risk-taking, and (d) Additional value to company and benefit to society as a whole. Any innovative idea should be: (a) new, (b) evolutionary or revolutionary, (c) useful, and (d) implementable (Dundon 2002).

Technological innovation can be defined as "the process by which any industry generates new and improved products and production processes". Generally, the activities include generation of an idea, research, development, and commercialization of the product or process for the benefit of society. The driving force for innovation is invariably a technology push or market conditions. Technology push arises from the need to be competitive by reducing costs, improving quality, and increasing market share (Walcoff et al. 1983).

V.I. Lakshmanan (✉) • R. Roy
Process Research ORTECH Inc.,
2350 Sheridan Park Drive, Mississauga, ON, Canada, L5K 2T4
e-mail: llakshmanan@processortech.com

R. Ramachandran
Consulting Engineer, 9650 E. Peregrine Place, Scottsdale, AZ 85262, USA

V.I. Lakshmanan et al. (eds.), *Innovative Process Development in Metallurgical Industry*,
DOI 10.1007/978-3-319-21599-0_1

As ore grades become lower, processing costs increase and environmental regulations tighten, there is more and more pressing need for innovative process development. Innovation involves not merely a collection of data, but its interpretation to generate knowledge in order to gain market advantage. As industrial processes become more complex, it is imperative to think ahead and look at the whole process in an integrated manner. For example, we need to understand the behavior of not only the value metals, but also the impurity elements through subsequent process steps. Failure to do so may result in a product that is not of intended market grade due to the presence of impurity elements, or it may result in tailings or discharges that don't meet the environmental guidelines for safe disposal.

Many times a process innovation results in process compression, which is elimination of a process step to make the process more economical. Some examples of process compression are the development of the Carbon-in-Pulp (CIP) process for gold processing and the Resin-in-Pulp (RIP) process for uranium ore processing. In CIP process, cyanide leach solution slurry containing gold is mixed with activated carbon in an agitation tank or in a countercurrent manner through a series of tanks. The gold cyanide complex is adsorbed on the activated carbon, which is separated from leach slurry by screening. RIP process is similar to CIP process except that activated carbon is replaced by solid spherical polystyrene resin beads and has been practiced in uranium industry. Both of these processes are now well accepted in industry as these processes minimize costly solid/liquid separation steps. The development of in situ and heap leaching minimizes or eliminates the need for crushing, grinding, and solid/liquid separation. In addition, in situ leaching eliminates tailings management while heap leaching minimizes issues related to tailings management. However, the plant design must make sure that leach solution doesn't reach the water table and also plan for reclamation of land after end of operation.

In its annual report on mining "Tracking the trends 2014" Deloitte calls for a radical rethink in mining sector with emphasis on innovation in the following words: (www.deloitte.com).

> To turn the current tide of poor productivity and hostile sentiment, companies may need to embrace innovation more aggressively. For starters, they must alter their approach to cost cutting by adopting more sustainable cost management practices. They must embrace new forms of technology to gain greater insight into their underlying performance metrics—not only from an operational perspective but also pertaining to their health, safety and talent management programs. Engaging in more strategic scenario planning, similarly, will help mitigate the risks of commodity price volatility, regulatory uncertainty and water/energy availability. Finally, they must take more innovative approaches in their dealings with communities, governments, shareholders, regulators and staff—including explaining their strategies to investors and analysts in more direct and compelling ways. While mining companies cannot change global economic trends, they can change the way they operate. As the mining industry works to attract more visionary leaders, the creativity and forward—thinking that characterize innovation and progress will cease to be industry anomalies and become the norm.

Hence the need for process innovation in meeting the challenges being faced by metallurgical industries is of the utmost importance. This book brings together highly experienced engineers and scientists from the metallurgical industry who take a comprehensive look at innovative process development and propose a way forward for the industry.

Technology/process innovation consists of many steps—from thinking of a new concept to selling it. Major steps involved are shown in Fig. 1.1.

These three phases also identify techniques used in managing innovation in any given organization. Inside any given organization, four factors influence how an innovative idea is brought to commercialization, viz., culture for innovation, technical expertise, availability of finance, and marketing capabilities. External factors—normally from governmental agencies—that may influence innovation are: (a) regulations (anticipated and/or real), (b) tax structure, (c) subsidies, and (d) inadequate patent protection (Walcoff et al. 1983).

A simplified model of the general innovation process as defined by Lager consists of (a) idea generation, (b) problem-solving, and (c) imple-

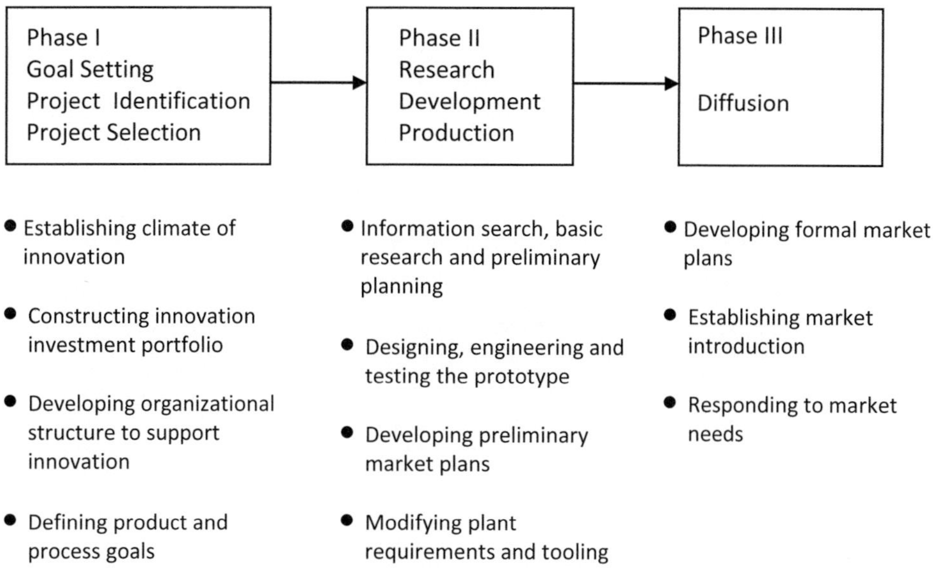

Fig. 1.1 Three phases of innovation process (from: Walcoff, C., et al., "Techniques for Managing Technological Innovation," page 2, Ann Arbor Science, Butterworth Group, 1983. [Reproduced with permission from Elsevier Publishers]

mentation (Lager 2011). Process innovation in the metallurgical industry is necessary to: (a) gain competitive advantage, (b) improve existing processes to reduce cost, and (c) improve environmental conditions in operations and benefit society.

To stay competitive in any given process industry, process innovation—in contrast to product innovation—is always an ongoing process, viz., reduce operating cost, improve environmental atmosphere and work place, and eliminate process steps resulting in process compression. This needs an ongoing investment in Research and Development as a corporate policy. The Research and Development effort also acts as a resource for production-related problem-solving.

Process innovation can happen in following ways:

1. Identifying internal process needs or process development needs.
2. Conceiving a totally "new" process idea for the production of a "new" product or a total replacement of an existing process.
3. Improve an existing process to: (a) reduce operating cost and (b) make a product with better quality and properties.
4. Improve workplace and environmental conditions in the plant.

Identifying internal process needs or process development needs:

The identification of internal process needs is the "ideas" generation phase. A conceptual model of the process development work process is shown in Fig. 1.2.

The phase two area represents all activities shown in Fig. 1.3.

In this case, there are three options to get new process technology:

1. Develop new technology within the internal organization.
2. Acquire new technology from external organization—already existing and can be licensed or purchased.
3. Combination of options one and two.

Option two essentially eliminates the development phase of the process. Option one is described below and shown in some detail in Fig. 1.4.

The work process is divided into four phases (Lager 2011, p. 186). They are:

1. Exploratory work
2. Development
3. Technology transfer

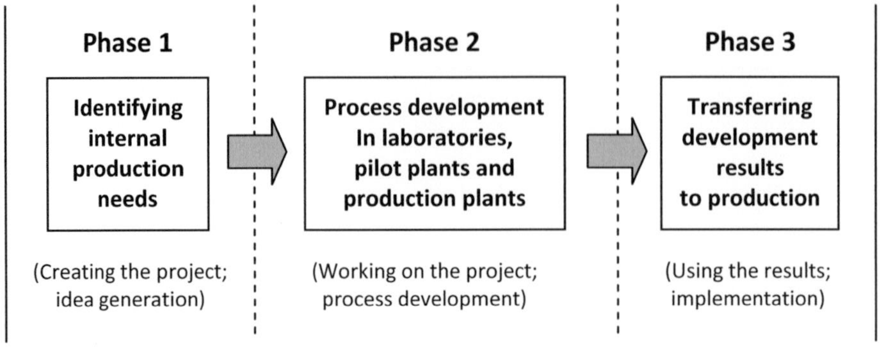

Fig. 1.2 A conceptual model of the process development work process. From: Lager, T., "Managing Process Innovation"—from Idea Generation to Implementation, Series on Technology Management, Vol. 17, page 167, Imperial College Press, 2011. [Reproduced with permission from Imperial College Press, UK.]

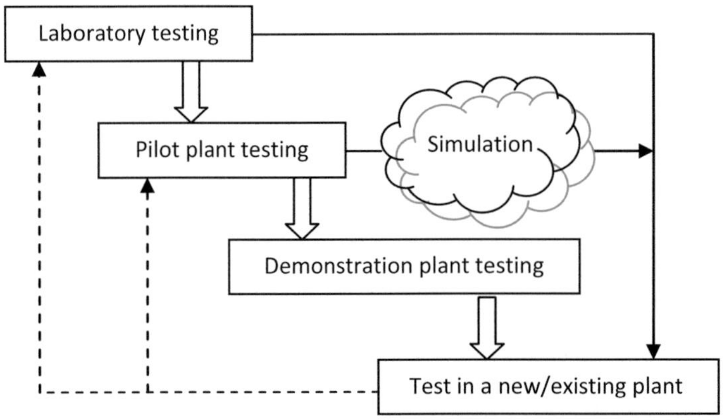

Fig. 1.3 Different kinds of test environments in the process industries. [From: Lager, T., "Managing Process Innovation"—from Idea Generation to Implementation, Series on Technology Management, Vol. 17, page 162, Imperial College Press, 2011.] [Also: Lager, T., (2000) A new conceptual model for the development of process technology in process industry, International Journal of Innovation Management, 4, 319–346.] [Reproduced with permission from Imperial College Press]

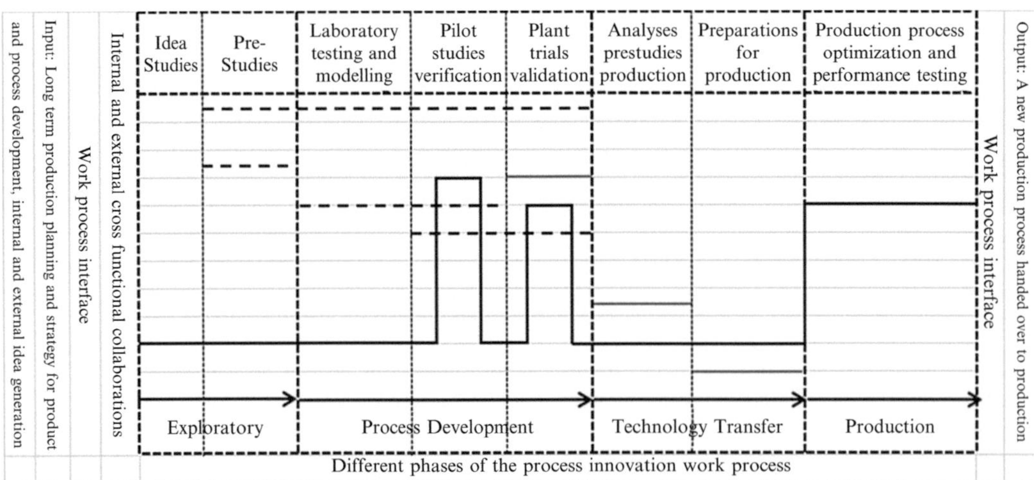

Fig. 1.4 Different phases of the process innovation work process. [From: Lager, T., "Managing Process Innovation"—from idea generation to implementation, Series on Technology Management, Vol. 17, page 185, Imperial College Press, UK, 2011.] [Reproduced with permission from Imperial College Press, UK]

4. Production (new process technology on stream) [Reproduced with permission from Imperial College Press, UK]

Details of these four phases have been described in detail by (Lager 2011). The building blocks in a process innovation work process have been identified. They can be used for both incremental and radical process development.

In summary:

1. An organization should develop and institutionalize a cultural mindset and a set of processes that support repeatable, *sustainable innovation*. This then becomes a foundation for ongoing competitive advantage.
2. The most successful and innovative firms have been characterized by strong in-house capability for Research and Development, generating a corporate research tradition and company-specific process of technology accumulation.
3. It is also important for Research and Development in the process industries to collaborate with external industries like process contractors, equipment manufacturers, and raw material suppliers.
4. Any new or improved technological process has to be sustainable, reduce operating cost, generate profit and meet or exceed environmental regulatory guidelines.

Finally, innovation is not an end-state. It is a combination of open-minded exploration, experimentation, results, and action. The cycle then repeats itself (Palmer and Kaplan 2013).

References

Dundon, E. (2002). *The seeds of innovation*. New York: American Management Association.

Lager, T. (2011). Managing process innovation: From idea generation to implementation. Series on Technology Management (Vol. 17, pages 160, 162, 167, 185, and 186). London: Imperial College Press.

Palmer, D., & Kaplan, S. (2013). *A framework for strategic innovation*. Whitepaper. San Fransisco: Innovation point LLC, 24p. www.innovation-point.com.

Perisco, J. (2011). Retrieved October 30, 2014, from http://www.innovationexcellence.com/blog/2011/07/20/process-innovation-or-product-innovation?.

Tracking the trends. (2014). www.deloitte.com. Retrieved April 28, 2015, from http://www2.deloitte.com/content/dam/Deloitte/ca/Documents/international-business/ca-en-ib-tracking-the-trends-2014.pdf.

Walcoff, C., et al. (1983). *Techniques for managing technological innovation* (pp. 1–2). Ann Arbor, MI: Ann Arbor Science.

Part I

Separation Processes and Process Selection

2 Physical Processing: Innovations in Mineral Processing

B.K. Gorain

2.1 Introduction

> *The mining industry, which dates back to the dawn of civilization, is often considered an ancient industry. Despite its age-old roots, natural resource exploration and exploitation have always been among the most active fields for renewal through innovation. Today, mining innovations address the growing demand for an ever-broadening range of mineral resources, while respecting communities and the environment.*
> (Extract from "100 innovations in the mining industry" by Minalliance 2012)

It is estimated that there are about 2500 metal producing mines, 2500 industrial minerals operations, and almost 100,000 quarries producing aggregates for construction purposes that are of "industrial scale" (not small scale or artisanal mining) operations (Ericsson 2012). Open-pit mining accounts for about 83 % of the total industrial metal mining production and underground producing the remaining 17 % of metal production.

Traditionally a mining operation is segmented into distinct core disciplines viz. geology, mining, and processing from a technical perspective with a focus on extracting an ore body to produce marketable concentrates or metal. There are various support roles that are critical to the success of these technical disciplines such as administration, maintenance, IT, supply chain, safety, environment, and community relations.

As mining is evolving, so are the challenges associated with mining. Some of the key challenges are complex ore bodies at greater depths, high capital and operating costs, water scarcity and quality issues, high energy costs, complex environmental issues including tailings management, worldwide constraints on resource availability, more and more elaborated regulations in emerging economies impacting the permitting processes, increasing stakeholder expectations, changing demographics of mining operations with ever-changing cultural or community expectations and the urgent need for a better assessment of project viability, this in very dynamic metals markets impacting the dynamics of mining projects development and sustainability.

These challenges are compelling mining companies to look at their mining business holistically and also from different points of views. An integrated approach to mining is gaining prominence not just from an economic perspective but also allows them to take into account the needs of various stakeholders. Mining companies are slowly tapping into some successful concepts that have transformed other industries such as manufacturing, chemical, pharmaceuticals along with oil and gas.

B.K. Gorain (✉)
Barrick Gold Corporation, 161 Bay Street, Suite 3700, Toronto, ON, Canada, M5J 2S1
e-mail: bgorain@barrick.com

V.I. Lakshmanan et al. (eds.), *Innovative Process Development in Metallurgical Industry*,
DOI 10.1007/978-3-319-21599-0_2

Some mining companies are already looking beyond the traditional disciplines of geology, mining, and processing to focus on the key value drivers that tend to integrate these disciplines and their interfaces. This is allowing mining companies to innovate and the focus is on step change in their cost structures and productivity.

Though this chapter is about innovation in mineral processing, the author strongly feels that this has to be seen in the light of mining value chain as a whole. Any innovation focusing on mineral processing alone will only provide part of the solution and will miss big opportunities. The question then is "What are these big opportunities that will allow a step change in profitability of a mining operation in a safe and sustainable manner?"

This chapter will emphasize on the following three key focusing questions related to the innovation in mineral processing in the present mining context:

1. What innovations are required to address the existing mining challenges?
2. What are the various mineral processing innovations that have already shaped and advanced the mining industry?
3. What can be done to bridge any innovation gaps for realizing the full potential?

2.2 The Hard Truth About Mining and Processing

Despite significant innovation, mining and processing still involves the old paradigm with significant amount of waste management including haulage, processing, and disposal as described below:

- The first step in a mining operation typically involves removal of overburden to access the ore body for an open-pit mine. This overburden in then stockpiled either as waste rock or to be re-used at closure.
- After drilling and blasting, haulage of blasted ore which is basically mineralized rock above a certain cutoff grade inclusive of waste dilution material that cannot be segregated further during the blasting and mucking activities. The ore is then carried out to the primary crusher or process plant. An ore after mining typically contains mostly gangue minerals. The amount of non-valuable gangue minerals could be more than 99.0 % for gold or PGM ores, ~97 % for a copper ore containing 1 % Cu as chalcopyrite.
- The next step is to crush and grind the ore in the process plant mainly to liberate the valuable from non-valuables. Comminution in the mining industry is intrinsically very energy intensive using 2–3 % of world's energy. Energy consumption could range from 10 to 25 kWhr/t treated. When treating 100,000 t/d, this equates to 1–2 GW/day treated by a comminution circuit drawing 50–100 MW power (Powell 2013).
- Intensive mineral processing steps are then carried out on the ore such as pre-concentration, beneficiation, or concentration to separate valuables from gangue.
- Once the valuables are recovered in the process plant, tailings management is a major step involving storage of tailings, water reclamation and detoxification of the mill and/or final site effluent prior to any environmental discharge, when applicable.

It is estimated that the overburden waste produced globally is about 10,000 million tonnes (Mt) per year, assuming a mining strip ratio of 2.5. This is a significant amount of waste generated by the mining industry. Generally, the lower the head grade, the larger the volumes of ore and waste that are produced. In the Kennecott mine in Utah alone, almost 100 Mt of ore and additional 200 Mt of waste are handled every year (Ericsson 2012). To meet the needs, bigger trucks and shovels are utilized, and the payload trucks have increased from around 200 t in 1990 to almost 350 t in 2012. At the same time, the installed power has increased by a similar factor. In just the past 5 years, the bucket volume of wheel excavators has increased from 25 to 40 m^3. The processing plants have followed suit with large 42 ft diameter mills,

500–700 m^3 flotation cells along with larger dewatering and tailings pond facilities.

The trend of increasing throughput and larger equipment is continuing despite all associated challenges along with an on-going trend of extremely high waste generation. Mining industry is still using the old paradigm. A radical shift is needed, starting with seeking alternatives to handling and processing large amount of waste.

2.3 The Big Opportunities

The basic extraction paradigm in the mining industry is "drill, blast, load, haul, dump, crush, grind, separate, leach, dewater, and tailings disposal." There are many variations, but fundamentally, the paradigm has not changed since ancient times. Almost all the innovations so far have made operations in this paradigm safer, more efficient, automated, and even autonomous (Dunbar 2014). What's needed is a major shift in this paradigm for a major breakthrough.

Since both mining and processing operations handle a significant amount of waste, the first consideration for any innovation should be to cut down on waste as early as possible in the mining value chain. Some of the major opportunities are:

1. *Minimal removal of overburden to access ore body*: Use of small diameter drill holes to access the ore body along with use of novel biotechnologies using to recover metals (Dunbar 2014). In situ recovery of metals such as uranium and copper oxides using solution mining is presently being pursued by the industry, which is a step in the right direction though new technologies and novel ideas are necessary to make this practical for most other ore types. It appears that companies like Rio Tinto are looking at opportunities with in situ recovery as key to the "mine-of-the-future" (Batterham 2008).
2. *Highly selective liberation of ores early in the mining process*: It is important to ask ourselves "Why create waste in the first place?" Comminution begins with mining and a new generation of drilling and blasting for selective mining of ores is critical to avoid or reduce waste removal early on in the mining process. This concept is referred to as "grade engineering" and is becoming a major focus of research and development (CRC-ORE 2014).
3. *Minimal haulage of waste*: This focuses on innovations that will allow processing to be closer to the mine site. There is an increasing interest in pre-concentration technologies such as sensor based mass sorting (ROM shovels/trucks), classification using screens, stream based ore-sorting, gravity and dense media separation, which is definitely a positive trend.
4. *Efficient comminution and processing*: Focus on highly selective separation of valuables utilizing efficient technologies resulting in minimal capital and lower operating costs along with significantly smaller footprint.
5. *Minimal generation of tailings and high reuse potential*: The important focus here is on recovery of by-products and non-conventional value-added products such as building materials with a generation of benign tailings for backfilling underground and open-pit mines.

Reflecting on these five opportunities suggests that the ideal scenario will be a "zero waste mining." Is this really possible? History tells us that the capability and ingenuity of mining and processing professionals have been phenomenal (Lynch et al. 2010). If there is a dire need with a strong vision and adequate resources, there is no reason why "zero waste mining" cannot happen. Various innovative technologies that have been developed by the mining industry definitely provide us the confidence. The area of "zero waste mining" is not the focus of this chapter but all trends in innovations must be seen in light of this ultimate goal.

2.4 Innovations in Mineral Processing

To understand the various innovations in mineral processing, we will need to understand the various challenges the mining industry has handled so far. In the early days, high grade ores and ore bodies were exposed on the surface. To extract these ores required minimal generation and treatment of waste. One could say, this was close to the ideal situation of "zero waste mining" scenario.

As head grades started to deteriorate with ore bodies located at a considerable depth, the need for overburden removal and therefore the generation of mining waste started to increase. With industrialization and the growth in demand of metals, the urgent need to treat low grade ores at much greater depth intensified. Innovation focused on how to mine higher throughputs and then process them mainly to meet the immediate needs. This paved the way for mass open-pit mining and larger mining machineries. Open-pit mines and unfavorable terrain resulted in building concentrators that are far away from mining. Also the consideration for a large tailings facility added to the complexity, resulting in a large mine footprint which increased in proportion with mill throughput.

In the early days when the environmental regulations were not as stringent and the access to power or energy sources relatively easier, the main focus of innovation activities was to build larger and more efficient equipment to treat the large amount of mined product (containing mostly waste) to meet the needs of metal production targets. Hence there was not much incentive to challenge the existing mining paradigm of mass scale waste generation, processing, and storage. This situation is still ongoing, although there is now a strong realization that this cannot continue for very long due to mounting energy costs, stringent environmental regulations, lack of water and other resources along with ever-increasing opposition by local communities.

The mining industry is now facing a dilemma and there is a strong push by companies, some governments and stakeholders to focus on innovation to address the challenges. The following are the main areas that have attracted serious attention for innovation in mineral processing:

- Energy consumption and GHG reduction
- Economy of scale (capital cost)
- Efficient use of consumables (media, liner, reagents)
- Instrumentation, control, and automation
- Use of alternative water sources
- Environmental management of tailings and waste
- Modeling and simulation tools (to reduce risk in design and operations optimization)
- Seamless integration with value chain
- Continuous improvement along with training and education to improve productivity
- Breakthrough technologies (focus on step change such as dry processing and in situ mining)

This chapter will focus on the key mineral processing innovations that have already resulted or have potential to make a significant contribution to the mining industry. Principles of operation and design of unit operations will not be the main focus of this chapter as they are readily available for reference in various publications (Fuerstenau and Han 2003; Wills 2006; Gupta and Yan 2006).

The following major areas in mineral processing will be subject of discussion in this chapter:

1. Ore body knowledge
2. Pre-concentration
3. Comminution and classification
4. Physical separation (gravity, magnetic and electrostatic separation)
5. Flotation
6. Dewatering and thickening
7. Tailings and water management
8. Automation and integration

2.5 Ore body knowledge

As ore bodies are becoming highly complex, it is more important than ever to understand the geology and mineralogy in details to enable the most appropriate design and operation of mining and mineral processing operations. Understanding of the geology and the uncertainties associated with the ore bodies is critical to the success of any mining operation. This is typically carried out using data generated through drill holes, geological mapping, geophysical surveys and the geologist's interpretation. Ore body modeling and resource estimation is the foundation on which the business case for future mine development and operation is intricately dependent.

One major input for resource estimation is quantitative mineralogy information, which is increasingly being recognized as more important than mere elemental assays. Interpretation of mineralogy through assays have been traditionally carried out but with increasing complexities and uncertainties associated with ore bodies, a direct way of determining mineralogy without the need to make certain assumptions is a positive trend. Having more mineralogy attributes for both valuables and gangue in the resource and reserve models allows for a better integration of geology and metallurgy, which is an emerging area widely known as "geometallurgy."

2.5.1 Quantitative Mineralogy

Advances in quantitative mineralogy technologies along with recent developments in computerization and data storage have been a major breakthrough for the mining industry. Process mineralogy is now a well-established field that provides ore characterization and better understanding of ore variability to reduce uncertainties and risks associated with ore body modeling, mineral processing plant design, and process plant optimization. Many demonstrated benefits have been realized by mining operations globally using quantitative mineralogy (Gu et al. 2014; Bradshaw 2014; Evans et al. 2011; Lotter 2011; MacDonald et al. 2011; Rule and Schouwstra 2011; Kormos et al. 2010; Kappes et al. 2009; Baum et al. 1989). Numerous plants have now onsite and in-house access to quantitative mineralogy analyzers and XRD or Near Infra-Red (NIR) analyses for carrying out planning, optimization, and for daily production needs, which is a major innovation in the mineral industry (Baum 2014).

Automated quantitative mineral analyzers like Mineral Liberation Analyzer (MLA) and QEMSCAN have been commercially available since the 1990s and have played a major role in redefining ore characterization in the mineral industry. In addition, advancement in XRD technologies (Rietveld) has allowed quantification of the gangue minerals. It is rare today to find any feasibility study carried out without quantitative mineralogy information. It is to be noted that these technologies are suited to visible minerals only and any elements that are in solid solution cannot be identified.

It is also common nowadays to use sophisticated tools and techniques for carrying out mass balances, low concentration precious metal and deleterious metals deportment, modal and liberation analysis, water chemistry and mineral surface-studies to obtain in-depth understanding necessary for metallurgical problem diagnostics (Smart et al. 2014; Buckley 2010; Grano 2010).

2.5.2 Quantitative Gold Deportment

An important development in gold mineralogy is the ability to carry out quantitative gold deportment for refractory and double refractory ores using state-of-the-art technologies. The characterization of these ores is challenging due to low gold ore grades, presence of variable proportion of preg-robbing total carbonaceous matters (TCM) with sulfide inclusions, visible and invisible gold in different sulfides phases (e.g., pyrite, chalcopyrite, chalcocite, and bornite) and also in iron oxide phases (e.g., magnetite, hematite, and goethite). This requires an integrated approach to process mineralogy involving various tools such as optical microscopy, mineral analyzer (MLA, QEMSCAN), X-ray diffraction (XRD), Dynamic secondary ion mass spectrometry (D-SIMS), Time-of-flight—secondary ion mass spectrometry (TOF-SIMS), X-ray photoelectron spectroscopy (XPS), and Laser ablation microprobe-inductively coupled plasma mass spectrometry (LAM-ICPMS). Details of these quantitative gold deportment techniques are presented elsewhere (Chattopadhyay and Gorain 2012, 2014; Chryssoulis and McMullen 2005). These measurements are immensely useful but could be expensive and time consuming. It is important to be prudent on representative sample selection with a clear purpose. Improper use of these tools could result in poor diagnosis and wrong interpretation.

2.5.3 Future of Process Mineralogy

The use of multiple mineralogical techniques to accurately identify element and mineral deportment is becoming important as the ore complexities are increasing and the tolerance for risks in plant design and operations is low. Bradshaw (2014) has highlighted the following four critical aspects for successful process mineralogy (the 4 I's):

- Information (appropriate measurement and accurate analysis)
- Interpretation (what does the information mean?)
- Implication (converting the interpretation into wisdom)
- Implementation (of a change to either the existing circuit or the process design)

Bradshaw has emphasized that if any of these steps are missing the whole process is threatened and much of the value is lost. The potential barriers to achieving the value can be attributed to short-term thinking, the imposition of inappropriate time or budget constraints, or to a lack of understanding of either the project aim or process mineralogy as a whole.

2.5.4 Geometallurgy

Recent advancements in process mineralogy have furthered the cause of integrating mine site geology and process plant to a great extent with a main focus on improving the value of the mine. Geometallurgy relates to the practice of combining geology and geo-statistics with extractive metallurgy to create a geologically based predictive model for mineral processing plants. It is used for risk management and mitigation during plant design and also to assist and enhance mine production planning (Kittler et al. 2011). From a comminution perspective, geometallurgy assists in optimizing the mine plan by mapping hardness and forecasting throughput into the mine block model using geo-statistics to achieve design tonnage over the life-of-mine of the deposit and eliminate the large throughput fluctuations caused by ore hardness (Brissette et al. 2014).

McCullough et al. (2013) have also highlighted the challenges associated with recognizing and embedding the value of geometallurgy in the mining value chain. The main challenge is to permanently change the behavior of people and the processes they follow. This requires executive sponsorship, technology, data integration along with automated intelligent analysis to realize the full value of geometallurgy. Kittler et al. (2011) has emphasized the need for proper sample selection with a clear focus on spatial and grade distribution. It is best to avoid blending of samples for geometallurgy samples to capture ore variability and to retain their spatial provenance. The requirements for operational geometallurgy have been presented by David (2010).

Bye (2011) has documented various industrial case studies that demonstrate strategies for gaining value from geometallurgical studies. Values include both operational benefits such as proactive fragmentation control and better strategic planning process along with building of geometallurgical domain models.

2.6 Pre-concentration

Removal of some waste as early as possible after blasting has the potential to reduce haulage costs to the mill in many situations. In addition, this provides an opportunity to upgrade the ore resulting in reduced energy and operating costs per unit of metal input to the plant. A higher head grade to the process plant typically results in better concentration ratio and better unit recoveries as well as the process now has to handle lower gangue content. This concept of pre-concentration is not new with application such as Dense Media separation and sorting on the surface since the 1930s (Munro et al. 1982) and also applied underground (Lloyd 1979).

It is worthwhile to note that not all ores are amenable to pre-concentration especially when the minerals are finely disseminated in the ore body. Pre-concentration of such ore bodies could result in high losses of valuables in the waste product. Often there is a resistance to pre-concentration because it is perceived as a process involving metal losses and costs, and this belief sometimes hinders any further discussion on this opportunity. It is therefore important to characterize the ore to assess the amenability of pre-concentration before this option is discarded.

Ores types that are amenable to pre-concentration can add significant economic benefit, such as at the Kroondal platinum mines in South Africa where about 50 % of bulk mining feed of UG2 ores to the mill is rejected with a PGM recovery of 95 % (Holloway et al. 2009). The value of pre-concentration has now been demonstrated for a wide range of ore types at a coarse size range (Mohanty et al. 2000; McCullough et al. 1999; Schena et al. 1990). The value of pre-concentration should be looked in a holistic way involving integrated mining, processing and waste disposal. Bamber (2008) has concluded based on his studies that the exploitation of a deposit with ore pre-concentration and waste disposal technologies integrated into the mining process prior to beneficiation on surface is superior to the conventional approach.

The following pre-concentration technologies appear to have significant potential and are slowly finding its way into different mining applications:

- Size classification
- Ore sorting
- Dense media separation
- Coarse particle flotation

2.6.1 Size Classification

Concentration by comminution and size classification alone has been found to be effective for some ore types (Burns and Grimes 1986; Sivamohan and Forssberg 1991). Size classification presents a low cost option for rejection of waste, with maximum economic benefit at coarse particle sizes either at naturally arising ROM particle size distribution or at a coarse crush size, as close as possible to the mining face (Klein et al. 2002).

2.6.2 Ore Sorting

Ore sorting in the mining industry has evolved from manual sorting in the early days to basic optical sorting first used in mining in the 1970s using camera technology and digital image processing for industrial minerals. Arvidson and Wotruba (2014) have provided a review of the various applications of ore sorting technologies. Some of the earliest applications of sorting in mining industry were in diamond and gem mining using X-ray fluorescence and X-ray transmission to identify the gems. Radiometric sorting has been successfully applied to the processing of uranium ores such as at Lodeve mine in France and at Northwest Kalimantan in Indonesia on ROM ores (IAEA 2000). The Indonesian ores typically responded well to ore sorting with about 50 % rejection with more than 98 % uranium recovery. X-ray sorting is presently used in various tungsten mining operations such as at the

Wolfram Bergbau in Australia and Wolfram Camp in Cairns, Australia. The importance of ore sorting in improving economics of marginal deposits is increasingly being realized by the mining industry (Lessard et al. 2014; Foggiatto et al. 2014; Walker 2014).

Bamber (2008) has presented different methods of ore sorting based on the following properties along with applications:

- Photometric: coal, sulfides, phosphates, oxides
- Radiometric: uranium, Witwatersrand gold ores
- Conductivity: metal sulfides, native metals
- Fluorescence: metal sulfides, limestone, iron ore
- X-ray luminescence: diamonds
- X-ray transmission: coal
- Electrostatic: salts, halite, sylvite
- Magnetic: iron ore, andalusite, quartz, kimberlites

At present, majority of automated ore sorters, outside of the diamond industry, are color or conductivity sorters (Bartram and Kowalczyk 2009). This is a significant innovation in mineral processing with an ability of pre-concentration and significantly improved economics. The optical sensors have quite a few applications but as the optical properties differ relatively little, laser based sensors could be useful. If the primary surface properties are distinct, NIR sensors are normally used. Novel sensor technologies are now necessary to extend the application to ore types beyond optical sensors that exploit material properties such as electrical conductivity, magnetization, molecular structure, and thermal conductivity. Combination of sensors also have potential such as optical/NIR, optical/inductive, or XRT/inductive (Arvidson and Wotruba 2014).

The recent development of LIBS (Laser-Induced Breakdown Spectroscopy), focused on pattern recognition appears to have potential for mass ore sorting. LIBS works by focusing a laser beam on the ore surface to create a plasma. When this plasma cools down it emits light that is representative for the elements in the plasma. LIBS provides elemental information of some ores such as copper without any radioactive sources. One major advantage is that there is no sample preparation or sizing required and the LIBS ore sorter can be positioned above the conveyor belt. The approach seems to make it possible to achieve fast sorting capabilities of 1.000 t per hour, while keeping good accuracy (SonicSampDrill 2014). Though there is potential, much work is required to make this technology commercial for large throughputs. Combination of LIBS with other sensor technologies may be necessary for some complex ores.

2.6.3 Dense Media Separation

Dense media separation is widely used in coal, diamond, chromite, and iron ore industries since its introduction by Dutch Sate Mines (DSM) in 1947. Though used occasionally in other industries, its application in precious and base metals viz. PGMs, gold, copper, and zinc are slowly gaining some momentum. Holloway et al. (2009) have estimated capital costs for DMS plants to be in the range from \$3.0 to \$4.5 per tonne of throughout per annum (excluding crushing), depending on the complexity of the plant design. Operating cost for heavy media (ferrosilicon or magnetite) is typically \$1.00/lb (delivered). Typical power consumption ranges from 2 to 3 kWh/t and cost of spare parts is estimated to be around \$0.50/t. These costs provide some rough guidelines only and could vary significantly depending on applications.

Napier-Munn et al. (2014) have reviewed various DMS technologies, with a focus on technologies that have been implemented in the industry. DMS technology has allowed Kroondal platinum mines to reject 50 % of bulk mining feed of UG2 ores with a PGM recovery of 95 %. Tati nickel plant in Botswana uses DMS to reject 70 % of mass with recovery of ~2/3rd of the contained nickel and has allowed bulk mining with a zero cutoff grade, thus simplifying the mine plan with increased resources. Another impressive application of DMS has been in recovery of fine grained gold from waste rock dumps at the Witwatersrand gold fields in South Africa. Despite a low feed grade (0.2–0.3 g/t), DMS was successful in recovering 70 % of the gold into 30 % of the mass.

Typically DMS technologies are compact processes with high-throughput of 300–1000 t/h. Holloway et al. (2009) have concluded that DMS seems to be well positioned with tremendous growth potential to address the need for reducing energy costs, improving mine-mill integration, and simplifying tailings handling through coarse ore pre-concentration. Bamber (2008) has carried out various studies and has highlighted the significant potential of DMS in underground applications.

2.6.4 Coarse Particle Flotation

Significant energy is needed to grind ores for conventional flotation which typically requires a feed that could range from 10 to 200 μm. Flash flotation applications can treat much coarser feed ranging from 600 to 1000 μm. Energy requirements magnify for regrinding applications with a target of less than 10 μm to liberate minerals in finely disseminated ores. From a pre-concentration point of view, the top size of flotation feed required is typically 1–3 mm.

Flotation of ore at a top size of 3 mm, followed by gravity concentration of the flotation tailings has been applied previously in South Africa to treat Witwatersrand gold ores resulting in production of a 40 % mass pull at an overall gold recovery of 98 % (Lloyd 1979). Coarse particle flotation of particle ranging from 3 to 5 mm is common in the potash industry. For many base metals and precious metals applications, coarse flotation is challenging in conventional and flash flotation machines for pre-concentration purposes.

Jameson (2014) has developed a fluidized bed flotation technology for coarse flotation focusing on pre-concentration at a coarse size (600–800 μm) with an estimated reduction in operating cost of 10–20 %. Though this is in experimental stage, nevertheless the concept has merit and deserves further attention. The Hydrofloat® technology from Eriez also uses fluidized bed for coarse particle flotation with commercial applications in grinding circuits similar to flash flotation, also with the possibility of producing a throwaway tails (Franco et al. 2015). These are interesting developments with implications for pre-concentration at a relatively finer size fraction.

Compared to other pre-concentration technologies, flotation offers the most potential especially for fine grained mineralogical complex ores as other technologies at a coarser grind will incur significant metal losses. A potential flow sheet involving HPGR product of 1–2 mm or SAG product of 800–1000 μm could be subjected to coarse particle flotation focusing on rejecting a significant amount of feed. The coarse flotation concentrate could then be ground to a finer size for conventional flotation as shown in Fig. 2.1.

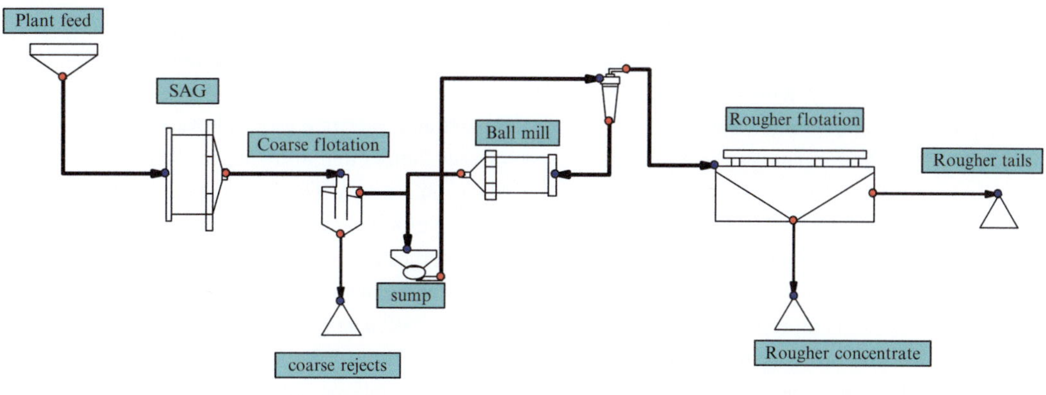

Fig. 2.1 Rejection of barren coarse gangue minerals using novel coarse particle flotation technology (based on Jameson 2014)

2.7 Comminution and Classification

The word comminution is derived from the Latin *comminuere*, meaning "to make small." Breaking of rocks from a mining perspective encompasses the following unit operations (Napier-Munn et al. 2005):

- *Crushers*: Gyratory crushers, jaw crusher, cone crushers, rolls crusher, high pressure grinding rolls (HPGR), impact crusher
- *Tumbling mills*: Autogenous grinding (AG), semi-autogenous grinding (SAG), rod and ball mills
- *Stirred mills*: Tower mills, vertical pin mills, horizontal mills
- *Sizing*: Screens, sieve bends, hydrocyclones, other classifiers

It is important to note that sizing processes are integral part of any comminution circuit and have a significant influence on comminution performance. It is estimated that about 30–70 % of the total power draw in a concentrator is consumed by comminution. It is reported that only 3 % of the total energy input is used in particle size reduction, though most of it is lost as mechanical energy and heat (Alvarado et al. 1998). The efficiency of comminution is important because the cost of breakage is a key factor that determines whether low grade ore bodies could be economic. There is a significant effort globally to develop more efficient comminution technologies. Johnson (2006) has carried out a review of various eco-efficient comminution devices. The coalition for Eco-Efficient Comminution (CEEC) has been recently established with support from various companies in the mining industry with an aim to promote awareness and knowledge transfer mainly to reduce energy requirements in comminution.

2.7.1 The Bond Work Index

Bond's work index is an innovative concept now widely used in the industry to size crushers and grinding mills since the publication of Fred Bond's approach in 1952. The work index is a comminution parameter that expresses the resistance of the material to crushing and grinding. Numerically, it is expressed as kWh per tonne required to reduce the material from theoretically infinite feed size to 80 % passing 100 μm. Grinding power calculated using work indices obtained from Bond grindability tests applies reasonably well in the range of conventional rod mill and ball mill grinding conditions, and can be corrected for other conditions (Napier-Munn et al. 2005). Several limitations to the Bond methods have been highlighted by many authors (Austin and Brame 1983; Yap et al. 1982; Herbst and Fuerstenau 1980), but the criticism of Bond's method as a "model" is not really warranted as its real value is to distinguish the grindability of different ores in design and as a yardstick to check the energy utilization of current operations. In addition, Bond methodology is simple and it does work for many circumstances to a reasonable degree of accuracy, since the corrections are based on a large industrial data base (Napier-Munn et al. 2005).

The Bond method, however, cannot be used in the case of autogenous and semi-autogenous milling because of the large size of ROM feed, making the Bond model for work input almost independent of feed size. MacPherson utilizes standard Bond rod mill and ball grindability tests, together with a dry autogenous test procedure, to obtain the autogenous work index and determine AG/SAG milling suitability, power requirement, and circuit configuration (Knight et al. 1989).

2.7.2 Selection and Design of Comminution Circuits

Barratt and Sherman (2002) have summarized the various factors that influence the selection of comminution circuits. The list of these factors is quite extensive depending upon the nature of the project, whether it is greenfields plant or an expansion, as well as on a thorough understanding of the ore characteristics and scoping of test work at each of the study. Table 2.1 shows the various comminution circuit options used in the industry.

Table 2.1 Options for various comminution circuits (based on Barratt and Sherman 2002)

Earlier circuits	Crusher, rod mill, ball mill	
	Crusher, single stage ball mill	
	Crusher, rod mill, pebble mill	
	Crusher, multistage ball mill	
Conventional circuits	*With autogenous (AG)*	*With semi-autogenous (SAG)*
	Single stage AG	Single stage SAG
	AG, ball mill	SAG, mill
	AG, pebble mill	SAG, ball mill, pebble crusher
	AG, pebble mill, pebble crusher	Pre-crushing, SAG, ball mill, pebble crusher
Emerging trends		
		HPGR, SAG, ball mill, pebble crusher
		HPGR, ball mill

2.7.3 Drilling and Blasting

A first step in comminution Drilling and blasting is an established technology evolved over the last century. This is a batch process and often closely embedded in the macho culture of a traditional tough and rough miner (Ericsson 2012). Hard rock cutting, a viable alternative to enable continuous mining is gaining wider acceptance in metal mining since its success with softer and nonabrasive bedrock and minerals such as potash and coal. This innovation seems to be attractive for narrow veins and reef mining allowing a more selective mining with less waste rock and dilution with positive impact in processing along with the economic benefits associated with continuous mining.

Mine-to-Mill optimization in various operations over the years have shown significant benefits such as high mill throughput rates from reduced top size from mining through increased powder factor or blast energies (Kanchibotla 2014). There are however some safety and environmental areas where special care and attention is needed to prevent any High wall damage, fly-rocks, noise, and vibrations from using higher blast energies. The introduction of electronic detonators and other techniques however allow mitigating some of these risks. New blasting technologies are emerging that have the potential to significantly increase powder factor through innovative blasting practices such as dual blast layers within a single blast event (Brent et al. 2013).

2.7.4 Crushing

- Crushing is an important step in comminution that prepares the ROM ore for primary grinding and other downstream processes such as heap leach. The selection of the right crushers and crusher flow sheet is dependent on the upstream mining method and blasting characteristics and the downstream process requirements.

As process plant throughputs are increasing, the need to reduce operating costs and capital cost per tonne of ore processed is critical. Both crusher and circuit design have evolved with larger crushers employing more horsepower and speed to treat higher throughputs at a reduced cost.

With the advent of SAG milling, cone crushers are now more or less eliminated from comminution flow sheets. Instead the cone crushers are employed in a SABC circuit to crush the recirculating pebbles. These pebbles tend to be more resistant to impact breakage in the SAG mill and the crusher creates ore surfaces more conducive to breakage in the grinding mills (Major 2002). Major also reiterated that a crusher has a more efficient transfer of applied power to the breakage of rock than a grinding mill. This has been seen with the development and acceptance of the SABC circuit and also led to Nordberg's development of the "waterflush" cone crusher. The application of waterflush technology has resulted in the production of a finer product than normally achievable with a shorthead cone crusher. SAG pre-crush is getting more and more popular, which support the findings of the Mine-to-Mill with finer fragmentation on the bench via blasting.

The earliest US patent for a crushing machine was issued in 1830, which was a crude device incorporating a drop hammer principle. The Blake jaw crusher was then patented in 1858, which is based on powerful toggle linkage, and is the standard by which all jaw crushers are judged. In 1881, Philters W. Gates was granted a patent on a machine that included in its design all of the essential features of the modern gyratory crusher. For some years after these machines were developed, mining was still manual with very low tonnages and the product specifications simple and liberal (Utley 2002). With the emergence of steam shovel, the mining industry grew treating large tonnages and adopted larger machines. In 1919, Traylor engineering built the largest 60 in. gyratory crusher and by 1969 they had already built the 72 in. gyratory crusher, which served the industry until 2001. Typical gyratory crusher capacities are 350–10,000 tph, known for their high capacity and low maintenance. Disadvantages are highest installed capital cost of any crusher design.

In the early 1980s, low speed sizers were introduced. They represent one of the only fundamental developments in primary crushers in three-quarters of a century. The main feature of the low speed sizers is that it exploits the fact that the ratio of compressive strength to tensile and shear strength in the majority of rocks is around 10:1. The low speed sizers break the rock in tension or in shear by its "snapping" and chopping action rather than in compression as conventional crushers do. Additionally, the position of the teeth on the rolls allows undersize to fall directly through the machine resulting in high throughputs at very low rotational speeds, which leads to greatly reduced wear and greatly reduced fines.

2.7.4.1 Gyratory Crushers

During the last 20 years, the size of gyratory crushers has not changed significantly, though their installed horsepower have increased allowing these crushers to treat higher throughputs. An example of the largest gyratory crusher is FLSmidth's 60′×113″ UD design with an installed power of 1 MW. Another feature of the new design of Gyratory crushers is the development of top service gyratory crusher which appears to enhance safety, reduce maintenance downtime, and lower cost of crusher installation (Erickson 2014). New monitoring tools such as CrusherMapper (Scanalyse 2012), which is a laser scanner tool, survey the physical dimensions of the crushing chamber and allows plant personnel to make informed decisions on crusher maintenance.

2.7.4.2 Cone Crushers

Cone crushers are traditionally used for secondary and tertiary crushing. With the advent of SAG milling circuits, however, cone crushers are increasingly being used for pebbles crushing. The principle of cone crusher remains the same but their sizes have increased over the years with two main manufacturers presently viz. Metso Minerals and FLSmidth. They manufacture large cone crushers with installed horsepower ranging from 800 to 2000. The new design of large cone crushers emphasize on increasing capacity without significantly increasing operating costs along with lower maintenance and high availability.

The "WaterFlush" crusher technology is sometimes used in conjunction with a cone crusher to allow the addition of water with the ore feed.

2.7.4.3 SELFRAG Technology

This is an emerging technology showing some promise in pre-weakening of mineral ores by high voltage pulses. This was pioneered in the 1950s but it is attracting some serious attention during the last few years. The key benefits of this technology compared to conventional comminution machines is that this is highly selective and its ability to weaken the ore with a potential for significant energy savings in further comminution (Shi et al. 2013).

Studies on a Newcrest's Cadia east Cu-Au ore samples by JKMRC showed an energy reduction of 5 kWh/t, along with a potential to remove 2 MP 1000 pebble crushers and one 10 MW ball mill. Simulation studies showed that Selfrag has the potential to reduce operating cost by A$ 19 million (Shi et al. 2014). Though this study showed some potential, a detailed capital cost estimate is needed to better evaluate the potential of this technology.

2.7.4.4 IMP Super Fine Crusher Technology

This "IMP" superfine crushing concept evolved from the simple premise that the probability of particle breakage increases as the number and intensity of forces simultaneously contacting the particles increase. This is quite different from conventional comminution machines where breakage rate decreases with increased energy inefficiencies as feeds become finer and harder. The IMP superfine crusher is designed to provide a mechanism that could effectively deal with fine hard feeds by maintaining high breakage rates as hard particle became finer and single particle population rapidly expands (Kelsey and Kelly 2014).

The "IMP" superfine crusher consists of a rotating compression chamber and an internal gyrating mandrel. The axis of rotation of the shell is displaced relative to the axis of the vertically mounted mandrel. Crusher feed is accelerated into a compacted particle bed, which lines the inside of the compression chamber. The depth of the compacted bed progressively increases until the mandrel is engaged and the counter rotating mandrel gyration drive is triggered. This delivers multiple compression at high pressure to the rotating particle bed allowing a high power intensity and extreme compression forces. The axially displaced rotation of the compression chamber effectively distributes the breakage forces throughout the particle bed. A detailed description of the mechanism can be found in Kelsey and Kelly (2014).

The "IMP" superfine crusher technology is presently in the pilot scale testing phase but has the potential for developing a broad range of application options as the technology matures.

2.7.4.5 High Pressure Grinding Rolls

Schonert (1982, 1987) first introduced the HPGR technology to the cement industry as an energy efficient alternative to ball milling, demonstrating about 25 % reduction in energy consumption for compressing and dis-agglomeration of the limestone and cement clinker. Since then many equipment manufacturers have developed HPGR technology for use in cement and minerals industry viz. KHD Humboldt Wedag AG (Germany), Krupp Polysius AG (Germany), FLSmidth (Denmark and USA), Maschinenfabrik Koppern GmBH (Germany), and Alpine AG (Germany).

Over the last two to three decades, HPGR is slowly being adopted in a wide range of mineral industry application such as diamond, iron ore, tin, copper, chrome, and gold (Erickson 2014). HPGR size ranges from installed power of 440–6800 kW.

Recent successes of HPGR in hard rock application include Freeport's Cerro Verde in Peru (Vanderbeek et al. 2006; Koski et al. 2011) and Newmont's Boddington in Western Australia (Dunne et al. 2007; Hart et al. 2011) with a target of grinding energy savings of 15–20 %. Both these operations have two stages of coarse crushing using gyratory and cone crushers followed by a third stage of crushing involving HPGR and a single stage ball milling circuit as shown in Fig. 2.2.

2.7.5 Primary Grinding

- Since the advent of AG and SAG milling technologies in the late 1950s, they have established themselves as the present standard and are commonly used in the industry now. These technologies have replaced the previous "conventional" comminution circuits involving crushing-ball mill or rod mill–ball mill circuits at least in the precious and base metals sectors. At present, there is a wide spectrum of proven milling technologies to a plant designer than say 20–30 years ago. Although such choice may be seen as an improvement, but also poses challenge in the selection of the most appropriate technologies for an application.

Pokrajcic et al. (2010) have demonstrated using their sustainability analysis tool "EcoNomics"

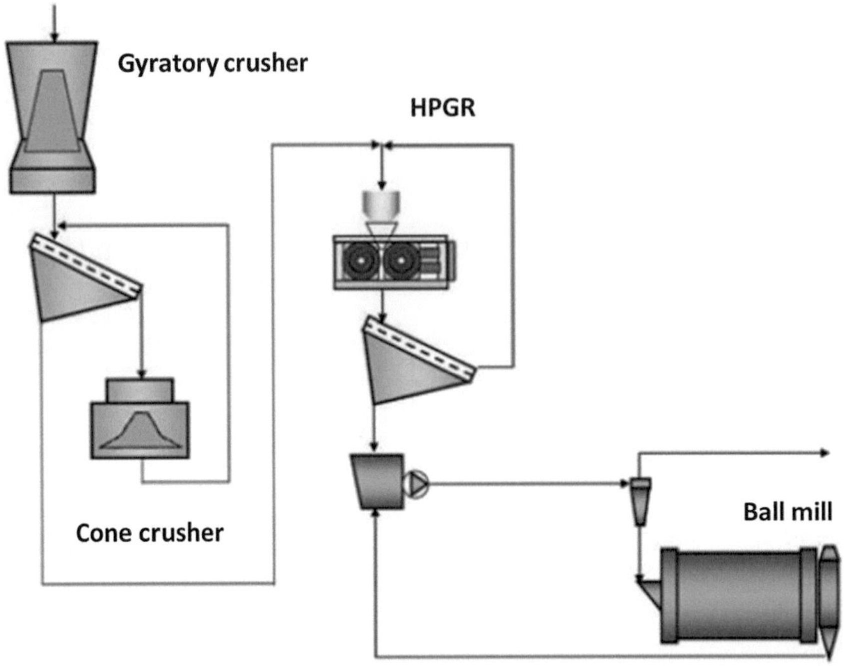

Fig. 2.2 Comminution flow sheet using HPGR at Boddington and Cerro Verde (Kirsch and Daniel 2009)

that significant savings in energy consumption could be achieved through the use of more efficient comminution equipment along with appropriate circuit design. They emphasized that there are significant benefits to business profitability when the energy efficiency features in comminution circuit design.

Grinding mills have increased from 4.5 MW in the late 1960s to 8 MW in 1980, 13 MW in 1996 and 20 MW in 1997. Large 28 MW, 12.2 m × 11 m long AG have been manufactured by CITIC recently for an iron ore operation in WA (Meka and Lane 2010).

2.7.5.1 Autogenous Milling

The largest autogenous (AG) mill (40′ × 33″ @ 28 MW GMD) is now operational at the Sino Iron mine at Cape Preston, about 100 km south west of Karratha in Western Australia. This is the largest magnetite mining and processing facility with design capacity of 24 million tons of iron concentrate produced through six processing lines (Tian et al. 2014).

2.7.5.2 Semi-Autogenous Milling

Metso designed and supplied the world's largest semi-autogenous grinding mill for Newmont Mining's Yanacocha Conga copper-gold project in Peru. The 12.8 × 7.6 m SAG mill will be the first to exceed 12.2 m in diameter. The SAG is equipped with an ABB gearless motor to allow operation at variable speeds. The Conga project is located approximately 900 km north of Lima at an altitude of 4100 m above sea level at the Yanacocha gold mine. The project reached the feasibility stage and while full funding was expected from the Newmont board in the first half of 2011, the project development has been on hold since then.

A brief comparison of AG and SAG mills given by (Tian et al. 2014):

- AG milling circuit offers higher power efficiency with high SG ores
- AG doesn't need grinding media, a major operating cost item for SAG
- Recent development of gearless mill drive (GMD) allows the application of larger AG mill
- Relatively insensitive to F80
- Lower operating cost
- Safer mill access conditions because of no steel
- Simpler layout, operation and potential lower maintenance cost

2.7.5.3 Comparison of SAG and HPGR

SAG milling technology has now successfully replaced conventional multistage crushing and rod/ball mill comminution circuits gradually over two to three decades mainly for base metals and precious metals applications. The main reasons for the attraction for SAG milling are circuit simplicity and typically lower capital and operating costs to meet the needs for operations with higher throughputs. SAG mills work well for handling clay rich sticky ores, which is a challenge for multistage crushing circuits requiring washing plants. Figure 2.3 shows a conventional SAG and Ball mill circuit (SABC).

SAG mills, however, are less energy efficient in handling harder and abrasive ores. As the ore bodies are steadily becoming more competent along with increasing energy costs, the industry has been looking for alternatives to SAG. It is interesting to note that the traditional crusher and ball milling technology has returned this time albeit with a more innovative approach to crush rocks using HPGR.

Morley and Staples (2010) carried out economic trade off studies of SAG vs HPGR and concluded that at the current stage of development of HPGR, the capital costs are generally higher than for the equivalent SAG based circuit. For highly competent ores and where power and grinding media costs are high, HPGR can offer substantial operating cost benefits with around 15 % energy savings.

It appears that HPGR has a niche application for very hard ores but further validation is required to demonstrate its advantages over SAG milling technology for a range of ore hardness and abrasiveness.

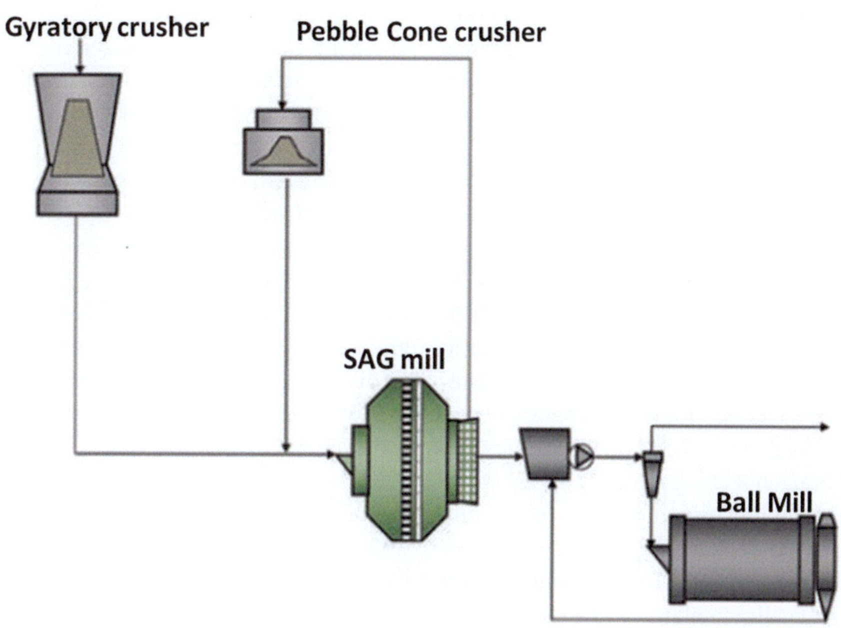

Fig. 2.3 Conventional semi-autogenous and ball milling circuit (SABC)

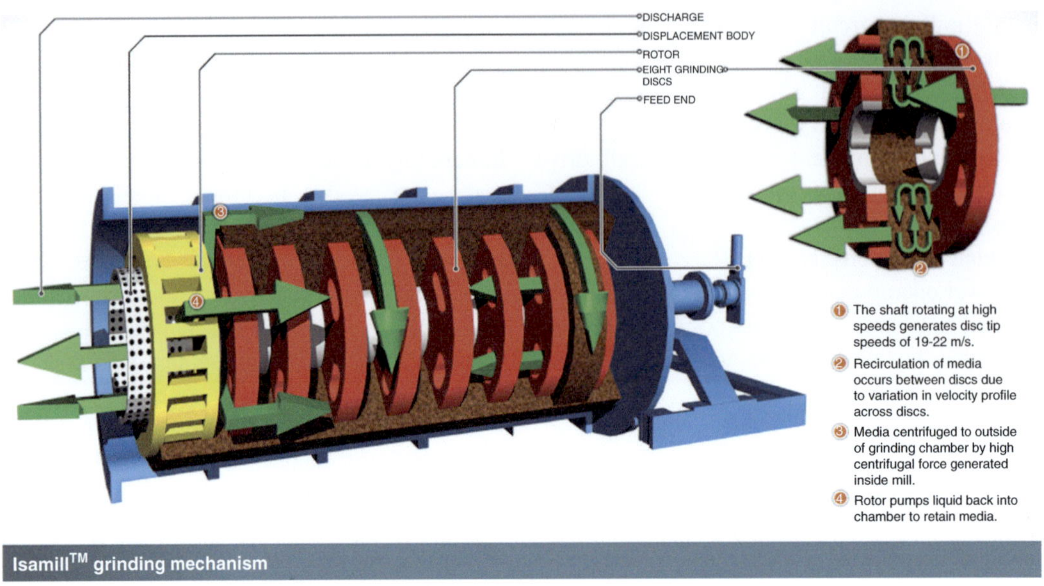

Fig. 2.4 A schematic of IsaMill™ (with permission from Glencore technologies)

2.7.6 Ultrafine Grinding

- The emergence of ultrafine grinding technologies is a response for treating low grade ore bodies with complex mineralogy requiring the need for grinding ores to as low as 5–7 μm to enable sufficient liberation for an efficient separation. There are a wide range of ultrafine grinding technologies worldwide. This section will discuss select few technologies that have made a significant impact in the industry.

2.7.6.1 IsaMills™

The IsaMill technology is an outcome of the significant efforts made in the early 1990s to develop a process for Mount Isa Mines (MIM) McArthur River deposit requiring regrinding to 80 % passing seven microns to produce a saleable zinc concentrate. IsaMill technology was jointly invented by MIM of Australia and Netzsch Feinmahltechnik GmbH of Germany for ultrafine and fine grinding duties in the minerals industry (Gao et al. 2002). Since the original installation at the MIM Pb-Zn concentrator in 1994, IsaMills totaling 130–140 MW of installed power have been installed in various facilities around the world (Rule 2011). IsaMills presently marketed by Glencore Technologies range from M500 (250 kW) to M50,000 (8 MW) though the most common installed unit is M10,000 (3 MW). Figure 2.4 shows a detailed schematic of a large IsaMill.

2.7.6.2 Stirred Mills (VertiMills® and Detritors)

There are two categories of stirred mills offered by Metso Minerals viz. gravity-induced and fluidized. The gravity-induced stirred mills, known as Vertimill®, use rotational movement of screw to initiate a ball charge motion; whereas the fluidized stirred mills, Stirred Media Detritors (SMD), use rotational energy to fluidize a media-slurry mixture for providing a grinding action. Typical feed size for Vertimill® is from 6 mm to produce grind sizes to sub 15 μm, using high density media such as steel or ceramic. Typical feed size for SMD ranges from 250 μm with product sizes to sub 5 μm.

Lichter and Davey (2006) have provided examples of many applications of stirred mills in different operations. Newcrest's Cadia, Ridgeway Concentrator in New South Wales have recently installed the largest unit VTM-3000-W (3000 hp or 2240 kW) supplied by Metso Minerals. Figure 2.5a, b shows a schematic of the Metso SMD and Vertimill®, respectively.

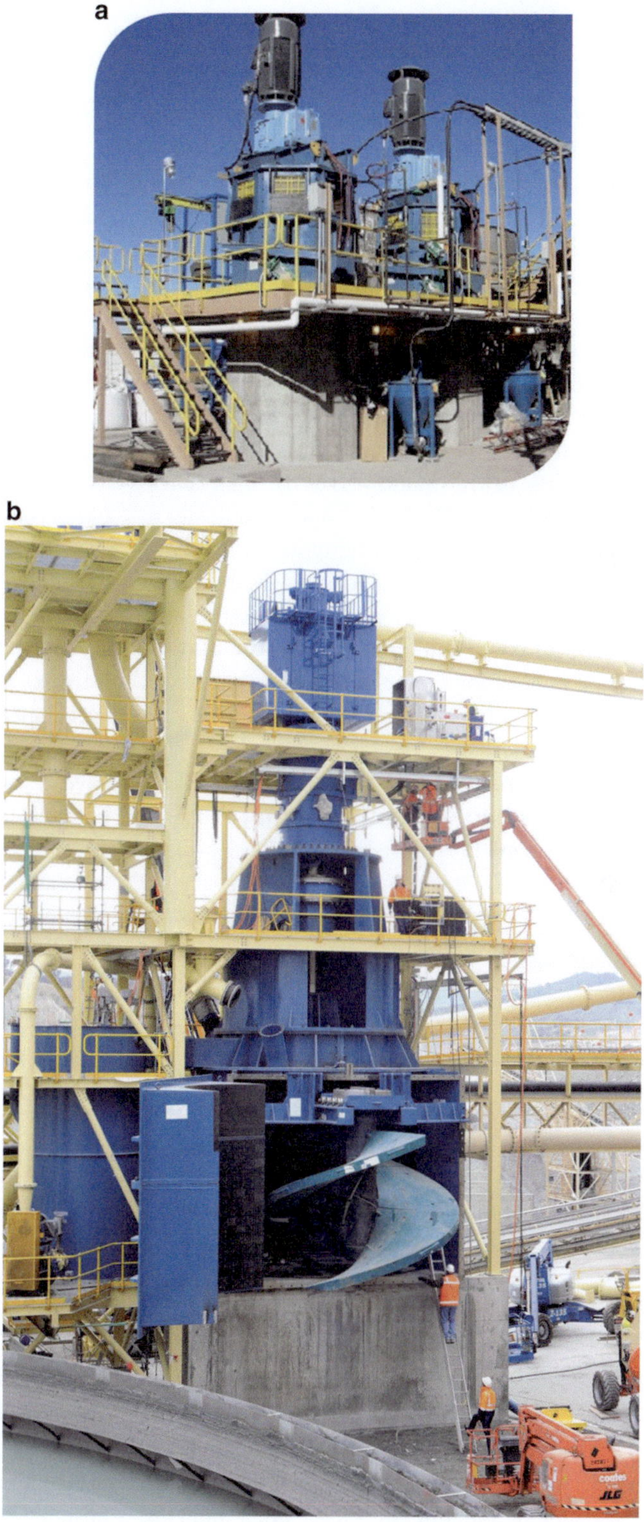

Fig. 2.5 (**a**) Application of Metso's Stirred Media Detritor (SMD), (**b**) VertiMill® in mineral processing plants (with permission from Metso Minerals)

2.7.7 Comminution Modeling and Ore Characterization

- One of the major advances in comminution innovation is the development of models as well as modeling and simulation techniques. Morrell (2014) has iterated that JKMRC has been in the forefront with over 30 years of research and development resulting in acceptance of JKSimMet and the Drop-weight test as being a commonplace, reliable design and optimization tool for comminution circuits. Development and commercialization of bench-scale ore characterization tests such as the SPI (TM), Starkey SAGDesign, Drop Weight Test, and SMC Test that are relevant to AG/SAG mill grinding circuits which have come to dominate circuit design. As discussed earlier, Bond's model is still an industry standard for comminution modeling even after its first publication more than 60 years ago.

2.7.8 Development of Classification Technologies in Grinding Circuits

- It is well known that a closed circuit grinding circuit with a classification device results in significant benefits to the comminution process. In the early days, mechanical rake and spiral classifiers were used in closed circuit grinding circuits. Due to capacity limitations and high operational costs because of wear and tear, these technologies are not commonly used in the present comminution circuits. Hydrocyclones became popular since the early 1950s as they are high capacity devices that allowed the industry to meet the demands of increasing throughputs to treat low grade ores. Since hydrocyclone separation is a function of both size and density, a sharp size separation is not always possible for many ore containing high specific gravity minerals and metals such as gold, PGMs and lead. This results in building of these small heavies, which are mostly liberated, in the recirculation stream in cyclone underflow leading to unnecessary grinding with poor gravity and flotation responses.

The inefficiency of the hydrocyclone separation requires the use of high circulating loads to minimize the mill residence time. High circulating loads increase the power requirements per tonne of ore (Albuquerque et al. 2008).

The key for great improvements in capacity and in energy consumption in closed circuit grinding is improved sharpness of classification (Hukki and Allenius 1968). This is where screen separation has a significant advantage because of its sharp separation. However, fine particle separation has been challenging due to blinding issues, low capacity, and high screen panel consumption (Albuquerque et al. 2008).

2.7.8.1 Derrick Stack Sizer®

Derrick Stack-Sizer® is a recent innovation (developed in 2005) in fine screening which allows high separation efficiency and high tonnage capacity on a much smaller footprint than possible using conventional screens (Clark 2007). The Derrick Stack-Sizer® is now available for separations from 10 mm down to 44 μm using Derrick Corporation's patented urethane screen surface technology, which provides an alternative to conventional hydrocyclones. There are many cases of application of the Derrick Stack-Sizer® replacing hydrocyclones in a closed circuit grinding applications for a large variety or minerals with significant metallurgical benefits reported (Valine et al. 2009). There are now over 400 installations around the globe in grinding circuits, as an alternative to hydrocyclones, for both metal and nonmetal beneficiation applications (Perkins 2015). Figure 2.6 shows an installation of the Derrick Stack-Sizer® in a grinding circuit.

2.7.8.2 Cavex Recyclone®

This is a double classification unit in one stage, which seems to increase the sharpness of separation by reducing the bypass of fines to the underflow. This technology uses wash water injection mainly to rupture the viscous layer to release the

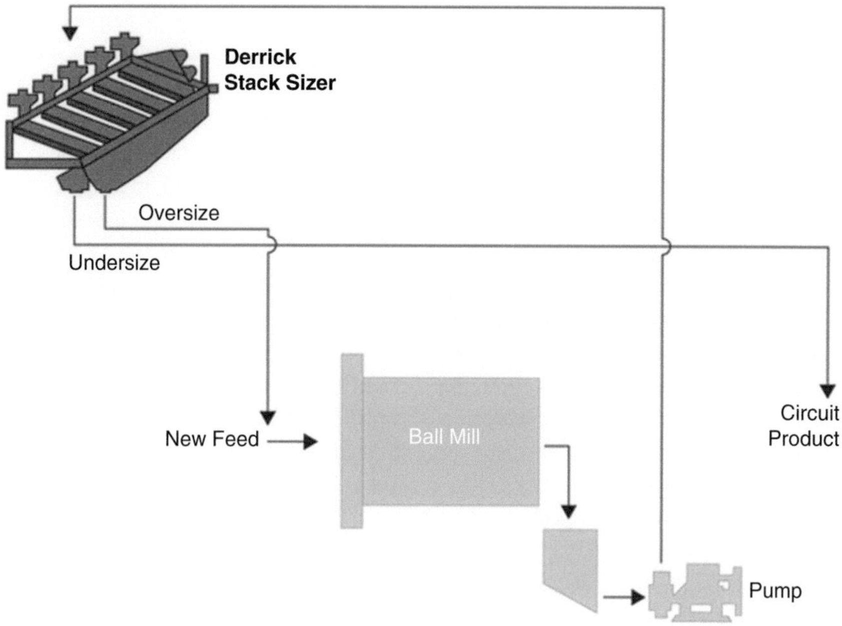

Fig. 2.6 Use of Derrick Stack-Sizer® in a grinding circuit (with permission from DerrickCorp)

trapped fines for proper classification. Recent trials and installations in grinding circuits have shown to significantly reduce fines misreporting to mill from 30 to 13 % resulting in reduction of circulating loads by around 50–60 % (Castro et al. 2009).

2.8 Froth Flotation

No metallurgical process developed in the twentieth century compares with that of froth flotation and the profound effect it had on the mineral industry (Fuerstenau 2007).

The first commercial process using froth flotation, patented by Adolf Bessel and his brother in 1877, was a graphite flotation process utilizing the natural hydrophobicity of graphite along with addition of oil and boiling the pulp. They also developed another patent in 1886, referred to as generation of bubbles by decomposition of limestone in an acid solution. It is interesting to note that they used oil, a nonpolar reagent, for improving the hydrophobicity of graphite. Lynch et al. (2010) have provided some interesting insights into the history of flotation developments before 1900s and also the breakthrough in flotation for the silver-lead-zinc ore deposit at Broken Hill in Australia during 1902–1920. Since then interest in flotation has exploded globally with further innovations and development in the USA, Canada, and Europe, and has now become a versatile process to concentrate billions of tonnes of ore annually, and is rightly regarded as one of the great technical achievements of the twentieth century.

The froth flotation process can be considered as an *interactive engineering system* consisting of three broad groups of variables for selective separation of a mineral from a liquid–solid suspension composed of both valuable and unwanted gangue mineral particles. The three key variables for the flotation process are

1. Chemical
2. Operational
3. Machine

The control of surface chemistry of the minerals to provide the potential conditions for bubble particle attachment is fundamental to the flotation process. The operating conditions in a flotation machine first develop the attachment of particles with bubbles and then permit levitation of particle bubble aggregates to the froth. The manner in which bubbles and particles interact with each other depends on the type of flotation machine used. Flotation machines used in the mineral industry can be classified into four types: mechanical, pneumatic, column, and froth separators. Of these, mechanical flotation machines have dominated the mineral industries since the early days of froth flotation all over the world.

2.8.1 Mechanical Flotation Cells

The first mechanical flotation cell was introduced by the Minerals Separation Company in 1905 which led to the development of Hoover's frothing machine used first at the Central Mine in Broken Hill, Australia. The first generation of sub-aeration machines came around 1915 and by 1927 many manufacturers introduced their own design of sub-aeration machines such as Fahrenwald (Denver sub-aeration), Fagergren, Agitair, Ruth, Kraut, Weinig, and Geco. From 1930 to 1960, the sub-aeration machines dominated the market with cell sizes increasing up to 100 cubic feet. The focus in the 1960s and 1970s shifted towards the design of large capacity machines ranging in cell sizes from 200 to 500 cubic feet in response to the falling ore grade and higher plant throughput. By early 2000, the cell sizes had increased up to 200 m^3, a 100-fold increase over 1940s volume. Detailed description of the different cell sizes and design offered by different manufacturers during the early 2000s has been provided elsewhere (Gorain et al. 2007). Cells as large as 700 m^3 are being designed and characterized by some flotation cell manufacturers (Grau et al. 2014). At present there are three major manufacturers of mechanical flotation cells representing most of the mechanical cells installations around the world viz. FLSmidth, Metso Minerals, and Outotec. Nelson et al. (2009) have provided a detailed description of the different mechanical flotation cell sizes, design, and operating principles of these three different cell manufacturers.

2.8.1.1 Flash Flotation

Flash flotation technology uses specialized mechanical cell design mainly to remove high specific gravity liberated minerals or metals from within the grinding circuit to reduce the overgrinding and consequent loss as slimes in the flotation circuit. Flash mechanical flotation normally treats cyclone underflow containing a high recirculating load of heavy minerals or metals that are significantly finer than the cyclone cut-size. Flash flotation has many applications in copper-gold and gold flotation circuits with an aim to recover free gold and gold associated with sulfides such as pyrite or copper minerals, in a high grade concentrates that could either be sold as copper or pyrite concentrates or processed onsite using enhanced gravity separators (Falcon or Knelson) and cyanide leaching. Newcombe (2014) has carried out an extensive study of flash flotation cells in Barrick operations and has provided some significant insight into optimization of flash flotation circuits. Outotec's SkimAir® cells are commonly used in the grinding circuit as an unit cell operation to float valuable liberated particles from the circulating load of the grinding circuit while still relatively coarse. There are over 200 SkimAir® cells installed worldwide, many in gold and copper-gold operations with recent installation of four SK2400 flash units at Esperanza operation in Chile (Lamberg and Bernal 2008). Figure 2.7 shows a conventional flash flotation circuit typically used in the grinding circuit for gold and copper-gold ores.

2.8.1.2 Operation of Large Mechanical Flotation Cells

The large cells provide significant advantages such as reduced plant footprint, lower power consumption, reduced maintenance and operating costs, and easier control. The economic benefits of these large cells have not been fully realized by the industry due to lack of understanding of large cell operation. The problem of machine design and metallurgical scale-up has increased with larger cells. Effective froth transportation

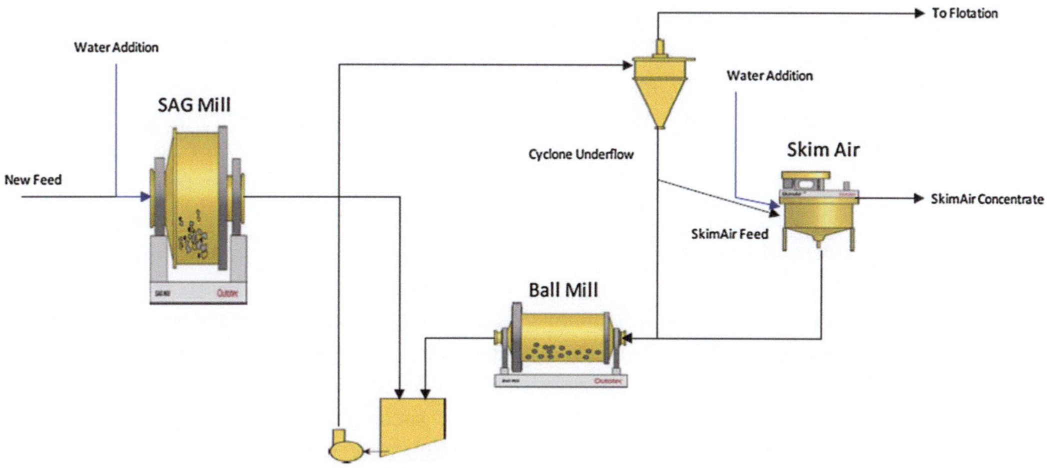

Fig. 2.7 Conventional flash flotation circuit in gold and copper-gold operations (with permission from Outotec)

and recovery is critical for efficient operation of large cells. Research efforts over the last two decades under the AMIRA P9 project and other work have provided significant insight into the understanding of hydrodynamics of large cell operations (Gomez and Finch 2009; Schwarz and Alexander 2006; Nesset et al. 2006; Gorain 2000, 2005). Different flotation manufacturers have also reported some operating data of their machines (Nelson et al. 2009). These contributions have undoubtedly improved our understanding of large cell operation, still more work is required to obtain a rational basis for selecting cell operating conditions and to understand the reasons for poor froth recovery in large cells, especially for treating ultrafine ($P_{80}<10$ µm) and coarse particles ($P_{80}>150$ µm).

Recent studies suggest that significant improvements in metallurgical performance of flotation banks with large cells are possible through a three-level cell optimization program. The first level is aimed at identification of the range of cell operating conditions for optimum mixing, gas dispersion, entrainment, froth flow behavior, and metallurgical performance in individual cells. The second level is aimed at identification of the optimum bank operating profile for cells down the bank in a circuit. The third level is focused at controlling these cell operating conditions and bank operating profiles for tighter control of concentrate grade and recovery. Higher metallurgical benefits are obtained once the findings of the three-level studies are integrated to develop an effective strategy for control of the optimum bank operating profile (Gorain and Stradling 2002).

Installation of few large cells in a circuit needs better cell control. The trend in many operations is the use of instrumentation like air flow control, variable speed control, as well as online measurement equipment for monitoring bubble size, superficial gas velocity, gas holdup, and bubble surface area flux. Froth vision system with an expert control is gaining prominence for better control of cell operating conditions. The development of flotation cells will continue as more and more fine particle processing and higher throughputs will be necessary in future. The large cells will have to be designed to generate very small bubbles and a high degree of micro-turbulence for effective bubble-particle collision. Entrainment will be a major issue in concentrators which will need refinement of froth washing technologies in mechanical flotation cells. An integrated approach to flotation cell operation will be required for efficient utilization of flotation cell capacity with focus on optimization of both Mine-to-Mill and Mill-to-Market. The dominance of mechanical cells over other types is expected to continue until the economies of scale with very large mechanical cells get exhausted as the disadvantages of these large cells outweigh the overall benefits. At this juncture, more innovative cell design is needed to treat a wide range of low grade complex ore bodies.

2.8.2 Nonmechanical Flotation Cells

- Nonmechanical cells are increasingly being accepted by the industry mainly in applications where the traditional mechanical cells fall short. Some examples are flotation of ultrafine particles below 10 μm and coarser particle beyond 180–300 μm depending on ore types. Also for certain ores requiring intensive rejection of non-sulfide gangue minerals, nonmechanical cells are sometimes preferred. Considerations such as energy savings, capital costs, and footprint are important as well in flotation cell selection. Details of the various nonmechanical cells developed for the mineral industry have been provided elsewhere (Yoon et al. 1992; Finch 1995; Zhou et al. 1997; Young et al. 2008; Kohmuench et al. 2009; Sanchez-Pino et al. 2014). Only a select few that are popular or becoming popular in the industry is described below.

2.8.2.1 Column Cells

Column flotation cells were first successfully used in the industry during the early 1980s (Cienski and Coffin 1981), though the concept has been patented by Tremblay and Boutin (1964). Unlike mechanical flotation cells, this technology doesn't use any agitation mechanism to generate bubbles or for solids suspension. As the name suggests, this technology utilizes tall columns fitted with a froth washing system on top of the columns mainly to reduce entrained gangue minerals. Columns use the countercurrent principle with feed introduced via a distributor located at about two-thirds of the height of the column and the tailings removed from the bottom of the column. Bubbles are generated independently using a sparging system and introduced near the base of the column. The principles of column operation have been discussed by Finch et al. (2007).

Industrial column height could range from 9 to 14 m with a diameter up to 3–4 m, and the froth depths are significantly deeper than mechanicals ranging from 0.5 to 2 m. Columns normally operate with a positive bias, where there is a net downward flow of liquid allowing a stable froth and better rejection of entrained gangue. Process control and automation of column flotation operation, using sensors to measure the main operating parameters such as air flow rate, froth depth, bias rates, gas holdup, feed, and tailings flow rate, is considered critical for its success in any operation (Bergh and Yinanatos 2003; Bouchard et al. 2009). Mechanical cells, in general, are easier to control than columns and therefore preferred by many design engineers and operators though with the advent of large mechanical tank cells, control is increasingly becoming important for their optimum performance.

There are two important developments in column technology over the last two decades, one is better froth washing system design and the other is generation of fine or ultrafine bubbles for improving flotation recovery of fine particles. There are two fine bubble generation technologies that have gained popularity viz. Microcel and Cavitation device, and have helped columns to compete with the new generation of mechanical tank cells with regards to better unit recoveries in cleaner applications. Development of the concept of bubble surface area flux and its relationship with flotation performance and operating conditions has helped in better cell control and performance optimization (Gorain et al. 1997; Desbiens et al. 2009).

2.8.2.2 Microcel™

Microcel™ technology was developed at the Virginia Polytechnic Institute by Professor Yoon and his team (Yoon et al. 1992). This involves an in-line mixer where slurry drawn from the base of the column is mixed with pressurized air at high intensity to produce microbubbles outside the column. The operation of Microcel has been described in other publications (Brake 1998; Phillips et al. 1997).

There are at present more than Microcel™ 100 installations worldwide, initially popular in coal cleaning but later came to base metals as well. Microcel™ spargers generate finer bubbles and have been shown to improve metallurgical performance. Installation of Microcel™ spargers on the zinc retreat columns at the Red Dog Mines reduced the mean bubble diameter from 3.4 to 1.9 mm, with the metallurgical improvements with a payback period of 1.5 months (Pyecha et al. 2005). Installation of Microcel™ technology at Antamina showed reduction in bubble size from 2.7 to 3.7 mm down to 1.6–2.6 mm in copper column cleaners and molybdenum circuits. This resulted in increased unit copper recovery by 6 % points and molybdenum recovery by 20 % points, along with improved selectivity against insols (Lizama et al. 2008). Figure 2.8 shows a schematic of a Microcel™ column installation.

2.8.2.3 CavTube™

This technology uses hydrodynamic cavitation to generate micron size (picobubbles) ultrafine bubbles through controlled pumping of slurry with a sparger of specific geometry to induce cavitation. Picobubbles are known to naturally exist in liquids such as seawater and distilled water and are found to be more readily attached to particles than larger bubbles. Studies have shown that combining cavitation-induced bubbles with mechanically generated bubbles produce higher flotation recoveries than by either method alone. This synergistic effect is caused by the deposition of picobubbles on the particle surfaces, which promotes attachment of larger bubbles (Fan et al. 2010; Krasowska and Malysa 2007). More details on the principles and applications of cavitation tube devices for columns have been provided by Wasmund (2014). Figure 2.9 shows a model of the CavTube™.

2.8.2.4 Imhoflot™

Imhoflot™ technology has been developed by Dr. Rainer Imhof of Germany based on previous work on pneumatic flotation cells at Clausthal

Fig. 2.8 Schematic of the Microcel™ column flotation cell (with permission from Metso minerals)

Fig. 2.9 A model of the CavTube™ sparging system (with permission from Eriez)

and Berlin universities and is now marketed by Maelgwyn Mineral Services Limited. The development of Imhoflot™ pneumatic flotation has been well documented in many technical papers (Battersby et al. 2011; Sanchez-Pino et al. 2008; Imhof et al. 2003).

There are two different types of self-aspirated Imhoflot™ pneumatic flotation cells viz. vertically fed (V-Cell) and tangentially fed (G-Cell). Pneumatic flotation differs from conventional flotation in that the bubble particle contact takes place outside of the cell itself, within the aerator. The associated pulp then enters a froth separation chamber to separate the froth from slurry. There are no moving parts in the Imhoflot™ cell where all the energy for mineral collection comes from the feed pump with energy focused into the restricted volume of the aerator. The net effect of this is very high air utilization rates up to an order of magnitude higher than conventional flotation. In the vertical type—Imhoflot V-Cell, the separating zone is essentially a vertical cone. Slurry is injected through nozzles upwards from the base of the cone, i.e., cocurrent with bubble movement. Sizing of the Imhoflot V-Cell vertical type is dictated by the time required for efficient froth separation.

The largest cells can treat around 1200 m^3/h with a diameter of 5 m. There are many successful applications reported in base metals, coal and iron ore applications (Sanchez-Pino et al. 2014; Imhof et al. 2005). Figure 2.10 shows an Imhoflot G-cell in an operation.

2.8.2.5 Jameson Cell

The Jameson cell flotation cell was developed at Mount Isa Mines in the late 1990s. This is a pneumatic flotation cell, and the fundamentals of its operation have been described by Clayton et al. (1991). The first installation of Jameson was in the lead flotation cleaners at Mt. Isa Mines. Jameson and Manlapig (1991) showed significant advantages of Jameson cells compared to mechanical and column cells including kinetics, footprint, and cost. During the 1990s, Jameson cells had great success in coal fines flotation and in SX-EW for removal of organics. With better understanding during various installations and through design improvements, Jameson cells are slowly being accepted in the lead, zinc, copper, and copper-gold operations (Young et al. 2008). Schematic of a Jameson cell is shown in Fig. 2.11.

2.8.2.6 The Woodgrove Staged Flotation Reactor

The Staged Flotation Reactor (SFR) was developed by Glenn Kosick and Glenn Dobby, who are also well known for their development of Contact Cell and jet sparging technologies for columns. The SFR technology uses different chambers for

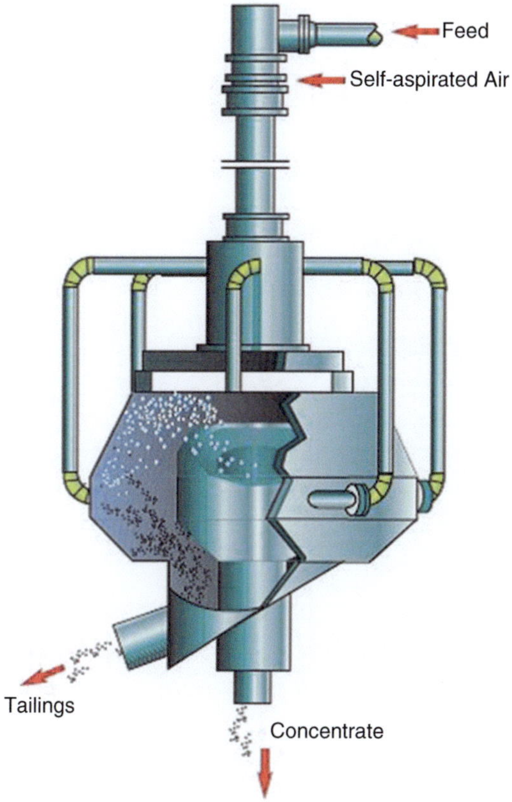

Fig. 2.10 Schematic of the Imhoflot™ G-cell (with permission from Maelgwyn Mineral Services)

particle collection and froth separation and appears to leverage the best features of flotation columns and mechanically agitated flotation cells within its design. The first commercial SFR was installed during July 2010 in the final stage of roughing at Anaconda's Pine Cove operation in Newfoundland. Their second installation was at Dundee's Chelopech copper-gold mine in Bulgaria during 2012 in cleaner applications with significant benefits (Dundee 2012). This is an emerging technology with ongoing pilot testing in various operations. This technology is presently being developed and evaluated further in full scale applications under a consortium sponsored by major mining companies. An installation of a large scale SFR in an operation is shown in Fig. 2.12.

2.8.3 Flotation Reagents

- A number of recent publications have provided details on the history of flotation reagent development since the early days (Nagaraj and Farinato 2014; Fuerstenau 2007; Nagaraj and Ravishankar 2007). This section will highlight only the key developments that have made a major impact in the industry.

A very important innovation during the early 1900s was the introduction of air or in situ generation of CO_2 which removed or reduced the need for large quantities of oils as buoyant separation medium used previously. With this reduction in oil consumption, inherent differences in the frothing and collecting powers of various oils were observed, which promoted the development of selective or differential flotation. Several important modifiers such as lime, sulfur oxide, sodium carbonate, sodium sulfide, and copper sulfate were used to enhance separation. It was also discovered that alkaline circuits provided much better flotation performance with lower reagent consumption compared to the popular sulfuric acid circuits. Perkins and his coworkers revealed numerous small organic compounds such as thiourea, thio alcohols, dyes, diazo, and amine compounds, which were vastly different from the fatty oils used then. There was a gradual realization that organic flotation reagents provided two different functions which led to the classification of collectors and frothers.

The period of 1921–1950 established itself as an era of *chemical flotation* due to phenomenal innovation and development in flotation chemistry (Nagaraj and Farinato 2014). This period was one of expanding on the concept of using small amount of short-chain (C_2–C_6) organic molecules, many of them were water soluble, as collectors for froth flotation of sulfide ores to replace large amounts of long chain (>C12) that are relatively insoluble oils. The reagents named in many patents all contained either sulfur or nitrogen or both. Most notable of these collectors were short chain amines, naphthol, chelating agents, xanthates,

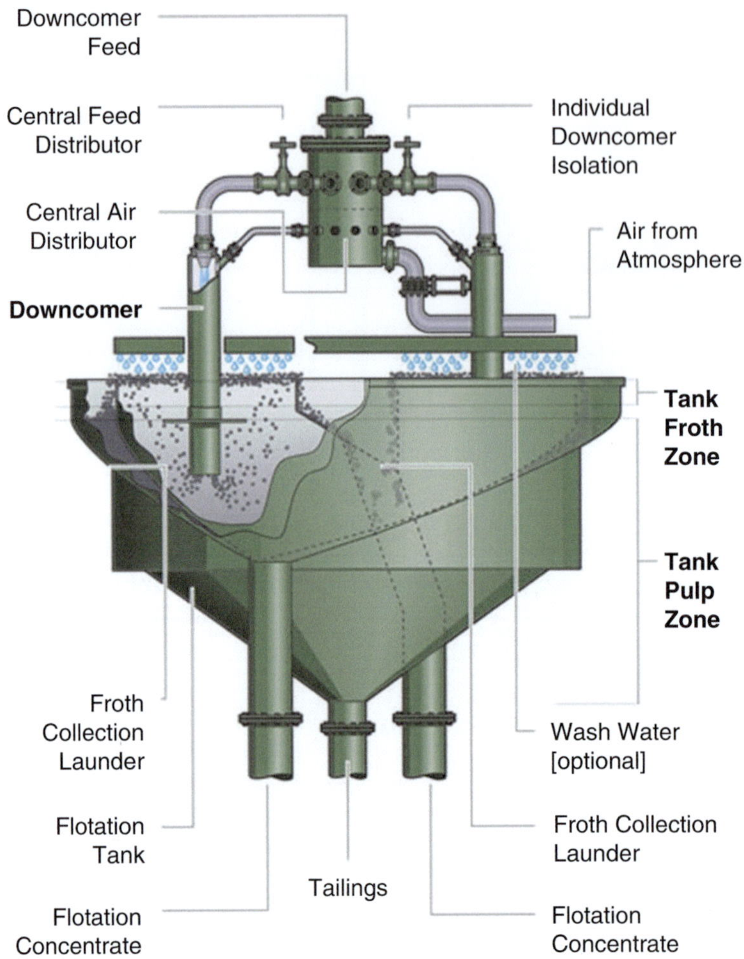

Fig. 2.11 Schematic of a Jameson™ flotation cell (with permission from Glencore technologies)

potassium salts of dithiocarbamic acid, and dithiophosphates (especially the cresylic acid derivative). Both xanthates and dithiophosphates (DTP) with applications in sulfide flotation are still commonly used in the industry. Dithiocarbamates, on the other hand, were not widely used and their usage is still relatively small. The importance of sulfur compounds in xanthates and DTPs was recognized, which led to the further search for these compounds from natural sources such as rubber and agricultural industries that were prevalent. Bolton (1928) patented mercaptobenzothiazole, which became popular and has applications even today. Synthetic mixtures of different collectors were also developed and commercialized during the 1930s.

Taggart and his coworkers laid the foundation for identifying and establishing chemical functional group features such as collectors, depressors, and frothers. This work along with contributions from Gaudin, Wark, Sutherland, and others led to significant advancement in further reagent development for both sulfide and non-sulfide mineral systems (Nagaraj and Farinato 2014). Work at U.S. Bureau of Mines and other institutions beginning in the late 1920s led to development of collectors, such as fatty acids, for oxide, carbonate, and phosphate minerals. Taggart (1951) has discussed development of

Fig. 2.12 Installation of staged flotation reactor (SFR) in an operation (with permission from Woodgrove technologies)

new reagent and schemes for oxides during the 1930s and 1940s. New collectors for non-sulfide ores such as hydroxamates and chelate forming reagents were also developed during this time. During this *chemical flotation* era, it was also realized that addition of certain modifiers, such as NaCN, dextrin, tannin, starch, dyes, and quebracho, work synergistically with collectors to facilitate differential flotation.

The next phase of innovation in flotation reagents during 1950–1990 came from better understanding of mechanisms rather than through screening and exploration (Nagaraj 1988; Jensen 1980). Like the 1920s, the 1980s was a period of intense activity in flotation reagent development with many chemical companies involved such as American Cyanamid (now Cytec), Dow Chemicals, Phillips Chemicals, Clariant (formerly Hoechst), Henkel, Montan Chemicals, and Elf Aquitaine. Some examples of new collectors developed were alkoxycarbonyl thionocarbamates, alkoxycarbonyl thioureas, dithiophosphinates, monothiophosphinates and monothiophosphates, dialky thionocarbamates, allyl alkyl thionocarbamates, allyl xanthate esters, alkyl hydroxamates, phosphonic acids, and sulfosuccinamates (Nagaraj and Farinato 2014). Examples of modifiers include synthetic polymers with mineral-specific functional groups containing the donor N, O, and S for both sulfide and non-sulfide systems.

Enhancements were also made during this period by merely changing the hydrocarbon (nonpolar) part of the reagent molecule without changing the functional group. Some remarkable developments of efficient and economical manufacturing processes for certain flotation reagents had a profound impact in the industry by catalyzing the much needed widespread applications of the specific reagents in many innovative ways (Nagaraj and Farinato 2014). This period also saw new developments of frothers for sulfide flotation such as polyglycol ethers and glycols, which are still in use along with alcohol frothers. Some new modifiers were also developed to address specific industry challenges such as development of zinc cyanide, guar gum, carboxymethyl cellulose (CMC), and sodium-ferrocyanide.

The intense R&D efforts and innovations of the 1980s in reagent development slowed down during the 1990s mainly due to lower metal prices and reduced mining activities. A holistic approach to reagent selection was becoming more important due to increased ore complexities and also the advent of quantitative mineralogy tools and better understanding of flotation cell hydrodynamics. In addition, due to stringent environmental regulations, there is now a focus on developing greener flotation reagents to meet sustainability goals (Somasundaran 2012). More recent application of electrochemical and surface analytical techniques has allowed better understanding of the fundamentals of flotation mechanism and chemistry (Smart et al. 2007). This is an important development that is already helping in better understanding of flotation behavior of complex ores and in developing solutions to complex flotation problems (Ralston et al. 2007; Fuerstenau 2007; Gorain 2013).

2.8.4 Flotation Circuit Design and Optimization using Modeling and Simulation

- The conventional approach to designing flotation circuit focuses on the use of safety factors in scaling-up residence time obtained from bench scale test work. Typically the safety factors range from 2 to 4 depending on ore type, personal preference, and inventive guesswork. There is no scientific basis for selection of this scale-up factor although several practitioners have hypothesized the scale-up factor to relate to power input and froth recovery factors. There are several instances of underestimation of flotation capacity leading to the operating plant not being capable of meeting the metallurgical design target. This could be a very risky proposition leading to lost opportunities due to production losses and the need for further capital expenditure for additional flotation capacity. The reason for high risks in conventional design is that the safety factors are based on previous experience on simpler ores whereas most of the present ores deposits are metallurgical complex requiring a deeper understanding of the mechanisms that drive flotation performance.

Flotation modeling and simulation techniques have recently emerged as important toolset to provide a more rational basis to design and optimization of flotation circuits (Manlapig et al. 1997; Gorain and Stradling 2002; Harris et al. 2002; Dobby and Savassi 2005; Herbst and Harris 2007; Connolly and Dobby 2009). This initiative has been driven by some of the major mining companies mainly to increase the confidence level in designing flotation circuits with minimal risks. This also allows simulation of "what-if" scenarios mainly to understand trade-off between incremental recovery and flotation capacity requirements or capital expenditure which helps in optimization of flotation circuit. Once a robust model has been developed for a deposit, simulations can be done to understand the effect of ore variability and circuit configuration on flotation performance. Flotation modeling and simulation techniques are not perfect due to difficulties in modeling complex ore types. The assumptions made for simulations should be carefully judged for better confidence in model predictions.

2.9 Physical Separation

2.9.1 Gravity Separation

Gravity separation is one of the oldest concentration techniques, with the use of jig screens and sluices for the separation of heavy metals as early as the sixteenth century. With industrialization and the need for large scale production of minerals, continuously operating Jigs and shaking tables were developed and commercialized in the late nineteenth century. Inventions that provided enhanced gravity separators in the mechanically applied centrifugal field were patented in the 1890s and served the basis of recently commercialized technology (Honaker et al. 2014). The early twentieth century brought continued development of high capacity separators for coarse particle concentration such as Baum jig and the Chance Cone. As the mining started to shift towards lower grade ores, the development activities shifted towards fines treatment. Flowing film separators such as spiral and Reichert cone concentrators were developed with wide range of commercial applications in the mid-twentieth century. The late twentieth century was the era of enhanced gravity separators to recover ultrafine gold and tin as fine as 10 μm. A detailed overview of commercially available density based separators can be found in Burt (1999).

2.9.1.1 Pulsed Devices

The Baum jig was a major development in 1892 which replaced the plungers used in older versions of jigs such as Harz with compressed air. This allowed better control of the pulsed cycle in larger compartments (bed width of around 2.5 m) for more efficient separation over a wide range of particle size at higher throughputs (Sanders et al. 2002). With the need for higher throughput and more efficient machines to maintain a uniform pulse across the bed, Batac jig was introduced in 1947, which uses compressed air injection in a chamber directly below the fixed jig screen instead of a separate chamber. Sanders et al. (2002) have summarized the benefits of Batac jigs, which included the possibility of larger beds with commercial units of 7 m wide along with the ability to achieve separation density values as high as 7.85.

A major development in the 1980s was the Humboldt ROMJIG, which utilizes a moving jig screen providing the jigging through mechanical motion provided by a hydraulic arm. The operational characteristics include mass feed rates up to 350 tph, with maximum and minimum feed particle sizes being 400 and 30 mm, respectively, with low water requirement of 185 lpm and no air requirements (Honaker et al. 2014; Sanders and Ziaja 2003). The In-Line Pressure jig (IPJ) was another interesting development in the 1990s, which also utilizes a moving jig screen attached to a hydraulic ram, and can treat particles up to 30 mm with a maximum throughput of 250 tph (Gray 1997).

2.9.1.2 Flowing Film Separators

During the 1960s, Reichert cone concentrator was developed for heavy mineral sand applications with a capacity of 45–75 tph and can treat minus 1.5 mm feed with an effective concentration to a particle size of 45 μm. Reichert concentrator was a significant improvement over the earliest forms of flowing film separators such as buddle used in the early nineteenth century to treat metallic ores such as tin, lead, and zinc. During the late nineteenth century, the vibrating flowing film separators were developed mainly to improve the efficiency of the earlier separators. Shaking tables became popular for recovery of gold, tin, and other heavy minerals in the particle size range of 1.65×0.074 mm and for coal in the 6.7×0.15 mm size range. Wilfley table is the most common shaking device, which has an oblong table with riffles that extend from the table deck and run across most of the deck length.

The most widely used flowing film separator is the Spiral Concentrator, based on Humphreys spiral introduced in 1943. The advancement in construction material such as fiberglass and spray coated with polyurethane during the 1980s allowed significant alternations to the spiral trough geometry and design to address the needs for more difficult separation applications (Honaker et al. 2014). This resulted in higher throughput spirals with a wide range of applications. The compound spiral was a later modification in the 1990s to further improve separation efficiencies for coal and heavy mineral applications (Luttrell et al. 2007).

2.9.1.3 Fluidized Bed Separators

Fluidized bed separators have been used in coal and mineral processing operations traditionally for particle size separation based on teetered bed or hindered bed settling principles. Stokes unit is one of the oldest fluidized bed separators that also provides a very efficient density based separation by using an upward flow of fluidization to suspend the finest density particles in the feed stream (1 × 0.15 mm), thus creating an autogenous dense medium in the separation unit. This creates a density bed that doesn't allow low density particles to penetrate and are carried by the fluidization water in an overflow stream of the separator. When the bed level reaches a certain height and bed pressure, the high density particles are allowed to fall through a discharge spigot in the underflow stream.

There are various commercial units based on Stokes unit such as Floatex, Lewis hydrosizer, Linatex hydrosizer, Allflux separator, and the Hydrosort (Honaker et al. 2014). The Allflux separator, however, is unique which allows two stages of density separation in one unit.

Several developments have taken place over the last two decades to resolve some of the issues with conventional fluidized bed separators such as high turbulence due to feed injection into the center of the unit, disturbing the density separation. The Reflux classifier is one key development consisting of a lower fluidization zone and an upper system of parallel inclined channels, which results in enhanced segregation due to the development of a phenomenon called "laminar-shear mechanism" (Gavin et al. 2010). The advantage of Reflux classifiers is that it can carry out significant upgrading along with de-sliming in only one stage for separating much finer or coarser particles than possible in a spiral separator. The inclined channels provide significant capacity advantage and the system is insensitive to low pulp density with no need for thickening cyclones compared to that for spirals.

The HydroFloat separator is another development that uses air bubbles that are injected into the fluidized particle bed to float the coarse low density particles to allow the bubble-particle aggregates to move upward to the overflow stream. Hence this technology can be used to treat a higher particle size ratio of 6:1 compared to 3–4:1 for a conventional fluidized bed separator (Luttrell et al. 2006).

2.9.1.4 Enhanced Gravity Separators

The enhanced gravity separators are based on applying centrifugal forces to enhance the particle settling behavior by G times the normal acceleration due to gravity (g). This is widely referred to as the G force, which significantly increases the terminal velocity of particles and the dependence of particle velocity on the particle size decreases. Though these separators have been in practice since the 1800s, but their recent resurgence over the past three decades can be attributed to the need to economically concentrate minerals such as gold at particle sizes approaching 1 μm at higher throughput capacities (Honaker et al. 2014).

A review of the various enhanced gravity separation technologies has been recently provided by Cole et al. (2012). These technologies use various levels of G forces along with different separation mechanisms and are applied even for particle size ranging from 1 to 10 μm. The semi-continuous machines are normally used for less than 1 % concentration of heavy minerals with high throughput capacity machines of 1000 t/h. The continuous machines are more effective for greater than 1 % heavy mineral concentration but are available at a lower capacity of 400 t/h.

The Falcon concentrator utilizes a G force up to 300 g's for their continuous machines, which is a vertically aligned, open topped bowl shaped as a truncated cone which is mounted on a rotating shaft. The semicontinuous unit uses G forces ranging from 50 to 200 g's, which however has a fluidization base added to the upper zone of the bowl. Recent development is a unit with an ability to deliver a G force of 600 g's and targets particle sizes of 37–3 μm.

The Knelson concentrators employ fluidized particle bed separation in a mechanically applied centrifugal force for both their continuous and semicontinuous machines. They operate by introducing water through a series of fluidization holes located in rings that circle the circumference of the bowl. The bowl is truncated cone shape and can deliver a G force of 200 g's. The semicontinuous machines are more popular in the industry. Figure 2.13a, b shows photographs of the popular Knelson and Falcon concentrators, respectively.

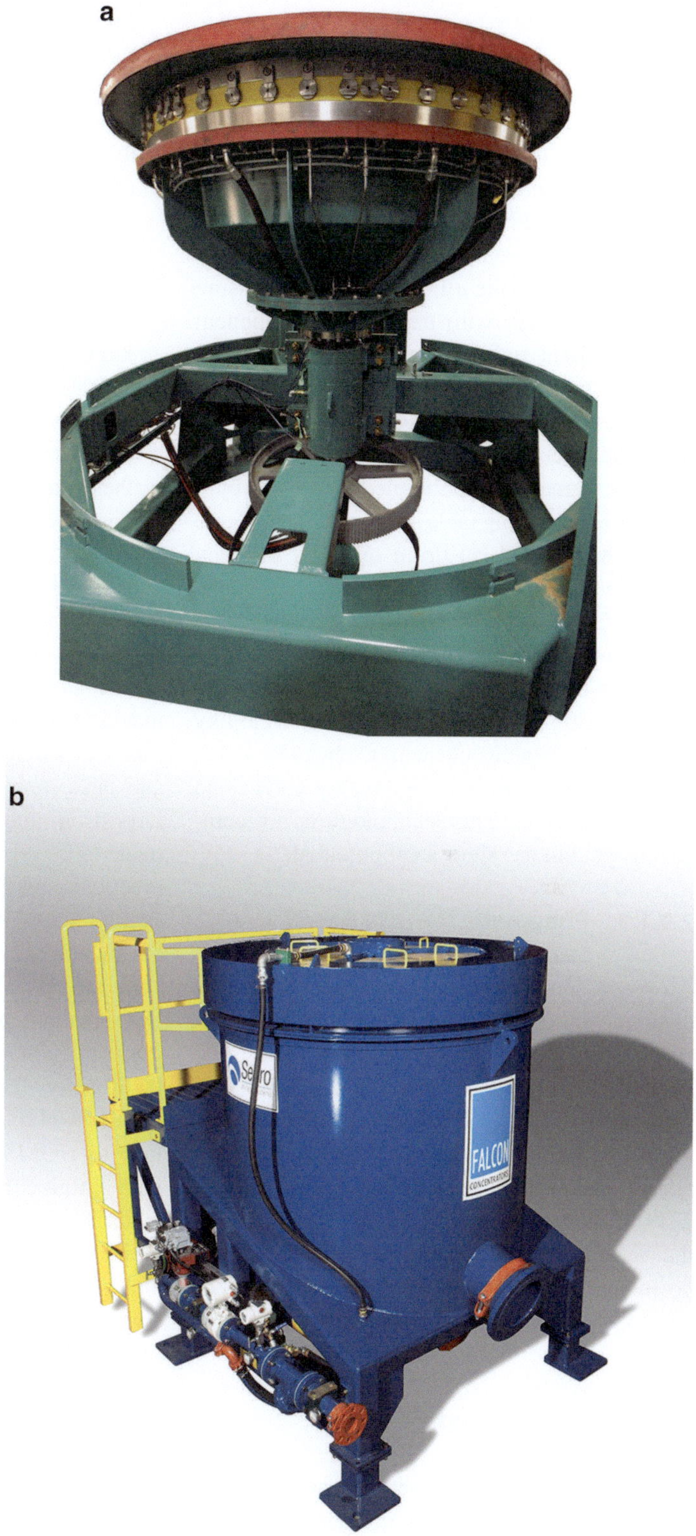

Fig. 2.13 (**a**) Knelson continuous variable-discharge (CVD) concentrator (with permission from FLSmidth). (**b**) Falcon SB concentrator (with permission from Sepro systems)

Kelsey jig is also an enhanced gravity device that can deliver a G force of up to 100. This technology has been successfully applied to many heavy metal applications with a 30–40 units installed globally. However, Knelson and Falcon separators are more popular with a few thousand installations worldwide.

Multi-Gravity Separator is another technology that seems to provide efficient separation, using riffling technology with three main component viz. cylindrical rotating drum, internal scraper network, and variable speed differential drive. Selective separation of fine particles are achieved along the internal surface of the rotating drum using the same basic principles employed by a conventional shaking table but replacing the table surface with a rotating drum resulting in many times the normal gravitational pull.

The Graviton is a centrifugal device that incorporates typically eight Reflux Classifier units with a system of parallel inclined channels aligned at 20° to the horizontal. This technology leverages the benefits of inclined channels and G forces based on the key finding by Gavin and Dickinson (2012). This technology is not commercial yet but has significant potential.

2.9.1.5 Pneumatic Density Based Separations

Air based density separation has been the focus of significant development in the early twentieth century, mainly focusing on coal cleaning. The Pneumatic technologies incorporate the same basic mechanisms used in wet separators including dense media separations, air jigging, riffled tabling, and air fluidized launders (Honaker et al. 2014). Due to water shortages in certain parts of the world, pneumatic separators have significant potential especially in conjunction with HPGR and optical sorting for dry processing.

2.9.2 Magnetic separation

- The largest use of magnetic separation is for iron ore processing in the minerals industry. There are tens of thousands of low-intensity magnetic separation (LIMS) and thousands of high-intensity magnetic separations (HIMS) used in the minerals industry today. Continuous HIMS for dry processing of industrial minerals were developed in the 1940s, but had limited unit capacities and could only process fine particles typically less than 1 mm. The continuous Wet High-Intensity magnetic separators (WHIMS) were introduced later in the 1960s to remove moderately magnetic particles from slurries, with larger Jones type machines introduced in the 1970s. During this time, the High-Gradient magnetic separators (HGMS) were also introduced mainly for kaolin clay processing and these machines were further developed that continued until the 1990s.

A significant magnetic separation development that made a major impact in the minerals industry was the drum separator for magnetite ores, mainly to treat fine grained and low grade magnetite ores for producing high grade concentrates for sinter feed, pellets for blast furnace feed and also for direct reduction iron processes (Arvidson and Norrgran 2014). The German made large capacity Jones type WHIMS became practically an industry standard for hematite and limonite/goethite iron ores around the world. In the industrial minerals area, the HGMS became an industry standard following a new concept of high-gradient generating matrix of filamentary material, which originated from Massachusetts Institute of Technology (MIT) in collaboration with J.M. Huber Company in the USA. The continuous HGMS made it possible to process ores with a high content (+10 %) of weakly magnetic material. The magnetic intensity for typical magnetite ore applications vary from around 800 to 1800 G, whereas for removing weakly magnetic components from kaolin clays requires a minimum of 18,000–20,000 G.

During the 1980s, the Rare-Earth Roll (RER) magnetic separators, using permanent magnet alloys involving rare-earth elements, came into the market with rapid acceptance by the industry. These separators are usually the high-intensity separator of choice for new operations now. The Rare-Earth Drums (RED) are now becoming prevalent for the concentration of some paramagnetic minerals such as hematite and ilmenite at relatively high capacity (Gover et al. 2011). The old induced magnetic rolls (IMR) are still in use but are increasingly being replaced by RER and RED as the latter have better separation efficiency and lower operating costs.

The Wetherill separator type is still in use today in various forms for separating high value minerals, typically after calcination of ore feed. In addition, other applications are removal of tramp iron or metals removal from conveyor belts, tin ores, removing wolframite and other moderately magnetic materials from cassiterite (Arvidson and Norrgran 2014).

Due to declining grades along with the presence of higher detrimental elements in iron ores, the need for sophisticated high capacity magnetic separators is becoming important. Drums for low-intensity magnetic separators in iron ore processing are becoming larger, with drum diameters around 1.5 m and widths exceeding 4 m. Use of RER and RED have allowed to extend the needed magnetic field despite increased operating gaps between the rolls to meet the high capacity requirements. The same trend of increasing size also applies to WHIMS. Other than the trend of increasing size, the use of cryogen-free superconducting coil system is also becoming prevalent, which is the most recent technology for superconducting magnetic separators using a niobium-titanium conductor to generate the magnetic field without the need for helium to cool the coil (Norrgran et al. 2009).

2.9.3 Electrostatic Separation

- Electrostatic separation (ESS) is mainly used for mineral sands application, though earlier patents suggest applications in iron ore and phosphate separations. The relative electric conductivity property of the mineral surfaces is exploited to make the separation. For example, conductive minerals such as ilmenite, monazite, and rutile are separated from non-conductive silica and zircon present in mineral sands. ESS works best on clean surfaces and in many cases the feed needs to be washed and classified for efficient separation. ESS was also successfully implemented at several precious metals smelters recovering metal prills from slag, typically after the Merrill Crowe process. In addition, this technology has been successfully applied in gold mines to recover gold and silver from slag (Hearn 2014). Dry grinding is typically used to liberate metal prills, de-dusting and then ESS to recover these prills for direct smelting rather than the conventional wet milling, gravity tabling followed by drying before sending for smelting (Maki and Taylor 1987).

The company that dominates the electrostatic mineral separation market is Ore Kinetics' Corona Stat, and their focus has been on fundamentals of good operating practices such as moisture control, feeding systems, roll, and electrode cleaning (Gates 2004).

The use of Triboelectric separation for separating hard salts was practiced since the late 1940s. This technique relies on electrical charge differences inducted in nonconductor materials by surface contacting, whereas minerals with a lesser tendency to acquire these electrons will tend to become positively charged and this differential charge is exploited to effect a separation. Gasiorowski et al. (2012) have described the STI technology, which is based on triboelectric separation but with a horizontal belt system with open mesh belts moving in opposite directions so that the material is collected by each discharges at opposite ends of the separator. These machines are mainly used to separate fly ash from carbon in the form of unburnt coal, and also in recycling and limited applications in potash.

Over the last few decades, the main development focus has been on bigger, more power, and better feed systems along with state-of-the-art process control systems. Corona Stat is still dominating the mineral sands but still uses the same principles employing similar grounded drums, charging and static electrodes as used previously in the Carpco separators in the 1950s. It is the innovation of applications and cleaning of these electrodes that has led to Corona's success (Hearn 2014).

The two dominant forms of electrostatic separators in the mineral sands have been a combination of high tension rolls (HTRs) and electrostatic plate separators (ESPs) and are in use mainly in Australia and South Africa. Recirculation of middlings is an important issue facing ESS and has significant implications on plant throughputs (Spiller and Norgren 2014; Germain et al. 2003). The Carrara HTR was a new development with the application of plate electrode into an HTR's body, which resulted in a vastly more efficient process with a sharper separation with noticeable reduction in middling streams and better capacity utilization (Germain et al. 2003).

Cytec is active in the field of surface coating of minerals to augment natural conductivity or non-conductivity properties, though its industrial application is not widely established yet (Ravishankar and Kolla 2009; Maki and Taylor 1987).

2.10 Dewatering and Thickening

Dewatering is an important process in mineral processing. The purpose of dewatering is to remove water absorbed by the particles which increases the pulp density. This is done for a number of reasons, specifically, to enable ore handling and concentrates to be transported easily, allow further processing to occur and to dispose of the gangue. The water extracted from the ore by dewatering is recirculated for plant operations after being sent to a water treatment plant. The main processes that are used in dewatering include dewatering screens, sedimentation, filtering, and thermal drying. These processes increase in difficulty and cost as the particle size decreases.

Dewatering screens operate by passing particles over a screen. The particles pass over the screen while the water passes through the apertures in the screen. This process is only viable for coarse ores that have a close size distribution as the apertures can allow small particles to pass through.

Sedimentation operates by passing water into a large thickener or clarifier. In these devices, the particles settle out of the slurry under the effects of gravity or centripetal forces. These are limited by the surface chemistry of the particles and the size of the particles. To aid in the sedimentation process, flocculants and coagulants are added to reduce the repulsive forces between the particles. This repulsive force is due to the double layer formed on the surface of the particles. The flocculants work by binding multiple particles together while the coagulants work by reducing the thickness of the charged layer on the outside of the particle.

Thermal drying is usually used for fine particles and to remove low water content in the particles. Some common processes include rotary dryers, fluidized beds, spray driers, hearth dryers, and rotary tray dryers. This process is usually expensive to operate due to the fuel requirement of the dryers.

- *Thickener technologies*

2.10.1 Flocculants

Development of synthetic polymers is a major advancement in physical separation during the last 50 years and they have largely replaced the organic polymers (Ballentine et al. 2011; Schoenbrunn 2011). Thickener applications often require handling particles finer than 0.1 μm and these particles are difficult to settle down due to electrostatic charge causing them to repel each other and hinder gravity settling. Economical use of flocculants requires constant attention due to dynamic changes in feed properties and the continuous need for controlling dosage, addition points and the type of flocculants used (McCaslin et al. 2014).

2.10.2 High Capacity Thickeners

Innovation in flocculants was the key driver for high capacity thickeners, which allowed increased tonnage in a smaller footprint. High capacity thickeners are now the norm in minerals industry. They require much stronger rakes than conventional thickeners especially for high specific gravity solids. Other than simply building robust rakes, other innovations include minimizing surface area in high-scale applications such as alumina refining and including spikes on blades to re-suspend heavily thickened materials like magnetite. Streamlined designs reduce torque on both drive and rakes, and posts create space between the blades and rake arms for sticky and/or viscous applications (Schoenbrunn and Laros 2002).

2.10.3 High Density and Deep Cone Thickeners

Outokumpu (now Outotec) and EIMCO developed competing High Density Thickener designs in the late 1980s. In 1996 EIMCO licensed the Alcan technology, leading to marketing of Deep Cone thickeners outside of coal and alumina (Schoenbrunn 2011). Higher under flow densities have also driven robust rake designs and have lower profiles to reduce load. Another universal modification extends above the rake arms to create channels that aid in water removal (Schoenbrunn 2007). Advancements in flocculants, feed dilution, and robust drives along with better tank design have made high density and deep cone technologies practical. Tall sidewalls in High density Thickeners permit 2–3 m bed depths, as compared to 1 m bed depth in conventional and high capacity thickeners. Floor slopes are in the range of 1:6, which is similar to traditional thickeners. Deep cone thickeners use high sidewalls and steep cone bottoms in the range of 30–60° (McCaslin et al. 2014).

2.10.4 Thickener Feed Dilution

Contrary to beliefs and practices in the earlier days, dilution of thickener feed has resulted in numerous advantages emanating from decreased interference between adjacent solid particles without reducing settling velocity. Dilution has improved thickening with greater underflow concentration resulting in the need for relatively smaller unit sizes. Different pumping arrangements have evolved in which the thickener overflow water is used for the dilution needs. EIMCO's patented dilution method utilizes an eductor in the feed line that draws in diluting supernatant from the upper reaches of the thickener. Flocculant addition and thorough mixing occurs before the feed reaches the feedwell. Proper eductor design considers the amount of dilution required but operators adjust with feed velocity (McCaslin et al. 2014). Thickener manufacturers are still looking at new ways of feed dilution along with optimization of equipment costs.

2.10.5 Thickener Drives

The adoption of large diameter precision bearings for the main gear is one of the most significant advancements in thickener drives in the early days. The present ring gears are as large as 3.56 m (140 in.) from a single forging process. This has greatly improved concentricity and permits higher loads. Advances in variable frequency drives, and the ability to link them, eliminated any concerns about imbalance and brought better control. This allows the rakes to be reversed easily and automatically speed up as torque increases (McCaslin et al. 2014)

- *Filtration technologies*: The horizontal pressure filters has been in use since the 1800s but they eventually lost ground to vertical presses in the metallurgical processing. While horizontal filters have not caught up in the number of installed unit on concentrates and tailings, they have caught up in process performance with the advent of better control and automation systems. Many operators and designers consider them a better value for many applications. Current trends are for larger units that require less floor space per area of filtration.

Vertical press filters were among the many innovations in processing and Larox was widely successful with many installations in copper, lead, zinc, nickel, and PGMs. Vertical presses generally have the highest capital cost per unit area of filtration, however, when evaluated from a total life-cycle cost, they are often second only to automated horizontal filter presses and can be a better option with filter cakes that are difficult to discharge (McCaslin et al. 2014).

Rotary vacuum disc filters are most frequently used in the minerals industry with an advantage that they offer continuous filtration with high solids throughput with a low cost and smaller footprint per filtration area. These filters are now available in disc diameters as large as 6 m with total filtration area approaching 200 m^2 operating with up to 20 discs. A recent innovation is simplified bayonet mounting the disc sectors. Rim retaining devices that permit quick and easy removal and installation are also a recent advancement. Rotary disc filters have been largely replaced by pressure filters in the 1980s and 1990s for concentrate filtration. Rotary discs are still used widely in tailings filtration where final cake moisture is less of a factor than many concentrate applications (McCaslin et al. 2014).

Horizontal belt vacuum filters are typically used for tailings dewatering, hydrometallurgical and zinc processing. They have the advantage of high wash efficiency utilizing only minimal wash liquor. The initial challenges of tearing and abrasion of the filter cloth were overcome through technology improvements that resulted in explosion of installations in the 1960s and 1970s. The indexing belt filters developed in the 1980s further improved its prospects with Delkor in 1986 installing 26 belt filters at the Nchanga tailings leach plants in Zambia for 80 m^2 filtration area (Chisakuta et al. 2005). This was a milestone in the race to build larger machines. Innovations in support deck design followed with introduction of wear belt decks and roller decks in the 1990s.

2.10.6 Filter Media

The trend of growing use of synthetic fibers since the 1940s is continuing, with materials such as nylon, polyester, polypropylene, polytetrafluroethylene (PTFE), and polyetheretherketone (PEEK). They offer significant advantages over the natural fibers such as cotton, wool, and paper in that they have better wear characteristics, improved strength, and greater stability along with improved filter cake release and reduced blinding depending on the fibers used. They also offer superior life leading to reduced down time and lower operating costs. Advances in weaving and finishing during the 1990s have allowed improved filtration of finer particles using the synthetic fibers. Plain weaving was common in the early days but developments in satin weaves and double layer weaves greatly affected filter performance. The former offered high airflow and a slick surface on the cake side for exceptional cake release, whereas the latter provided better filtration and high strength. The quality of finishing significantly improves filter media's performance (Hoijer and Grimm 2011).

2.11 Tailings Retreatment and Water Quality

The generation of mine tailings is likely to increase as the industry is processing low grade ores with high throughputs to meet the demands of metals and minerals. This is posing a significant environmental and community relations issue. The volume ratio between concentrate and tailings is suggested to be as high as 1:100. It is estimated that there are at least 3500 tailings dams and ponds globally. In European Union alone, 34 % of the total waste is generated by mining and about 2–5 tailings dam accidents occur with significant environmental impact, despite some stringent regulations (Riina et al. 2014). For every ton of copper metal produced, it is estimated that about 128 t of solid copper tailings is generated (Gordon 2002). About 230,000 t of dry copper tailings is produced daily at the Escondida mines in Chile (Boger 2009).

It is interesting to note that the Chilean Environment Impact Assessment System (EIAS) has evaluated 1257 mining projects during 1997–2013, with an investment of US$ 103 billion, of which 78 % of the projects were approved. It is important that mining companies have a clear understanding of similar regulations in various countries and their evaluation methodologies along with the critical steps involved for all the mining projects to ensure a "license to operate."

Mine tailings presents a very important source of future ores; as an example, the content of copper and precious metals in copper mine tailings is few times higher than the primary copper ores (Han et al. 2014). Retreatment of mine tailings is gaining attention and there are a number of investigations being carried out to recover base and precious metals along with some environmental remediation efforts (Riina et al. 2014; Han et al. 2014; Ebell and Capstick 2013; Dudeney et al. 2013; Primeau and Gilbert 2012; Pal et al. 2010; Newell and Bradshaw 2007; Xie et al. 2005). There is also a growing interest in using mine tailings for construction and building materials (Onuaguluchi and Eren 2012).

2.11.1 Seawater Processing

With the trend of declining head grades requiring high throughputs to improve economics, the need for water has increased significantly over years. A review of water needs in the mining industry has been presented by Global Water Intelligence (Arowoshola et al. 2011).

In temperate and equatorial latitudes, there is no shortage of water in general, although the disposal of used water is a significant issue. In other arid parts of the world such as in Chile, Western Australia, and South West USA, water is scarce or unavailable requiring the need to use sea, hypersaline or brackish water for mining. The situation has become so dire in Chile that, in March 2012, authorities declared a state of emergency on the Copiapó River watershed and it is impossible to capture any more water from the aquifers.

Desalination of seawater is expensive as new desalination plants and associated pumping and pipeline systems can easily have a capital cost exceeding $500 million along with high operating costs (Blin and Dion-Ortega 2013). In addition, the impact of raw seawater and hypersaline water on processing such as grinding, gravity, flotation, leaching, and electrowinning is also not favorable for some ore types due to issues with high density, viscosity, and pH buffering issues (Aral et al. 2010). Seawater is used in Chile, Indonesia, and Australia using hypersaline and raw seawater but the processes need modifications. KCGM in Western Australia uses hypersaline water with 30–200 g/L of dissolve solids, which results in high cyanide consumption as the cyanide leaching process is operated at pH 9 due to buffering issues. Minera Michilla Copper in Chile carries out washing of the leached solution before electrowinning to avoid generation of dangerous chlorine gas. Use of raw seawater in copper and copper-molybdenum processing plants in Chile requires innovative processing such as use of the patented AMBS flotation process at Antofagasta Minerals Esperanza operation and dilution of seawater with tailings water in the process plant at Las Luces (Blin and Dion-Ortega 2013; Moreno et al. 2011).

The Escondida mine, also located in the Atacama Desert, has been desalinating seawater since 2006, while Freeport MacMoRan has nearly completed a desalinating plant and pipeline to feed its Candelaria project, located South of Copiapó. Other major mines have made similar choices in Australia, such as Newmont's Boddington, which has contracted Osmoflo to process the brackish water coming out of its dam. It is estimated that there are about 20–30 mines globally that use desalinated water right now, 10 of which are located in Australia or Chile alone. About 15 more desalination projects have already been commissioned, and several dozen are under evaluation (Blin and Dion-Ortega 2013).

Barrick Gold has developed a new process viz. Air-Metabisulfite treatment (AMBS), which allows use of seawater or brackish water for pyrite depression in copper flotation with minimal metallurgical impact compared to that with the conventional lime based process (Gorain 2012). The AMBS treatment has also resolved the issue of molybdenum flotation in copper-molybdenum ores using seawater.

2.12 Automation, Control, and Integration

2.12.1 Automation and Control

- The mineral industry has made significant progress in instrumentation and control with contributions from different players, with major benefits in improving productivity, increase in equipment availability and utilization along with safety. The early efforts were focused on regulatory control, driven mainly by the introduction of mini-computers and computer savvy process engineers. These successes helped drive development in new analysis techniques and the early adoption of new generation of control hardware. Regulatory control improvements led to an interest in advanced process control and this sparked the exploration of advanced instrumentation. Over the past few years, a new innovation path has emerged—peripheral applications that harmonize well with the goals of process control (Flintoff et al. 2014).

The first step in any process control is to have a deeper understanding of the key variables (or drivers) of the process and the relationship between these drivers with metallurgical performance. The importance of these variables will vary depending on ore types and the target performance for each ore type. As an example, some of the key variables in a flotation operation could be:

1. Grind size P80 (primary and regrinding)
2. Slurry flow rate (feed, concentrate, and tailings)
3. Slurry density or % solids by weight (mill, rougher, cleaners)
4. pH (mill, rougher, cleaners)
5. Eh (mill, rougher, cleaners)
6. Dissolved oxygen levels (rougher, cleaners)
7. Assays (feed, concentrate and tailings)
8. Reagent addition rates (mill, roughers, and cleaners)
9. Particle size distribution (concentrates and tailings)
10. Mineralogy (feed, concentrate, and tailings)
11. Air flow rates (rougher and cleaner cells and banks)
12. Froth depth (rougher and cleaner cells and banks)

It is best to target a limited number of these variables to simplify the control system otherwise it becomes an expensive undertaking and overwhelming for many plant operators resulting in loss of trust in the system and eventually results in implementation failure.

There are different levels of process control one can utilize in a process plant. It is important to note that the effectiveness of controllers depends on the foundation upon which they are built. Reliable measurement of control variables is critical for effective process control. The best control system also incorporates a strong interaction between process operators, metallurgists, and process control personnel to ensure the system is actually utilized. The need of the operation must be met through leveraging the experience of operating personnel along with the capabilities of the process control system.

Technology advancements have now made it possible to monitor several key drivers with a much higher level of accuracy. Instrumentation for process control should be based on the value they bring in to make control decisions. More instrumentation is not always better. Reliable sensors are always better than virtual sensors or model estimates and model estimates are better than no sensor at all. Issues with sensor reliability force the need to include virtual sensors for higher data reliability in better control (McKay et al. 2009).

2.12.2 Online Monitoring Systems

- Technology advancements have now made it possible to monitor several important plant measurements with a much higher level of confidence than possible earlier. It is important that we utilize these technologies as much as possible to assist in better control of our operations along with improved reconciliation between geology, mining, milling, and metal production. In the early 1960s, online chemical analysis using X-ray fluorescence (XRF) allowed continuous analyses of process streams in mineral processing operations. This allowed more productive work than testing of routine samples.

2.12.2.1 On-Stream Analysis

Development of on-stream analysis (OSA) systems has been critical for flotation circuit control and optimization. This technology enabled online measurement of assays of key flotation streams such as plant feed, tailings, and final concentrates, which allowed estimation of recovery every few minutes. The development of radioisotope based on-stream analysis system was due to a decade long work by different players, which led to the commercialization of a radioisotope OSA system in Australia (Lynch et al. 2007).

2.12.2.2 Split Online Fragmentation Analysis

Optical sizing technology for measuring the size distribution of fragmented rocks has been used for many years now. The WipFrag fragmentation sizing system was originally designed using a roving camera and operator assisted analysis. Using optical sizing technology, mining blast professionals could evaluate, reassess, and redesign their blasts, while understanding the effect of their design on their final product. In addition, they could begin to quantitatively evaluate the effect of geological structure on their blasts. Although the accuracy of this method was low, it was found to be useful as the alternative method of screening large masses of rocks which is almost impractical.

More recently, the new Split-Online digital image analysis has been applied for accurate, continuous, and rapid measurement of rock fragmentation. Cameras are installed along key stages of crush, convey, and milling processes to determine rock fragmentation size. The critical data for mining operations, including particle size, shape, color, and texture, are calculated by the advanced Split algorithms and can be reported to a centralized database to enable real-time evaluation by mine operators and management. Figure 2.14 shows installation of the Split-Online camera

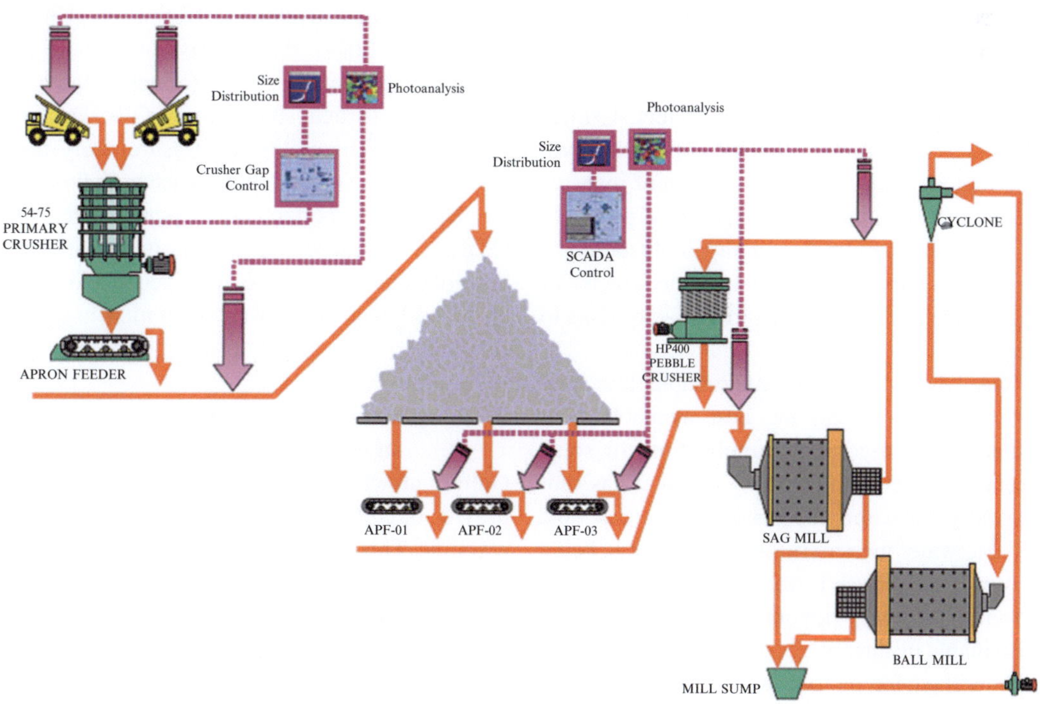

Fig. 2.14 Installation of the Split-Online camera systems at the Morila gold mine in West Africa (Gillot 2006)

systems in various stages of the comminution process at the Morila gold mine in West Africa. A 10 % mill throughput improvement was achieved with this system. Split-Online cameras along with ore tracking system have significant potential for operations and can provide information on the less understood interface between mine and mill, which is key to successful implementation of an integrated operations strategy.

2.12.2.3 Slurry Flow Meters and Density Gauges

Accurate and sustainable measurement of flows in a process plant is fundamental to successful process control in any operation. Also this is critical to reliable solids, water, and metal balance along with recovery calculations needed to meet the criteria for an operational metal accounting standards (MAS). Many of the electromagnetic flow meters used in operations have several limitations to accurate flow measurements over a long period of time. The coupling of electromagnetic flow meters with nuclear density gauges is prone to errors due to entrained air and scale build-up on the pipe inner wall. Scaling is common due to corrosive environment in our metallurgical processing, which causes eventual measurement deterioration and the need for maintenance and scheduled replacement of both electromagnetic and ultrasonic meters. Critical lines are costly to shut down because that will interrupt plant operation. Installation of invasive meters in old piping carries a risk of pipe cracking that will require costly repair. Installing a large, heavy invasive electromagnetic flow meter is logistically difficult and carries safety risks to personnel.

One key to an accurate and reliable flow measurement is selecting the best available flow meter technology for the application. In some operations, increasing scarcity of water has substantially increased the need for accurate and reliable water measurements. This need is being driven by water use restrictions imposed by the government along with our desire to operate in a sustainable manner as good corporate citizens. It is important for us to demonstrate to both the communities and the government that we are operating within their agreed-upon consumption limits, which may even be reduced in the future.

The SONARtrac® flow meters from CiDRA® have recently achieved wide acceptability in the mining industry with some significant benefits compared to the traditional flow meter technologies such as Electromagnetic, Ultrasonic Doppler, Differential Pressure, or Coriolis (Markoja 2011). The CiDRA® flow meters are light weight, noninvasive as they clamp on to the existing pipes and have no wetted parts, which maintains the full integrity of the piping system and eliminates the possibility of leakage associated with flow meter fittings. These flow meters are easy to install, unaffected by internal scale, work on pipe of any material, both lined and unlined, and are maintenance free. In addition, the *SONARtrac®* flow meters have demonstrated better accuracy compared to the existing magnetic meters. Figure 2.15 shows some of the typical applications of CiDRA® flow meters in a process plant.

These CiDRA® flow meters have recently been used for better quantification of recirculating loads in a grinding circuit and also for detecting the presence of coarse or even rock particles in cyclone overflow, which allows operators to optimize cyclone performance (Cirulis and Russell 2011). Also these flow meters are being used to monitor the performance and condition of centrifugal slurry pumps (Van der Spek et al. 2009). The ability to measure slurry flow rates in pipes with air holdups such as in flotation has been an important development using the CiDRA® flow meters. It is important that these new robust flow meters be evaluated to allow better metal balancing and reconciliation in a process plant.

2.12.2.4 Online Monitoring of Mineralogy and Assays

Online mineralogy and assays measurements provide opportunity to optimize process plant in real time. This will allow a tighter control of tailings losses, which otherwise is difficult to achieve in an off-line based mineralogy and assays measurements that we use presently. Not all assays such as gold or platinum can be measured online accurately, but for every ore type a proxy for gold

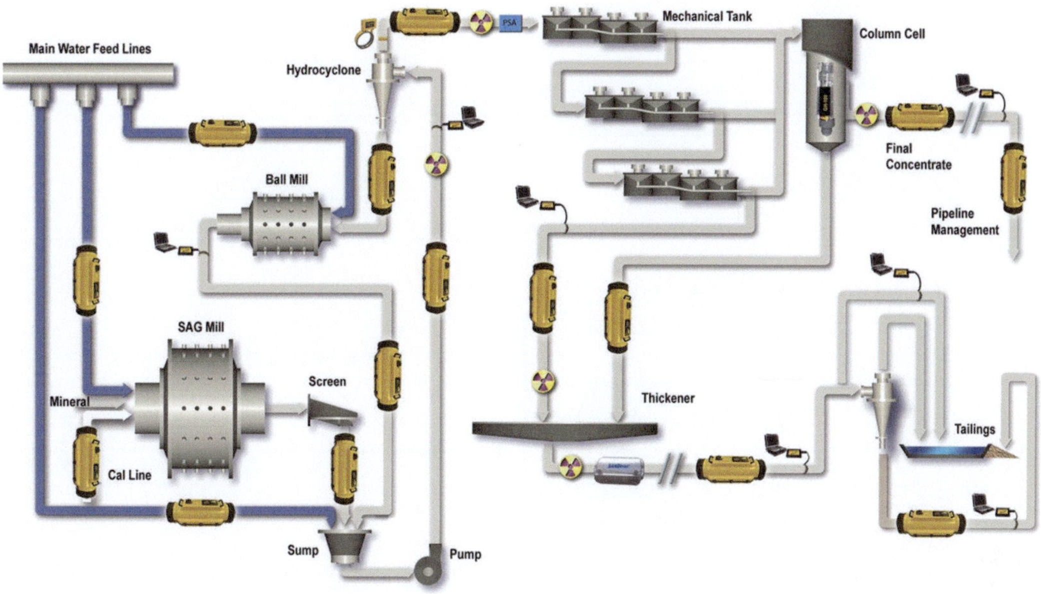

Fig. 2.15 SONARtrac® flow meters (depicted in *yellow*) in various duties in a process plant for accurate and sustainable measurements (with permission from CiDRA®)

can be used with a higher level of confidence through frequent calibration using the centralized database. For refractory gold ores, this proxy could be pyrite or iron assays.

The recent development of Laser-Induced Breakdown Spectroscopy (LIBS) provides real-time online assay measurement on belt conveyor and has potential for ore sorting and also for obtaining online assays of mill feed for plant control (SonicSampDrill 2014).

Online mineralogy measurement from BlueCube Systems (Pty) Ltd. is a major breakthrough and has recently been successfully implemented in many precious and base metals industries (Mumbi 2012). This technology is based on diffused reflective spectroscopy combined with propriety chemometric techniques.

BlueCube's MQi Slurry analyzer is used for an in-line real-time measurement of slurry stream composition viz. mineral grade, elemental grade and particle size. With the availability of real-time data, the effect of process changes can be tracked without having to do a full plant survey. This technology along with slurry flow measurements such as with CiDRA® flow meters can be effectively used to optimize recovery. This is an important tool for maximizing revenue from a process stream. In a flotation application, this technology can be used in conjunction with froth cameras for better control of flotation banks.

BlueCube's MQi Hydromet Analyser is used for an in-line measurement of the composition of solutions in a hydrometallurgical process plant. Ionic species in clear solutions are measured in-line and in real time. The technology is based on absorption spectroscopy combined with proprietary chemometric techniques. The technology appears to be suitable for installation in any plant environment as long as the feed to the system is filtered and free of suspended solids. This is an important development for optimization of pressure oxidation and leaching systems. Figure 2.16 shows a BlueCube® system for controlling a flotation circuit.

2.12.2.5 Ore Tracking for Mine to Mill Integration

- Ore tracking system is becoming a necessity as the mining operations are becoming immensely complex. This system provides an effective

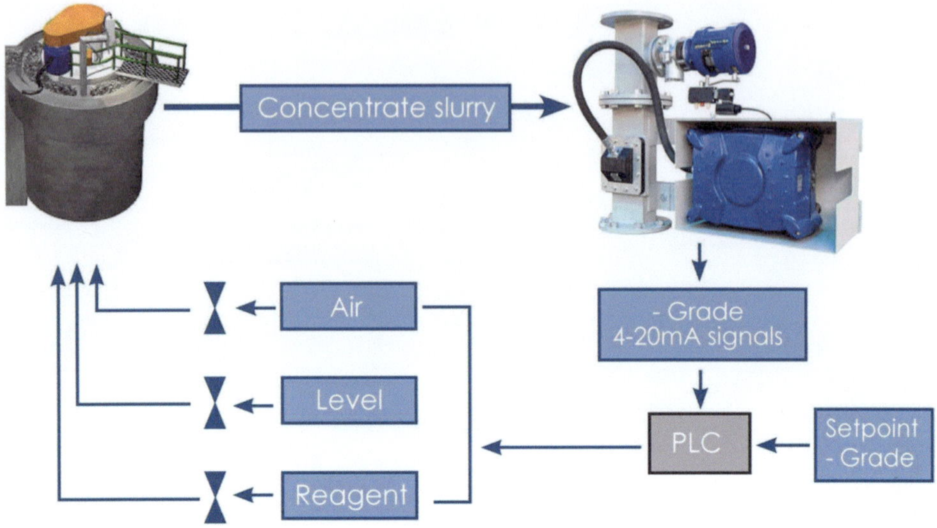

Fig. 2.16 Control of a flotation circuit using inline real-time measurement of mineralogy and assays using the BlueCube® system (with permission from BlueCube Systems Pty Ltd.)

integration tool between mine and the mill. It is essentially a system to track ore types from the mine to the mill and through the process plants using RFID tags. Physical RFID tags are now available from a number of suppliers, with some tags specially designed for mining applications. They can be passive or active RFID tags which affects how close the reader needs to be to detect the presence of the tag. Also active tags require a battery power source and therefore may not be appropriate for long-term stockpile applications.

This system helps to identify the origin of the ore and its behavior throughout the mining process. Metso's SmartTag™ ore tracking has recently been successfully implemented in process integration and optimization (Isokangas et al. 2012). These tags travel through a mine and process plant in a series of steps. Initially the tag and insertion location is logged using a handheld computer or PDA, and then it is inserted into the rock mass in the same holes where blasting explosives are placed. The tag travels with the ore through digging, transport, and processing before being detected by sensors that are positioned along conveyor belts after crusher and in the stockpile reclaim belt or SAG feed, when the time and specific tag is recorded. The RFID tag data is then loaded into a centralized database and analyzed as required.

The use of a SmartTag™ system allows development of relationships between ore characteristics (available in the mine block models) and operating parameters in the mine and processing plant (such as ore dilution, fragmentation, stockpile residence times, segregation, energy consumption, and ore grade) along with actual mine production and metal recovery. With this knowledge, operating parameters can be optimized to respond rapidly to changes in ore characteristics, thus resulting in controlling operating costs and improving productivity.

Some of the key benefits of using SmartTag™ include linking of spatial mine data to time-based processing data, increased confidence in measuring ore blend, proactive process changes for known ore types, identifying material handling logistics issues and accurate measurement of residence times in stockpiles and bins thus assisting in better stockpile management (Jansen et al. 2009).

A significant advantage of using SmartTag™ is that every tag that travels from the mine to the plant is a geometallurgical data point. By utilizing this data and process knowledge, one can create models and refine existing mine block models to predict plant performance based on geological data. Figure 2.17a, b shows some for the advantages of using SmartTag™ systems.

Ultimately a link between the spatial understanding of the mine and the temporal understanding of the process plant has immense long-term benefits such as Grade Engineering (McKee 2013). Having a central database with estimates

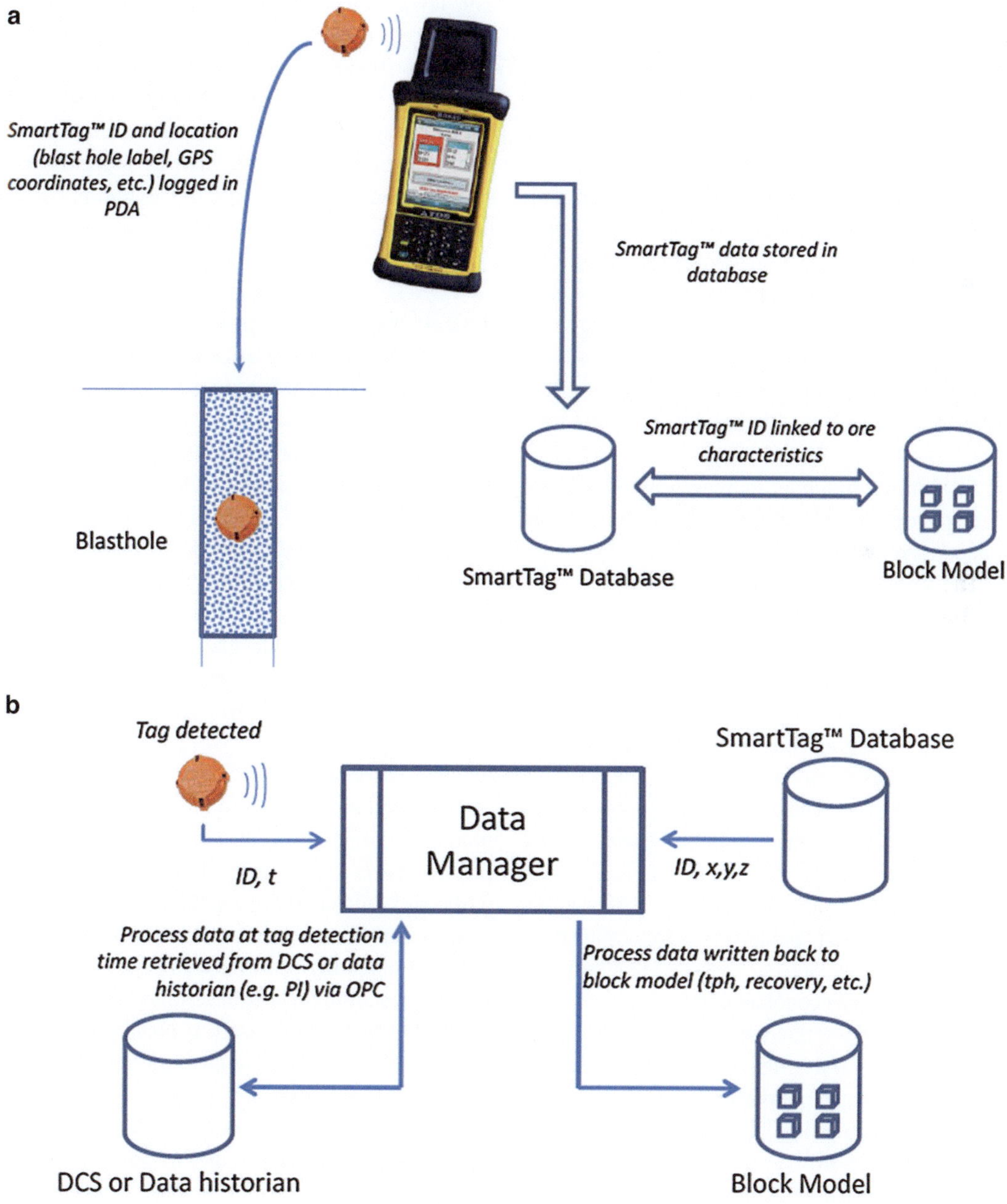

Fig. 2.17 (**a**) A snapshot of the Metso SmartTag ore tracking system to characterize ores from different blast holes (Wortley and La Rosa 2013). (**b**) The use of SmartTag for mine to mill reconciliation and continuous refinement of the mine block models (Wortley and La Rosa 2013)

of key value drivers and KPIs allows benchmarking of an operation in terms of its unique set of KPIs. For example, a relatively simple throughput forecasting model can be developed which considers the range of ore types and how they will interact as blends. This can provide benchmarking estimates and KPIs can then be developed as a percentage of actual vs. benchmark.

2.12.3 Integration: mine-to-mill strategy: optimization of blasting costs vs. milling costs

Mine-to-Mill is an operating strategy for mining operations to enhance the performance of mining and downstream processing activities. One of the key objectives of the Mine-to-Mill strategy is to maximize operations profitability through a holistic approach to the optimization of ore fragmentation. Generation of fines by blasting can have a significant impact on the following:

- SAG mill throughput increase (10–30 %)
- Reduced overall energy costs (up to 30 % decrease in kWh/t)
- Higher excavator productivity
- Higher truck loading
- Better primary crusher productivity
- Better heap leach permeability

The key task for Mine-to-Mill optimization is to identify the optimum feed size distribution for the crusher or the mill. The optimum feed size distribution is typically generated through blasting (ROM size distribution), crusher settling along with stockpile and feeder management. Figure 2.18 demonstrates that the size reduction starts from mine itself and the mine must deliver a product that will produce the optimum economic outcome.

Table 2.2 shows that the energy costs for milling are significantly higher than blasting costs, which provides an opportunity in maximizing value from the blasting. This is a strong justification for Mine-to-Mill.

Mine-to-Mill now is a proven methodology and has been applied at many large open-pit operations around the world (McKee 2013).

Table 2.2 A typical comparison of the blasting and milling costs (JKTech 2004)

	Cost ($/tonne)	Energy (kWh/T)
Blasting	0.20	0.2
Crushing	0.80	2.0
Grinding	4.00	20.0

This involves rock characterization, benchmarking, and process modeling. Many of the projects have achieved 10–30 % increase in mill throughput. The issues facing success with this approach are not just technical, but involve cultural change and sustained implementation, such as departmental silos and lack of proper systems to retain knowhow, that exist in many mining operations.

2.12.4 Integrated Operations (integration with upstream and downstream processes)

One of the key approaches that is of immense value in this new operating philosophy is the concept of "Integrated Operations." The goal is to drive improvement through better integration of operations. It is about optimizing the individual parts but it is also about optimizing the value across the whole value chain (Sherring 2013). This model has proven to be successful in many industries including oil and gas, manufacturing and recently in mining by bringing many fragmented solutions together.

Some mining companies are increasingly integrating their operating data scattered across multiple locations, into centralized database with dashboards and analytics capability with an aim to make a step change in improving productivity and reducing operating costs through better collaboration between different business units. This strategy is proving to be immensely beneficial for many operations resulting in improved mill throughputs (10–30 %), reduced grinding energy costs (10–20 %), improved metals production, better reconciliation between mine to mill and also between reserve model, mine production and bullion production along with many intangible benefits such as improved communication and cooperation

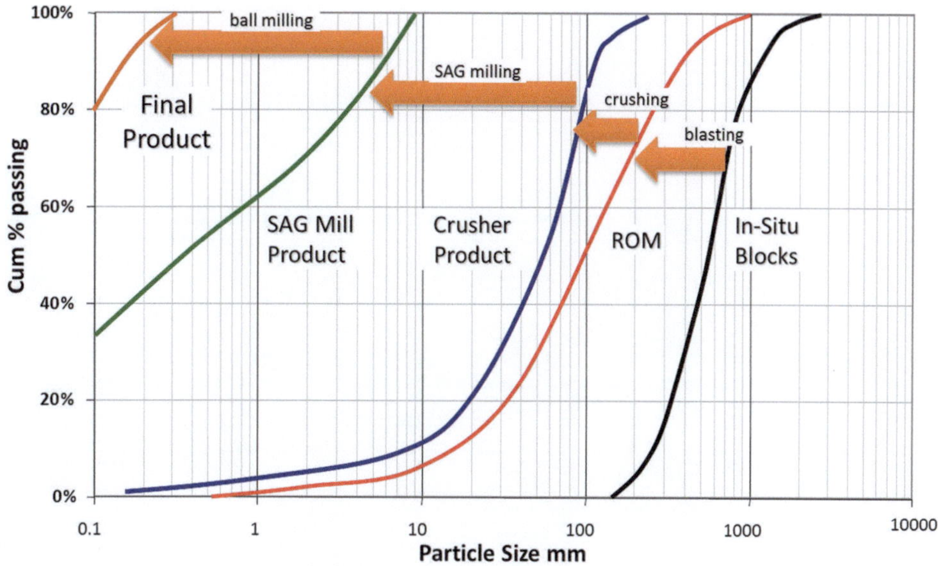

Fig. 2.18 A key strategy for size reduction in Mine-to-Mill optimization (Dance 2013)

through data transparency and knowledge sharing, effective problem solving through quick root cause analysis and ultimately better decision making.

This "optimization of the parts," no doubt, allows improving efficiency of individual units and is an important part of the business. But this results in potential economic benefits to be missed as the focus is on constant execution to meet the targets of the individual business units, and the improvement opportunities associated with a systemic approach come only as an afterthought. This is typical of operators working in various business unit silos resulting in limited conformance to life-of-mine (LOM) plan and looking beyond the plan is often challenging.

To address these challenges and to improve profitability of operations, it is becoming imperative to look at a mining operation holistically requiring a strong interaction of professionals in various business disciplines involving Geology, Mining, Processing, Maintenance, Environment, Health and Safety, Finance, Supply Chain and Logistics, IT, HR, Community and Public Relations.

There are many examples that suggest that integrating different business disciplines such as Mine-to-Mill involving true integration of geology, mining, and processing functions have led to significant productivity gains and cost savings along with intangible benefits including a more cohesive, satisfied, and performance oriented workforce transcending the traditional boundaries (McKee 2013).

A photograph of Kairos Mining's Collaboration Centre (a joint venture of Codelco and Honeywell) in Santiago, Chile, is shown in Fig. 2.19. This center provides real-time monitoring of plant operating data and automated control systems for multiple sites with a focus on knowledge management and long-term process performance improvements (Zamora et al. 2010).

Based on recent successes in operations of some companies, this Integrated Operations strategy provides an opportunity for companies to make a paradigm shift in our approach to productivity improvements and operating cost savings. Figure 2.20 shows an integrated operation strategy with a stage wise focus.

2.12.5 4D-BIM (Building Information Model)

4D-BIM is an emerging socio-technical communication tool using time as an add-on to 3D mod-

Fig. 2.19 Kairos Mining's Integrated Collaboration Centre in Santiago, Chile; a joint venture of Codelco and Honeywell (with permission from Honeywell)

els. This is defined as digital representation of physical and functional characteristics of an operation, and is a shared knowledge resource for information about a facility forming a reliable basis for decisions during its life cycle (from project conception to operation closure). This technology will allow mining construction to be more efficient, less costly and can bridge potential information loss caused by hand over to production. The best use of this tool is to improve communication of the proposed project phasing to all stakeholders. This will allow stakeholders to visualize how the project affects them throughout the life cycle of a mining operation along with any environmental and community issues. This is communication tool for the public to provide feedback and therefore improves dialogue and engagement between stakeholders (Huls 2014).

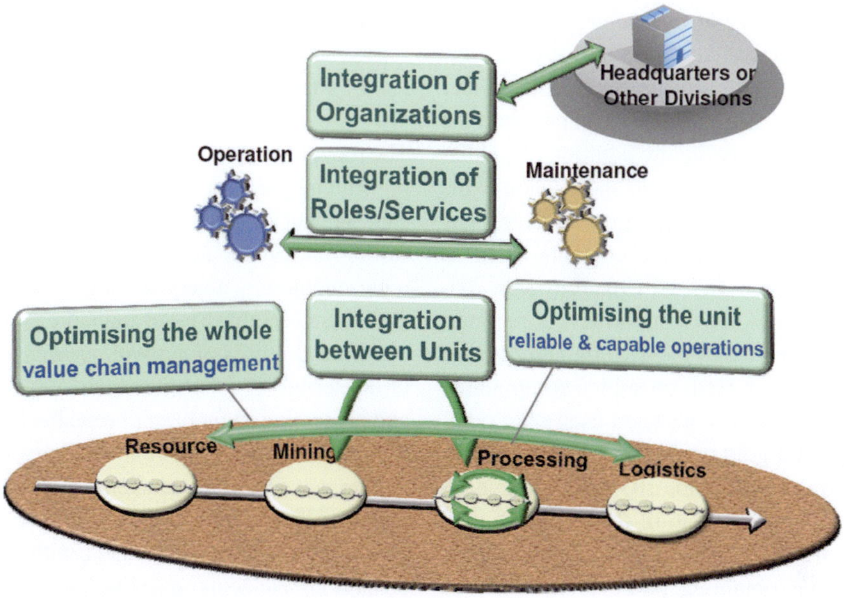

Fig. 2.20 An integrated operation strategy with a stage-wise focus: first on value chain optimization for an operation followed by integration of multiple operations (Sinclair 2012)

2.13 Shifting Paradigm in Mining and Processing

2.13.1 In situ leach recovery (The merging of mining and processing)

In situ recovery (ISR), also called in situ leaching (ISL) or solution mining, is a mining process used to recover minerals such as copper and uranium through boreholes drilled into a deposit. The process initially involves drilling of holes into the ore deposit. Explosive or hydraulic fracturing may be used to create open pathways in the deposit for solution to penetrate. Leaching solution is pumped into the deposit where it makes contact with the ore. This process allows the extraction of metals and salts from an ore body without the need for conventional mining involving drill-and-blast, open-cut or underground mining.

In 2013, 47 % of world uranium mined was from ISL operations. Most uranium mining in the USA, Kazakhstan, and Uzbekistan is now by in situ leach methods. ISL mining of uranium is undertaken in Australia, China, and Russia as well (World nuclear association website 2014). ISL is seen as a cost-effective and environmentally acceptable method of mining, and is now getting some attention due to various challenges with the conventional mining.

In situ leaching of copper is usually carried out using acid (sulfuric acid or hydrochloric acid), then recovered from solution by solvent extraction electrowinning (SX-EW) or by chemical precipitation. The San Manuel copper mine, owned by BHP Billiton, was a successful operation that integrated ISR methods with open-pit and underground mining and produced approximately 3.25 billion pounds of copper in 14 years of production. Two additional projects owned by BHP Billiton, Pinot Valley and the Miami Unit, also used ISR to extract copper (combined with SX-EW in the case of the Miami Unit). The Florence Copper Project, owned by Curis Resources Ltd., is located roughly 150 km to the northwest of Gunnison and is host to a resource of 2.84 billion pounds of oxide copper (Excelsior Mining Website 2014).

Ores most amenable to leaching include the copper carbonates malachite and azurite, the oxide tenorite, and the silicate chrysocolla. Other copper minerals, such as the oxide cuprite and the sulfide chalcocite may require addition of oxidizing agents such as ferric sulfate and oxygen to the leachate before the minerals are dissolved. The ores with the highest sulfide contents, such as bornite and chalcopyrite, will require more oxidants and will dissolve more slowly. Sometimes oxidation is expedited by the bacteria *Thiobacillus ferrooxidans*, which feeds on sulfide compounds.

In situ leaching has not been used on a commercial scale for gold mining. A 3-year pilot program was undertaken in the 1970s to in situ leach gold ore at the Ajax mine in the Cripple Creek district in the USA, using a chloride and iodide solution. After obtaining poor results, perhaps because of the complex telluride ore, the test was halted (Chamberlain and Pojar 1984).

Figure 2.21 shows a pictorial representation of the ISL process for recovering uranium.

Some of the main advantages of ISL mining are no creation of open holes, waste dumps and leaching or leach pads, with minimal visual disturbance, resulting in lower capital and operating costs as no waste or ore is moved. Fewer permits are required for ISR compared to other mining processes. Despite these advantages, one of the key barriers to this technology is the environmental concern associated with ground water contamination from injection of hazardous lixiviants such as cyanide or sulfuric acid. Lixiviant production recovery is an issue with the existing ISL technology for many ore types due to poor rock permeability and inadequate understanding of lixiviant flow control.

Recent developments in discrete fracturing network (DFN) site characterization, analysis and modeling, have the potential to address both the production efficiency and environmental issues, with improved well-design and through transfer of learnings from the oil and gas industry (Dershowitz 2011). Dershowitz has also reiterated that global experience in ISL has better understanding of hydraulic cage control of lixiviant to ensure all fluids are captured and regulated.

2.13.2 Zero waste mining (The future of mining and mineral processing)

The root cause of most of the problems in the existing mining paradigm is the generation of significant amount of waste and the need for handling, storing, and processing of this waste with valuables representing only a very small fraction. As the ore bodies are becoming more and more complex with lower head grades along with greater proportions of deleterious elements, the innovations in mining and processing so far hasn't been able to keep pace with the increasing degree of challenges confronting the industry. These challenges include suboptimal metallurgy, high operating and capital costs, environmental issues associated with tailings and deleterious elements, ever increasing mine footprint, lack of resources including energy, quality water, skilled personnel, and ever-growing community issues. The more waste we generate, the more these challenges magnify.

Significant efforts have been made by many players in the industry to address these challenges with some remarkable successes, but unless we address the root cause of the problem, which is waste generation, it will be difficult to find a robust sustainable solution. The ideal scenario will be the case for Zero Waste Mining, in which mining will target only the valuables without the need to remove the host rock and also all mined material will be used to create value added products with no waste dumps or tailings disposal.

Zero waste mining is best approached in stages as follows:

- Stage 1: Improve efficiencies of unit operations in the present mining value chain paradigm to maximize recovery of valuables presently being lost to tailings. Synergies of various processes with integrated goals to optimize the overall mining process have been recognized as key to success (USNMA 2000). Retreatment of old tailings to recover valuables previously lost due to inefficient processes is being pursued (Riina et al. 2014). This stage also includes recovery of by-products, not previously considered due to lack of technolo-

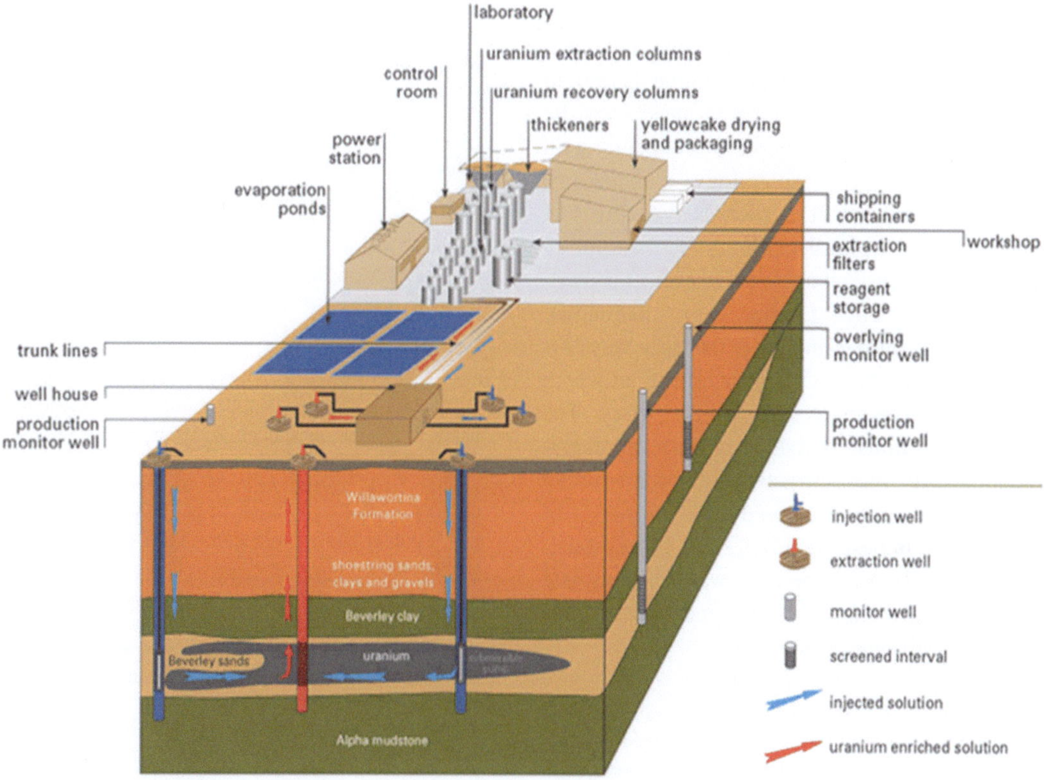

Fig. 2.21 Pictorial representation of the ISL process for uranium ore (with permission from Heathgate Resources, www.heathgate.com.au)

gies, such as rare earth elements and iron metal from base and precious metal mining tailings ponds (PRO 2014). This stage also involves reduced wastage of limited resources such as water, energy, and technical skills.

- Stage 2: Focus on bringing the process plant as close as possible to the mine with pre-concentration either underground or close to the mining phase in open-pit with an aim to significantly reduce transportation, processing, and storage of mine waste. This is also known as near-to-face processing. Studies have suggested that integrated mining and waste rejection processes has high potential for deep underground mining (Dammers et al. 2013; Bamber 2008; Batterham 2003). Other concepts such as rejection of waste as coarse as possible with an aim to reduce energy and water requirements along with tailoring this coarse waste for value added products such as building materials needed for local community and markets are also being pursued (Onuaguluchi and Eren 2012). Stage 2 entails a much smaller mining and processing footprint with a focus on high recovery of several by-products along with the main valuable minerals and a step-change in capital and operating costs for a robust economics.
- Stage 3: This stage is the ultimate zero waste scenario where the conventional boundaries of mine and mill merge with no overburden or waste removal to start-with, resulting in a complete shift in mineral extraction paradigm. In situ recovery (ISR) definitely falls into this category and has significant potential, but requires further development and novel technologies to address environmental and lixiviant production issues for different ores types. New technologies involving bacterial leaching, chemical comminution, and discrete fracturing network (DFN) appear to have potential and are subjects of further research and development (Follink 2010; Dershowitz 2011; Muir 2014).

The concept of zero waste mining is not new, but to make it a reality requires a phased approach working through the existing mining paradigm for most commodities. Different phases will be required for different commodities depending on geology, ore types, mineralogy associations, location, and other factors such as availability of technologies. Multidisciplinary collaboration involving various stakeholders is a must to realize its full benefits. During these challenging times for the mining industry, the only way forward is to transcend to a higher level than where the problem was created in the first place. The underlying premise is that zero waste mining is economically very attractive, environmentally friendly, and fully integrated with the needs of local communities, societies, and other stakeholders. A vision for the zero waste mining is a subject of another discussion and will be presented by the author elsewhere.

Acknowledgments The chapter is based on the experience gained by the author over 25 years, working with various mining companies, centers of excellence, commercial laboratories, engineering companies, equipment vendors, reagent suppliers, and consultants working on many plant operational issues and greenfield projects, and would like to thank them all for their support, discussion, partnership, and collaboration. The author would like to thank colleagues at Barrick Gold Corporation in corporate and operations for their input in discussion on selection, development, design, and implementation of technologies in various ways during the last 10 years.

References

Albuquerque, L., Wheeler, J., Valine, S., Ganahl, B., & Barrios, G. (2008). Application of high frequency screens in closing grinding circuits. In *Proceedings of the Vth International Mineral Processing Seminar*, Procemin 2008, 22–24th October, Santiago, Chile.

Alvarado, S., Alguerno, J., Auracher, H., & Casali, A. (1998). Energy-exergy optimization of comminution. *Energy Oxford, 23*(2), 153–158.

Aral, H., Mead, S., Cuevas, J., Davey, K., & Bruckard, W. (2010). Desperate times call for desperate measures—The use of sea water in mineral processing. In *Proceedings of Sustainable Mining 2010* (pp. 148–163). The Australasian Institute of Mining and Metallurgy: Melbourne.

Arowoshola, L., Gasson, C., Gonzalez-Manchon, L. H., Reemeyer, L., & Uzelac, J. (2011). *Water for mining: Opportunities in scarcity and environmental regulation*. Global water intelligence report, published by Media Analytics Ltd., Oxford, England.

Arvidson, B. R., & Norrgran, D. (2014). Magnetic separation. In C. G. Anderson, R. C. Dunne, & J. L. Uhrie (Eds.), *Mineral processing and extractive metallurgy—100 years of innovation* (pp. 223–233). Englewood, CO: Society for Mining, Metallurgy & Exploration.

Arvidson, B. R., & Wotruba, H. (2014). Ore sorting. In C. G. Anderson, R. C. Dunne, & J. L. Uhrie (Eds.), *Mineral processing and extractive metallurgy—100 years of innovation* (pp. 209–211). Englewood, CO: Society for Mining, Metallurgy & Exploration.

Austin, L. G., & Brame, K. (1983). A comparison of Bond method for sizing wet tumbling mills with a size-mass balance simulation model. *Powder Technology, 34*, 261–274.

Ballentine, F., Lewellyn, M.F., & Moffatt, S.A. (2011). Red Mud flocculants used in the Bayer process. In S. J. Lindsay (Ed.), *Light metals* 2011 (pp. 107–108). Orlando, FL: TMS.

Bamber, A. S. (2008). *Integrated mining, pre-concentration and waste disposal systems for the increased sustainability of hard rock metal mining*. Ph.D. thesis, University of British Columbia, April.

Barratt, D., & Sherman, M. (2002). Factors that influence the selection of communition circuits. In A. L. Mular, D. N. Halbe, & D. J. Barratt (Eds.), *Mineral processing plant design, practice and control* (Vol. 1, pp. 539–565). Englewood, CO: SME.

Bartram, K., & Kowalczyk, M. (2009). New developments in sensor based ore sorting. In *Proceedings of 48th Conference of Metallurgists* (pp. 421–432). Sudbury, ON, Canada, August 23rd to 26th.

Batterham, R. B. (2003). The mine of the future, Will it be visible? In *Proceedings of the XXII International Mineral Processing Congress,* Cape Town, South Africa, 28 September–3 October.

Batterham, R. B. (2008). Current practices shaping the science and practice of mineral processing. In *Proceedings of the XXV International Mineral Processing Congress* (Vol. 1, pp. 55–59). Beijing, China, 24–28th September.

Battersby, M., Battersby, R. M., Flatman, S., Imhof, R., Sprenger, H., & Bragado, T. (2011). *Recovery of ultra fines using imhoflot pneumatic flotation—Two pilot plant case studies recovering nickel and zinc from tailings streams. Presented at the MEI Conference*, Cape Town, South Africa.

Baum, W. (2014). Innovations in process mineralogy and laboratory automation. In C. J. Anderson, R. C. Dunne, & J. L. Uhrie (Eds.), *Mineral processing and extractive metallurgy: 100 years of innovation* (pp. 569–575). Englewood, CO: SME.

Baum, W., Sanhueza, J., Smith, E. H., & Tufar, W. (1989). The use of process mineralogy for plant optimization at the El Indio gold-silver-copper operation. *Erzmetall, 42*(N9), 373–378.

Bergh, L. G., & Yinanatos, J. B. (2003). Flotation column automation. *Control Engineering Practice, 11*, 67–72.
Blin, P., & Dion-Ortega, A. (2013, June/July). High and dry: With water in short supply, some are looking to the ocean for answers, *CIM Bulletin*.
Bolton, E. K. (1928). CA281647
Bouchard, J., Desbiens, A., del Villar, R., & Nunez, E. (2009). Column flotation simulation and control: An overview. *Minerals Engineering, 22*, 519–529.
Boger, D. V. (2009). Rheology and the resource industry. *Chemical Engineering Science, 64*, 4525–4536.
Bradshaw, D. (2014). The role of process mineralogy in improving the process performance of complex sulphide ores. In *Proceedings of the XXVII International Mineral Processing Congress* (pp. 1–23). Santiago, Chile (Chapter 14).
Brake, I. R. (1998). The development and commissioning of a new Microcell column flotation circuits for BHP Coal's peak Downs coal preparation plant. In A. C. Partridge & L. R. Partridge (Eds.), *Proceedings of the XIII International Coal Preparation Congress* (pp. 767–776). Brisbane, NSW: Australian Coal Preparation Society.
Brent, G. F., Dare-Bryan, P., Hawke, S., & Rothery, M. (2013). Ultra-high-intensity blasting—A new paradigm in mining. In *Proceedings of World Gold Conference* (pp. 395–400). Brisbane, 26–29 September.
Brissette, M., Mihajlovic, V., & Sanuri, S. (2014). Geometallurgy: New accurate test work to meet required accuracies of mining project development. In *Proceedings of the XXVII International Mineral Processing Congress*, Santiago, Chile (Chapter 14).
Buckley, A. (2010). Surface chemical characterisation of identifying and solving problems within base metal sulphide flotation plants. In *Flotation plant optimisation* (pp. 137–153). AusIMM, January (Chapter 8).
Burns, R., & Grimes, A. (1986). The application of Preconcentration by Screening at Bougainville Copper Limited. In *Proceedings of the AUSIMM Mineral Development Symposium*, Madang, Papua New Guinea, June.
Burt, R. O. (1999). Selection of gravity concentration equipment. In A. L. Mular, M. A. Anderson (Eds.), *Design and installation of concentration and dewatering circuits* (p. 194). Englewood, CO: SME.
Bye, A. R. (2011). Case studies demonstrating value from geometallurgical studies. In *Proceedings of the First AusIMM International Geometallurgy Conference* (pp. 9–30). Brisbane, 5–7 September.
Castro, E., Lopez, J., & Switzer, D. (2009). Maximum classification and efficiency through classification using the recyclone, In D. Malhotra, P. R. Taylor, E. Spiller, & M LeVier (Eds.), *Recent advances in mineral processing plant design* (pp. 444–454). Englewood, CO: SME.
Chamberlain, P. G., & Pojar, M. G. (1984). Gold and silver leaching practices in the United States. In *US Bureau of Mines* (p. 24). Information Circular 8969.
Chattopadhyay, A., & Gorain, B. K. (2012). Gold deportment studies on a copper-gold ore—A systematic approach to quantitative mineralogy focusing on diagnostic metallurgy, paper #615. In *Proceedings of the XXVI International Mineral Processing Congress* (pp. 778–788). New Delhi, 24th–28th September.
Chattopadhyay, A., & Gorain, B. K. (2014). Gold deportment studies on a copper-gold ore—A novel approach to quantify gold deportment for double refractory gold ores. In *Proceedings of the XXVII International Mineral Processing Congress* (pp. 160–180). Santiago, Chile (Chapter 14).
Chryssoulis, S. L., & McMullen, J. (2005). Mineralogical investment of gold ores. In *Advances in Gold Processing, Developments in Mineral Processing* (Vol. 15, pp. 21–67). Amsterdam: Elsevier.
Chisakuta, G. C., Banda, M., & Chabinga, M. (2005). Process developments at tailings leach plant. In *Presented at the 3rd South African Conference on Base Metals*, SAIMM (p. 380).
Cienski, T., & Coffin, V. (1981). Column flotation operation at Mines Gaspe molybdenum circuit. In *Proceedings of the 13th Canadian Mineral Processors Conference* (pp. 240–262). Ottawa.
Cirulis, D., & Russell, J. (2011). CiDRA CYCLONEtrac at Kennecott Utah Copper. *Presented at the E&MJ Processing Conference*.
Clark, B. H. (2007). The Derrick stack sizer: Evolutionary advancements in wet screening. In *Proceedings of the 39th Canadian Mineral Processors Conference* (pp. 413–428). January 23 to 25th.
Clayton, R., Jameson, G. J., & Manlapig, E. V. (1991). The development and application of the Jameson cell. *Minerals Engineering, 4*, 925–933.
Cole, J., Dunne, R., & Giblett, A. (2012). Review of current enhanced gravity separation technologies and applications. In C. A. Young & G. H. Luttrell (Eds.), *Separation technologies for minerals, coal and earth resources*, 9th. Englewood, CO: SME.
Connolly, J., & Dobby, G. (2009). Benchmarking of flotation plants: A key element of flotation modelling, In D. Malhotra, P. R. Taylor, E. Spiller, & M LeVier (Eds.), *Recent advances in mineral processing plant design* (pp. 211–219). Englewood, CO: SME.
CRC-ORE. (2014). Transforming mining productivity with grade engineering. www.crcore.org.au.
Dammers, M., Schropp, C., Martens, P. N., Rattmann, L., & Robben, C. (2013). In *6th International Conference on Sustainable Development in the Minerals Industry* (pp. 1–6). 30th June to 3rd July, Milos Island, Greece.
Dance, A. (2013). Personal communication.
David, D. (2010). Operational geometallurgy. In *Flotation plant optimisation* (pp. 201–209). AusIMM, January (Chapter 12).
Dershowitz, B. (2011, November). Discrete fracturing network modeling in support of in-situ leach mining. *Mining magazine*, pp. 46–50.
Dobby, G. and Savassi, O. (2005). An advanced modelling technique for scale-up of batch flotation results to

plant metallurgical performance, Centenary of Flotation, AusIMM, Brisbane.
Desbiens, A., del Villar, R., Maldonado, M., & Bouchard, J. (2009). Towards the optimization of flotation column operation. In *COM 2009*, Laurentian University, August 25th.
Dudeney, A. W. L., Chan, B. K. C., Bouzalakos, S., & Huisman, J. L. (2013). Management of waste and waste water from mineral industry processes, especially leaching of sulfide resources: State of the art. *International Journal of Mining, Reclamation and the Environment, 27*, 2–37.
Dunbar, W. S. (2014). Will the mine of the future be a mine at all? In *Globe and Mail*, August 18th. .
Dundee, (2012). Dundee Precious Metals Annual Report, pp 7.
Dunne, R., Hart, S., Parker, B., & Veilette, G. (2007). Boddington gold mine—An example of sustaining gold production for 30 years. In *World Gold 2007 Conference* (pp. 213–230). AusIMM.
Ebell, M. B., & Capstick, D. (2013). Improving extraction efficiency from Witwatersrand gold tailings using flotation and ultrafine grinding of concentrates. In *World Gold 2013* (pp. 301–306). AusIMM.
Ericsson, M. (2012, March). Mining technology—Trends and development. *Polinares Consortium*, Working paper # 29.
Erickson, M. T. (2014). Innovations in comminution equipment: Crushers, high pressure grinding rolls, semi-autogenous grinding, ball mills and regrind mills. In C. J. Anderson, R. C. Dunne, & J. L. Uhrie (Eds.), *Mineral processing and extractive metallurgy: 100 years of innovation* (pp. 65–75). Englewood, CO: SME.
Evans, C. C., Wightman, E. M., Manlapig, E. V., & Coulter, B. L. (2011). Application of process mineralogy as a tool in sustainable processing. *Minerals Engineering, 24*(12), 1242–1248.
Excelsior Mining Website. (2014). http://www.excelsiormining.com/index.php/in-situ-recovery.
Fan, M., Tao, D., Honaker, R., & Luo, Z. (2010). Nanobubble generation and its application in froth flotation (part II): Fundamental study and theoretical analysis. *Mining Science and Technology, 20*, 159–177.
Finch, J. A., Cilliers, J., & Yianatos, J. (2007). Column flotation. In M. C. Fuerstenau, G. J. Jameson, & R.H. Yoon (Eds.), *Froth flotation: A century of innovation*. (pp. 681–737). Englewood, CO: SME.
Finch, J. A. (1995). Column flotation: A selected review—Part IV: Novel flotation devices. *Minerals Engineering, 8*(6), 587–602.
Flintoff, B., Guyot, O., McKay, J., & Vien, A. (2014). Innovations in comminution instrumentation and control. In C. J. Anderson, R. C. Dunne, & J. L. Uhrie (Eds.), *Mineral processing and extractive metallurgy: 100 years of innovation* (pp. 91–113). Englewood, CO: SME.
Foggiatto, B., Bueno, M., Lane, G., McLean, E., & Chandromohan, R. (2014). The economics of the large scale sorting. In *Proceedings of the XXVII International Mineral Processing Congress* (pp. 11–20). Santiago, Chile (Chapter 16).
Follink, B. (2010). Bio-mineral processing—Its potential to revolutionize the industry. In *Proceedings of the XXV International Mineral Processing Congress* (p. 61). Brisbane, Australia: AusIMM.
Franco, J. J. C., Castillo, M. F., Concha, J., Christodoulou, L., & Wasmund, E. B. (2015). Coarse gold recovery using flotation in a fluidized bed. In *Proceedings of the 13th Canadian Mineral Processors Conference* (pp. 439–447). Ottawa.
Fuerstenau, D. W. (2007). A century of developments in the chemistry of flotation processing. In M. C. Fuerstenau, G. J. Jameson, & R. H. Yoon (Eds.), *Froth flotation: A century of innovation* (pp. 3–64). Englewood, CO: SME.
Fuerstenau, M. C., & Han, K. (2003). *Principles of mineral processing*. Littleton, CO: SME.
Gao, M., Young, M., & Allum, P. (2002). IsaMill fine grinding technology and its industrial applications at Mount Isa Mines. In *Proceedings of the 34th Canadian Mineral Processors Conference* (pp. 171–188). Ottawa: CIM.
Gasiorowski, S. A., Bittner, J. D., & Hrach, F. J. (2012). Triboelectric belt separator for beneficiation of fine minerals. In *Proceedings of the 44th Canadian Mineral Processors Conference* (pp. 399–408). Ottawa: CIM.
Gates, P. (2004). US patent office, 6723938, pp. 1–6.
Gavin, K. P., Walton, K., & Zhou, J. (2010). Application of closely spaced inclined channels in gravity separation of fine particles. *Minerals Engineering, 23*, 26.
Gavin, K. P., & Dickinson, J. (2012). Particle transport and separation in inclined channels subject to g forces. *Chemical Engineering Sciences, 87*, 294.
Germain, M., Lawson, T., Henderson, D. K., & MacHunter, D. M. (2003). The application of new design concepts in high tension electrostatic separation to the processing of mineral sands concentrates. *Heavy Minerals 2003*; SAIMM, p. 101.
Gillot, P. (2006). Pit-to-Plant optimization at Morila Gold mines. In *Proceedings of the SAG 2006 Conference*, Vancouver.
Gomez, C. O., & Finch, J. A. (2009). Gas dispersion measurements: Uses and opportunities. In D. Malhotra, P. R. Taylor, E. Spiller, & M. LeVier (Eds.), *Recent advances in mineral processing plant design* (pp. 60–70). Englewood, CO: SME.
Gorain, B. K., Franzidis, J. P., & Manlapig, E. V. (1997). Studies on impeller type, impeller speed and air flow rate in an industrial-scale flotation cell—Part 4: Effect of bubble surface area flux on flotation performance. *Minerals Engineering, 10*(4), 367–379.
Gorain, B. K. (2000). Selection of cell operating conditions to optimise performance of flotation circuits with large cells. In *Proceedings of the 7th Mill Operators' Conference* (pp. 179–187). Kalgoorlie: AusIMM.
Gorain, B. K., & Stradling, A. W. (2002). An integrated approach to modelling, design and optimisation of the

flotation plant at the Red Dog concentrator. In *Proceedings of the 34th Canadian Mineral Processors Conference* (pp. 279–296). Ottawa.

Gorain, B. K. (2005). Optimization of flotation circuits with large flotation cells. In Jameson (Ed.), *Centenary of Flotation Symposium*, p. 843.

Gorain, B. K., Oravainen, H., Allenius, H., Peaker, R., Weber, A., & Tracyzk, F. (2007), Mechanical froth flotation cells. In M.C. Fuerstenau, G. J. Jameson, R.H. Yoon (Eds.), *Froth flotation: A century of innovation* (pp. 637–680). Englewood, CO: SME.

Gorain, (2012). Separation of copper minerals from pyrite using air-metabisulfite treatment, patent application global.

Gorain, B. (2013). Developing solutions to complex flotation problems. In *Proceedings of the 34th Canadian Mineral Processors Conference* (pp. 293–312). Ottawa, Jan 22nd–24th.

Gordon, R. B. (2002). Production residues in copper technological cycles. *Resource Conservation Recycling, 36*(2), 66–87.

Gover, D., Berton, A., & Norrgran, D. (2011). Magnetic concentration of Hematite at Wabush Mines—Plant scale application of Rare Earth Permanent Magnetic Separators. *SME Annual Meeting*, R-H Yoon symposium, Magnetic and Electrostatic separation, Feb 27—Mar 2nd, Denver, CO.

Gray, A. H. (1997). InLine Pressure Jig—An exciting, low cost technology with significant operational benefits in gravity separation of minerals. In *Proceedings of the AusIMM Annual Conference* (p. 259).

Grano, S. (2010). Chemical measurements during plant surveys and their interpretation cells. In *Flotation Plant Optimisation* (pp. 107–121). AusIMM, January (Chapter 6).

Grau, R., Nousiainen, M., & Yanez, A. (2014). Gas dispersion measurements in three Outotec flotation cells: Tankcell1, e300 and e500. In *Proceedings of the XXVII International Mineral Processing Congress* (pp. 204–214). Santiago, Chile (Chapter 5).

Gu, Y., Schowstra, R. P., & Rule, C. (2014). The value of automated mineralogy. *Minerals Engineering, 58*, 100–103.

Gupta, A., & Yan, D. S. (2006). *Mineral processing design and operations*. Amsterdam: Elsevier.

Han, B., Altansukh, B., Hag, K., Ishiyama, D., Shibayama, A., Stevanovic, Z., et al. (2014). The investigation of copper recovery process from mine tailings by flotation and high pressure leaching, In *Proceedings of the XXVII International Mineral Processing Congress* (pp. 141–158). Santiago, Chile (Chapter 19).

Harris, M., Runge, K. C., Whiten, W. J., & Morrison, R. D. (2002). JKSimFloat as a practical tool for flotation process design and optimization, In *Proceedings of Mineral Processing Plant Design, Practice and Control* (Vol. 1, pp. 461–478). Englewood, CO: SME.

Hart, S., Parker, B., Rees, T., Manesh, A., & Megaffin, I. (2011). Commissioning and ramp up of the HPGR circuit at Newmont Boddington Gold, *SAG 2011*.

Hearn, S. (2014). Innovations in electrostatic separation. In C. J. Anderson, R. C. Dunne, & J. L. Uhrie (Eds.), *Mineral processing and extractive metallurgy: 100 years of innovation* (pp. 235–242). Englewood, CO: SME.

Herbst, J. A., & Fuerstenau, D. W. (1980). Scale-up procedure for continuous grinding mill design using population balance models. *International Journal of Mineral Processing, 7*, 1–31.

Herbst, J. A., & Harris, M. (2007). Modeling and simulation of industrial flotation processes. In M. C. Fuerstenau, G.J. Jameson, R.H. Yoon (Eds.), *Froth flotation: A century of innovation* (pp. 757–777). Englewood, CO: SME.

Hoijer, J., & Grimm, R. (2011). The Black art of filter media, Issue 28, *Outotec Australia quarterly E newsletter*.

Holloway, A. R., Jones, L., & Lawrence, S. R. (2009). Pre-concentration of platinum group and base metal ores by dense medium separation. In *Proceedings of 48th Conference of Metallurgists* (pp. 373–384). Sudbury, Ontario, Canada Aug 23rd to 26th.

Honaker, R., Dunne, E. C., & Galvin, K. (2014). Density-based separation innovations in coal and mineral processing application. In C. J. Anderson, R. C. Dunne, & J. L. Uhrie (Eds.), *Mineral processing and extractive metallurgy: 100 years of innovation* (pp. 243–264). Englewood, CO: SME.

Hukki, R. T., & Allenius, H. (1968). A quantitative investigation of closed grinding circuit. *SME AIME Transactions, 241*, 482–487.

Huls, B. (2014). Drivers that will change the design and plant operation in the mining industry. In *Proceedings of the XXVII International Mineral Processing Congress* (pp. 16–37). Santiago, Chile (Chapter 1).

Imhof, R. M., Battersby, M. J. G., Brown, J. V., Lotzien, R. M., & Kleefeld, J. (2003). Development of pneumatic flotation incorporating centrifugal separation. *Presented at the XXII International Mineral Processing Congress*, 28 October–3 September, Cape Town, South Africa.

Imhof, R. M., Battersby, M., Parra, F., & Sanchez-Pino, S. (2005). The successful application of pneumatic flotation technology for the removal of silica by reverse flotation at the iron ore pellet plant of Compania Minera Huasco, Chile, *Centenary of Froth Flotation*, Brisbane.

International Atomic Energy Agency. (2000). Methods of exploitation of different types of uranium deposits. In *International Atomic Energy Report*, IAFA-TECDEC-1174, p. 84.

Isokangas, E., Sonmez, B., Wortley, M., & Valery, W. (2012). Using SmartTag to track ore in process integration and optimization projects: Some case studies in a variety of applications. *Platinum 2012, SAIMM Conference* (pp. 871–882).

Jameson, G. (2014). Experiments on the flotation of coarse composite particles, In *Proceedings of the XXVII International Mineral Processing Congress* (pp. 102–111). Santiago, Chile (Chapter 5).

Jameson, G. J., & Manlapig, E. V. (1991). Flotation cell design—Experiences with Jameson cell. *Presented at the 5th AusIMM Extraction Metallurgical Conference*.

Jansen, W., Morrison, R., Wortley, M., & Rivett, T. (2009). Tracer-based mine-mill ore tracking via process hold-ups at NorthParkes. In *Proceedings of the 10th Mill Operators Conference* (pp. 345–356). Adelaide, October 12–14th.

Jensen, W. B. (1980). *The lewis acid-base concepts: An overview*. New York: Wiley-Interscience.

JKTech. (2004). www.jktech.com.

Johnson, N. W. (2006). Review of existing eco-efficient comminution devices, in project 2B1 Energy (Eco)—Efficient liberation and comminution—report to sponsors and providers (pp. 1–29).

Kanchibotla, S. (2014). Mine to mill value chain optimization—Role of blasting. In C. J. Anderson, R. C. Dunne, & J. L. Uhrie (Eds.), *Mineral processing and extractive metallurgy: 100 years of Innovation* (pp. 51–64). Englewood, CO: SME.

Kappes, R., Brosnahan, D., & Gathje, J. C. (2009). Utilizing mineral liberation analysis (MLA) to determine pyrite, arsenopyrite and arsenian pyrite floatabilities for Carlin trend ores. Annual SME meeting, Denver, CO.

Kelsey, C., & Kelly, J. (2014). Super-fine crushing to ultra-fine size the "IMP" super-fine crusher, IMPC proceedings. In *Proceedings of the XXVII International Mineral Processing Congress* (pp. 239–251). Santiago, Chile (Chapter 9).

Kirsch, S. W., & Daniel, M. J. (2009). The optimization of semi-autogenous and ball mill based circuits for mineral processing by means of versatile and efficient high pressure grinding roll technology. In *Proceedings of 10th Mill Operators Conference* (pp. 81–87). Adelaide 12–14th October.

Kittler, P., Liebezeit, V., Ehrig, K., Macmillan, E., & Lower, C. (2011). 'It seemed like a good idea at the time….'common mistakes in geometallurgy, *Presented at the First International AusIMM Geometallurgical Conference*, Brisbane, 5–7 September.

Klein, B., Dunbar, W. S., & Scoble, M. (2002). Integrating mining and mineral processing for advanced mining systems. *CIM Bulletin, 95*(1057), 63–68.

Knight, D. A., MacPherson, A. R., Medina, V. G., & Spiller, D. E. (1989). Case histories of using small-scale tests to design SAG mill circuits. SAG'89. In A. L. Mular & G. Agar (Eds.), Advances in autogenous and semiautogenous grinding technologies (pp. 105–117) UBC Vancouver.

Kohmuench, J. N., Mankosa, M. J, Wyslouzil, H., & Luttrell, G. H. (2009). Column and non-traditional flotation. In D. Malhotra, P. R. Taylor, E. Spiller, & M LeVier (Eds.), *Recent advances in mineral processing plant design* (pp. 220–231) Englewood, CO: SME.

Kormos, L. J., Oliveira, J., Fragomeni, D, Whiteman, E., & Carrion De la Cruz, J. (2010). Process diagnosis using quantitative mineralogy. In *Proceedings of 42nd Canadian Mineral Processors Conference* (pp. 97–120) Ottawa, paper 8.

Koski, S., Vanderbeek, J. L., & Enriquez, J. (2011). Cerro Verde concentrator-four years operating HPGR's, *SAG 2011*.

Krasowska, M., & Malysa, K. (2007). Kinetics of bubble collision and attachment to hydrophobic solids: 1. Effect of surface roughness. *International Journal of Mineral Processing, 81*, 159–177.

Lamberg, P., & Bernal, L. (2008). Flash flotation—From pilot size to full size installation at Esperanza. In *Proceedings of V International Mineral Processing seminar (Procemin 2008)* (pp. 251–256). Chile, 22–24 October.

Lessard, J., deBakker, J., & McHugh, K. (2014). Development of ore sorting and its impact on mineral processing economics. *Minerals Engineering, 65*, 88–97.

Lizama, H. M., Carrion, J., & Estrella, D. (2008). Improvements in column flotation through the use of Microcel spargers at Antamina. In *Proceedings of the 40th Annual Canadian Mineral Processors Conference* (pp. 363–375). Ottawa, January 22–24.

Lichter, J. K., & Davey, G. (2006). Selection and sizing of ultrafine and stirred grinding mills. In S. K. Kawatra (Ed.), *Advances in comminution* (pp. 69–85). Littleton, CO: SME.

Lloyd, P. J. D. (1979). An integrated mining and extraction system for use on Witwatersrand gold mines. *Journal of the SAIMM, 79*(1), 135–148.

Lotter, N. O. (2011). Modern process mineralogy: An integrated multi-disciplined approach to flow sheeting. *Minerals Engineering, 24*(12), 1229–1237.

Luttrell, G. H., Honekar, R. Q., Bethell, P., & Stanley, F. (2007). Design of high-efficiency spiral circuits for preparation plants. In B. Arnold, M. Klima, & P. Bethell (Eds.), *Designing the coal preparation plants for the future* (pp. 73–88). Littleton, CO: SME.

Luttrell, G. H., Westerfield, T. C., Kohmuench, J. N., Mankosa, M. J., Mikkola, K. A., & Oswald, G. (2006). Development of high efficiency hydraulic separators. *Mineral and Metallurgical Processing, 23*(1), 33.

Lynch, A. J., Watt, J. S., Finch, J. A., & Harbort, G. E. (2007). History of flotation technology. In M. C. Fuerstenau, G. J. Jameson, & R.-H. Yoon (Eds.), *Froth flotation: A century of innovation* (pp. 65–91). Littleton, CO: SME.

Lynch, A. J., Harbort, G. J., & Nelson, M. G. (2010). *History of flotation*. Melbourne: AusIMM.

Manlapig, E. V., Gorain, B. K., Harris, M. C., & Franzidis, J. -P. (1997). Designing flotation plants in the 21st century. In *Proceedings of XX International Mineral Processing Congress* (pp. 235–245). Aachen, Germany.

MacDonald, M., Adair, B. J., Dunn, M., & Latti, D. (2011). Learnings from five years of on-site MLA at Kennecott Utah Copper Corporation (myth busters through quantitative evidence). In *Proceedings ICAM*, Norway, 1–5 August.

Major, K. (2002). Types and characteristics of crushing equipment and circuit flowsheets. In A. L Mular, D. N.

Halbe, D. J. Barratt (Eds.), *Mineral processing plant design, practice and control* (pp. 566–583). Littleton, CO: Englewood, CO: SME.

Maki, T. D., & Taylor, J. B. (1987). Precious metal slag treatment using an electrostatic separator. In *Small mines development in precious metals* (pp. 245–252).

Markoja, B. (2011). Cidra SONARtrac flowmeters: An alternative flow measurement technology. Presented at *ISTOG Winter Conference*, Clearwater, FL, December 5–9th.

Meka, Z., & Lane, G. (2010). Recent metallurgical developments and their impact on minerals project execution. In *Proceedings of XXV International Mineral Processing Congress* (pp. 21–28). Brisbane, Australia, 6–10th September.

McCaslin, M. L., Mounteer, K. J., & Phillips, M. (2014). Innovations in liquid/solid separation for metallurgical processing. In C. J. Anderson, R. C. Dunne, & J. L. Uhrie (Eds.), *Mineral processing and extractive metallurgy: 100 years of innovation* (pp. 333–344). Littleton, CO: SME.

McCullough, G., Mutsaers, B., & Jessop, J. (2013). Obstacles to embedding geometallurgy decision-making into operations and how to increase maturity of this critical capability. In *2nd AusIMM International Geometallurgy Conference*, Brisbane, 30 Sept to 2nd Oct.

McCullough, W. E., Bhappu, R. B., & Hightower, J. D. (1999). Copper ore preconcentration by heavy media separation for reduced capital and operating costs, Presented in *Cobre 99*.

McKay, J., Guyot, O., & Broussaud, A. (2009). Purposeful instrumentation for effective flotation control. In D. Malhotra, P. R. Taylor, E. Spiller, & M LeVier (Eds.), *Recent advances in mineral processing plant design* (pp. 204–210). Littleton, CO: SME.

McKee, D. (2013). Understanding mine to mill. *A monograph published by the Co-operative Research Centre for Optimizing Resource Extraction (CRC ORE)*, Brisbane, Australia.

Minalliance. (2012). 100 innovations in the mining industry, ebook from www.minalliance.ca.

Mohanty, J. K., Rao, D. S., Das, B., & Rao, R. B. (2000). Mineralogy and pre-concentration of the chromite overburden of the Sukinda Ultramafic Belt. *CIM Bulletin, 93*(1038), 37–43.

Moreno, P. A., Aral, H., Cuevas, J., Monardes, A., Adaro, M., Norgate, T., et al. (2011). The use of sea water as process water at Las copper-molybdenum beneficiation plant at Taltal (Chile). *Minerals Engineering, 24*, 852–859.

Morley, C., & Staples, P. (2010). SAG or HPGR?—The current dilemma. In *Proceedings of the 42nd Annual Meeting of the Canadian Mineral Processors* (pp. 491–508) CIM.

Morrell, S. (2014). Innovations in comminution modelling and ore characterization. In D. Malhotra, P. R. Taylor, E. Spiller, & M. LeVier (Eds.), *Recent advances in mineral processing plant design* (pp. 77–90). Englewood, CO: SME.

Muir, A. (2014). Evolution in comminution. Keynote address in *Comminution 2014*, organized by MEI, 7-1-April, Cape Town, South Africa.

Mumbi, M. (2012). Process optimization of mixed copper ores through real-time mineralogical analysis. Presentation from Quantum Minerals, Zambia.

Munro, P. D., Schache, I. S., Park, W. G., & Watsford, R. M. S. (1982). The design, construction and commissioning of a heavy media plant for silver-lead-zinc ore treatment—Mount Isa Limited. In *Proceedings of the XIV International Mineral Processing Congress* (pp. 1–20) Toronto, VI-6..

Nagaraj, D.R. (1988). The chemistry and applications of chelating or complexing agents in mineral separations, In *Reagents in Mineral Technology*, Edited by P. Somasundaran and B.M. Moudgil, SME.

Nagaraj, D. R., & Ravishankar, S. A. (2007). Flotation reagents—A critical overview from an industry perspective. In M. C. Fuerstenau, G. J. Jameson, & R. H. Yoon (Eds.), *Froth flotation: A century of innovation* (pp. 375–424). Littleton, CO: SME.

Nagaraj, D. R., & Farinato, R. S. (2014). Major innovations in the evolution of flotation reagents. In C. J. Anderson, R. C. Dunne, & J. L. Uhrie (Eds.), *Mineral processing and extractive metallurgy: 100 years of Innovation* (pp. 159–175). Englewood, CO: SME.

Napier-Munn, T. J., Morrell, S., Morrison, R. D., & Kojovic, T. (2005). *Mineral comminution circuits*. JKMRC monograph.

Napier-Munn, T. J., Bosman, J., & Holtham, P. (2014). Innovations in dense media separation technology. In C. J. Anderson, R. C. Dunne, & J. L. Uhrie (Eds.), *Mineral processing and extractive metallurgy: 100 years of Innovation* (pp. 265–275). Englewood, CO: SME.

Nelson, M. G., Lelinski, D., & Gronstrand, S. (2009). Design and operation of mechanical flotation machines. In D. Malhotra, P. R. Taylor, E. Spiller, M. LeVier (Eds.), *Recent advances in mineral processing plant design* (pp. 168–189) Englewood, CO: SME.

Nesset, J. E., Hernandez-Aguilar, J. R., Acuna, C., Gomez, C. O., & Finch, J. A. (2006). Some gas dispersion characteristics of mechanical flotation machines. *Minerals Engineering, 19*, 807–815.

Newcombe, B. (2014). Characterising and predicting the performance of an industrial flash flotation cell, PhD thesis, University of Queensland.

Newell, A. J. H., & Bradshaw, D. J. (2007). The development of a sulfidisation technique to restore the flotation of oxidized pentlandite. *Minerals Engineering, 20*, 1039–1046.

Norrgran, D., Ashton, D., & Fears, P. (2009). Installation and operation of the Eriez cryogen-free superconducting magnet at Sibelco. In *Mineral Processing Plant Design—An update Conference*, September 30–October 3rd, Tucson, Arizona.

Onuaguluchi, O., & Eren, P. (2012). Recycling of copper tailings as an additive in cement mortars. *Construction and Building Materials, 37*, 723–727.

Pal, A., Samanta, A. N., & Ray, S. (2010). Treatment of iron ore slimes for value addition. *Hydrometallurgy, 5*, 30–35.

Perkin, D. (2015). Private communication.

Phillips, D. I., Yoon, R., Lutrell, G. H., Fish, I., & Toney, T. A. (1997). Installation of 4-meter diameter Microcel flotation columns at LadyDunn preparation plant. In *14th International Coal Preparation Congress* (pp. 115–132). Lexington, Kentucky.

Pokrajcic, Z., O'Haaloran, R., & Jones, C. (2010, September). Evaluation of comminution circuit design for sustainability using economics. In *Proceedings of the XXV International Mineral Processing Congress* (pp. 3597–3609). Brisbane, Australia.

Powell, M. (2013). Utilising orebody knowledge to improve comminution circuit design and energy utilization. In *2nd AusIMM International Geometallurgical Conference*, Brisbane, 30 September to 2 October.

Primeau, D., & Gilbert, M. M. (2012). Remediation of red mud using an innovative approach. In Proceedings of *ICSOBA 2012*, International committee for study of Bauxite, Alumina and Aluminium, Belem, Brazil.

PRO. (2014). Process Research Ortech chloride leach technology. Private Communication.

Pyecha, J., Sims, S., Lacouture, B., Hope, G., & Stradling, A. (2005). Evaluation of a Microcel sparger in the Red Dog column flotation cells. In *Proceedings of Centenary of Flotation Symposium* (pp. 1017–1085). Brisbane.

Ralston, J., Fornasiero, D., & Grano, S. (2007). Pulp and solution chemistry. In M. C. Fuerstenau, G. J. Jameson, R. H. Yoon (Eds.), *Froth flotation: A century of innovation* (pp. 227–281). Englewood, CO: SME.

Ravishankar, S. A., & Kolla, H. (2009). Chemically enhanced ESS. In *Proceedings of 7th International Heavy Mineral Conference* (pp. 203–206) SAIMM.

Riina, S., Maria, M., & Antti, H. (2014). Recovery of metals from tailings. In *Proceedings of the XXVII International Mineral Processing Congress* (pp. 224–232). Santiago, Chile (Chapter 19).

Rule, C. (2011, February). Stirred milling—New comminution technology in the PGM industry. *SAIMM Journal* (pp. 101–107).

Rule, C., & Schouwstra, R. P. (2011). Process mineralogy delivering significant value at Anglo Platinum Concentrator Operations. In *Proceedings ICAM*, Norway, 1–5 August.

Sanchez-Pino, S., Sanchez-Baquedano, A., & Imhof, R. M. (2008). Imhoflot G-cell pneumatic flotation technology for fine particle application. *Presented at the V International Mineral Processing Seminar, Procemin 2008*, 22–24 October, Santiago.

Sanchez-Pino, S. E., Sanchez-Baqueano, A., Imhof, R., & Sprenger, H. (2014). Industrial application progress of pneumatic flotation imhoflot technology in Chile. *XXVII International Mineral Processing Congress* (p. 245) Santiago Chile (abstract only).

Sanders, G. J., Ziajoa, D., & Kottman, J. (2002). Cost efficient beneficiation of coal by Romjigs and Batac jigs. In *Proceedings of XIV International Coal Preparation Congress* (p. 395) SAIMM.

Sanders, G. J., & Ziaja, D. (2003). The role and performance of Romjig in modern coal preparation practice. In *Proceedings of the 20th International Coal Preparation Conference* (p. 207). Lexington, KY.

Scanalyse. (2012). CrusherMapping—Introducing the CrusherMapping for gyratory crushers, Now under Outotec.

Schena, G. D., Gochin, R. J., & Ferrara, G. (1990). Preconcentration by dense media separation—An economic evaluation. *Transactions of the Institute of Mining and Metallurgy, 99*, C21–C31.

Schoenbrunn, F., & Laros, T. (2002). Design features and types of sedimentation. In A. L. Mular, D. N. Halbe, & D. J. Barratt (Eds.), *Mineral Processing Plant Design, Practice and Control Proceedings* (1st ed., pp. 1331–1341). Englewood, CO: SME.

Schoenbrunn, F. (2007). A short history of deep cone thickener development. In *Australian Centre for Geomechanics* (pp. 19–24).

Schoenbrunn, F. (2011). Dewatering to higher densities: An industry review. In *Australian Centre for Geomechanics* (pp. 51–56).

Schonert, K. (1982). Method of fine and very fine comminution of materials having brittle behavior. US Patent 4357287.

Schonert, K. (1987). A first survey of grinding with high compression roller mills. *International Journal of Mineral Processing, 22*, 401–412.

Schwarz, S., & Alexander, D. (2006). Gas dispersion measurements in industrial flotation cells. *Minerals Engineering, 19*, 554–560.

Sherring, A. (2013). Personal communication.

Shi, F., Zuo, W., & Manlapig, E. (2013). Characterization of pre-weakening effect of ores by high voltages electric pulses using single particle tests. *Minerals Engineering, 50–51*, 69–76.

Shi, F., Krishnan, N., von der Weld, F., van-der-Wielen, K., Zuo, W., & Manlapig, E. (2014). A potential application of high voltage pulse technology in a gold-copper ore grinding circuit. In *Proceedings of the XXVII International Mineral Processing Congress* (pp. 68–77) Chile, Santiago (Chapter 9).

Sinclair, M. (2012). Remote Collaboration Centres (Honeywell), Presentation from the website https://www.posccaesar.org/svn/pub/PCA/MemberMeeting/201210/Day2/RemoteCollaborationCenters-Mike Sinclair-Oct2012.pdf

Sivamohan, R., & Forssberg, E. (1991). Electronic sorting and other pre-concentration methods. *Minerals Engineering, 4*(7–11), 797–814.

Smart, R. St. C., Skinner, W. M., Gersen, A. R., Mielczarskim, J., Chryssoulis, S., Pratt, A. R., et al. (2007). Surface characterization and new tools for research. In M. C. Fuerstenau, G. J. Jameson, & R. H. Yoon (Eds.), *Froth flotation: A century of innovation* (pp. 283–337). Englewood, CO: SME.

Smart, R. St. C., Gerson, A. R., Hart, B., Beattie, D. A., & Young, C. (2014). Innovations in measurement of mineral structure and surface chemistry in flotation: Past, present and future. In C. J. Anderson, R. C. Dunne, & J. L. Uhrie (Eds.), *Mineral processing and extractive metallurgy: 100 years of innovation* (pp. 577–602) Englewood, CO: SME.

Somasundaran, P. (2012). A new paradigm for sustainable minerals extraction. Presentation at the *International Mineral Processing Congress*, New Delhi, September 24th–28th.

SonicSampDrill. (2014). http://www.sonicsampdrill.com/news/sorting-ore-with-the-speed-of-light-1000-tons-per-hour-.htm.

Spiller, D. E., & Norgren, A. A. (2014). Mineral sand separation. In C. J. Anderson, R. C. Dunne, & J. L. Uhrie (Eds.), *Mineral processing and extractive metallurgy: 100 years of innovation* (pp. 317–332). Englewood, CO: SME.

Taggart, A. F. (1951). *Elements of ore dressing*. New York: Wiley.

Tian, J., Zhang, C., & Wang, C. (2014). Operation and process optimization of sino iron ore's autogenous milling circuits: The largest in the world. In *Proceedings of XXVII International Mineral Processing Congress* (pp. 141–153). Santiago, Chile.

Tremblay, R., & Boutin, P. (1964). Method and apparatus for the separation and recovery of ores. Canadian patent No 694547, 15th September.

USNMA. (2000, September). Mineral processing technology roadmap, *United States National Mining Association workshop on Mining Industry of the Future*.

Utley, R. W. (2002). Selection and sizing of primary crushers. In A. L. Mular, D. N. Halbe, & D. J. Barratt (Eds.), *Mineral Processing Plant Design, Practice and Control Proceedings* (1st ed., pp. 584–605). Englewood, CO: SME.

Valine, S. B., Wheeler, J. E., & Albuquerque, L. G. (2009). Fine sizing with the Derrick stack sizer screen. In D. Malhotra, P. R. Taylor, E. Spiller, & M. LeVier (Eds.), *Recent advances in mineral processing plant design* (pp. 433–444). Englewood, CO: SME.

Vanderbeek, J. L., Linde, T. B., Brack, W. S., & Marsden, J. O. (2006). HPGR implementation at Cerro Verde, *SAG2006* (Vol. 4, pp. 45–61).

Van der Spek, A. M., Maron, R. J., O'Keefe, C. V., & Rothman, P. J. (2009). Use of sonar flow measurement for performance and condition monitoring of slurry pumps. In *MAPLA 2009*, Antofogasta, Chile.

Walker, S. (2014). Ore sorters and analyzers: the technology develops. *Engineering and Mining Journal*, 16th April.

Wasmund, E. B. (2014). Flotation technology for coarse and fine particle recovery. In *Congreso Internacional de Flotacion de Minerals*, Lima Peru, August.

Wills, B. A. (2006). *Minerals processing technology* (7th ed.). Amsterdam: Elsevier.

World Nuclear Association Website. (2014). http://www.world-nuclear.org/info/Nuclear-Fuel-Cycle/Mining-of-Uranium/In-Situ-Leach-Mining-of-Uranium/.

Wortley, M., & La Rosa, D. (2013). Technical Description of SmartTags, Metso Process Technology and Innovation, Personal Communication, June 19th.

Xie, Y., Xu, Y., Yan, L., & Yang, R. (2005). Recovery of nickel, copper and cobalt from low grade Ni-Cu sulphide tailings. *Hydrometallurgy, 80*, 54–58.

Yap, R. F., Sepuvelda, J. L., & Jauregui, R. (1982). Determination of the Bond work index using an ordinary laboratory batch mill. In A. L. Mular & G. V. Jergensen (Eds.), *Design and installation of comminution circuits* (pp. 176–203). New York: SME/AIME.

Young, M. F., Barnes, K. E., Anderson, G. S., & Pease, J. D. (2008). Jameson cell: The "comeback" in base metals applications using improved design and flowsheet. In *Proceedings of the 38th Canadian Mineral Processors Conference* (pp. 311–332). Ottawa.

Yoon, R.-H., Luttrell, G. H., Adel, G. T., & Mankosa, M. J. (1992). The application of Microcel column flotation to fine coal cleaning. *International Coal Preparation Journal, 10*, 177–188.

Zamora, C., Schifferli, G., Castelli, L., & Bonomelli, A. (2010). Kairos mining: A comprehensive approach to sustained long term benefits in copper processing plant automation. Presented at the 42nd Annual *Canadian Mineral Processors Conference*, Ottawa, June 19–21st.

Zhou, Z. A., Xu, Z., Finch, J. A., Hu, H., & Rao, R. (1997). Role of hydrodynamic cavitation in fine particle flotation. *International Journal of Mineral Processing, 51*(1–4), 139–149.

Thermal Processing: Pyrometallurgy—Non-ferrous

3

Innovations in Non-ferrous Pyrometallurgical Processing: Case Study of the Peirce–Smith Converter

Nathan M. Stubina

The Merriam-Webster dictionary defines “innovation” as:

- A new idea, device or method
- The act or process of introducing new ideas, devices or methods

Many people use the words “innovation” and “research” interchangeably, but in reality, there is a world of difference. There is an interesting quote that has been attributed to Will Westgate of 3 M: “Research is the transformation of money into knowledge, and innovation and imagination are the transformation of knowledge into money” (CIM). Innovation is much more than a breakthrough or the “lightbulb” moment–innovation is a process. It is the entire process of transforming the initial creative idea into a new product that has commercial value.

The history of pyrometallurgy can be traced back at least 6000 years to the simple copper smelters of present-day Israel (Themelis 1994) and likely some time later in China (Mackey 2014). Copper, silver and gold were the very first metals to be used by mankind, since they were originally found in their native metallic form. Those original chunks of metal led us to use fire (pyrometallurgy) in order to process various minerals into metals and alloys. Metals and metallurgy have played such a pivotal role in the development of our civilization that major periods in our history are marked by such names as the Bronze Age (3300–1200 BC) and the Iron Age (1200–500 BC).

Pyrometallurgical processing has many inherent advantages over other ambient temperature processes (Themelis 1985):

- High reaction rates due to elevated temperatures
- High concentration of metals in processing streams
- Easy phase separation
- Favourable shift of equilibrium at high temperatures

Most of the world’s copper and nickel are currently produced by smelting sulphide concentrates and then converting the matte into metal. In this chapter, we will examine some of the innovations that have appeared over the years in the pyrometallurgical processing of these metals. We will be using the Peirce–Smith converter as a case study.

There are many articles in our trade journals that decry the paucity of innovation in mining and metallurgy. It is assumed that our industry does not embrace change very quickly. We have

N.M. Stubina (✉)
McEwen Mining, 150 King Street West, Suite 2800, Toronto, ON, Canada, M5H 1J9
e-mail: nstubina@mcewenmining.com

V.I. Lakshmanan et al. (eds.), *Innovative Process Development in Metallurgical Industry*,
DOI 10.1007/978-3-319-21599-0_3

been using some equipment, for example, the flotation cell, the ball mill or the Peirce–Smith converter for over a century. Is industry not generating enough novel ideas? Are we not funding enough R&D? Is there insufficient capital available? Or perhaps the problem relates to the type of people who are attracted to our industry? Why do some industries, such as cell phone manufacturers, produce devices that are obsolete every two years, whereas the non-ferrous pyrometallurgical industry is still using technologies that are over 100 years old? We will see, however, that hidden in the 100-year history is a remarkable story of innovation and regeneration, thus bringing the technology into modern times; the 100-year-old processing concept now has "high-tech" features, making it very state-of-the-art indeed.

An interesting psychometric study was done in South Africa where they investigated the personality traits of the 2010 final year mining engineering students. The Herrmann Brain Dominance Instrument (HBDI) tool was used to identify the thinking preferences of the students (Webber-Youngman and Callaghan 2011). The HBDI test depicts the degree of preference individuals have for thinking in each of the following four brain quadrants:

- Rational
- Practical
- Feeling
- Experimental

The thinking behaviour of the students fell into the following sectors as shown in Fig. 3.1.

Typical words that describe the BLUEs include: factual, quantitative, logical and analytical. For the GREENs: controlled, sequential, detailed and conservative. The REDs are emotional and intuitive

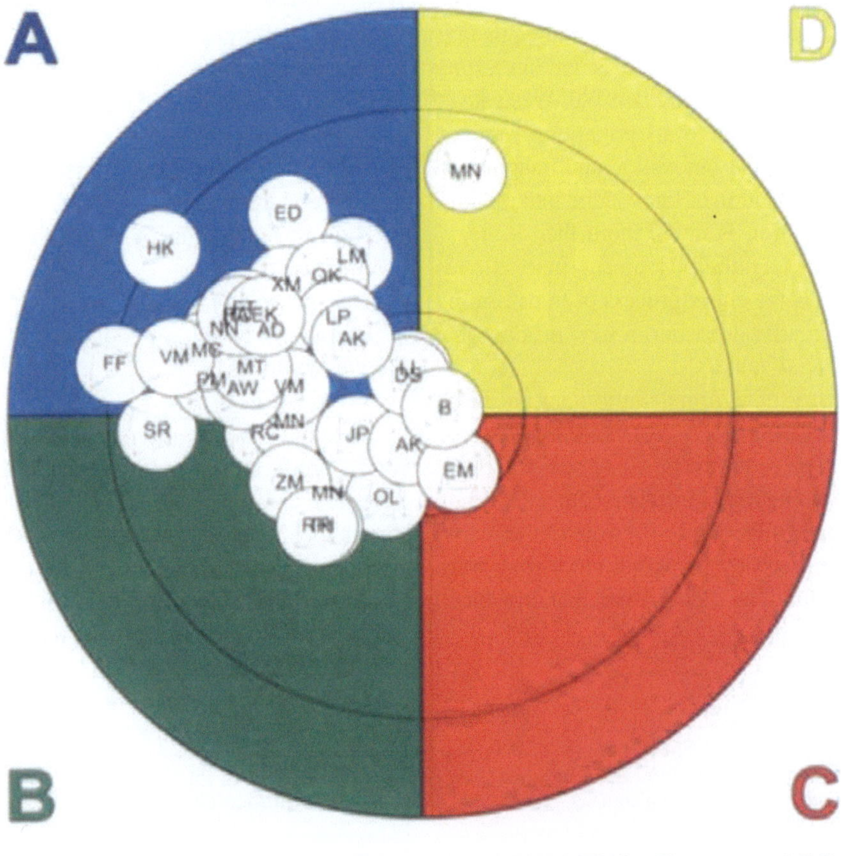

Fig. 3.1 The personality traits of the final year mining engineering students (Webber-Youngman and Callaghan 2011)

while the YELLOWs are described as imaginative and artistic. It is interesting to note that the majority of the students fell into the blue (engineering) quadrant. Only one student fell into the yellow (imaginative) quadrant. Perhaps this goes a long way to explaining why the industry is so conservative and reluctant to try new concepts and technologies? Why are the innovators not attracted to mining? What is required to shift this paradigm? Perhaps another aspect to this is that recently, unlike in the past, many mining companies have tended to shun innovation, further alienating nouveau innovators in the industry.

Let us begin our journey into innovative ideas in non-ferrous pyrometallurgy, using the Peirce–Smith converter as a case study. We will describe some of the breakthrough concepts that were steadily introduced over the years. This is not an exhaustive survey, but a few of the more important discoveries will be presented.

A patent for the Peirce–Smith converter was issued in 1909 (Peirce and Smith 1909). Peirce–Smith converting is currently used in the copper, nickel and platinum industries to remove iron and sulphur from a molten matte phase. Although the converter has recently celebrated its 100th birthday, very little has changed in its fundamental concept since its inception. In its simplest form, low pressure air is supplied by a blower. The air is introduced into the vessel using a number of tuyeres. Inside the vessel, the oxygen in the air is used to oxidize iron and sulphur to iron oxides and sulphur dioxide (Davenport et al. 2002). The bubbling injection regime is inherently inefficient from an energy perspective (Wraith et al. 1999). Inefficiencies in the design of the vessel include: unreacted oxygen in the air leaves the vessel, causing productivity losses; dilution of sulphur dioxide in the off-gas due to the nitrogen from the injected air and energy consumed by the main blower, punching machines, etc. The introduction of air into the vessel results in the formation of accretions that block the air channel through the tuyeres. These required mechanical "punching" in order to reopen the tuyeres (Kapusta et al. 2012). A historical photograph of this back-breaking work is shown in Fig. 3.2 (Southwick 2008; EMJ 1914). The punchers

Fig. 3.2 Manual punching of Peirce–Smith converter (Southwick 2008; EMJ 1914)

worked in an extremely noisy and dusty work environment.

One of the first great innovations in the converting practice occurred in the area of punching at Noranda's Gaspé Smelter, bringing to the world a far superior mechanical puncher than had hitherto been in use. Management at Gaspé realized that converter capacity and converter blowing rates limited higher smelter throughput. At that time, the converters were hand-punched, a very physically demanding job and unsafe job. It should be noted that mechanical punchers were in use at the time, for example, at the Kennecott puncher (Larson 1950). It was felt, however, that the Kennecott design had some limitations. By having the punch bar remaining inside the tuyere pipe at all times, the airflow would be impeded and an alternative was sought (Diaz et al. 2011). The new punching system developed at the Gaspé Smelter consisted of an externally mounted, hydraulically operated punch bar that was pushed into the tuyere with great rapidity and force and then quickly removed. This enabled the mechanical puncher to clean the tuyeres more efficiently and as the bar was removed, the tuyere pipe remained unrestricted (Fowler et al. 1968). One of the inventors, Albert Pelletier, described how a prototype model of the new puncher was constructed from wood and metal and was initially evaluated in the garage of one of the smelter personnel. A working experimental model was soon built and tested in the smelter during 1962. Later that year, an improved version was designed, built and tested. It consisted of two punch bars on a cradle; the bars were activated by an air-operated pneumatic cylinder with the assembly mounted on a track set in the floor along the converter length. The puncher had an upwards angle of approximately 5°; this approximated the stroke angle of a hand puncher. During 1964, a more robust unit was built essentially along the same lines and put into service. A photograph of this device is shown in Fig. 3.3.

The converter equipped with this new puncher design immediately showed higher average blowing rates. As the puncher performance improved, the device was installed on the two Gaspé converters and the average blowing rates gradually increased to over 40,000 Nm^3/h, thus approaching the maximum capacity of the blower. This helped drive the copper throughput to more than the smelter's design capacity. Worldwide patents for the new puncher, now known as the Gaspé puncher, were granted and the decision was made to establish an exclusive commercial and

Fig. 3.3 Mechanical punching of Peirce–Smith converter (Diaz et al. 2011)

Fig. 3.4 Gaspé puncher for Peirce–Smith converter (Diaz et al. 2011)

marketing arrangement with Heath and Sherwood Ltd. of Kirkland Lake, Ontario.

A photograph of a more modern Gaspé puncher is shown in Fig. 3.4.

The next innovation that we will examine is the Noranda Tuyere Silencer which was developed at Noranda's Horne Smelter during the 1970s by minimizing air leakage at bar entry. It also lessened the noise of the punch bar entering the tuyere assembly (Pelletier 1976). The silencer consisted of four spring-loaded segments fitted into a tuyere block that was positioned just ahead of the ball valve. This device gripped the moving punch bar and virtually eliminated any air losses and resulting blast noise caused by the punch bar entering the ball valve. It was estimated that 4 % of the air that was previously lost during punching could be utilized in the vessel. This led to a direct and immediate increase in productivity. In addition, the silencer was found to lower the puncher noise level to within 2 dBA of ambient. The Gaspé puncher and silencer transformed the Peirce–Smith converter by removing many of the uncertainties associated with the tuyere line operation, by stabilizing and maximizing blowing rates and by creating a safer and quieter work environment.

In copper and nickel smelting, reliable and continuous measurement of the melt temperature is critical in order to achieve effective control over the process. An innovative approach to measuring the temperature was developed at the Noranda Research Centre. Prior to this invention, a hood-mounted pyrometer was the technique commonly used by the industry. Stationary pyrometers have many limitations, such as the need to have direct line of sight to the bath, the requirement for frequent cleaning and not providing the true temperature. A number of unsuccessful attempts using thermocouples were tried in the past. A novel approach using sighting the melt through a submerged tuyere was developed by John Lucas and Greg Wint (Lucas 1987). The new design incorporated the following:

- The new pyrometer was to be sighted through a tuyere
- Punching could not be restricted
- Instrument electronics would be mounted away from the vessel
- The measurement was not to be affected by changes in received light due to tuyere blocking or tuyere pipe burn-back

Following initial studies, a prototype was built in the mid-1980s. The unit used a retractable pyrometer periscope that sighted the melt through an operating tuyere. It employed a fibre-optic cable to convey the radiation emitted by the bath to a specially designed two-wavelength

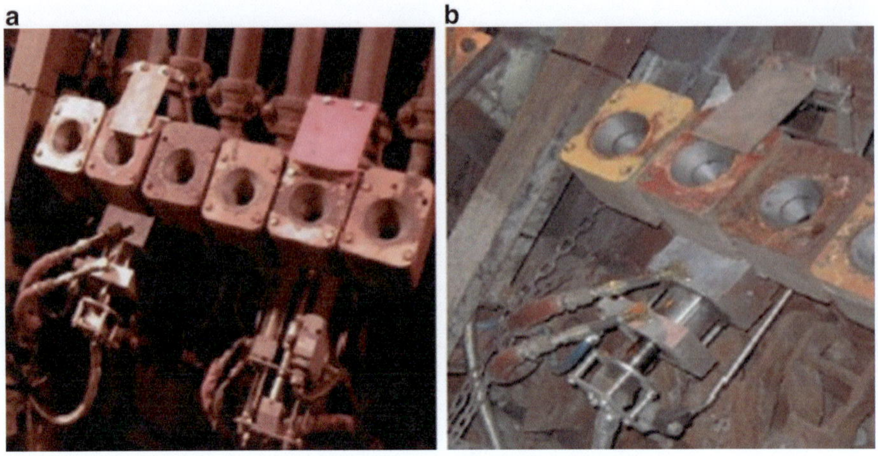

Fig. 3.5 First and second generation Noranda Tuyere Pyrometer (Diaz et al. 2011)

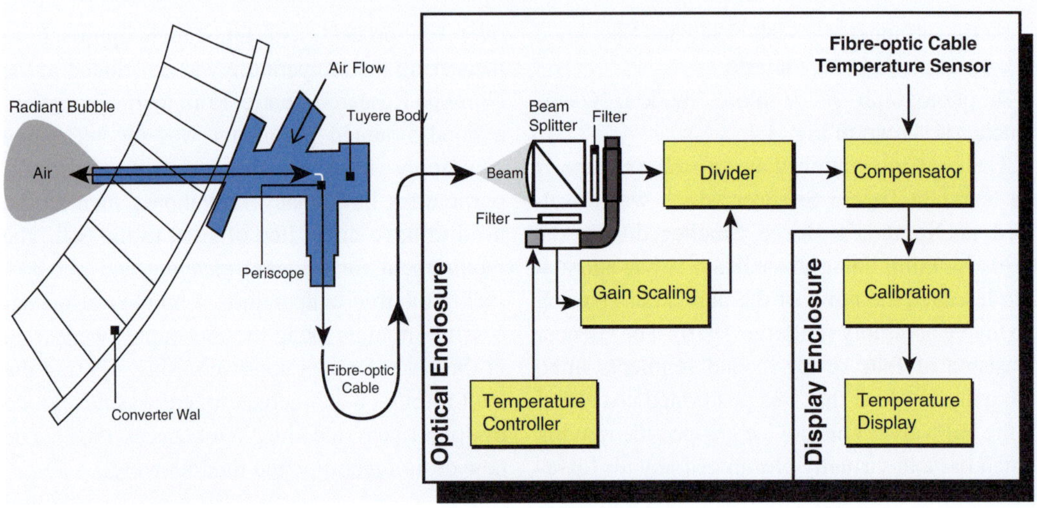

Fig. 3.6 Optical features of the Noranda Tuyere Pyrometer (Diaz et al. 2011)

pyrometer located away from the vessel. The tuyere body was fitted with a heavy guard plate in order to protect the periscope from accidental puncher damage. Initial calibration using a manual "Temptip" thermocouple was very successful. A first and second generation Noranda Tuyere Pyrometer are shown in Fig. 3.5.

The important optical features of the pyrometer are illustrated in Fig. 3.6.

In theory, it should be possible to attach an LIBS (Laser Induced Breakdown Spectroscopy) analyser next to the pyrometer in order to get instantaneous assays for key elements (e.g. As, Bi, Te) during the converting cycle. This would greatly improve the operation of the vessel. By knowing the temperature and composition of the bath on a continuous basis, the operator would be more assured of meeting smelter specifications. This concept was tested by Noranda in the 1990s.

The next major innovation to occur in the non-ferrous pyrometallurgical field is sonic injection of air/oxygen. Kapusta (2013) wrote an award winning paper on this topic. He opined on the fact that the steelmaking and non-ferrous metals industries have behaved as "two solitudes" in the way that they used different approaches to solve similar processing issues. Submerged gas injection is one such example. Whereas sonic injection has revolutionized steelmaking (e.g. Q-BOP, AOD), it has found limited applications in sulphide bath smelting and converting. Mackey and Brimacombe (1992) suggested that the reason was more than just economic. It might be related to the fact that the growth in oxygen usage in non-ferrous pyrometallurgy has been at a much slower pace compared to the steel industry. The steelmaking converter requires a high oxygen tuyere in order to overcome the productivity limits that had been reached. The non-ferrous converter, however, did not reach a similar barrier. Enriching the blast air, a few percent of oxygen had generally been sufficient for debottlenecking non-ferrous processes. This situation is rapidly changing and a new paradigm is emerging. Non-ferrous smelters are being squeezed by lower operating margins.

A wealth of knowledge focused on understanding gas injection started to emerge during the 1960s. Metallurgical research laboratories around the world investigated features of this process leading to many excellent technical papers on this topic that even today remain a benchmark. The steel industry used this knowledge to their advantage and great strides in productivity gains were made. The non-ferrous industry lagged behind.

Sir Henry Bessemer developed the first inexpensive industrial process for the mass production of steel from molten pig iron using air injection. Kapusta (2013) noted that although Bessemer included the use of oxygen in his patent, it was not possible to use oxygen at that time due to the severe erosion at the bottom of the vessel. In fact, submerged oxygen injection was not used for over a century. This required the genius of Savard and Lee (1958) of Canadian Liquid Air who developed the concentric tuyere. In the original Savard–Lee concentric tuyere, an oxygen stream at sonic velocity is shrouded with a hydrocarbon gas that cracks at steelmaking temperatures, and this phenomenon provides local cooling at the tuyere's tip. By thus shrouding the oxygen jet with a medium that is non-reactive with the melt, but reactive with oxygen, the injector is protected, especially at the critical zone near the injector–refractory interface where refractory wear occurs.

Keith Brimacombe and his group at the University of British Columbia (Hoefele and Brimacombe 1979) conducted new research into gas injection. They clearly demonstrated that the properties of the bath had a major influence on the gas jet penetration. They found that converting operations, in particular in copper and nickel converting, were characterized by large discrete bubbles of oxygen containing gas rising vertically above the tuyere tips. This suggested that the high refractory erosion at the back wall of the vessel was directly related to the dynamics of the gas injection process. They determined that above a critical back pressure, air injected into the converter becomes underexpanded and discharges as a steady jet stream with a much greater penetration into the bath (jetting regime). This work suggested that a jetting regime at high injection pressures could offer major benefits to converting operations. They speculated that underexpanded jets could lower the need for punching due to the greater momentum of the air jet.

The next major breakthrough in this area came in 1989, when Alejandro Bustos, who completed his Ph.D. work under Keith Brimacombe, joined the same Air Liquide group pioneered by Messrs. Lee and Savard. Bustos applied his experience in converting and sonic injection to the development of a technology for high oxygen injection into non-ferrous converting vessels. This work led to the Air Liquide Shrouded Injector, which is also more commonly known as the ALSI. This was a truly innovative concept designed to take full advantage of the benefits of operating at high oxygen enrichment levels without increasing the rate of refractory wear (Bustos 1995). A schematic diagram of the ALSI injector is shown in Fig. 3.7 (Kapusta 2013; Kapusta and Lee 2013).

The injector consists of an inner pipe through which oxygen enriched air is injected. This pipe is surrounded by an annulus through which nitrogen, or another inert gas or a hydrocarbon, flows. Both gas flows are injected at pressures such that

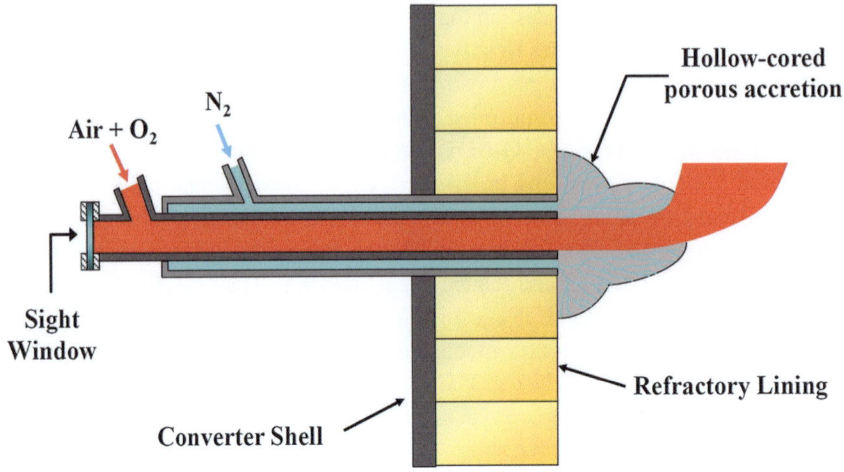

Fig. 3.7 A schematic diagram of the ALSI injector (Kapusta 2013; Kapusta and Lee 2013)

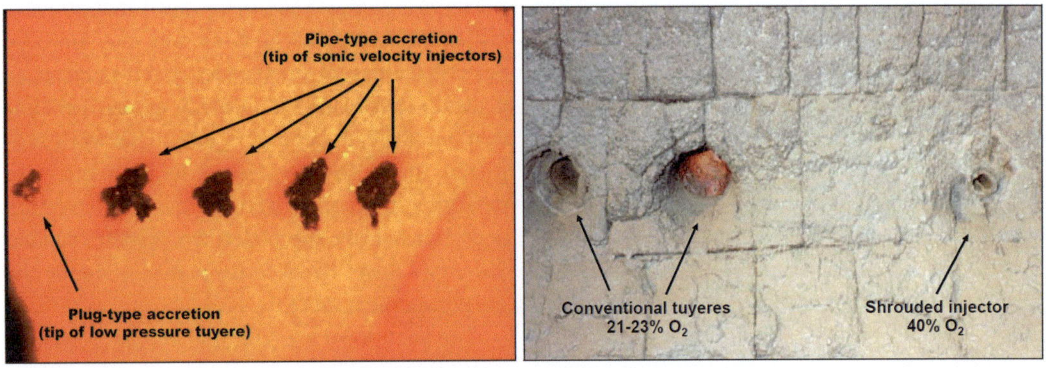

Fig. 3.8 Reduction of refractory wear by shrouded tuyere injectors (Bustos et al. 1999)

the flow through the inner pipe and the annular space is choked. This technology reduces wear by two mechanisms: (a) a protective accretion forms at the injector tip, thus providing chemical protection and (b) the elimination of punching, thus reducing mechanical damage. This technology has been tested in copper converters at the Union Minière Hoboken Smelter (now closed, company now known as Umicore) and in nickel converting at the Falconbridge (now Glencore) Smelter. Some spectacular shots from the Falconbridge tests are shown in Fig. 3.8 (Bustos et al. 1999). The photograph on the left clearly shows an example of a protective accretion that formed at the tip of the injector. The photograph on the right shows a comparison of refractory wear that is observed in "conventional" operation versus the minimal wear observed using shrouded tuyere injectors. More recently, the ALSI technology has been tested at Thai Copper (Pagador et al. 2009) and at the Lonmin Platinum Smelter (Kapusta et al. 2012).

3.1 Epilogue

It is interesting to note that although the Peirce–Smith converter is over 100 years old, the technology concept has not changed very much since its original introduction. What has changed is that as a result of the relentless innovation and perseverance by operators and company managers to improve the way the process was operated, it now performs very well and in an environmentally friendly way at the world's best smelters. Peirce–Smith converting now accounts for approximately 90 % of copper matte processing (Davenport et al. 2002). The process, however, suffers from several drawbacks:

- It operates batch-wise, giving an uneven flow of SO_2 gas to the sulphuric acid plant
- It leaks SO_2 gas into the workplace during charging and pouring activities
- Air leaks into the off-gas between the converter mouth and the gas-collection hood, thus producing a relatively weak SO_2 gas

Some of these deficiencies have been addressed over the years by such technologies as the Hoboken or siphon converter; this is essentially a Peirce–Smith converter with an improved gas-collection system. The Mitsubishi continuous top-blown converter which blows oxygen enriched air onto a molten matte surface using vertical lances, the Outokumpu flash converting process and the Noranda continuous submerged tuyere converter are all examples of continuous processes. Continuous operations provide a more uniform gas strength to the acid plant.

Some of the innovations discussed here, such as the Gaspé puncher, improved workplace safety immensely and improved productivity. Other inventions such as the tuyere pyrometer helped the converter operator (skimmer) to better control the process. Prior to that, the operators would rely on "indicators", such as colour of the off-gas flame and slag fluidity, to determine "endpoints". New technologies helped to standardize the operation from blow to blow. This helped turn the converting operation from an "art" to more of a "science".

There is no doubt that some amazing inventions have emerged over the years, such as the ALSI shrouded tuyeres, but the non-ferrous pyrometallurgical industry as a whole has been reluctant to embrace new technologies. As new cost and environmental constraints are placed on the industry, new (and some old) ideas will need to be implemented. It is hoped that the innovative spirit that was so evident in the work at the Gaspé, Noranda, Falconbridge and Hoboken plants, to name a few, will be encouraged by present and future mine and smelter owners so that this technology will be continually improved and modernized well into the twenty-first century.

References

Bustos, A. A. (1995). Process to convert non-ferrous metal such as copper or nickel by oxygen enrichment. U.S. Patent No. 5,435,833.

Bustos, A. A., Kapusta, J. P., Macnamara, B. R., & Coffin, M. R. (1999). High oxygen shrouded injection at Falconbridge. In C. Diaz, C. Landolt, & T. Utigard (Eds.), *Proceedings of the Copper99—Cobre 99 International Conference, Vol. VI—Smelting, Technology Development, Process Modeling and Fundamentals* (pp. 93–107). Warrendale, PA: TMS.

CIM Website: http://www.cim.org/en/Publications-and-Technical-Resources/Publications/CIM-Magazine/August-2011/cim-news/the_power_of_imagination.aspx.

Davenport, W. G., King, M., Schlesinger, M., & Biswas, A. K. (2002). *Extractive metallurgy of copper* (4th ed., pp. 131–172). Oxford: Elsevier.

Diaz, C. M., Levac, C., Mackey, P. J., Marcuson, S. W., Schonewille, R., & Themelis, N. J. (2011). Innovation in nonferrous Pyrometallurgy—1961–2011. In J. Kapusta, P. Mackey, N. Stubina (Eds.), *The Canadian metallurgical & materials landscape 1960 to 2011* (pp. 333–360). Montreal, Qubec: MetSoc of CIM.

EMJ. (1914). The original Peirce-Smith converter. *Engineering and Mining Journal, 4*, 718–720.

Fowler, P. L., Mills, L. A., & Balogh, A. G. (1968). The Gaspé mechanical tuyere puncher and converter performance. *Journal of the Minerals, Metals & Materials Society, 20*, 43–47.

Hoefele, E. O., & Brimacombe, J. K. (1979). Flow regimes in submerged gas injection. *Metallurgical Transactions B, 10*, 631–648.

Kapusta, J. P. T., Davis, J., Bezuidenhout, G. A., Lefume, S., & Chibwe, D. K. (2012). Industrial evaluation of sonic injection in a Peirce-Smith converter at the Lonmin smelter. In R. H Schonewille, D. Rioux, S. Kashani-Nejad, M. Kreuh, & M. E. S. Muinonen (Eds.), *Towards clean metallurgical processing for profit, social and environmental stewardship. 51st Conference of Metallurgists*. Montreal, Qubec: MetSoc of CIM.

Kapusta, J. P., & Lee, R. G. H. (2013). The Savard-Lee shrouded injector: A review of its adoption and adaptation from ferrous to non-ferrous pyrometallurgy. In *Proceedings of Copper 2013* (03. Pyrometallurgy and Process Engineering) (pp. 1115–1151).

Kapusta, J. P. T. (2013). Sonic injection in bath smelting and converting: myths, facts and dreams. In *Ralph Lloyd Harris Memorial Symposium, Proceedings of the Materials Science & Technology 2013* (pp. 267–318).

Larson, L. (1950). Development of mechanical puncher at McGill smelter: Trans. AIME. *Journal of the Minerals, Metals & Materials Society, 180*, 929–932.

Lucas, J. M. (1987). A fibre-optic pyrometer for tuyere temperature measurement. In *Proceedings of the Instrument Society of America, International Conference and Exhibit* (pp. 1299–1313). Anaheim, CA.

Mackey, P. J., & Brimacombe, J. K. (1992). Savard and Lee—Transforming the metallurgical landscape. In J. K. Brimacombe, P. J. Mackey, G. J. W. Kor, C. Bickert, & M. G. Ranada (Eds.), *Proceedings of the Savard/Lee International Symposium on Bath Smelting* (pp. 3–28). Warrendale, PA: TMS.

Mackey, P. J. (2014). China copper output climb continues. *Copper Worldwide (UK), 4*(3), 14–16.

Pagador, R. U., Wachgama, N., Khuankla, C., & Kapusta, J. P. T. (2009). Operation of the air liquide shrouded injection (ALSI) technology in a Hoboken siphon converter. In J. Kapusta & T. Warner (Eds.), *International Peirce-Smith Converting Centennial* (pp. 367–381). Warrendale, PA: TMS.

Peirce, W. H., & Smith, E. A. C. (1909). Method of and converter vessel for Bessemerizing copper matte. U.S. Patent No. 942,346.

Pelletier, A. (1976). *Development and operation of the Noranda puncher-silencer for converters.* Paper presented at the 15th Conference of Metallurgists, Montreal, Qubec: MetSoc of CIM.

Savard, G., & Lee, R. (1958). Method and apparatus for treating molten metal with oxygen. U.S. Patent No. 2,855,293.

Southwick, L. M. (2008). William Peirce and EA Cappelen Smith and their amazing copper converting machine. *Journal of the Minerals, Metals & Materials Society, 60*(10), 24–34.

Themelis, N. J. (1994). Pyrometallurgy near the end of the 20th century. *Journal of the Minerals, Metals & Materials Society, 46*(8), 51–57.

Themelis, N. J. (1985). Pyrometallurgical frontiers and challenges. In: J. F. Spisak, & G. V. Jergensen II (Eds.), *Frontier Technology in Mineral Processing, Chapter 5—Advanced Pyrometallurgy.* New York: SME.

Webber-Youngman, R., & Callaghan, R. (2011). Educating the future mining engineering practitioner. *SAIMM Journal, 111*, 815–820.

Wraith, A. E., Mackey, P. J., Levac, C. A., & Element, P. (1999). Converter and bath smelting vessel design—Blast delivery and tuyère performance: A reassessment of design characteristics. In C. Diaz, C. Landolt, & T. Utigard (Eds.), *Proceedings of Copper 99—Cobre 99 International Conference, Vol. VI—Smelting, Technology Development, Process Modeling and Fundamentals* (pp. 67–82). Warrendale, PA: TMS.

4 Thermal Processing: Pyrometallurgy—Ferrous

Some Perspectives on the Development of Converter Steelmaking Within Japan

Toshihiko Emi and Alexander McLean

4.1 Introduction

This review is intended to provide a concise but panoramic understanding of the past developments with respect to converter steelmaking and stimulate further revolutionary technological advances for the future development of the steel industry.

4.2 The Early Years

During the nineteenth century the process of steelmaking was revolutionized. In 1856, the converter furnace with an acidic refractory lining and air blown bottom tuyeres was developed. In 1857, the Siemens open hearth furnace was introduced and in 1864 this technology was combined with the Martin furnace for scrap melting to create the Siemens-Martin open hearth. It required about 20 more years however, before the converter and open hearth furnaces, both lined with basic refractory material, successfully dephosphorized hot metal for mass production of steel with acceptable quality. Hot metal in Europe had high phosphorus (P) content, since the available iron ores were high in phosphorus.

Mass production of steel in those days was dominated by Britain, Germany, Belgium, and the USA. In Japan, the acidic open hearth and the acidic Siemens-Martin open hearth furnaces were installed in 1890 and 1896 respectively at navy and army steelmaking factories. Annual production of these factories was 2.4 k tons, which was about 1.1 % of annual steel imports of 220 k tons.

From the end of the Russo-Japanese War in 1905 until the outbreak of World War I in 1914, the Japanese steel industry suffered a recession, facing plant shutdowns, and mergers. Inadequacy of equipment and lack of good quality coal and iron ore were problematic. Despite the unfavorable circumstances, efforts by Yawata Works made it possible in 1914 to convert steel ingot production from remelting iron made by the traditional Tatara method with charcoal (17 k tons/year) to steel made with the Basic Open Hearth (BOH) (275 k tons/year) and the acidic Bessemer converter (106 k tons/year).

Productivity of the acidic Bessemer converter was much greater than the BOH, but the slopping loss of metal out of the converter mouth and erosion wear of the bottom tuyeres resulted in 8 % lower metal yield on an ingot basis and 3.4 % higher cost of production. In addition, the Bessemer converter with an acidic lining could not make steel with phosphorus contents lower than about 0.1 wt% and hence the application of

T. Emi
Formerly with Tohoku University and Kawasaki Steel, Takaido Higashi 1-31-6-603 Suginami, Tokyo 168-0072, Japan

A. McLean (✉)
University of Toronto, Toronto, ON, Canada, M5S-3E4
e-mail: amclean16@cogeco.ca

V.I. Lakshmanan et al. (eds.), *Innovative Process Development in Metallurgical Industry*,
DOI 10.1007/978-3-319-21599-0_4

steel products was limited to commodity wire, rod, and medium gauge rail. The high phosphorus content was a consequence of unavoidable use of phosphorus-bearing iron ore (0.2–0.3 wt%P) in the blast furnace due to the war-time shortage of low phosphorus ore imports. To improve dephosphorization (deP), refining the converter melt in the BOH was attempted by the Converter-BOH Combination process (double refining). However, this process resulted in 6 % lower metal yield, 14 % higher ingot cost and could not compete with the advanced operation of the BOH, and was therefore terminated in 1927.

NKK introduced three basic refractory lined, bottom air blown 20 t Thomas converters from Germany and started operation with high phosphorus hot metal in 1938. Two more Thomas converters were installed by 1941 to produce 0.35 million tons of steel in 1942. However, due to war damage, all converters were shut down in 1945. When World War II ended, the number of BOHs was 197 with a production of 1.23 million tons. In 1947, when post-war disturbances still remained, only 20 BOHs were in operation, producing a very limited amount of steel, 0.4 million tons. However in order to promote economic self-reliance within Japan and meet the needs of the General Head Quarters (GHQ) of the United Nations forces (UNf) to supply materials in support of the Korean war which lasted from 1950 until 1953, steel production was increased. (Compilation Committee for Yawata Steel Historical Documents 1980; Harabuchi and Imoto 2009; Shimokawa 1989; Kawasaki 1982; Yawata 1980; Philbrook et al. 1951; - The Iron and Steel Institute of Japan 1964; The Iron and Steel Institute of Japan 1992; Sugita 1995).

4.3 Steelmaking with the Oxygen Top Blown Converter

4.3.1 Evolution of Oxygen Top Blown LD Converter Steelmaking

The oxygen top blown Linz Donawitz converter (LD) began operation at Vöest Linz with three 30 t vessels in 1952 and at ÖMAG (Alpine) Donawitz with two 30 t vessels, all with BOH hot metal. In Europe, the Thomas converter process was in operation with high phosphorus hot metal smelted from European iron ores with high phosphorus content. However, phosphorus-bearing iron ores were not available in Austria where only low phosphorus hot metal was produced and this composition was not appropriate to sustain the heat balance required for the Thomas converter. In addition, there were increasing demands for steels low in phosphorus and nitrogen in order to obtain steels with improved cold formability. It was difficult for Thomas converter steel to meet such demands. Attempts were made to decrease nitrogen in the converter with an oxygen enriched air blow, but this was not successful due to the heavy tuyere erosion. Scrap supply was insufficient in Austria to sustain BOH operations despite the fact that BOH steel with lower phosphorus and nitrogen contents has better cold formability.

To solve these problems, Dürrer and colleagues successfully carried out experiments with pure oxygen, blown through a water-cooled top lance onto the surface of the molten metal. Vöest and ÖMAG adopted this new technology, went into a joint development, and succeeded in operating a 2 t vessel, and subsequently vessels of 12–15 t capacity in 1949, making steels with phosphorus, nitrogen, and oxygen contents below those found in BOH steel (Sugita 1995; Doi 1969; Lynn 1986; The Iron and Steel Institute of Japan 1982; Iron and Steel Handbook 2002).

4.3.2 Transfer of the LD Steelmaking Process to Japan

Regarding pneumatic converter operations in Japan, Yawata had operational experience with the Bessemer converter which lasted until 1927 and with a 5 t top blown converter from 1954. NKK revitalized the basic Thomas converter operation in 1954 with an oxygen enriched air blow. The two companies had exposure to information on the LD operation in Austria via the Ministry of International Trade and Industry (MITI) and a trading company within Japan. They were stimulated by the operational advantages of the new LD technology and its suitability for low phosphorus hot metal, compared to the Thomas converter or BOH, lower scrap ratio, lower refractory consumption, and no need for heavy oil as fuel. Although the process technology was still in its infancy with respect to industrial mass production, it was speculated that there would be improvements in productivity, installation cost, labor cost, operating cost, and steel quality with the LD compared to the Thomas converter or BOH (Doi 1969; Lynn 1986; The Iron and Steel Institute of Japan 1982).

The two companies sent their representatives to confirm these aspects on site and were convinced of the future potential of the LD process. They attempted to negotiate independently with Alpine however in view of national interest, it was concluded, under the auspices of MITI, that NKK would be the sole representative licensee for the use and granting of sublicenses of the technology in Japan. Accordingly, a licensee contract was struck between NKK and Brassert Oxygen Technik AG (BOT, licensing company) via Alpine in 1956. Yawata struck a sublicensee contract with NKK which was kept open for domestic third parties. Milestones in the history of LD steelmaking in Japan were established with the installation of a 50 t LD at Yawata and a 42 t LD at NKK in 1957 and 1958, respectively.

The aforementioned advantages combined with financial support by MITI as part of the second rationalization program of the Japanese steel industry prompted other steel companies to install LD converters (Fig. 4.1 (Baker et al. 1998), upper left). The share of LD steel production which was around 5 % in 1958 surpassed BOH steel share in 1965 with 55 % of the steel produced, and in 1970 achieved 79 % with a production of 73.51 million tons, the highest in the world. New installations of LDs, conversion of BOHs to LDs, and revamping of obsolete LDs continued until 1973, just before the oil shock crisis. With a total of 92 LDs including eleven 250 t vessels and six 300–340 t vessels, Japan held the leading position in both the amount and share of the world's production of LD steel, exceeding West Germany, the USA, Britain, and the USSR.

In contrast, BOH steel production declined sharply. In 1970, BOHs in major steel companies were closed down, and in 1971 only 1.99 million tons were produced within Japan. The last BOH was shut down at Tokyo Steel Okayama works in 1977, bringing a long and glorious history to a close.

4.3.3 Development of the LD Steelmaking Process Within Japan

The success of the LD process can be ascribed to advantages such as greater productivity, lower production cost (~10% for carbon steel, ~30% for HSLA steel), and a lower requirement for scrap than the BOH process. These advantages were achieved due to the following improvements and developments in equipment and operation after the transfer of the LD technology to Japan:

1. Enlargement of the shell volume, implementation of a supporting and tilting mechanism for the vessel with a concentric tulip profile and taphole, and elimination of a detachable bottom. For example, in the case of a 300 t vessel, the inner steel shell volume and height/diameter ratio were set at 553 m^3 and 1.3, respectively. The top cone angle was optimized, and a trunnion ring support together with a stepless variable speed tilting system was installed.
2. A water-cooled main lance with multiple hole nozzle tip was invented to prevent spitting and bottom refractory erosion during the blow (1962–1970).
3. An automatic exchange system for a sublance equipped with temperature sensor and carbon meter was made fully operational in 1966.
4. A non-combustive off-gas recovery system (OG) was developed by Nippon Steel (now Nippon Steel Sumitomo Metal, NSSM) to retain off-gas energy and minimize off-gas volume and particulate emission (1962–1969). This system was upgraded and

subsequently implemented by steel companies throughout the world.

5. A considerable reduction was achieved of vessel refractory consumption to ~7kg/t of steel with the development of tar bonded dolomite, stabilized dolomite bricks, and magnesia-carbon bricks.
6. Significant improvement in the hit-rate of the carbon content and temperature (C-T) window at the end of blow was achieved with static and dynamic computer control of the main lance height and oxygen flow rate, utilizing accumulated blow data calibrated with C-T values measured on-line with the sub-lance.
7. Hot metal low in silicon and phosphorus for charging to the LD was produced by advanced blast furnace operations using low phosphorus iron ores imported from Brazil and Australia, accompanied by a decreased return of phosphorus-bearing BOH slag to the blast furnace.
8. Development and implementation of hot metal pretreatment processes and secondary refining processes prior to and after the LD operation.
9. Advances in single slag and catch carbon blowing techniques for medium and high carbon steels, low alloy steels and stainless steels to the extent that the resulting steel quality met or exceeded the standards required by JIS specifications.
10. Prolongation of vessel life beyond 5000 heats/campaign by developing zone lining of refractory, hot gunning refractory repair technology, and enrichment of magnesia in the LD slag during the blow, all of which increased LD productivity.
11. Progress in (a) measurements at elevated temperatures of thermodynamic quantities and physical properties of the materials relevant to the process, (b) equilibrium calculations for steelmaking reactions, (c) modeling of heat, mass, and energy transport phenomena for process analysis, and (d) fluid dynamic simulation of the process.
12. Development of the theory and appropriate systems for the control and automation of the process.

Many of the above factors depended on the progress achieved in Europe and the USA. Items (2) and (4) however were developed and implemented within Japan. Considerable progress was achieved with items (5) through (12) after they were introduced from abroad. Cooperation among the licensee companies in sharing relevant information on timely technological advances at the meetings and plant visits of the LD committee (later Discussion Meetings for Japan LD Technology), promoted by NKK and Yawata for Japan BOT members (1958–1966), contributed greatly to the technological advances. These meetings were succeeded by the Steelmaking Division of the Joint Study Committee of the Iron and Steel Institute of Japan (ISIJ). In addition, collaborative activities between academia and industry facilitated by the Joint Study Committee for the Fundamentals of Iron and Steel, by the Melt Refining Division and at bi-annual Meetings, all held under the auspices of ISIJ, and those by the 19th Committee (1934-) of the Japan Society for the Promotion of Science (JSPS) greatly enhanced progress. It must also be noted that these activities were supported either directly or indirectly by the Japan Iron and Steel Federation (JISF) and MITI (The Iron and Steel Institute of Japan 1982; Iron and Steel Handbook 2002; Tetsu-to-Hagané 1975, 1985).

4.3.4 Maturation of the LD Process for High Productivity

During the favorable economy period which prevailed after 1967, the LD operation continued to move toward higher productivity with extended flexibility for hot metal ratios. Notable technologies that enhanced productivity included:

1. Charge time reduction with large capacity torpedo/ladle cars to transport hot metal to the steelmaking shop with parallel improvements in scrap loading.
2. Thinner refractory lining to enlarge the vessel volume with the development of magnesia-carbon brick.

3. Hot gunning of magnesia or dolomite with vessel profile monitoring and slag coating to cutback the repair time and prolong service life of the vessel refractory.
4. Computer control of the main lance height and oxygen flow rate and optimization of the multiple hole lance tip design to minimize slopping and enhance dephosphorization under much increased oxygen flow rate.
5. Simultaneous operation of all three vessels in a three vessel shop.
6. Development of "Direct tapping" which eliminated the time consuming end point sub-lance measurement of C and T, as the ultimate aim in dynamic blow control.

As a consequence of the improved equipment and operations, average productivity of LDs reached 240 t/h in 1974. In many plants, the end point hit-rate for the C-T window was about 90 % for low C steels, even though the scrap rate was limited. Vessel refractory life was very much extended, with a record of 10,110 heats/campaign established at the Kimitsu Works of Nippon Steel Corporation (now NSSM).

Since that time, noteworthy progress has involved further prolonged refractory life with a slag splashing practice developed in the USA (Baker et al. 1998). After tapping, dolomite is added to the remaining slag on the bottom of the vessel, the main lance is lowered to about 70 cm above the bottom, and the magnesia enriched molten slag is splashed with nitrogen gas blown from the main lance to weld coat the inner surface of the refractory lining. No particular equipment is necessary, and the operation is simple, lasting only a short period of time before proceeding to the next heat. For these reasons, the splashing technique has been widely adopted throughout the world. Refractory life from a fresh lining to the first relining has exceeded 20,000 heats. With the slag splash coating practice, however, clogging of tuyeres installed at the bottom of mixed blown converters, as discussed later, was an initial concern. This was resolved by controlling the flow rate of the nitrogen flushing gas. It has been reported (Liu et al. 2012) that with delivery of inert gas through concentric bottom tuyeres combined with an appropriate gas flushing practice, the refractory life was extended up to 50,000 heats/campaign. Caution is, however, necessary to ensure that the proper blow characteristics within the vessel are not adversely affected with an off-design, distorted inner profile which may result from an excessive number of splashing cycles.

With the automated blow operation, the C-T hit-rate was improved with a narrower target window. Trimming of the blow pattern was refined to reduce spitting and slopping with amended material additions. These improvements were made possible on the basis of integrated computer analysis of the data acquired with sensors for the volume and composition of the off-gas, temperature and composition of the steel melt, as well as vibration and acoustic wave characterization associated with the refining process.

4.4 Steelmaking with Bottom Blown and Mixed Blown Converters

4.4.1 Birth of the Bottom Blown Converter, OBM/Q-BOP

In spite of the advantages mentioned above, the LD process had some inherent disadvantages:

1. Insufficient mixing of the metal bath, resulting in heterogeneity in temperature and chemistry within the bath.
2. Sluggish formation of molten slag with the added lime.
3. Loss of iron to the slag phase caused by excessively high temperature and over-oxidation at the impinging points of the oxygen jet on the metal bath.
4. Difficult to prevent slopping when abrupt evolution of CO bubbles occurred due to the over-oxidation and heterogeneity of carbon distribution in the bath. The multiple hole lance tip was to some extent effective in reducing slopping by decreasing the heterogeneity within the bath. Slopping happens when

bubbles are retained in the slag under an unfavorable combination of temperature and slag properties and the decarburization rate is high.

In the case of the bottom blown Thomas converter, heterogeneity within the bath was avoided and hence slopping was prevented. As mentioned previously however, the Thomas converter was limited in the amount of oxygen that could be used in the blowing gas due to the serious incidence of tuyere erosion. In the mid 60s, after many difficult trials, Savard and Lee (1992) of Canadian Liquid Air overcame the erosion problem with a new concept based on the use of concentric, double tube, annular tuyeres. Hydrocarbon gases (propane, methane) were passed through the outer annulus of the concentric tuyeres as a coolant which endothermically decomposed to cool the tip of the tuyere tube. Oxygen gas and lime powder were injected into the melt through the inner tube. A porous accretion called a "mushroom" was formed at the tip which prevented direct contact of the steel melt with the tuyere, and the cool decomposed gas passed through the pores (Guthrie et al. 1992).

In 1967, the Savard and Lee technology was validated on an industrial scale when Brotzmann of Maxhütte conducted trials with a 20 t Thomas converter in which the conventional tuyeres were replaced with the new annular tuyeres. In 1968, this innovative technology was commercialized under the name of Oxygen Bottom Blown Maxhütte (OBM) (Brotzmann 1992). In 1973, US Steel implemented the annular tuyere technology on a large scale at Gary works in a 200 t basic oxygen furnace (BOF), which was similar in concept to the LD, and in 1974 at Fairfield works in 160 t BOFs. US Steel gave the new technology the name Q-BOP which stands for Quick refining, Quiet blowing, Quality Basic Oxygen Process.

A major problem encountered with the Q-BOP was poor endurance of the bottom tuyeres. For long life of the tuyere bottom, it was necessary to have mushroom accretions of similar size on all tuyeres. If imbalance in the cooling allowed a mushroom to melt away, the tip of the tuyere is burned back causing damage to the bottom. In an extreme case, the burn back proceeded to the outer side of the bottom, caused burn-through of the tuyere and burned the connecting oxygen piping, resulting in leakage of molten steel from the vessel (Nozaki 2000a; Ueda et al. 1995; Savard and Lee 1992).

4.4.2 Blowing Characteristics of the Q-BOP

In 1977, Kawasaki Steel introduced the Q-BOP technology to Japan. Two 230 t Q-BOPs with 18 and 22 tuyere bottoms were installed at Chiba (Fig. 4.1, upper right) (Baker et al. 1998). Parameters for control of the blowing process were established through extensive investigations with a water model and a 5 t Q-BOP. Aspects investigated included melt flow in the vessel, mode of lime injection, characteristics of steelmaking reactions, and wear of refractory (Nozaki 2000a, b).

The Q-BOP is distinctly different from the LD in that all of the oxygen together with lime powder is injected through the bottom tuyeres. This has the following advantages:

1. Quick melting of the charged scrap.
2. Very fast homogenization of the steel melt is sustained into the low C range. In terms of the time required for uniform mixing of the melt, the Q-BOP takes only about 10 % of that required by the LD (Fig. 4.2) (Nakanishi et al. 1978).
3. The rate of decarburization (deC) is also very fast, proceeding to near equilibrium for the C–O reaction.
4. Loss of Fe and Mn in the slag caused by over-oxidation is smaller, and hence the yield of Fe and Mn is higher. For example, the total Fe content in the slag (T. Fe) at 1630 °C and 0.04 %C is ~12 % compared to ~23 % for the LD.
5. Accordingly, the oxygen consumption for the Q-BOP is lower. Decarburization oxygen efficiency for the LD decreases from unity at 0.8 %C down to 0.6 at 0.2 %C, whereas for

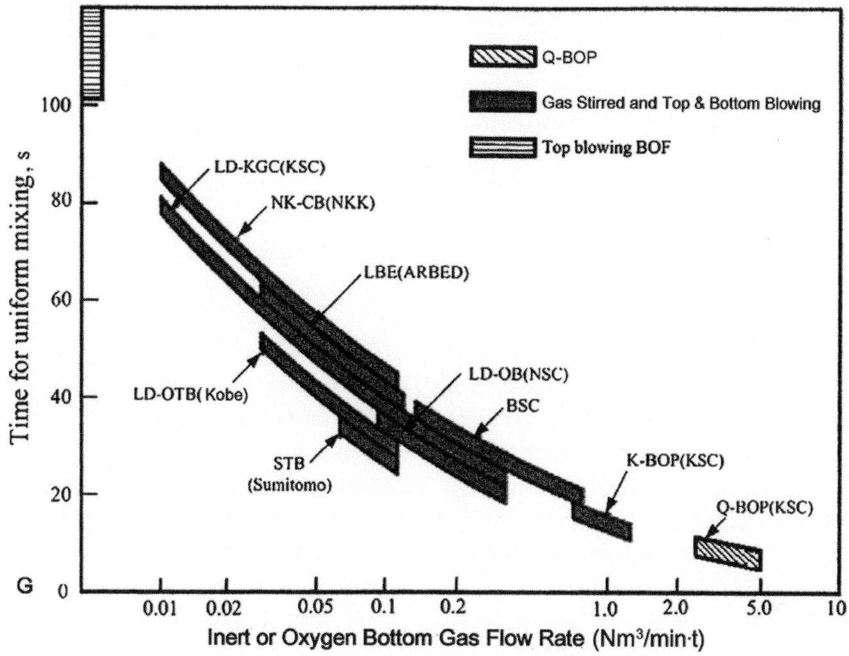

Fig. 4.2 Mixing time of melt in various BOFs with bottom gas flow rate (Nakanishi et al. 1978)

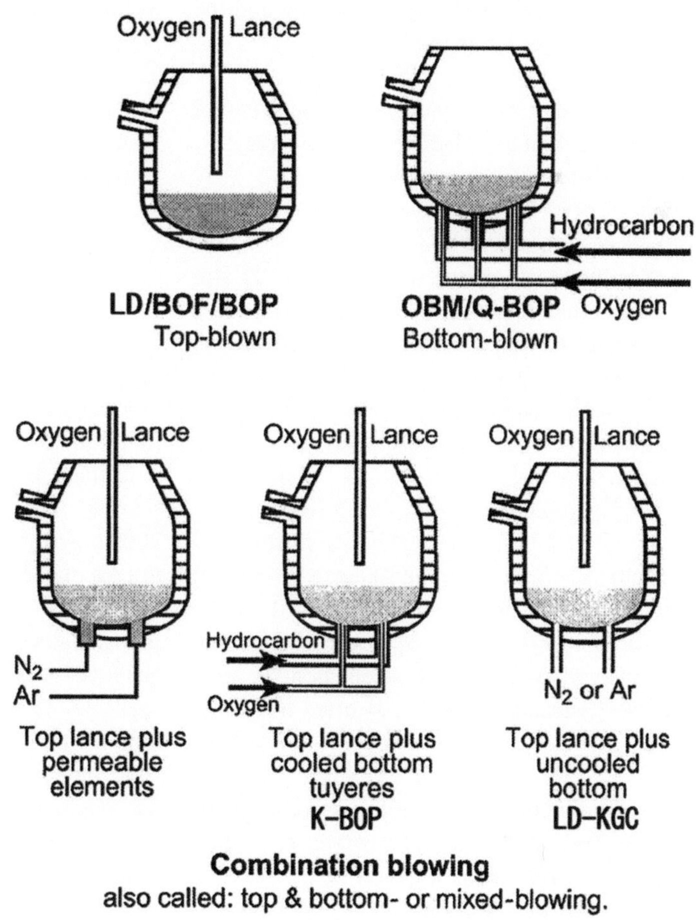

Fig. 4.1 LD (BOF), mixed blowing BOFs and bottom blown BOF (Q-BOP) (Baker et al. 1998)

the Q-BOP, the efficiency remains at unity until 0.4 %C and then decreases only to 0.9 at 0.2 %C. In addition, the decarburization limit for the Q-BOP is lower than that for the LD and can be less than 0.02 %C.

6. Slag formation is quicker, over-oxidation and slag amount are much smaller, and hence samples taken with the sub-lance are a better representation of the C and T of the bulk melt. Consequently, the hit-rate at the end of blow is close to 99 % and the reblow rate less than 1 % for a target window of 0.05 ± 0.015 %C and 1610 ± 10 °C with a 10% scrap operation.
7. Slopping is much decreased and off-gas recovery with OG can reach 1.4 GJ/t.
8. The desulfurization (deS) ratio is improved, and despite the lower (T. Fe) in the slag, the deP ratio is not much different from the LD, provided the injection of lime is properly distributed between the initial period and the later period of blowing.

On the other hand, there are some inherent disadvantages of the Q-BOP and these are outlined below:

1. Higher cost of investment.
2. Refractory life of the tuyere bottom is still shorter than that of the vessel, calling for 2–3 bottom exchanges during a campaign of the vessel refractory.
3. The scrap charge ratio is lower to the extent equivalent to the heat loss caused by use of the coolant gas.
4. The hydrogen content in the Q-BOP melt at the end of blow, even after Ar flushing, is higher (4–7 ppm) than an LD melt (2–3 ppm) due to the hydrogen input generated by the decomposition of the hydrocarbon coolant. However, in a modern steelmaking system where vacuum degassing is often available, this may not be a serious issue.

In a similar process to Q-BOP, kerosene coolant was used instead of hydrocarbon gases in a 240 t LWS converter at Sollac in 1978.

The previously mentioned blowing characteristics were well explained on a semi-empirical basis by Nakanishi et al. (1978) in terms of an Index for Selective Carbon Oxidation (ISCO). Derived from the results of a model study combined with operational data, the ISCO consists of the product of two terms:

1. A thermodynamic term to account for the partial pressure of CO which defines the C–O equilibrium at the melt surface.
2. A relative mass transfer term that represents the ratio of mass flux of oxygen supplied from the bottom tuyeres (or the main lance) to the mass flux of C supplied from the bath to the gas/melt interface. Here, the mass flux of C was approximated by the average melt flow rate, q, in the bath, and q was defined as the inverse of the uniform mixing time, τ, of the melt ($q = 1/\tau$). τ was determined to be proportional to about −0.4 power of the mixing energy, ε, supplied to the bath ($\tau \propto \varepsilon^{-0.4}$). Term (2) shows the predominance of either the oxidation of Fe or the oxidation of C from the melt under a given CO pressure. The smaller the ISCO value, the better is the preferential oxidation of C to a lower C range without much loss of Fe by oxidation into the slag phase.

The ISCO value successfully described for the first time in an integrated way the relation between the degree of melt stirring and the oxidation into the slag of constituent elements in the metal bath for a variety of primary and secondary refining furnaces (Figs. 4.2 and 4.3) (Nozaki 2000a; Nakanishi et al. 1978), and is considered to be a major contribution to process metallurgy. Later, Kai et al. (1982) proposed an amended index, Balance of Oxygen and Carbon Feeding Rate (BOC), which eliminated term (1) and replaced the metal flow rate in term (2) in ISCO with the mass flux of C. For the LD process, the BOC index was reported to give a slightly better correlation than ISCO in describing the behavior of (T. Fe) in the slag phase within the range of 0.02–0.22 %C in the melt.

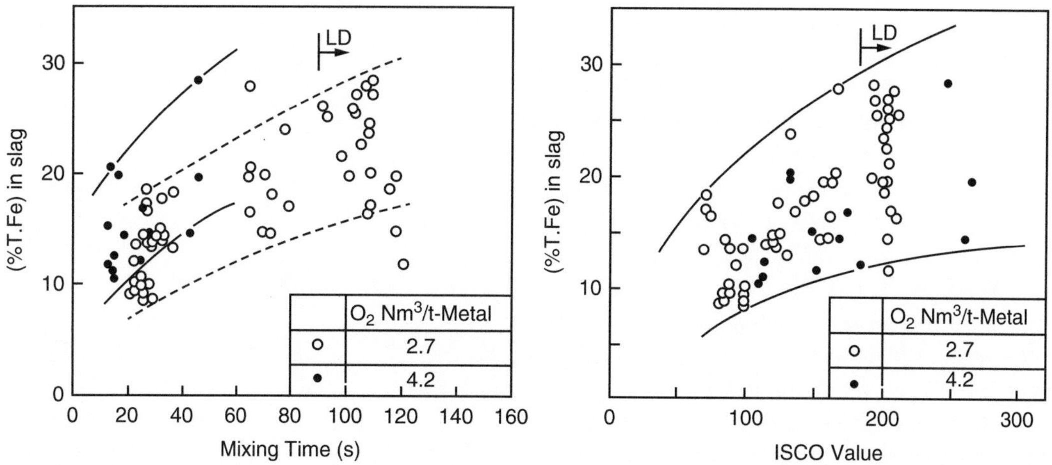

Fig. 4.3 Oxidation of iron in slag in various BOFs with mixing time or ISCO (Nozaki 2000a; Nakanishi et al. 1978)

4.4.3 Development of Mixed Blowing Converters

A mixed blowing converter which incorporated aspects of the LD and Q-BOP, with 70 % oxygen blown from a top lance and 30 % oxygen injected through bottom tuyeres, was put into operation at Mizushima (three 250 t vessels) and at Chiba (two 85 t vessels) of Kawasaki Steel during 1980–1981. The top and bottom blowing converter was named K-BOP (Kawatetsu-BOP where the prefix means Kawasaki Steel in Japanese, Fig. 4.1 bottom center). The ISCO value for the K-BOP was determined to be 64, which is very close to 58 for the Q-BOP, and much smaller than about 230 for a 160 t LD, despite the limited fraction of bottom injected oxygen. With the K-BOP only oxygen is blown from the top and bottom until a critical C content of 0.10–0.15 % is reached, at which point inert gas is injected through the bottom tuyeres to promote deC in the range below the critical C content (called Inert gas Decarburization, ID). With the K-BOP, it is possible to change the main lance height to prevent slopping, thus acquiring better freedom to form slag and achieve better control of the deP operation.

Similar mixed blowing or top and bottom blowing LDs with the bottom blowing through the concentric tuyeres, called LD-OB (LD-Oxygen Bottom Blowing), has also been implemented with a total of ten installations at Nippon Steel Yawata, Oita, Kimitsu, and Nagoya in 1984. There are now over 100 top blowing converters with oxygen bottom blowing (OBM, Q-BOP, K-BOP, LD-OB, etc., all inclusive) installed in steel plants throughout the world. Sumitomo Metals (now NSSM) commercialized a top and bottom blowing process, STB (Sumitomo Top and Bottom), with CO_2 as the bottom tuyere coolant in place of hydrocarbon gases.

It was known that in the Thomas converter and in secondary refining furnaces, bottom injection of gases promoted stirring and enhanced flow of the steel melt. However, the considerable improvement in blowing characteristics for the LD with the bottom injection of oxygen was a revelation to the steelmaking community, triggering the evolution of a variety of top and bottom blown converters.

As noted before, uniform mixing time, τ, is inversely proportional to 0.4 power of energy ε of stirring imposed and dissipated in the melt. The average flow rate q (ton/s) of the melt is inversely proportional to τ, i.e., $q = 2W/\tau$, with W (ton) being the mass of the steel melt.

For an inert gas bottom stirred 250 t LD (LD-KGC) at Mizushima, τ was calculated to be 37 s with a relatively small 10 % injection of Ar, which is much shorter than about 75 s for an LD without bottom injection, and not too much longer than 14 s for the Q-BOP at Chiba. The concentration product of C and O in the melt in mixed blowing converters is found to decrease below the equilibrium concentration product of Ceq and Oeq with a shorter τ. In former times, this anomaly was interpreted to have been caused by lower partial pressure of CO than unity in the converters due to decomposed hydrocarbon coolant. However, Kishimoto et al. (1989) showed that the O content in the melt did not change with the species of hydrocarbon coolant, but decreased with increased q which implies a shorter τ. They considered that the anomaly is caused by the following: At the impinging points of the oxygen jet on the steel melt, the activity of (Fe_tO) formed by the oxidation of the melt is unity and oxygen in the melt may come close to the equilibrium value corresponding to the C and T at that location. When melt stirring or melt flow rate is large enough, however, oxygen in the bulk melt should be controlled by (Fe_tO) which is generated at the impinging points but diluted in slag which covers a much wider surface area of the melt than the area of the fire spots. The stronger the stirring or the greater the flow rate, the lower the activity of (Fe_tO) which in turn controls the oxygen in the melt, and hence oxygen decreases with q.

4.4.4 Development of LDs with Bottom Injection of Inert Gases

In view of the considerable improvement in blowing characteristics of the LD with a small amount of inert gas injection from the bottom, the injection of inert gas soon replaced oxygen and lime injection through the concentric tuyeres, thus eliminating the exchangeable bottom with the tuyeres and reducing the investment and running costs.

The following versions of LDs with gas stirring have been commercialized by developing different types of Ar or N_2 bottom injection, resolving the erosion problems associated with the gas inlet plug and increasing the gas flow rate: LBE by IRSID-ARBED, LD-BC by CRM, and UBDT by Krupp employ either slit brick or

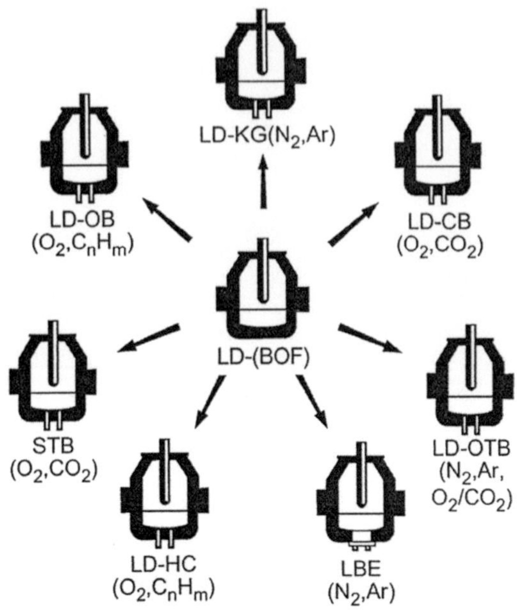

Fig. 4.4 Variants of mixed blowing BOFs as developed from LD/BOF and Q-BOP/OBM (Baker et al. 1998)

permeable brick for the plug, while a metal tube (generally a single one) is used in the LD-AB by Nippon Steel, LD-KGC by Kawasaki Steel, LD-OTB by Kobe Steel, and NK-CB by NKK. Injected gas flow rates range from 0.01–0.50 Nm^3/min.ton (Fig. 4.4) (Baker et al. 1998), with the majority less than 0.2 Nm^3/min.ton.

With inert gas stirred LDs, there is a decrease in the critical C content at which the rate controlling step for deC changes from mass transfer of O to mass transfer of C, and hence the decarburization oxygen efficiency in the lower C range is improved and the (T. Fe) in the slag is reduced to below 20 % which corresponds to increased iron yield. For these reasons, conventional LDs were rapidly converted into the gas stirring variants.

To avoid the confusion associated with using the names, LD in Europe and BOF in the USA, the vessels are collectively denoted here as BOFs. Gas stirred BOFs with bottom injection of inert gas are designated as "Inert Gas Stirred BOFs," BOFs with bottom oxygen blowing as "Top and Bottom Blowing BOFs," and the two types of BOFs are identified as "Mixed Blowing BOFs or Combination Blowing BOFs."

In recent years, even better productivity was demanded from the mixed blowing BOFs. However, when greater oxygen gas flow was supplied, there was interference between the fire spots where the top-gas jet impinged onto the melt surface, generating spitting of molten metal droplets out of the vessels. The spitting effects were particularly severe with a lean slag operation when the use of pretreated hot metal low in Si, P, and S was introduced. Fukagawa (2006) designed a new main lance nozzle tip with six nozzle holes inclined in a particular way. Oxygen jets from the holes did not interfere with each other at the impinging points, thus preventing the formation of spitting. Using the new nozzle design on the main lance for 250 t mixed blowing BOFs at Wakayama, it was possible to produce 300 k tons/month with a 1 out of 2 converter operation at a blowing time of 9 min and tap-to-tap time of 20 min.

4.4.5 Utilization of Mixed Blowing BOFs for Scrap Melting and Smelting Reduction

In an integrated steel plant equipped with an OG system, top and bottom blowing BOFs are capable of melting scrap with less energy consumption (~3.9 GJ/t, after deduction of the recovered off-gas credit) than the electric arc furnace (EAF) (~4.5 GJ/t). At Hirohata works in 1993, a BOF was converted into a top and bottom blowing vessel and operated as a coal- based, scrap melting unit (Oonuki 1998) using the existing off-gas recovery system. When the combustion rate of off-gas in the vessel was increased from 0 to 30 %, the scrap ratio is increased by only 10 %. Addition of carbon-bearing fuel at the furnace mouth was not successful either, since much of the material was simply blown away. These shortcomings were resolved by using a hot heel operation that consisted of a high C melt into which oxygen and pulverized coal with low volatile materials were blown through the bottom tuyeres with nitrogen as the carrier gas. Scrap was charged from a shoot on top of the vessel, oxygen was blown from the top lance to enhance scrap melting by secondary combustion of the off-gas in the vessel. High speed and stable low temperature melting practice increased the utilization efficiency of the coal, prolonged refractory lining life, and sustained both the heat for melting and the energy contained in the recovered off-gas by controlling the degree of secondary combustion. In addition, evolution of dust, slopping caused by bursting bubbles, enhancement of deP by controlling the (T. Fe) content of the slag, maintaining optimum C content of the high C melt, and stabilizing heat compensation were all optimized by controlling S and ash content in the coal, the ratio of oxygen blown from the top lance, and the amount of addition of iron oxide pellets. The resulting hot metal was decarburized in another BOF to produce high quality steel which was reported to be of better quality than EAF steel.

During the 1980s, top and bottom blowing BOFs were also developed for the smelting reduction of Cr ore to produce stainless steel. Okuyama et al. (2013) have reported that at Chiba, 50 % of deP hot metal, Cr ore pellets that had been sintered in a rotary kiln, coke and stainless steel scrap were charged into a K-BOP to obtain Cr and Ni bearing hot metal. The hot metal was separated from slag, poured into a large size mixer equipped with a channel heater, and stainless scrap added to trim the Ni and Cr contents. The hot metal was then decarburized and dephosphorized in a second K-BOP to produce SUS 304 and 430 stainless steels. The smelting reduction process provided Chiba with greater freedom to choose Cr sources depending on the market situation.

Recently, an additional lance has been installed to add Cr ore fines with burner heating into the hot metal in the first stage K-BOP. Hydrogen-bearing fuel is combusted with oxygen, and granular Cr ore is preheated in the combusting flame to compensate for the 20 %/(unit mass Cr ore) required for the endothermic heat of reduction. Emission of CO_2 is naturally decreased, together with erosion of the refractory lining. The top addition of coal to the hot metal combined with oxygen blown through the auxiliary lance has been in operation since the latter half of the 2000s. The coal disintegrates and disperses as fine particles in the slag, thus providing favorable conditions for reduction of the preheated Cr ore granules which are also suspended in the slag. The auxiliary lance is equipped with a multiple hole tip, designed to combust the CO gas evolving in the vessel space from the Cr ore reduction to fully compensate for the heat required for the reduction reactions.

To obtain high Cr ferritic stainless steels with extra low C and N contents, the melt from the second K-BOP is further refined under vacuum with the Vacuum Oxygen Decarburization (VOD) process or SS-VOD. The total amount of stainless steel produced with the duplex K-BOP process has been about 0.7 million tons/year.

Traditionally, stainless steels have been produced from stainless steel scrap, Ni from various sources, and Fe-Cr alloys using EAFs in combination with converters such as the AOD (Argon Oxygen Decarburization) and CLU (Creusot Loire Uddeholm), both of which control oxygen partial pressure within a wide range during the blowing to minimize the loss of Cr. The steels are then further processed in ladle refining furnaces (LRF) depending on the quality requirement. The smelting reduction type duplex K-BOP route is reported to consume about 35 % less energy than the traditional EAF-AOD route.

Steelmaking and refining processes for stainless steels are quite diverse (Stainless Steel Handbook1995). For example, Muroran installed the RH-OB process in 1972, Yawata put into operation a 150 t LD-VAC (VOD) in 1979 and hot metal pretreatment with soda ash combined with LD-OB in 1980. Wakayama started an oxygen top blowing AOD in 1982, the same year that Nippon Metal began operations with an AOD-VOD, Wakayama continued with an AOD-VOD/VOD-PB in 1990. In the same year, Fukuyama installed a top and bottom blowing BOF type Smelting Reduction Furnace (SRF) for processing Ni- and Cr ore. In 1991, Daido Specialty Steel introduced a vacuum AOD (VCR), in 1995 Yawata initiated REDA and in 1996, a VOD was installed at Hikari. The different choices of process are made based on the local conditions for each plant.

4.5 Future of Technological Innovation in the Steel Industry

With respect to the future, the high productivity steelmaking system will be one which is cost competitive and meets the requirements for high quality steels such as Interstitial-Free, High-Strength Low Alloy, and line pipe grades. Minimizing the paths for the flow of materials among different unit processes, selecting appropriate equipment to ensure quality processing, and decreasing labor costs by automation combined with robotics, all aimed at reducing production costs, would appear to provide a foundation for future progress. As a final challenge, these advanced systems will require to remain competitive for many years within the world market in terms of productivity, quality, and cost.

In reviewing the history of steelmaking, the enthusiasm, endeavor, and accomplishments of our predecessors in industry, academia, and government who developed and improved the core technologies are most praiseworthy and highly commendable. The steel industry is a huge equipment-based organization. After the commissioning, construction, and start up of a new facility, the development of strategies for effective energy utilization and environmental protection is essential. The accumulation of these innovative developments and comprehensive improvements as a whole can certainly be viewed as revolutionary.

4.6 Concluding Comments

With the passage of time, current steelmaking equipment, systems, and operations have come to maturity. On the other hand, some facilities which are not at the leading edge call for refurbishment with the most modern systems, but this requires heavy investment. It is worth noting however, that in the past, whenever there were major refurbishments, there were also many associated developments. Enthusiastic discussions held cooperatively between the engineers of the steel industry and those of the machine building industry to determine specifications, install and start up the equipment, brought forth unexpected new ideas to further develop the facilities. Unfortunately, major refurbishments have become limited in recent years. Motivation for developments and improvements during a mature period usually comes externally from the demands of the market for enhanced quality and reduced costs, and internally from the demands for increased productivity and cost savings. Despite such circumstances, dedicated team work with high motivation has traditionally been a characteristic of those involved in steel production. This concerted team work is still effective in providing significant results.

Within the steel industry, the period from invention to industrialization usually takes a long time. For this reason, this is now the time to develop new core production technologies in order to secure high ROE and compensate for more expensive raw materials, increased labor costs, and expenses associated with new tax laws. This is appropriate not only for the high quality steel sector but also within the commodity steel sector. A critical factor in the future development of the steel industry is the importance of the personnel involved. High quality products require high quality processing and both require high quality people. Motivated capable people are the key to success. Of course, people may not be well motivated or capable in the beginning. Nevertheless, they should be motivated continually, in particular during hard times, and become more capable through challenges and persistent efforts. Researchers and engineers who are well trained and are actively engaged in R&D and who participate in cooperative projects between academia and industry are mandatory requirements for the survival of the industry. On site cooperation means experiments, data acquisition, and discussion within modern world class plants and academia. To ensure such endeavors are not limited to success only in reports, young engineers and researchers must traverse the "Sea of Darwin" and address the technological challenges pertaining to new equipment, novel instrumentation, and sensor-based automation systems, implementing them within innovative processes in order to ensure the continuing prominence of advanced steel production technologies in the industrialized world of tomorrow.

Acknowledgements Sincere appreciation is expressed to the Iron and Steel Institute of Japan for granting permission to publish this material, some of which was reproduced from a previous paper by one of the authors, T. Emi, in ISIJ International, Vol. 55 (2015), No. 1, pp. 36–66, as part of the Centennial activities of ISIJ.

References

Baker, K. J., Paules, J. R., Rymarchyk, N., & Jancosko, R. M. (1998). The making, shaping and treating of steel. 11th ed., Steelmaking and Refining Vol., AISE, Pittsburgh, PA, Ch. 8, CD-ROM: 433, 470.

Brotzmann, K. (1992). In J. K. Brimacombe, P. J. Mackey, G. J. W. Kor, C. Bickert, & M. G. Ranade (Eds.), *Savard/Lee International Symposium on Bath Smelting* (p. 29). TMS, USA.

Compilation Committee for Yawata Steel Historical Documents: Yawata Steel 80 Year History, Divisional History, Vols. 1 & 2 (1980) Yawata Seitetsusho Shoshi Hensan Committee. Yawata, Fukuoka, Japan (in Japanese).

Doi, J. (1969). *Converter steelmaking* (3rd ed., Vol. 2, p. 241). Tokyo, Japan: Nikkan Kogyo Shinbun (in Japanese).

Fukagawa, S. (2006). *CAMP-ISIJ, 19*, 125.

Guthrie, R. I. L., Lee, H. C., & Sahai, Y. (1992). In J. K. Brimacombe, P. J. Mackey, G. J. W. Kor, C. Bickert, & M. G. Ranade (Eds.), *Savard/Lee International Symposium on Bath Smelting* (p. 445). TMS, USA.

Harabuchi, K., & Imoto, T. (2009). Kitakyusyu-Shi industrial technology investigation. In *Changes of Equipment and Technology at Yawata Steel: Vol. 3, Steelmaking Equipment. Kitakyusyu.* Japan: Kitakyushu-Shi Center for Conservation and Inheritance of Industrial Technology (in Japanese).

Iron and Steel Handbook, 4th ed. (2002), 2(1), ISIJ, Tokyo, Japan (CD-ROM).

Kai, K., Okohira, K., Hirai, M., Murakami, S., & Sato, N. (1982). Influence of bath agitation intensity on metallurgical characteristics in top and bottom blown converter. *Tetsu-to-Hagane, 68*, 1946.

Kawasaki, T. (1982). *Japanese steel industry—Its trajectory* (p. 7). Tokyo, Japan: Tekko Shinbun.

Kishimoto, Y., Kato, Y., Sakuraya, T., & Fujii, T. (1989). Effect of stirring force by bottom blown gas and partial pressure of CO on characteristics of metallurgical reaction in a converter. *Tetsu-to-Hagané, 75*, 1300.

Liu, I. L., Wang, I. M., & Su, T. -S. (2012). *Proceedings of Fifth International Congress on Science and*

Technology of Steelmaking (*ICS2012*), Dresden, CD-ROM.

Lynn, L. H. (1986). *Essence of innovation—A comparison between Japan and USA of steel technology implementation process*. Translated by Y. Onda: Toyo Keizai Shinpo, Tokyo, Japan (in Japanese).

Nakanishi, K., Kato, Y., Suzuki, K., & Katsuki, J. (1978). Abstract for ISIJ Meeting. *Tetsu-to-Hagané*. 64, S169, also Nakanishi, K., Nozaki, T., Kato, Y., Suzuki, K., & Katsuki, J. (1979). *7th Japan-USSR Joint Symposium on Physical Chemistry of Metallurgical Processes*, Moscow. Also, *Tetsu-to-Hagané* (1995) 81: 99, 134.

Nozaki, T. (2000a). *Bottom blowing converter*. Tokyo, Japan: ISIJ (in Japanese).

Nozaki, T. (2000b). *Bottom blowing converter* (p. 121). Tokyo, Japan: ISIJ (in Japanese).

Okuyama, G., Uchida, Y., Ogawa, N., Miki, Y., & Kishimoto, Y. (2013). *Materia Japan, 52*, 119.

Oonuki, K. (1998). Ph.D. Thesis. Tohoku University.

Philbrook, W. O., Bever, M. B., Emerick, H. B., & Larsen, B. M. (Eds.). (1951). *Basic open hearth steelmaking* (2nd ed., p. 21). New York: AIME.

Savard, G., & Lee, R. G. H. (1992). In J. K. Brimacombe, P. J. Mackey, & G. J. W. Kor, C. Bickert, & M. G. Ranade (Eds.), *Savard/Lee International Symposium on Bath Smelting* (p. 645). TMS, USA.

Shimokawa, Y. (1989). *History of iron and steel technology in Japan*. Tokyo, Japan: AGNE Gijyutsu Center.

Sugita, K. (1995). *Refractory for ironmaking and steelmaking*. Tokyo, Japan: ISIJ (in Japanese).

Stainless Steel Handbook, 3rd ed.,1995, Sutenresu Kyokai,Tokyo, Japan, p. 752 (in Japanese).

The Iron and Steel Institute of Japan. (1964). Progress of open-hearth process. Third Report of the Steelmaking Committee of ISIJ,Tetsu-to-Hagané 50 (9).

The Iron and Steel Institute of Japan. (1975). Progress of steel technology, 2. Steelmaking. Steelmaking Committee of ISIJ,Tetsu-to-Hagané 61: 505.

The Iron and Steel Institute of Japan. (1985). Progress of steel technology, 3.Steelmaking. Steelmaking Committee of ISIJ, The 70th Anniversary Special Issue, Tetsu-to-Hagané 71: 371.

Ueda, Y., et al. (1995). Progress of steel technology in the last decade, 3. Steelmaking. The 80th Anniversary Special Issue.Tetsu-to-Hagané 81: 39.

The Iron and Steel Institute of Japan. (1982). History of oxygen steelmaking in Japan. Special Report, No. 33, ISIJ, Tokyo, Japan (in Japanese).

The Iron and Steel Institute of Japan (1992). Development of Steel Technology in Japan during Post World War II Period, ISIJ, Tokyo (1992), and JISF: History of Iron and Steel during Post World War II Period, JISF, Tokyo, Japan (1959) (in Japanese).

Yawata. (1980). First appeared in Yawata Seitetsusyo 80nenshi Sougoushi, in Japanese (Yawata Steel 80 year Comprehensive History), Yawata Seitetsusho Shoshi Hensan Committee, 48, For details, see: Kozasu I (2000) Bull. Iron Steel Inst. Jpn. Vol. 5: 578 (in Japanese).

5 Chemical Processing: Hydrometallurgy

V.I. Lakshmanan, M.A. Halim, and Shiv Vijayan

5.1 Introduction

Technological innovations in the hydrometallurgical industry in the last 50 years have consisted largely of changes that allowed companies to exploit lower grade ores and to continually reduce the cost of metal production. Traditional pyrometallurgical processing of complex ores and concentrates is becoming increasingly expensive because of environmental constraints due to undesirable gaseous emissions, production of dusts, and the losses of valuable metal by-products in slags and residues that are discharged as effluents from the plants. Hydrometallurgical metal recovery processes were introduced with the objective of meeting improved environmental requirements and reducing capital and operating costs by introducing innovative reagents, process chemistry, equipment, and operating modes. However, the hydrometallurgical processes have their share of challenges in managing the generation and disposal of liquid effluents.

In hydrometallurgical processing, the first step is metal extraction. Extraction is generally performed by leaching the target metal from the feed material using a variety of chemical reagents at optimum process chemistry conditions. In addition, accelerated extraction of different metals can involve the application of elevated temperatures and pressures. The solution containing the metals of interest, referred to as the pregnant leach solution, can also contain some undesirable metals that were originally present in the feed material and are leached with the target metal. The solid leach residue or solid effluent, which contains most of the gangue or waste minerals, gets disposed in tailings dams or dumps. The extraction step is followed by metal purification and recovery steps to obtain the final purified product. The purification step in chemical processing utilizes compounds that have high affinity for specific target metals to achieve effective separation. Processing techniques such as solvent extraction, ion exchange, carbon adsorption, and precipitation are commonly used for the separation of dissolved metals from the leach solution. Electrolysis and precipitation are used for the final recovery of metals from purified solutions.

Innovative development of a competitive and viable process factoring economical and environmental considerations would require the application of knowledge and experience gained in one or more of the chemical processes. Such innovations are critical for a sustainable metallurgical industry. This section discusses the state-of-the-art and major technological innovations in chemical separation processes that recover metals

V.I. Lakshmanan (✉) • M.A. Halim • S. Vijayan
Process Research ORTECH Inc., 2350 Sheridan Park Drive, Mississauga, ON, Canada, L4K 2T4
e-mail: llakshmanan@processortech.com

V.I. Lakshmanan et al. (eds.), *Innovative Process Development in Metallurgical Industry*,
DOI 10.1007/978-3-319-21599-0_5

through leaching of feed materials followed by solvent extraction, ion exchange, carbon adsorption, precipitation, and electrowinning.

5.2 Leaching

Leaching of material is performed using different process configurations. Some configurations involve leaching at atmospheric pressure or at higher pressures whereas others use extraction in situ or by assembling the raw material in a heap. Atmospheric leaching involves reaction vessels operating at atmospheric pressure containing the leaching solution and ground feed material. The pressure leaching has been used for ores containing metals such as uranium, zinc–lead, copper, nickel–cobalt, gold–silver, and platinum group metals (PGM). This technique is often practiced on crushed and ground ores and concentrates to accelerate dissolution at elevated temperatures in pressurized vessels (autoclaves).

Heap leaching is accomplished by introducing a chemical solution to a bed of crushed low to medium grade ores and/or tails, which are often agglomerated. In response to the ever increasing price of base metals like copper, nickel, cobalt, and precious metals, heap leaching has become a major extraction technique used for large tonnage-low grade (generally 0.5–1 g/t) ores which cannot be economically beneficiated into a concentrate or processed through atmospheric or pressure leaching. In-situ leaching (ISL) process is applied to extract different metals from the ore deposit where the target metals are located within permeable materials, such as sands or sandstones, is saturated and confined above and below by low permeability materials. Under these conditions, leaching solutions are pumped into the ore deposit, which dissolve the metals of interest. These metal-loaded solutions are then pumped to the above ground surface facilities for metal recovery.

5.2.1 Atmospheric Leaching

Based on sulfate and chloride media, several innovative atmospheric leaching processes have been developed for the recovery of base and precious metals from ores and concentrates as alternative to the pressure leaching and the traditional roasting processes. The innovative aspects include (a) the production of elemental sulfur instead of sulfuric acid, (b) elimination of the addition of surfactant, (c) reduction of environmental risks, and (d) lower capital and maintenance costs as a result of leaching in vessels operating at atmospheric pressure.

Due to the aggressive nature of chloride leaching and stability of chloro-complexes, atmospheric chloride leaching has been gaining increased attention as an option in hydrometallurgical process applications. Several innovative processes such as Duval CLEAR Process, Canadian Titanium Limited (CTL) Process, the PRO's Nickel Laterite Process, HydroCopper™ Process, and the SMM New Copper Process have been developed for the recovery of value metals from their feed materials. Among these processes, Duval CLEAR Process has been commercialized and others are under consideration for commercialization. Considering the advantages of the chloride chemistry, an innovative mixed chloride ($HCl + MgCl_2$) atmospheric leaching process has been developed for the recovery of value metals including titanium from ilmenite ores and nickel, cobalt, and iron from low-grade laterite ores (Lakshmanan et al. 2013). The process offers many advantages: (a) direct extraction of nickel from cobalt raffinate at lower pH (around 1.0) without adding any neutralization reagents, (b) conversion of ferric chloride to hematite and production of concentrated HCl, which is recycled to the leaching step, and (c) recycling of $MgCl_2$ in the nickel raffinate to the leaching step.

5.2.2 Pressure Leaching

Pressure leaching is an innovative process step to reduce the environmental concerns related to traditional smelting processes and to decrease the operating costs by simplifying the process flow sheet. Initially, this process step was successfully applied in commercial operations for the recovery of copper, nickel, and cobalt from their feed materials. This success has allowed several applications of the technique for the recovery of several other metals including uranium, gold, and zinc.

During 1993, a two-stage pressure leaching of zinc sulfide concentrates was commissioned to replace the traditional roast-leach-electrowinning (RLE) process at Hudson Bay Mining and Smelting Co. Ltd. (now HudBay Minerals Inc.) with the goal of meeting stringent environmental requirements (Barth et al. 1998). This process is now operated in different countries including Kazakhstan since 2003 and China since 2009. The first stage leaching is performed at low acidity to produce a low-iron pregnant zinc solution, which is suitable for iron removal followed by purification. The second leaching stage is carried out at a higher acidity to achieve a high overall zinc extraction. The feed to the second leaching stage comprises the unleached zinc sulfides and zinc-ferrite from the first stage leaching and the balance of the return spent electrolyte. The process has several merits over RLE including minimization of SO_2 production, conversion of sulfide sulfur to elemental sulfur, removal of iron as jarosite with generation of acid, and overall cost reduction. Some of the limitations of the process are autoclave scaling and molten sulfur production.

The recovery of gold from sulfide ores and concentrates by pressure oxidation of sulfide minerals under the acidic media is now practiced more than the traditional roasting process. In the beginning, the pulp is treated with acid to avoid the generation of CO_2 during pressure oxidation. Antifoaming agents such as lignosulfonates are used to avoid foaming of the pulp. Air is supplied to completely remove the carbon dioxide gas generated in the process. The pulp is washed before neutralization to minimize cyanide consumption in the gold leaching circuit. The cost saving components of the process include higher gold recovery, minimization of gas emission, removal of arsenic as ferric arsenate, precipitation of iron as hematite or as jarosite, and reduction of lime addition to neutralization circuit. The carbon-in-pulp process configuration (similar to resin-in-pulp and carbon-in-leach) provides an attractive and economical option by eliminating the problematic solid–liquid separation step, which as a result also reduces the volume of wash water and the loss of gold.

5.2.3 Heap Leaching

Heap leaching is another innovative process that has revolutionized the metal recovery industry in the 1970s and 1980s from uranium to gold in the United States, Peru, and other countries (Brunk 1997). It has several favorable attributes: (a) elimination of several steps in regular leaching process such as crushing, grinding, agitation leaching, solid–liquid separation, and countercurrent decantation circuits; (b) simple process and equipment configuration; (c) lower environmental risks due to significant reductions in the generation of liquid and solid effluents, and virtual elimination of particulates/dust emissions; (d) low energy consumption; (e) applicability to low-grade complex ores and tailings; and (f) low capital and operating costs. The rapid expansion of the heap leaching combined with solvent extraction and electrowinning for copper production occurred in Chile during 1990s. In the past decade, heap leaching of nickel laterites and mixed metal sulfide ores (Taylor 2009) were practiced. Because some less desirable features such as (a) leaching kinetics is slow; (b) leaching cycles are long; (c) metal recovery is low; (d) footprint is large; and (e) heaps cannot be fixed after construction, several heap leaching plants were shut down. Currently, a sulfide heap leaching plant is in operation in Finland.

It is important for any ISL operation to implement a well-designed groundwater monitoring system that can detect the escape of any process solutions (excursion). Recent advances in electronic monitoring may be incorporated as part of the bore-well permanently to monitor pH, salinity, or concentration of specific metals of interest. Efficient continuous data collection and evaluation would allow more rapid detection of process excursions and take corrective actions to minimize any adverse impacts.

5.3 Separation Processes

5.3.1 Solvent Extraction

The increasing demand for high purity metals for industrial and domestic uses has provided impetus to the development of efficient and cost-effective solution purification steps to recover metals from leach solutions. Solvent extraction is one such process step for the separation and purification of metals deployed in the hydrometallurgical industry. This process has its roots in analytical chemistry, where it is used to separate metals from solutions selectively and quantitatively. The first large-scale process was deployed for the recovery of uranium and other isotopes from uranium ores and irradiated nuclear fuels. The key factors in the success of SX process are (a) the innovative development of cost-effective and less hazardous organic extractants that are specific to the metal being extracted and (b) a detailed understanding of the chemistry of the feed solution and the coordination chemistry of the particular metal ion in the solution. The first commercial extractant specific to copper was LIX 64N®, developed by General Mills Corporation. Commercial solvent extraction processes have been expanding considerably since the 1960s. In this section, major innovations in the solvent extraction process are discussed related to the separation of value metals, particularly, separation of cobalt, nickel, copper, zinc, titanium, niobium, tantalum, rare-earth elements, and PGM.

5.3.1.1 Cobalt and Nickel Separation

The separation of cobalt from nickel in aqueous solution has been found to be difficult because of their very similar chemical behavior. Although cobalt and nickel exist as divalent hexahydrated ions in aqueous solution, the rate of water exchange on the cobalt ion is much higher than for nickel. Thus, the formation of ion complex often proceeds much more readily with divalent cobalt than with nickel. The trivalent cobalt ion is much less labile and forms in preference to nickel even though the redox potentials for the cobalt(II)-cobalt(III) and nickel(II)-nickel(III) couples are nearly identical. In the divalent state, cobalt has a strong tendency to form tetrahedral complexes in concentrate electrolytes instead of the hexa-coordination complexes, as in such aqueous solutions nickel(II) retains a hexagonal configuration. These general differences in chemical behavior provide the basis for the solvent extraction processes currently used or proposed for cobalt–nickel separation. In commercial solvent extraction plants, anion exchangers and acidic extractants are mainly used for the separation of cobalt from nickel in chloride solutions, ammoniacal solutions, and sulfate solutions.

In strong chloride solution, cobalt can readily form anionic chloro-complex such as $CoCl_3^-$ and $CoCl_4^{2-}$, whereas nickel does not form such complexes. This difference with cobalt(II) allows the separation of cobalt from nickel in chloride medium with high selectivity. Falconbridge Nikkelverk in Norway has applied the principle to two feed solutions: (1) Adogen 281 (a secondary amine) is used to extract copper and cobalt from chloride solutions generated from leaching nickel matte and (2) tri-isooctylamine is used to extract cobalt from a nickel chloride solution. Société Le Nickel (now Minemet) and Nippon Mining use Adogen 381 (a tertiary amine) and trioctyl amine, respectively, to extract cobalt from impure nickel electrolyte. In these operations, cobalt is readily stripped from the metal-loaded organic with a dilute hydrochloric acid solution.

In 1975, Nippon Mining's Hitachi began applying LIX® 64N for Ni extraction from a sulfate solution at a pH value of 9–10 using ammonia for pH control. Nickel was stripped with spent electrolyte from the organic and recovered by electrowinning. Later, it was revealed that LIX® 64N was losing its oxime group (C=NOH) by conversion to ketone (C=O) through the oxidation of co-extracted cobalt(II) to cobalt(III) in the organic phase. A similar problem has been observed in the Cawse laterite project in Western Australian and the modified Yabulu refinery operations at Queensland Australia, where LIX 84I and LIX®87QN (a mixture of LIX®84 and Escaild 110) are used, respectively, for nickel extraction from ammoniacal solutions. The ammoniacal leach liquor is aerated to oxidize cobalt(II) to cobalt (III) to prevent co-extraction in the subsequent nickel solvent extraction process in Yabulu operation. However, nickel extraction most likely proceeds through a cation-exchange reaction with nickel(II) replacing the loaded NH_4^+ from the extractant. The problem of organic degradation has been solved by the reoximation of the ketone group through the original oxime contact with hydroxylamine salts in alkaline solutions. It should be noted that because small amounts of copper in the feed to extraction would accumulate in the organic due to high pH conditions in both extraction and stripping, the copper removal would require separate H_2SO_4 stripping (Bacon and Mihaylov 2002).

The solvent extraction process for the separation of cobalt from nickel using di(2-ethylhexyl) phosphoric acid (D2EHPA) has been used in the mid-1960s at different laboratories including Eldorado Nuclear refinery, CANMET in Canada and Rustenburg Base Metals Refinery in South Africa. Some of the disadvantages of using D2EHPA are due to strong iron extraction, requiring a reductive or HCI stripping, and low degree of cobalt–nickel selectivity requiring many extraction stages. In Eldorado Nuclear refinery, the sieve-plate pulsed column was used to avoid multistage units like mixer-settlers in cobalt extraction. A reductive re-leach of the solids produces a nickel–cobalt sulfate solution, which is fed, after removal of impurities such as iron, copper, and lead with NaOH, to the cobalt solvent extraction circuit in Rustenburg operation. The organic is preloaded with sodium to avoid adding base for pH control (pH 5–6) during cobalt(II) extraction. Trace amounts of calcium and magnesium remaining in the feed solution are co-extracted with D2EHPA, and calcium is removed as gypsum from the cobalt pregnant strip solution. Nippon Mining has used an organophosphonic acid extractant, PC-88A/Ionquest 801, for selective extraction of cobalt over nickel (nickel extraction requires a higher pH) from feed solution produced by leaching of the sulfide cake at Queensland Nickel.

In early 1982, Cyanamid (now Cytec) Canada Inc. developed the di(2,4,4-trimethylpentyl) phosphinic acid (CYANEX® 272) for selective extraction of cobalt over nickel at pH 5–6 with a substantially higher cobalt–nickel separation factor by lesser extraction and scrub stages. The development of CYANEX® 272 expanded solvent extraction applications in cobalt–nickel refining and allowed the replacement of expensive and environmentally undesirable technologies. For example, the cobalt solvent extraction process with CYANEX® 272 at Harjavalta refinery replaced Outokumpu's older process of cobalt recovery from the $NiSO_4$ electrolyte by precipitation as $Co(OH)_3$ with $Ni(OH)_3$.

In Bulong operation, cobalt was extracted with CYANEX® 272 from the leach liquor after two-step precipitation of iron, aluminum, and chromium at pH 4.2–4.5. The nickel in the raffinate was then extracted with a carboxylic acid, Versatic 10 (Preston and du Preez 1995). Previously, SMM Niihama refinery and the Rustenburg refinery operations have also used Versatic 10 to transfer nickel and cobalt from sulfate solution into a concentrated chloride solution. Cobalt is then selectively extracted with a tertiary amine, and finally both Co and Ni are recovered by electrowinning. The cost of Versatic 10 is much lower than the organophosphorous extractants, but it is a weaker extractant, requiring approximately a pH of 7 for efficient Ni extraction, which exacerbates its other problems, namely, the high aqueous solubility (3–5 g/L) and relatively poor phase separation performance (Bacon and Mihaylov 2002).

In the late 1980s, Cytec developed CYANEX® 301, very similar to CYANEX® 272. The replacement of both oxygen atoms with sulfur resulted in a very different extractive strength and selectivity for base metals. Vale INCO has chosen this extractant for cobalt and nickel extraction in their Goro laterite project in New Caledonia. The use of CYANEX® 301 allows nickel and cobalt to be extracted quickly due to its fast extraction kinetics efficiently (>99.9 % extraction) from the large flow of a relatively dilute partially neutralized leach solution without the co-extraction of any of the major impurities (magnesium, calcium, and manganese) and without the need for pH control and the addition of expensive base reagents.

However, CYANEX® 301 is not stable and decomposes in two steps. The first step is reversible, where the formation of disulfide occurs, and the second step is irreversible, where disulfide decomposes to elemental sulfur (Flett 2005).

Several synergistic solvent extraction (SSX) reagents have been proposed (Flett 2004; Lakshmanan and Lawson 1973; Ritcey 2006; Cheng et al. 2011) to improve selectivity and stripping efficiency and to potentially create new, innovative applications. Lakshmanan and Lawson (1973) observed that cobalt(II) cannot be stripped from Kelex 100 in kerosene with 7N sulfuric acid but the stripping was easier with dilute acid when they used a mixture of Kelex 100 and Versatic 911 in kerosene as an extractant. The combination of LIX 63 and Versatic 10 has recently gained more importance due to the significant synergistic shifts of the ΔpH_{50} values for nickel, cobalt, copper, zinc, and manganese and an antagonistic shift for calcium (Cheng et al. 2011). For an example, the ΔpH_{50} values of nickel, cobalt, zinc, and manganese compared to Versatic 10 have been found to be 2.79, 3.50, 1.99, and 1.17 pH units, respectively, with a mixture of 0.5 M Versatic 10 and 0.28 M LIX 63. However, the hydroxyoxime is susceptible to degradation if the solvent extraction process is operated under improper conditions (Halford and Barnard 2011).

A further understanding of the chloride chemistry has enabled Process Research ORTECH Inc. (PRO) to develop recently an innovative solvent extraction process for the recovery of nickel, cobalt, and iron from low-grade laterite ores. The process involves atmospheric mixed chloride (mixture of hydrochloric acid and magnesium chloride) leaching of ore, solid–liquid separation, and nickel and cobalt recovery by solvent extraction. One of the major advantages of the PRO's proprietary process is the direct extraction of nickel from cobalt raffinate with a hydroxyoxime extractant at low pH values of 1.0–1.5 without adding any neutralization reagents (Lakshmanan et al. 2014). The process claims to minimize intermediate neutralization and precipitation, consumes significantly less energy, and to produce less waste for environmental release using a closed loop operation with recycle.

5.3.1.2 Copper Separation

Over 20 % of the global output of refined copper is produced by solvent extraction-electrowinning (SX-EW) hydrometallurgical plants and is generally suited for low-grade ores. The SX-EW process is superior to other processes such as cementation of copper in terms of improved product quality and costs. New organic extractants greatly reduce iron co-extraction and allow for a much wider range of copper concentrations (Rotuska and Chmielewski 2008). A noteworthy achievement in solvent extraction comes from extractants based on hydroxyoximes. In the beginning, the most important issue was the need to increase the strength of extractant and consequently to improve the stripping efficiency. For example, different mixtures of extractants were used based on LIX 64N (blend of LIX 64 and LIX 63). LIX 64 alone was able to extract copper selectively over iron in the pH range of 1.5–2.5, but with the disadvantage of slow rate of extraction (Cox 2004). This problem was alleviated by the addition of a small amount of LIX 63. LIX 64N mixture is the extractant of choice for commercial copper recovery by solvent extraction from acidic leach solutions (Szymanowski 1993).

The addition of modifiers to the organic phase minimized and/or eliminated problems related to stripping performance. Although the modifier blend resulted in a weaker extractant mixture, it was easier to strip with conventional electrowinning electrolytes. For example, Avecia with their Acorga reagents has showed that addition of nonyl phenol to hydroxyoxime, tridecanol, or alkyl esters improves the stripping performance with insignificant negative effects on extraction (Cox 2004). However, the addition of modifiers produced less desirable crud formation and entrainment, which resulted in greater contamination of the electrolyte with the leaching solution, increased losses of extractant and solvent from the extraction circuit, and poor selectivity. To remedy these problems, an innovative approach was established by Outokumpu Oyj in Finland through the Vertical Smooth Flow (VSF™) mixers. The basic idea of the VSF™ technology is to maintain smooth agitation throughout the solvent extraction plant to avoid oxidation of organic and development of overly small droplet size in dispersion (Outotec Oyj 2007).

The solvent extraction reagents can selectively extract copper over other metals from feed solution (Kordosky et al. 2006). The best reagent to achieve this for a number of years was Acorga P-5100, but in recent years, it has been replaced with new Acorga reagents such as M5640 and PT5050, both of which exhibit better selectivity and higher recovery characteristics (Rotuska and Chmielewski 2008). This new class of modified aldoxime extractants has greatly contributed to the growth trend in SX-EW technologies. The pyridine carboxylic ester derivative such as Acorga CLX 50, containing 50 % of the active ingredient, has been developed for selective removal of copper from chloride leaching of sulfide ores (Dalton et al. 1987). This extractant provides favorable selectivity for copper over other elements including iron, zinc, and lead present in the feed solution. The reduction of chloride concentration in the pregnant strip solution is crucial for the success of the electrowinning process.

5.3.1.3 Zinc Separation

Zinc electrowinning from sulfate medium is extremely sensitive to the presence of trace impurities and requires a highly purified electrolyte. The selectivity of modern solvent extractants, an improved understanding of the process chemistry, and engineering innovations, have enabled today the deployment of solvent extraction as a purification step ahead of the electrowinning step. The di(2-ethylhexyl)-phosphoric acid (D2EHPA) is used in most of the industrial solvent extraction for zinc recovery from sulfate media. This extractant has a high selectivity for zinc over most of the metal halides and the accompanying metals (copper, cadmium, cobalt, and nickel) that are harmful for electrolysis and has the ability to readily strip by the spent electrolyte. However, iron (III) is built up in the organic phase that needs to be controlled by a concentrated HCl treatment. For example, Skorpion mine (Namibia) uses the modified ZINCEX process, where a bleed of the organic phase is separately contacted with 4–8 M HCl solution to strip iron and aluminum (Filippou 2004). The HydroZinc process has been tested on pilot plant scale at Red Dog mine (Alaska, USA), where iron buildup in the organic phase is removed by the reductive stripping with zinc dust (O'Keefe 1993). The mixing of D2EHPA with other extractants such as tri-*n*-butyl phosphate (TBP), tri-*n*-octyl phosphine oxide (TOPO), CYANEX 923, and amines may enable the use of sulfuric acid as a stripping reagent for iron(III) (Deep and Carvalho 2008).

The co-extraction of iron(III) with D2EHPA, TBP, and CYANEX 923 has been a problem in the chloride media. To avoid this issue, iron (III) is reduced to iron (II) prior to the extraction of zinc (II) (Deep and Carvalho 2008). The reagent, bis-benzimidazole (ACORGA ZNX 50), is used for selective extraction of zinc(II) from iron, arsenic, calcium, chromium, lead, magnesium, manganese, and nickel containing chloride leach solutions. The co-extracted copper is removed by scrubbing with water followed by stripping of zinc(II) with a chloride solution containing 30 g/L Zn (as zinc chloride), 2 mol/L NaCl, and 5 g/L HCl (Dalton et al. 1992).

5.3.1.4 Titanium, Niobium, and Tantalum Separation

A recently developed process is being commercialized for the production of TiO_2 pigment (Lakshmanan et al. 2014). In this process, titanium(IV) is selectively extracted from the process solution at controlled pH and Eh conditions. The process promises to minimize CO_2 emissions, consume significantly less energy, and produce less secondary waste due to closed loop operation with recycle (Sect. 9.3).

The separation of niobium and tantalum is inherently difficult due to their very similar chemical properties. A commercial process based on solvent extraction performs separation of the metals as fluoride ions containing sulfuric acid or hydrochloric acid solution. Mainly niobium(V) and tantalum(V) are in stable states and form very stable complexes such as $NbOF_2^{5-}$ and NbF_6^-, TaF_7^{2-} and TaF_6^- with fluorides in solution (Zhu and Cheng 2011). Among the various extractants, methyl iso-butyl ketone (MIBK), tri-butyl phosphate (TBP), cyclohexanone (CHN), and 2-Octanol (OCL) are widely used in the industry for separating and purifying niobium and tantalum. The extraction of tantalum is more favorable than that of niobium in low HF concentration solutions with these extractants. For example, 2-octanol is used to extract tantalum(V) as $HTaF_6$ from a solution containing 0–0.5 M free HF and 1.5–2 M H_2SO_4 followed by niobium(V) as $HNbF_6$ in the presence of 7–10 M free HF and 5–6 M H_2SO_4 (Agulyansky et al. 2004; Zhu and Cheng 2011). After scrubbing the impurities, both tantalum and niobium are separately stripped with water from the loaded organics.

5.3.1.5 Rare-Earths Separation

Separation of rare earths (Lutetium or Lutecium plus yttrium) is a challenging task because of the close similarity of their chemical properties. Solvent extraction is generally accepted as the most appropriate process step commercially for separating these elements in hydrochloric and nitric acid media. The separation is based on the decreasing ionic size of these elements in the series, which gives an increasing acidity with increasing atomic number. This causes a variation in the formation coefficient of rare earths-extractant complexes, allowing preferential extraction of the complex into the organic phase (Cox 2004). A large number of extraction stages are required to reach the equilibrium due to separation factors between adjacent elements being small. All three major classes of extractants (acidic, neutral, and basic extractants) have been utilized for separating rare earths. However, the organophosphorus acids have recently gained more importance for individual separation of rare earths in hydrochloric acid medium. Separation factors between adjacent elements follow the order: phosphinic acid > phosphonic acid > phosphoric acid. The 2-ethylhexyl ester of 2-ethylhexylphosphonic acid (HEHEHP) (e.g., PC-88A, SME 418, Ionquest 801, and P-507) shows higher distribution coefficients for lanthanides than di(2-ethylhexyl)phosphate (D2EHPA) in both chloride and nitrate media. A process developed by Daihachi for separating rare earths using HEHEHP has been applied commercially at the separation plant in Baotou, China (Zhu 1991). Recently, CYANEX 272 has been found to provide significant advantages over other organophosphorus acid extractants for the extraction of heavy rare earths with greater average separation factors and easier stripping. A mixture of CYANEX 272 and PC 88A gives significant synergism for the separation of heavy rare earths with higher loadings than either CYANEX 272 or PC 88A separately in chloride solutions (Xie et al. 2014).

5.3.1.6 Palladium and Platinum Separation

Solvent extraction separation is an efficient and practical method for extraction and recovery of precious group metals from leach solutions obtained by chloride leaching of ores and spent catalysts. The separation and purification of palladium(II) and platinum(IV) are again difficult because of their similar chemical properties in acidic (e.g., hydrochloric acid) media. Understanding the solution chemistry of these elements is key to successful separation by solvent extraction. Palladium(II) and platinum(IV) have two common oxidation states and stereo chemistries of square planar and octahedral (Cox 2004; Lakshmanan and Ryder 1989). The octahedral $[Pt(Cl)_6]^{2-}$ is essentially inert, while the square planar $[Pd(Cl)_4]^{2-}$ has lower thermal stability but higher dynamic activity, which is favorable for $[Pd(Cl)_4]^{2-}$ to have reactions of hydration and ligand exchange in hydrochloric acid media. For example, phenolic oximes can be used to separate palladium(II) and platinum(IV). In this case, the former is extracted as a 2:1 complex, $[Pd(L)_n]$, at $pH \geq 3.0$ while the kinetic inertness of platinum(IV) ensures that it is only loaded very slowly and therefore remains in the aqueous phase.

In general, kinetically inert metals do not lose coordinated anions to generate lower charge during the extraction of more readily extractable species. These metal anions could be easily separated on the basis of their charge differences. For example, by maintaining the oxidation state of iridium(III) and rhodium(III) as their trianions $[Ir(Cl)_6]^{3-}$ and $[Rh(Cl)_6]^{3-}$, respectively, platinum(IV) as its dianion $[Pt(Cl)_6]^{2-}$ is selectively extracted with tri-*n*-butylphosphate over these metal anions from acid chloride solution. The key to this separation process is to ensure that $[Ir(Cl)_6]^{3-}$ and $[Rh(Cl)_6]^{3-}$ are not converted into their more readily extracted $[Ir(Cl)_6]^{2-}$ and $[Rh(Cl)_6]^{2-}$ complexes. Similarly, by considering the stripping advantage, weakly basic alkyl amides are used for the extraction of platinum(IV) from acidic chloride solutions.

5.3.2 Ion Exchange Technology

Ion exchange (IX) is another process step that is widely used in hydrometallurgy for metal separation and recovery. The natures of the media used for ion exchange, and the configuration and mode of ion-exchange column operation have evolved through many innovative developments. Ion exchange was first used to recover uranium in the 1960s (O'Malley 2002). In the 1970s, the Union of Soviet Socialist Republics (USSR) applied synthetic organic resins instead of activated carbon for the recovery of gold and silver from cyanide leach solutions (Bolinski and Shirley 1996). This process also plays an important role in recovering PGM, molybdenum, nickel, copper, and other valuable metals and for the treatment of process water (van Deventer 2011). They are superior to activated carbon with respect to gold loading kinetics and the maximum gold loading achievable. Other advantages include the ability for the metal ions to be eluted from the loaded resins at room temperature, media resistivity to poisoning by organic species, and strength to minimize attrition and breakage, in addition to resin's capability against blinding by fine clay particles (Fleming et al. 2003). The resins are relatively expensive, have a lower selectivity for gold over other base metal complexes, and are smaller in particle size than activated carbon.

The resin-in-pulp (RIP) and the similar carbon-in-pulp (CIP) technologies have been developed for the recovery of gold and uranium from low-grade ores (van Tonder and van Hege 2007). Most commercial operations for the recovery of gold use special strong-base or medium-base anion-exchange resins with an optimal balance between the weak- and strong-base groups on the resin. Operations in the USSR use mainly strong-base resins, such as AM2B, manufactured by the Pridneprovsky Chemical Plant, Ukraine. Other commercially available strong-base resins are Aurix 100 (Cognis), Dowex Minix, and Purolite A100/S992/2412 (van Deventer 2011). Similarly, strong-base anion-exchange resins, containing quaternary ammonium functional groups, are most commonly used for commercial uranium extraction both from alkaline and acidic leach liquors and pulps.

Recently, the RIP process has been developed for nickel and cobalt recovery from laterite ores (Littlejohn and Vaughan 2012). In this process, resin and pulp are contacted countercurrently in a series of agitated contactors. The loaded resin is separated from the pulp over screens, washed, and transferred to elution. The eluted resin is then returned to the ion-exchange circuit. The process provides an attractive and economical option by avoiding a pulp filtration step, which is usually problematic. Thus, the ion-exchange operation reduces the wash-water volume and minimizes the loss of value metals.

The growing environmental and occupational safety concerns over the use of cyanide in gold processing have resulted in increased interests towards alternative lixiviants such as thiosulfate, thiourea, thiocyanate, and halides. Among these reagents, considerable research has been aimed at thiosulfate leaching of gold. For such leach solutions, the activated carbon has not been found to be an effective medium for the adsorption of the gold thiosulfate complex. However, much success has been achieved commercially in the use of strong-base resins, such as Purolite A500C, in RIP operations (Fleming et al. 2003).

The separation and purification of Pt(IV), Pd(IV), and Rh(IV) ions in hydrochloric acid solutions are relatively difficult, primarily because of the complex chemical properties of these ions and their tendency to form different chemical complexes in hydrochloric acid solutions. By maintaining the conditions such as temperature, redox potential, and acidity, palladium has been selectively separated from these metal ions using a molecular recognition-type resin, SuperLig® 2, at the Impala Refineries in Springs, South Africa (Black et al. 2006). The molecular recognition technology (MRT) utilizes "lock and key" chemistry. It offers highly selective separation of the target metal ion in the presence of high concentrations of competing ions through selective complex formation with specially designed organic chelators (ligands) that are chemically bonded to solid supports such as silica gel or polymer substrates.

Similarly, through an understanding of parameters such as size, geometry, and coordination chemistry, the separation of cobalt is facilitated by oxidizing cobalt(II) to cobalt(III) on to the molecular recognition-type resin. For example, cobalt(II) was selectively separated over other ions including nickel(II), copper(II), and zinc(II) by oxidizing SuperLig®138-bound cobalt(II) to cobalt(III).

Purification of metal electrolytes is important prior to the production of metal by electrowinning. For example, the impurities such as copper, nickel, and zinc in cobalt electrolyte create problems in the electrowinning of cobalt. The removal of small amounts of nickel from cobalt electrolyte is not a trivial task because of their similar chemical properties. Here, ion exchange has proved to be an appropriate technology over precipitation and solvent extraction processes to remove trace quantities of these impurities from cobalt containing process streams. Chambishi Metals PLC in Zambia has successfully removed nickel from cobalt with bis-picolylamine functionality, DowM4195 resins. In a mildly acidic solution, it exhibits a selectivity order of (van Deventer 2011): $Cu^{2+} > Ni^{2+} > Fe^{3+} > Zn^{2+} > Co^{2+} > Cd^{2+} > Fe^{2+} > Mn^{2+}$. Similarly, the di(2-ethylhexyl) phosphoric acid (D2EHPA)-impregnated resin, Lewatit VP OC 1026, has been successfully used since the 1980s at Vale's Port Colbourne, Canada, refinery for the removal of zinc from a nickel electrolyte prior to electrowinning (Agnew et al. 1998) and is also used in several nickel plating bath systems for the removal of ferric iron and zinc (Kotze 2012).

For effective water management in hydrometallurgical operations, recycling the process waters is essential that often leads to cross contamination of the water sources. Also, finding good quality water for mining operations in remote areas remains a challenge. The Purolite's Cyclic Ion Exchange (CIX-RO) technology, using the inert-core Shallow Shell resin, Purolite SST65, has the ability to use very low concentrations of brine for regeneration. This technology provides savings in power consumption, water, and waste disposal and also makes the overall process more environmental friendly.

5.4 Metal Recovery

5.4.1 Precipitation

Precipitation is one of the important and widely used industrial process steps for the recovery of metals from different solutions. The development of innovative iron precipitation techniques such as hematite and jarosite allows the hydrometallurgical process to recover zinc commercially from leach solution obtained from sulfuric acid leach of roasted zinc concentrates. Similarly, the hematite process was commercially practiced by the Akita Zinc Company in Japan (Tsunoda et al. 1973).

In jarosite process, precipitation of iron occurs from acidic sulfate solution at a pH of about 1.5 and a temperature of about 95 °C. This process overcomes the gel formation in the conventional ferric hydroxide precipitation, which in turn prevents serious problems related to settling, filtration, and washing of the residue. The jarosite process was developed in the mid-1960s by Asturiana de Zinc S.A. of Spain, Det Norske Zinkkompani A/S of Norway, and Electrolytic Zinc Company of Australasia Ltd., independently (Arslan and Arslan 2003). It should be noted that the jarosite process is used for iron removal and not for iron recovery as the precipitated jarosite is sent for disposal.

The increased applications of nucleation or seeding and supersaturation control in precipitation processes may be viewed as innovative approaches. Nucleation is used to initiate the precipitation of chosen metal ions, particularly at low concentrations, and also to enhance the morphological character of the precipitate. It is generally accomplished through recycle of the precipitated solids, e.g., production of nickel from nickel ammine solutions by hydrogen reduction. The supersaturation control can result in significantly improved precipitation efficiency and controlling the particle size of the precipitated solids (Demopoulos 2009). Finally, these features can also improve the solid–liquid separation characteristics of the slurry.

Sulfide precipitation is another important process over hydroxide precipitation for the recovery of either intermediate or final product of base metals including nickel, copper, cobalt, and zinc. The main advantages of this process are the high recovery of metals at lower pH values (2–3) within a short reaction time, less metal sludge generation, and easy solid–liquid separation. In the nickel industry, the sulfide precipitation is a well-accepted process step since it rejects manganese and gives a clean, readily saleable product. For example, the process was successfully applied for the recovery of nickel from low-grade laterite ores (<1.5 % Ni) at Moa Bay operation in Cuba and for the purification of nickel at Freeport Nickel Company refinery in Louisiana. The process has also some limitations such as the potential for hydrogen sulfide (H_2S) gas evolution and the concern for sulfide toxicity in emission releases. To reduce H_2S contamination, different sulfur containing reagents such as sodium sulfide, sodium hydrosulfide, and ferrous sulfide have been introduced and are used.

Oxidative precipitation of iron and manganese from sulfate solutions using a gas mixture of sulfur dioxide (SO_2) and air (or O_2) is another innovative separation process for the recovery of base metals including nickel, cobalt, and zinc. The process offers many attributes, especially for plants running a sulfur burner or a sulfide roasting facility on site where off-gas is used as a source of SO_2 and mixed with air. This SO_2/air mixture is a relatively cheap and strong oxidant that allows the oxidative precipitation of iron and manganese at relatively low pH values. The oxidation precipitation process is very effective for the separation of cobalt from nickel in solutions. Cobalt is quite selectively oxidized and precipitated in presence of relatively large concentrations of nickel although their pH-Eh characteristics are very similar. In this case, a strong oxidant such as ammonium persulfate, chlorine, ozone, or Caro's acid is used to increase the redox potential at around +1.75 V. This process has been deployed at different installations including INCO and Falconbridge in Canada, and Jinchuan Group Ltd. in China for cobalt removal from nickel solutions.

Precipitation of metals by cementation offers high purity metal recoveries with low costs and eliminates the need for close monitoring of the waste stream composition and external control of the supply rate of the treatment reagents.

5.4.2 Electrowinning

Electrowinning is an important metal recovery process step, which allows selective recovery of metals such as copper, nickel, zinc, gold, and silver from different solutions. The process is typically used on drag-out recovery solutions or combined with ion exchange/solvent extraction. Innovations in the design of electrowinning devices include extending the usable current range by increasing the cathode surface area, and reducing the diffusion barrier using agitation or heating. For example, the use of rotating electrodes can provide attractive economics for the recovery of a variety of metals even at lower concentrations.

In the copper industry, a major breakthrough in electrowinning came when Bagdad cathode produced from a leach and SX process was registered on the Comex in 1975 followed by the registration of Anamax cathode on the London

Metal Exchange (LME) (Kordosky 2002). The copper quality and current efficiency were also improved by plating copper on stainless steel blanks. In the 1980s, Magma Copper began to use a manifold to consistently distribute the electrolyte to every cathode in a cell, which was a key factor in their ability to produce LME quality copper at current densities up to 320 A/m^2. A modern copper electrowinning tank house coupled with solvent extraction can achieve 90–93 % current efficiency while producing 60–80 % more copper per unit of the tank house area than the earlier electrowinning tank houses. Adding cobalt sulfate in the electrolyte reduces corrosion of lead anode in copper electrowinning. The strong electrolyte stream from SX is cleaned by filtration through garnet sand and anthracite. Special grades of water-soluble polymers are used as smoothing agents and acid mist suppressants (Kordosky 2002).

The recent innovations in copper electrowinning include advances in energy reduction and acid mist reduction and improvements in product quality and productivity. In 2007, a PGM-coated titanium mesh anode was commercially introduced in the electrowinning cell at Chino, New Mexico (Sandoval et al. 2010). Similar anodes have been installed at other electrowinning facilities in North and South Americas. The mesh anode has produced a 15 % reduction in cell voltage and eliminated lead completely from the electrowinning operations. In 2010, Freeport-McMoRan researchers claimed an alternative anode reaction technology, which allows the oxidation of ferrous ion to ferric ion at the anode, thereby reducing cell voltage by approximately 50 % and avoiding the generation of acid mist. Air sparging under the cathode in an electrowinning cell allows for the production of high quality copper cathode and results in a smaller tank house footprint. At Anglo Los Bronces (north of Santiago, Chile), fiber-reinforced plastic electrode frames manufactured by SELE were first installed in the electrowinning cells to produce a high quality copper product.

In 2011, Anglo American Platinum implemented several new technologies including permanent titanium cathodes, mechanical stripping, anode skirts, and cell hoods in nickel tank house at Rustenburg, South Africa. The permanent titanium cathodes have allowed the use of mechanical stripping. The cell hoods and anode skirts have been employed to reduce worker exposure to aerosol and mist exposure. Recently, Vale has designed a novel nickel tank house with coated titanium anodes, which allows the use of mixed sulfate/chloride electrolyte at Long Harbor, Newfoundland, Canada, where generated chlorine gas is recycled to the leaching stage.

The development of mechanical stripping of the zinc cathodes has resulted in increasing zinc productivity from 8.1 man-hour per tonne in 1993 to 5.1 man-hour per tonne in 2003. Optimization of the stripping machines to fit the needs of each zinc plant was conducted relating to the type of plastic edges as well as the insulators. Noranda Technology Centre developed an online sensor for detecting the quality of the purified solution entering the cell house (Houlachi et al. 1993). This sensor detects and allows for remedial actions during process upsets affecting the quality of the electrolyte.

5.5 Summary

In summary, innovations in chemical processing steps have made it possible to meet the challenges posed by economic and environmental needs with improved resource utilization. A detailed summary of the innovations in major separation processes used in the hydrometallurgical industry is presented in Table 5.1.

Table 5.1 A summary of major innovations in key separation process steps used for metal recovery in chemical processing employed in hydrometallurgy

Metal separation process step	Innovative components	Metal recovery and application
Pressure leaching	• Dilute sulfuric acid is used in the first stage of the two-stage leaching to extract zinc for the direct recovery by electrowinning, while the concentrated acid is used in the second stage for the extraction of the remaining zinc • Elemental sulfur is produced instead of sulfur dioxide • Iron is precipitated as jarosite while acid is liberated for recycling • Pressure vessel scaling and molten sulfur production may be considered as negative aspects	Zinc
	• In pressure oxidation leaching of sulfide ores and concentrates (e.g., gold), the carbonate in the ore is neutralized with acid and carbon dioxide is removed by injection of air; an antifoaming agent is used to minimize foaming • High carbonate flotation tailings are used to minimize the amount of lime addition for neutralization • Pulp washing before neutralization minimizes cyanide consumption • The use of carbon-in-pulp or resin-in-pulp configuration eliminates the need for pulp filtration steps and reduces wash-water volume and loss of gold	Gold
	• Sulfuric acid is produced by the conversion of sulfide sulfur, and iron is precipitated as hematite or jarsosite in the Moa Bay process. A possible drawback is increased lime consumption for neutralization of the excess acid in the process • The excess acid is neutralized with partially leached high acid consuming ore, which significantly reduces lime consumption and operating cost in the AMAX process • Cobalt and nickel are precipitated as hydroxides with magnesia instead of mixed sulfides using H_2S in the SURAL process, which reduces the operating cost and negative environmental impacts caused by the use of H_2S • Precipiation of iron as natrojarosite by using saline water instead of fresh water at the Cawse, Bulong, and Ravensthorpe operations releases more acid back into solution	Nickel and cobalt
	• The pregnant leach solution is used for leaching of copper oxide ore in heaps, which reduces the cost of heap leaching in the Phelps Dodge process • Chloride-assisted (5–10 g/L of NaCl) sulfate leaching is carried out in the PLATSOL process, where base metals are formed as metal sulfates, while the precious metals are converted to chloro-complexes . The copper sulfide "cements" the precious metals onto the solid surface, resulting in an enriched product for further processing • The by-product acid is recycled in the Freeport-McMoRan Inc. (FCX) process to the leaching vessel and to the heap in order to eliminate the need for sulfuric acid purchase for pressure leaching and heap leaching • In these processes, the precious metals are recovered from the leach residue by pressure cyanidation coupled with carbon-in-column (CIC)	Copper and PGM

(continued)

Table 5.1 (continued)

Metal separation process step	Innovative components	Metal recovery and application
Atmospheric leaching	• The ferric sulfate used in the Galvanox™ process to leach copper from concentrates is added with a pyrite catalyst to achieve near complete dissolution of copper within a short time	Copper
	• The atmospheric zinc sulfide leaching with ferric ion (Outotec process) is regenerated by pure oxygen injection in the leaching reactors. The process produces elemental sulfur instead of sulfuric acid and eliminates the need for the addition of a surfactant	Zinc
	• The reductive pretreatment of ilmenite and titanium slag is used in sulfuric leaching with low acid consumption to obtain a high recovery of titanium dioxide (the BHP titanium process)	Titanium dioxide
	• A mixed chloride ($HCl+MgCl_2$) lixiviant is used to recover nickel, cobalt, and iron from nickel laterite ores and TiO_2 from ilmenite ores in the PRO process. The nickel recovery is achieved at low pH values of 1.0–1.5 without neutralization. Other innovations include conversion of ferric chloride to a valuable hematite product, generation of concentrated HCl, and recycling of HCl and $MgCl_2$ to the leaching stage	Nickel, cobalt and iron from laterite ores, TiO_2 from ilmenite ores
	• A cupric chloride-brine solution has been effectively used to leach a chalcopyrite concentrate in the HydroCopper™ process. The process generates NaOH, and chlorine and hydrogen gases through the electrolysis of NaCl solution, and these materials are recycled to the leaching stage	Copper and PGM
	• In the Intec copper process, the $BrCl_2^-$ ion (known as Halex™) is used as an oxidant to obtain high recoveries of copper, gold, and other PGMs	Copper and PGM
	• The chlorine gas is used for leaching of copper sulfide concentrates, and copper and iron separation and purification are achieved without iron precipitation using solvent extraction	Copper and iron
Heap leaching	• The innovative aspect is its simplicity in physical configuration which eliminates commonly used steps such as crushing, grinding, agitation leaching, solid–liquid separation, and countercurrent decantation circuits. This configuration saves significant gains in energy consumption and insignificant issues with tailings management and offers flexibility for low-grade ore processing • Use of agglomerated ore to form a heap increases heap permeability and porosity and facilitates uniform distribution of the leachant throughout the heap, leading to provide high metal recoveries	Uranium, gold and copper

Solvent extraction	• In a concentrated electrolyte solution, cobalt(II) forms tetrahedral complex whereas nickel(II) retains hexagonal configuration. This property allows the separation of cobalt from nickel with anion exchange and acidic extractants (CYANEX® 272) • In a strong chloride solution, cobalt (II) easily forms anionic chloro-complexes such as $CoCl_3^-$ and $CoCl_4^{2-}$, whereas nickel does not. This property has been exploited to separate cobalt selectively from nickel with Alamine 336 in a chloride medium • The oxime functional group in LIX® 64N, LIX 84I, and LIX®87QN during Ni extraction from a sulfate solution at higher pH values gets converted to a ketone group through the oxidation of co-extracted cobalt(II) to cobalt(III) in the organic phase. The innovative component to maintain the functional group was reoximation of the ketone group through the original oxime contact with hydroxylamine salts in alkaline solutions • A sieve-plate pulsed column configuration for the extraction equipment has been used instead of the large footprint mixer-settler equipment at the Eldorado Nuclear refinery • The higher selectivity of CYANEX® 272 and CYANEX® 301 reagents allows solvent extraction use in copper and nickel extractions over a range of pH values	Nickel and cobalt
	• The slow extraction rate of LIX 64 for copper at pH of 1.5–2.5 is overcome by the addition of a small amount of LIX 63. The mixture, LIX 64N, is the choice for the commercial solvent extraction of copper • New Acorga reagents such as CLX 50, M5640, and PT5050 exhibit better selectivity and higher recovery characteristics for copper	Copper
	• The higher selectivity for zinc over most base metal halides is achieved by di(2-ethylhexyl)-phosphoric acid (D2EHPA) reagent for the recovery of zinc	Zinc
	• The preferential extraction of tantalum over niobium in low HF concentration solutions is achieved by the use of 2-octanol	Niobium and tantalum
	• The addition of a mixture of CYANEX 272 and PC 88A at a proper ratio gives preferred separation and synergism for heavy rare earths over lighter rare earths	Rare earth elements
	• The kinetically labile character of the chloro-complexes of palladium allows the formation of highly extractable and stable anionic complexes at room temperature. Platinum, which reacts slowly under the same conditions, is separated from palladium using the solvent	Palladium and platinum

(continued)

Table 5.1 (continued)

Metal separation process step	Innovative components	Metal recovery and application
Ion exchange	• The development of resin-in-pulp and carbon-in-pulp processes for the recovery of uranium and gold from low-grade ores economically by eliminating solid–liquid separation, reducing the wash-water volume, and minimizing the loss of value metals • The molecular recognition technology allows the selective separation of copper over other metals including iron, cobalt, and nickel with SuperLig®152 in a sulfate system	Uranium, gold and copper
Precipitation	• Precipitation of iron as hematite and jarosite instead of hydroxide overcomes the gel formation, thereby avoiding the serious problems related to the settling, filtration, and washing of the residue, reducing the solid wastes disposal costs, and enabling a significant recovery of value metals • Oxidative precipitation of iron and manganese with a gas mixture of sulfur dioxide and air for the recovery of base metals including nickel, cobalt, and zinc offers many advantages for plants running a sulfur burner or a sulfide roasting facility, where the off-gas is used as a source for sulfur dioxide • Initiatives on Nucleation or seeding and supersaturation control allow precipitation of metal ions at low concentrations, enhance the morphological character of the precipitate, and improve precipitation efficiency and solid–liquid separation characteristics	Iron, manganese, cobalt, nickel and zinc
Electrowinning	• Corrosion of lead anode in copper electrowinning is reduced by the addition of cobalt sulfate to the electrolyte • Development of PGM-coated titanium mesh anode allows a significant reduction of cell voltage and eliminates lead from cathode copper in copper electrowinning operations • Development of the novel nickel tank house with coated titanium anodes permits the use of mixed sulfate/chloride electrolytes that generate chlorine gas, which is recycled to the leaching stage • Air sparging under the cathode in an electrowinning cell allows for the production of high quality copper cathode and results in a smaller footprint tank house • Development of a mechanical stripping method for zinc cathodes results in significant increases in zinc production	Copper, nickel and zinc

References

Agnew, R., et al. (1998). Production of electrolytic cobalt rounds at Inco's cobalt refinery in Port Colborne, Ontario, Canada. In G. P. Tyroler & C. A. Landolt (Eds.), *Proceedings of the extractive metallurgy of nickel and cobalt* (pp. 555–561). Warrendale, PA: The Metallurgical Society.

Agulyansky, A., Agulyansky, L., & Travkin, V. F. (2004). Liquid–liquid extraction of tantalum, with 2-octanol. *Chemical Engineering and Processing: Process Intensification, 43*, 1231–1237.

Arslan, C., & Arslan, F. (2003). Thermochemical review of jarosite and goethite stability regions at 25 and 95 °C. *Turkish Journal of Engineering and Environmental Sciences, 27*, 45–52.

Bacon, G., & Mihaylov, I. (2002). Solvent extraction as an enabling technology in the nickel industry. *Journal of the Southern African Institute of Mining and Metallurgy, November/December,* 435–444.

Barth, T. R., Hairu, A. T. C., & Meier, T. P. (1998). The operation of the HBM&S zinc pressure leach plant. In J. A. Gonzalez, G. L. Bolton, P. Hancock, & J. E. Dutrizac (Eds.), *Zinc and lead processing* (pp. 811–823). Montreal, Canada: The Metallurgical Society of CIM.

Black, W. H., et al. (2006). The application of molecular recognition technology (MRT) in the palladium refining process at Impala and other selected commercial applications. In *IPMI Conference 2006* (pp. 507–519). Pensacola, FL: International Precious Metals Institute.

Bolinski, L., & Shirley, J. (1996). Russian resin in pulp technology, current status and recent developments. In H. von Michaelis (Ed.), *Randol Gold Forum 1996* (pp. 419–423). Squaw Creek, CA: Randol International.

Brunk, K. A. (1997). *Process technology: Its role as a component of a strategic business plan*. Paper presented at the symposium on–global exploitation of heap leachable gold deposits, Orlando, FL, 9–13 February 1997.

Cheng, C. Y., Barnard, K. R., Zhang, W., et al. (2011). Synergistic solvent extraction of nickel and cobalt: A review of recent developments. *Solvent Extraction and Ion Exchange, 29*, 719–754.

Cox, M. (2004). Solvent extraction in hydrometallurgy. In J. Rydberg, M. Cox, C. Musikas, & G. R. Choppin (Eds.), *Solvent extraction principles and practice* (2nd ed., pp. 455–505). New York: Marcel Dekker.

Dalton, R. F., Burgess, A., & Quan, P. M. (1992). ACORGA ZNX-50 a new selective reagent for the selective solvent extraction of zinc from chloride leach solutions. *Hydrometallurgy, 30*, 385–400.

Dalton, R. F., Prince, R., Henmana, E., et al. (1987). The CUPREX process—A new chloride-based hydrometallurgical process for the recovery of copper from sulphidic ores. In G. A. Davies (Ed.), *Separation process in hydrometallurgy* (pp. 466–476). Chichester, England: Eillis-Horwood.

Deep, A., & de Carvalho, J. M. R. (2008). Review on the recent developments in the solvent extraction of zinc. *Solvent Extraction and Ion Exchange, 26*, 375–404.

Demopoulos, G. P. (2009). Aqueous precipitation and crystallization for the production of particulate solids with desired properties. *Hydrometallurgy, 96*, 199–214.

Filippou, D. (2004). Innovative hydrometallurgical processes for the primary processing of zinc. *Mineral Processing and Extractive Metallurgy Review, 25*, 205–252.

Fleming, C. A., McMullen, J., Thomas, K. G., et al. (2003). Recent advances in the development of an alternative to the cyanidation process: Thiosulfate leaching and resin in pulp. *Minerals & Metallurgical Processing, 20*, 1–9.

Flett, D. S. (2004). Cobalt-nickel separation in hydrometallurgy: A review. *Chemistry for Sustainable Development, 12*, 81–91.

Flett, D. S. (2005). Solvent extraction in hydrometallurgy: The role of organophosphorus extractants. *Journal of Organometallic Chemistry, 690*, 2426–2438.

Halford, R., & Barnard, K. R. (2011). Factors affecting hydroxyoxime degradation in the LIX®63/Versatic 10 system under strip conditions. In F. L. Valenzuela, & B. A. Moyer (Eds.), *Proceedings of the 19th International Solvent Extraction Conference* (pp. 1–8). Santiago, Chile.

Houlachi, G., et al. (1993). On-line sensor for monitoring the quality of zinc sulphate electrolyte. In *Proceedings of the 23rd Annual Hydrometallurgical Meeting, Modelling, Simulation and Control of Hydrometallurgical Processes* (pp. 249–259). Quebec: CIM, 24 August–2 September 1993.

Kordosky, G. A. (2002). Copper recovery using leach/solvent extraction/electrowinning technology: Forty years of innovation, 2.2 million tonnes of copper annually. *Journal of the Southern African Institute of Mining and Metallurgy* 445–450.

Kordosky, G., Virning, M., & Boley, B. (2006). Equilibrium copper strip points as a function of temperature and other operating parameters: Implications for commercial copper solvent extraction plants. *Tsinghua Science and Technology, 11*, 160–164.

Kotze, M. H. (2012). *What are the major roles of ion exchange in hydrometallurgy*. Paper Presented at the International Conference on Ion Exchange. Cambridge: Society of Chemical Industry, 18–21 September 2012.

Lakshmanan, V. I., Bhowmick, A., & Halim, M. A. (2014). Titanium dioxide: Production, properties and application. In J. Brown (Ed.), *Titanium dioxide: Chemical properties, applications and environmental effects* (pp. 75–130). New York: Nova Science.

Lakshmanan, V. I., & Lawson, G. J. (1973). Extraction of cobalt by Kelex 100 and Kelex 100/versatic 911 mixtures. *Journal of Inorganic and Nuclear Chemistry, 35*, 4285–4294.

Lakshmanan, V. I., & Ryder, J. (1989). Recovery of precious metals from spent automotive catalysts.

In A. Torma & I. H. Guindles (Eds.), *Precious and rare metal technologies* (pp. 381–393). Amsterdam: Elsevier.

Lakshmanan, V. I., Sridhar, R., & Halim, M. A. (2013). *Recovery of nickel in leaching of laterite ores*. US Patent Application 20140294702.

Littlejohn, P., & Vaughan, J. (2012). Selectivity of commercial and novel mixed functionality cation exchange resins in mildly acidic sulfate and mixed sulfate–Chloride solution. *Hydrometallurgy, 121–124*, 90–99.

O'Keefe, T. J. (1993). *Method for stripping metals in solvent extraction*. US Patent 5228903, 20 July 1993.

O'Malley, G. (2002). Recovery of gold from thiosulfate solutions and pulps with anion exchange resins. http://core.kmi.open.ac.uk/download/pdf/11233893.pdf. March 2002.

Outotec Oyj. (2007). *Copper SX-EW technology*. http://www.outotec.com/ImageVaultFiles/id_713/cf_2/OTE_Copper_SX-EW_technology_eng_web.PDF.

Preston, J. S., & du Preez, A. C. (1995). The solvent extraction of nickel and cobalt by mixtures of carboxylic acids and pyridinecarboxylate esters. *Solvent Extraction and Ion Exchange, 13*, 465–494.

Ritcey, G. M. (2006). Processes. In *Solvent extraction: Principles and application to process metallurgy* (Vol. 2, pp. 356–359). Ottawa: Ritcey GM and Associates Incorporated.

Rotuska, K., & Chmielewski, T. (2008). Growing role of solvent extraction in copper ores processing. *Physicochemical Problems of Mineral Processing, 42*, 29–36.

Sandoval, S., et al. (2010). Development and commercialization of an alternative anode for copper electrowinning. In J. Harre (Ed.), *Copper'10: Electrowinning and refining, Hamburg, June 2010* (Vol. 4, pp. 1635–1648). Clausthal-Zellerfeld, Germany: GDMB.

Szymanowski, J. (1993). *Hydroxyoximes and copper hydrometallurgy*. Boca Raton, FL: CRC Press.

Taylor, A. (2009). *Trends in nickel-cobalt processing*. Paper presented at the ALTA 2009-nickel/cobalt, copper and uranium conference, Perth, 25–30 May 2009.

Tsunoda, S., et al. (1973). *The construction and operation of the Iijima Electrolytic Zinc Plant*. Paper presented at the 102nd annual meeting of AIME, TMS, Chicago, 21–24 August 1973.

van Deventer, J. (2011). Selected ion exchange applications in the hydrometallurgical industry. *Solvent Extraction and Ion Exchange, 2929*, 695–718.

van Tonder, D., & van Hege, B. (2007). Uranium recovery from acid leach liquors: The optimization of RIP/SX based flowsheets. In *Proceedings of the Fourth Southern African Conference on Base Metals* (pp. 241–260). Swakopmund, Namibia: Southern African Institute of Mining and Metallurgy, 23–27 July 2007.

Xie, F., Zhang, T. A., Dreisinger, D., et al. (2014). A critical review on solvent extraction of rare earths from aqueous solutions. *Minerals Engineering, 56*, 10–28.

Zhu, T. (1991). Solvent extraction in China. *Hydrometallurgy, 27*, 231–245.

Zhu, Z., & Cheng, C. Y. (2011). Solvent extraction technology for the separation and purification of niobium and tantalum: A review. *Hydrometallurgy, 107*, 1–12.

6 Biological Processing: Biological Processing of Sulfidic Ores and Concentrates—Integrating Innovations

Corale L. Brierley

6.1 Introduction

Biological processing of sulfidic ores and concentrates is a commercially proven hydrometallurgical process employing naturally occurring microorganisms to catalyze the oxidation of sulfide minerals to extract base metals, such as copper, nickel, and zinc, and to enhance the recovery of precious metals occluded (locked) within sulfide minerals, such as pyrite and arsenopyrite. Biological processing is also referred to as bioleaching, biomining, biohydrometallurgy, and minerals biooxidation, the latter when applied to precious metal ores or concentrates to enhance recovery of precious metals.

Biological processing has likely been unknowingly used for the recovery of copper from ores for over 2000 years (Rossi 1990), although the role of microorganisms in the leaching of sulfide minerals was unknown until the late 1940s and early 1950s (Colmer and Hinkle 1947; Colmer et al. 1950; Temple and Colmer 1951). Early records dating from 166 AD indicate the widespread practice of copper leaching in Cyprus and it is well known that the Romans recovered copper by leaching from a deposit located at what later became the Rio Tinto mine in southern Spain (Rossi 1990). This area of Spain is rich in metal sulfide minerals, so bioleaching of these sulfides most likely occurred during Roman times. Even in very early times copper, resulting from the biologically facilitated dissolution of copper sulfide minerals, was recovered from solution by displacement with iron. The Chinese king Liu-An (177–122 BC) described this electrochemical conversion process of cementation, or reduction of copper sulfate, using iron. Paracelsus (1493–1541), a Swiss physician and alchemist, also described copper cementation on iron, referring to it as a transmutation of elements (Rossi 1990).

Today biological processing is widely practiced at commercial scale for the extraction of copper and gold. Some 18–20 % of the world's mined copper production is now the result of biological processing. This estimate is obtained by examining world copper mine production, copper produced by solvent extraction/electrowinning (SX/EW) (about 22 %), and approximating percentages of oxide ores versus sulfide ores (ICSG 2013). Chile and the United States account for 93 % of the world's electrowon copper (Edelstein 2013). Identifying the major copper sulfide leach operations in these two countries and totaling these operation's annual production provide additional information to estimate production by biological processing.

C.L. Brierley (✉)
Brierley Consultancy LLC,
Highlands Ranch, CO, USA
e-mail: clbrierley@msn.com

V.I. Lakshmanan et al. (eds.), *Innovative Process Development in Metallurgical Industry*,
DOI 10.1007/978-3-319-21599-0_6

About 3 % of the global mined gold production is now the result of biological processing based on annual gold production from plants using biological processing (Biomin 2014) and comparing that with total annual mine production of gold (USGS 2014).

This chapter begins with a description of the chemistry and microbiology of biological processing of sulfide ores and concentrates. This is followed by an explanation of how the process is engineered and the innovations that have been introduced in commercial practices. The chapter concludes with a view of how biological processing technologies for ores may be applied in the future with new advances and the motivating factors for using biological processing.

6.2 The Microbiology and Chemistry of Biological Processing

It is important to understand the microbiological and chemical underpinnings of biological processing to appreciate the innovations that have revolutionized biological processing in the last several decades and continue to impact the technology.

6.2.1 Microbiology

The natural habitats of all microorganisms used in bioleaching and minerals biooxidation are natural outcroppings of sulfide minerals, mined areas where sulfide minerals have been exposed to air, acidic hot springs (e.g., Yellowstone National Park, Iceland and New Zealand), and volcanic areas.

All microorganisms used in biomining have several things in common. They:

- Are single-celled organisms that multiply by simple cell division.
- Derive energy for growth and cell functioning by oxidizing[1] ferrous iron (Fe^{2+}).

$$4Fe^{2+} + O_2 + 4H^+ \rightarrow 4Fe^{3+} + 2H_2O \quad (6.1)$$

- and certain reduced sulfur compounds such as elemental sulfur (S°).

$$S^\circ + H_2O + 1.5O_2 \rightarrow 2H^+ + SO_4^{2-} \quad (6.2)$$

- Obtain carbon for their cellular bodies from carbon dioxide (CO_2) in the atmosphere.
- Require oxygen (O_2) taken from the atmosphere; O_2 serves as the electron acceptor. Certain oxidized metals can also function as electron acceptors and this is discussed later.
- Need ammonium (NH_4^+) and phosphate (PO_4^{3-}) ions and certain trace elements as building blocks for amino acids, DNA, and other constituents. The trace elements required (Mg^{2+}, K^+, etc.) are often abundant in the ore or concentrate feedstock.
- Require a sulfuric acid environment to metabolize and multiply. The acidity should be less than pH 2.5 to ensure the metals of value that are being leached remain soluble in the leach solution and Fe^{2+} remains soluble and available as an energy source for the organisms.

Biomining microorganisms do not cause diseases in humans, animals, or plants. They can't survive in or on plants and animals, because their energy (food) source is inorganic and because they require a sulfuric acid environment.

Since the discovery in 1947 of *Thiobacillus ferrooxidans* (now *Acidithiobacillus ferrooxidans*), many more microorganisms have been discovered that are also involved in metals extraction (Norris 2007; Olson and Clark 2004; Schippers 2007; Brierley and Brierley 2013). The microorganisms used in biological processing are

[1] Oxidation involves the removal of electrons from a substance. In biological processing, the microbes remove electrons from dissolved ferrous iron (Fe^{2+}) converting it to ferric iron (Fe^{3+}) and from elemental sulfur (S°) and other chemically reduced sulfur compounds converting them to sulfuric acid (H_2SO_4).

often grouped within temperature ranges at which they grow and where they are found in the natural environment.

Ambient temperature microorganisms (Mesophiles): These cylindrical-shaped bacteria are about 1 μm long by 1/2 μm in diameter. They only metabolize and multiply from about 10 to 40 °C. If the temperature is too low, these bacteria become dormant. If the temperature exceeds 45 °C the organisms, with some exceptions, die when their proteins coagulate. There is great diversity among the mesophilic bacteria that are active in industrial biomining applications. Some members of this group are *Acidithiobacillus ferrooxidans*, various species of *Leptospirillum, Acidithiobacillus thiooxidans*, and various *Ferroplasma* species, the latter being archaea[2] not bacteria. Some of the mesophilic organisms only oxidize ferrous iron, some oxidize only reduced sulfur, while some oxidize both.

Moderately thermophilic (heat-loving) microorganisms: There is not a precise temperature that divides the mesophilic microorganisms from the moderately thermophilic microbes, because some mesophilic microorganisms are thermotolerant and some moderate thermophiles are active at 25 °C. However, biomining organisms, whose optimum temperature is at or above 45 °C and actively function in the temperature range of 40 °C to about 60 °C, are considered moderate thermophiles. Most moderate thermophiles die when the temperature exceeds about 65 °C. These microorganisms are similar to the "mesophilic" biomining bacteria, except they are somewhat larger in length—about 2–5 μm long. Some bacteria included in this grouping are *Sulfobacillus thermosulfidooxidans* and *Acidimicrobium ferrooxidans*. Like the mesophilic organisms, some moderate thermophiles oxidize iron and some oxidize sulfur.

Extremely thermophilic archaea[2]: While similar in size (one μm in diameter) to bacteria, archaea have a different molecular organization. In the tree of life, archaea occupy the lowest branch and are extant members of an offshoot of primitive microbes. They have a spherical shape and characteristically lack a rigid cell wall; rather the contents of the single cell are enclosed by a membrane. These microbes, nevertheless, are extremely robust and grow and perform only at temperatures between 60 and 85 °C. Several archaea important in commercial biological processing of sulfide minerals are *Acidianus brierleyi, Sulfolobus metallicus*, and *Metallosphaera sedula*.

In addition to the mesophilic, moderately thermophilic, and extremely thermophilic microbes, there are a variety of other acid-loving microorganisms present in biomining operations that use organic matter as their energy source. The exact role of these organisms in biological processing is unknown; however, it has been suggested that these organisms benefit biological processing by scavenging organic matter that may be toxic to the organisms using inorganics as energy sources (Johnson and Roberto 1997). Small amounts of organic matter are present in commercial operations from the death of microbes, from entrainment of organic reagents used in mineral processing operations that precede the bioleaching/minerals biooxidation step, and from ores that naturally contain organic matter.

Microorganisms in biological processing operations can be analyzed with varying success using several techniques. Counting cells under a microscope, even using stains, is unreliable because of difficulties in distinguishing microorganisms attached to mineral particles and because dead cells can't be differentiated from living cells. Cultivation techniques are not only labor intensive but also select for only a subset of the whole microbial community that is present. In recent years, nucleic-acid based molecular techniques have been increasingly applied to identify and quantify microorganisms present in commercial biological processing operations (Johnson and Hallberg 2007; Schippers 2007; Chávex et al. 2011). While these molecular techniques

[2] Archaea, also called archaebacteria, are single-celled microorganisms that are genetically distinct from bacteria. Archaebacteria evolved separately and are regarded as "living" fossils of an ancient group of organisms bridging the evolutionary gap between bacteria and eukaryotes, the latter being multicelled organisms (Biology Online 2014).

have provided much new information about microbial diversity in biological processing operations, they are most applicable to determining the microbes present in the leach solutions and to those microorganisms that can be effectively "washed" from the mineral particles. Organisms firmly bound to mineral particles are not as successfully identified or quantified as those in the leach liquor.

There are a number of factors that affect the performance of biological processing microorganisms (Brierley and Briggs 2002).

Temperature: Temperature impacts biological processing by selecting for the group of microorganisms that will predominate at a specific temperature range.

pH: All microorganisms currently used in commercial biological processing are acid-loving and perform best when the pH is between 1.2 and 2.3. Above pH 2.5 soluble ferric iron hydrolyzes (reacts with water) and precipitates from solution as various compounds. What this means in an operating plant is that the key microbial energy source (Fe^{2+}) and the ferric iron product (Fe^{3+}) of the microbial oxidation, which is the oxidant of sulfide minerals, becomes limited. The higher pH is also not favorable for the dissolution of metal cations, which may be the products of value, such as copper and zinc. The ultimate consequences of a pH that is too high in a commercial biological processing plant are a decline in PLS (pregnant leach solution) tenor for base metals and lower than anticipated extraction of precious metals.

Microorganisms are remarkably adaptable and slow changes in acidity and other operating parameters allow time for microbial populations to adapt to a range of adverse conditions without the loss of important members of the microbial population. Abrupt changes in pH and other conditions will often result in the loss of microorganisms in the biological process.

Oxidation-reduction potential: In biological processing, oxidation-reduction potential (redox) is controlled by the ratio of ferrous iron (Fe^{2+}) to ferric iron (Fe^{3+}) in the leach solution. The Fe^{2+}:Fe^{3+} ratio selects certain microorganisms in an operating biological processing plant (Rawlings et al. 1999). For example, if the redox potential is low, because of increasing Fe^{2+} in solution relative to Fe^{3+}, the bacterium, *Acidithiobacillus ferrooxidans* will predominate, because this organism has a faster growth rate than other iron-oxidizing microorganisms when an abundance of its ferrous iron energy source is available. However, as the redox potential increases due to a lower Fe^{2+}:Fe^{3+} ratio, *Leptospirillum ferrooxidans* will predominate, because these organisms have a higher affinity for Fe^{2+} than does *A. ferrooxidans*; in other words, *L. ferrooxidans* will oxidize available Fe^{2+} before *A. ferrooxidans. A. ferrooxidans* is also more sensitive to inhibition from high concentrations of Fe^{3+} in solution. Therefore, in a stirred-tank reactor, in which the redox potential remains relatively constant and is high, *L. ferrooxidans* is likely to be the predominating iron-oxidizing microorganism in the reactor.

Oxygen: Bioleaching microorganisms in commercial plants require O_2. O_2 accepts the electrons in the redox reactions catalyzed by the microorganisms. The surest way to cause a production problem in a biological processing operation is to limit O_2. Getting air into the circuits and distributing it efficiently are significant engineering challenges in the design of biological processing plants.

Nutrients: The leaching microorganisms have few nutritional requirements: PO_4^{3-}, NH_4^+, and a few trace elements. Trace elements, such as Mg^{2+} and K^+, are generally present in sufficient quantities from the degradation of rock in the acid leach. PO_4^{3-}, NH_4^+, and K^+ are added to stirred-tank biological processing operations (Brierley and Briggs 2002).

Carbon dioxide: Microorganisms require carbon for synthesis of cellular components. The microbes used in biological processing of sulfide ores and concentrates obtain carbon from atmospheric CO_2 and convert this to cellular constituents, such as enzymes, by chemical reduction of the CO_2 in a complex metabolic pathway. Microorganisms expend considerable energy in assimilation of this carbon, called "carbon fixation". CO_2 is generally available from the air or from the acid neutralization of limestone added for pH control in stirred tank bioreactors. CO_2 limitation in heaps can occur, if the ore has little or no carbonate mineralization or CO_2 consumption by the microbial population is high.

Energy (food) source: Microorganisms used in biological processing of sulfide ores and concentrates require an energy source and that energy source is ferrous iron (Fe^{2+}) for the iron-oxidizing microbes and chemically reduced sulfur compounds, such as S°, for the sulfur-oxidizing microorganisms.

Microorganisms obey the laws of thermodynamics; they do not perform any oxidation reactions that are not thermodynamically possible. Microbes are also referred to as "catalysts" because they speed up certain reactions. For example, the oxidation of Fe^{2+} to Fe^{3+} in an acid solution is extremely slow chemically; microorganisms increase the rate of this oxidation by some 500,000 times (Lacy and Lawson 1970). The reason the organisms are so good at iron oxidation is because they must oxidize a lot of it to obtain enough energy to fix CO_2 and synthesize complex proteins, carbohydrates, DNA, etc.

Salinity: The microorganisms involved in bioleaching are relatively intolerant to the chloride ion (Cl^-). The diversity of the population as well as rates of ferrous iron oxidation are negatively affected at Cl^- concentrations greater than about 3–4 g/L (Gahan et al. 2009). Attempts to adapt the biological processing microorganisms to higher Cl^- concentrations have been unsuccessful (Lawson et al. 1995).

Soluble cation and anion metal/metalloid concentrations: Leaching microorganisms are tolerant to high concentrations of most heavy metal cations and can readily be adapted to even higher concentrations. In stirred-tank bioleach plants, heavy metal cation concentrations can exceed 20 or 30 g/L; adaptation of the microbial culture to anticipated metal concentrations is an important design step. There are some cationic metals/metalloids, which can be toxic to the organisms. For these substances to be toxic, they must be soluble. Mercury and silver, though toxic, are usually not serious problems, because silver has a low solubility in acidic leach solutions and mercury adsorbs to rock, mitigating its toxic effect. Arsenate (As^{5+}) is not toxic, but arsenite (As^{3+}) is. It is important, particularly in heap leach operations, that the redox potential is sufficiently high to ensure that, when arsenic-bearing minerals such as realgar (AsS), orpiment (As_2S_3) and arsenopyrite (FeAsS) are solubilized either through acid addition or oxidation, that As^{3+} is oxidized to As^{5+}. Ferric iron effectively oxidizes As^{3+}, but particular attention may be necessary in commercial plant start-up to ensure adequate ferric iron is available to oxidize any As^{3+} in solution.

$$2Fe^{3+} + H_3As^{III}O_3 + H_2O \rightarrow 2Fe^{2+} + H_3As^{V}O_4 + 2H^+ \quad (6.3)$$

Nitrate anion (NO_3^-) presents toxicity issues; NO_3^- concentrations in excess of 200 mg/L slow the rate of Fe^{2+} oxidation by the microorganisms. Like other anions, such as Cl^-, the mechanism of toxicity is likely to be disruption of the cell membrane and uncontrolled transport of NO_3^- into the cell, which suggests that adaptation of the microbes to NO_3^- may not be effective.

Fluoride (F^-) can be problematic in bioleaching, but not because the anion is toxic. Acidophilic microbes maintain an intracellular pH near neutral, thus a huge proton gradient exists across their cell membranes when they grow in acidic environments of pH 2 or less. This gradient accounts for the organism's sensitivity to F^-. Below pH 3.45 fluoride (F^-) occurs predominately as HF. HF crosses the cell membrane as an uncharged molecule; once inside the cell HF disassociates, releasing protons that acidify and kill the cell (Brierley and Kuhn 2010). It is important to note that F^- toxicity levels are directly related to the presence of other ions in solution; fluoride complexes with aluminum and this complexation significantly reduces the toxicity of F^- to the biomining microorganisms, because the molecule is too large to cross the cell membrane of the microorganism.

Process reagents and materials: Process reagents, for example, flotation reagents, used in biological processing plants, must be tested to ensure they are not toxic to the microorganisms. Materials, such as rubber linings in tanks, leach pad liners, and all materials that microbes come in contact with in the process, should be evaluated in lab tests to ensure that there are no inhibitory effects.

Tailings waters, containing traces of cyanide (CN^-), thiocyanate (SCN^-), or cyanate (CNO^-), must not be used as process water or make-up water to bioleach circuits. These agents are respiratory inhibitors that deactivate microbial enzymes, and if they enter the circuit, the result is significant loss in plant performance at best (Bell and Quan 1997) and a total loss of microbial activity at worst. Obviously, the toxicity of cyanide has implications in the treatment of concentrates that have been previously cyanide leached.

Oils, greases, hydraulic fluids, water treatment chemicals, dust suppressors, and anti-scalant chemicals are common substances in metallurgical plants and are potential inhibitors to the leaching microorganisms. Some of these agents are surfactants, which damage the organism's cell membrane causing the membrane to break open. Little quantitative data are available on the exact concentrations that induce problems. Good housekeeping in metallurgical plants is necessary to avoid contaminating anything in which the microorganisms come in contact with (Brierley and Briggs 2002).

Biocides are used in stirred-tank reactor cooling circuits to eliminate microbial contamination. For obvious reasons, biocides must never be allowed in any part of the circuit in which the biological processing takes place.

6.2.2 Chemistry of Biological Processing

6.2.2.1 Oxidation of Sulfide Minerals by Microbially Produced Ferric Iron

Ferric iron (Fe^{3+}) is the reagent that oxidizes the sulfide minerals. The primary function of the microorganisms is to produce the strong oxidant (Fe^{3+}) for the dissolution of the sulfide minerals (see reaction 6.1); the other role of the organisms is to oxidize the elemental sulfur (S°) (see reaction 6.2) that is usually a by-product of the ferric iron oxidation of the sulfide minerals. Metal ions go into solution as metal cations (e.g., Cu^{2+}, Zn^{2+}, Ni^{2+}); iron is solubilized as Fe^{2+}, which is the energy source for the microorganisms; the corresponding anion is SO_4^{2-}. Reactions 6.4 and 6.5 illustrate the two-stage oxidation of chalcocite (Cu_2S), a common copper sulfide mineral found in supergene deposits.

$$Cu_2S + 2Fe^{3+} \rightarrow Cu^{2+} + 2Fe^{2+} + CuS \quad (6.4)$$

chalcocite blaubleibender covellite

$$CuS + 2Fe^{3+} \rightarrow Cu^{2+} + 2Fe^{2+} + S^{\circ} \quad (6.5)$$

The oxidation of chalcocite, as shown in reaction 6.4, occurs at a relatively low redox potential (370 mV Ag/AgCl electrode; ~600 mV SHE[3]) and is kinetically fast. Because of this, many operators of copper sulfide leach operations consider the first mole of copper produced from Cu_2S as an acid dissolution reaction. However, it is an oxidation-reduction reaction, which consumes ferric iron and must be considered in the air requirement calculations in plant design, because the re-oxidation of the resulting Fe^{2+} by the microbes requires oxygen (see reaction 6.1). The oxidation of Cu_2S also produces CuS (reaction 6.4), often referred to as blaubleibender covellite, but is a member of a series of nonstoichiometric copper sulfides (e.g., dijurleite, $Cu_{1.97}S$; digenite $Cu_{1.8}S$; yarrowite, $Cu_{1.12}S$, etc.) The oxidation of this "CuS" product (reaction 6.5) requires a much higher redox potential (650 mV Ag/AgCl electrode; ~870 mV SHE) than reaction 6.4 and "CuS" is kinetically slow to leach. In heap bioleaching of secondary copper sulfide ores, the slow rate of leaching of the "CuS" product is an important consideration in the design of the leach circuit. In both reaction 6.4 and reaction 6.5, one of the reaction products of the oxidation is ferrous iron, which is then re-oxidized to ferric iron by the iron-oxidizing microorganisms (reaction 6.1). Another reaction product

[3] Standard Hydrogen Electrode.

of the oxidation of "CuS" is elemental sulfur (S°), which is oxidized to sulfuric acid by sulfur-oxidizing microorganisms present in the leach circuit according to reaction 6.2.

As noted for chalcocite and its oxidation product "CuS", sulfide minerals oxidize at different redox potentials and at different rates. For example, sphalerite (ZnS) oxidizes rapidly at a relatively low redox potential

$$ZnS + 2Fe^{3+} \rightarrow Zn^{2+} + S^{\circ} + 2Fe^{2+} \quad (6.6)$$

Sphalerite

Chalcopyrite oxidation (reaction 6.7) is thermodynamically favorable in an acidic ferric sulfate system with an oxidation-reduction potential of 427 mV (SHE) (Basson 2010); however, the dissolution of the mineral is severely limited. This has been widely attributed to passivation of the chalcopyrite surface (Warren et al. 1982). The passivation is only overcome at solution redox potentials of greater than 1000 mV (SHE) at low temperatures—redox potentials that are well above those achieved in hydrometallurgy and bioleaching operations—or by operating at temperatures above 60 °C (Crundwell 2014). Crundwell (2014) describes the mechanisms of chalcopyrite dissolution in terms of the semiconductor properties of chalcopyrite and demonstrates how these mechanisms explain the observed passivation. Crundwell's (2013, 2014) theory differs radically from other proposed models of chalcopyrite passivation (Munoz et al. 1979; Dutrizac 1989; Hackl et al. 1995; Stott et al. 2001; Tshilombo et al. 2002; Parker et al. 2003; Majuste et al. 2012, 2013) and provides reasons for slow rates of dissolution irrespective of reaction products formed on the chalcopyrite surface.

$$CuFeS_2 + 4Fe^{3+} \rightarrow 5Fe^{2+} + Cu^{2+} + 2S^{\circ} \quad (6.7)$$

Chalcopyrite

To achieve the oxidation-reduction potential necessary to oxidize the targeted sulfide minerals, there must be sufficient iron in solution; 1 g/L or less of total iron in solution is enough to achieve a redox potential sufficiently high to oxidize most sulfide minerals, assuming most of the iron is efficiently maintained in the ferric form by the microorganisms (Wadsworth 1975). To avoid a decrease in the redox potential, plant conditions must be favorable for sustained oxidation of Fe^{2+} by the microorganisms. The only way to maintain the high redox potential is to ensure that everything the organisms require (O_2, CO_2, acidic conditions, nutrients, optimum temperature conditions, etc.) is optimized. This is normally achievable in stirred-tank reactors, but realizing optimum conditions for the organisms in dump (stockpile) and heap leach operations is much more difficult and usually not consistently attainable.

Precious metals (primarily gold and silver) are often associated with sulfide minerals, principally pyrite (FeS_2) and arsenopyrite (FeAsS). If the precious metals are occluded (embedded) within pyrite, arsenopyrite, or some other sulfide minerals, the ore or concentrate is referred to as being "sulfidic-refractory". In such sulfidic-refractory gold ores and concentrates the gold grains can be submicron in size. To effectively recover the precious metals, the sulfides must be oxidized. In biological processing, this oxidation is accomplished by microbially produced Fe^{3+} that oxidizes the pyrite and arsenopyrite accordingly,

$$FeS_2 + 14Fe^{3+} + 8H_2O \rightarrow 15Fe^{2+} + 2SO_4^{2-} + 16H^+ \quad (6.8)$$

Pyrite

$$FeAsS + 5Fe^{3+} \rightarrow As^{3+} + 6Fe^{2+} + S^{\circ} \quad (6.9)$$

Arsenopyrite

As^{3+} is oxidized by Fe^{3+} to As^{5+} (see reaction 6.3).

The oxidation of sulfide ores and concentrates by the microbially produced ferric iron is a heat-generating reaction. Different sulfide minerals have differing heats of reaction; pyrite, when oxidized, releases the most heat among the common sulfide minerals (−12,884 kJ/kg of FeS_2). In commercial stirred-tank reactor plants employing biological processing heat generation must be considered in the engineering design.

For example, when biologically processing sulfidic-refractory gold concentrates in which the gold is locked in pyrite, a large amount of heat is generated and the reactors must be cooled. A critical consideration is the design of the cooling circuit.

6.2.2.2 Microbial Attachment of Mineral and Biofilm Formation

Most microorganisms present in biological processing systems are firmly attached to the mineral surfaces while other microorganisms are suspended in the aqueous phase. Microbes initially attach via electrostatic interaction and the attachment is preferentially to sulfur-rich zones and dislocation sites such as cracks, crystal defects, and mineral grain boundaries (Fig. 6.1) (Sand et al. 1999; Noël et al. 2010). Once attached the microorganisms begin producing a biofilm, which is an extracellular polymer principally composed of sugars and lipids. The biofilm eventually consists of a community of different microbes involved in the biological processing of sulfidic ores and concentrates. The polymer constituents in the biofilm complex and concentrate large amounts of ferric iron creating a reaction zone between the biofilm and the mineral sulfide, which significantly enhances the dissolution of the metals beneath the biofilm (Sand and Gehrke 2006; Zhang et al. 2014). The biofilm also serves to protect the microbial community from inhibitory conditions that may exist in the bulk solution.

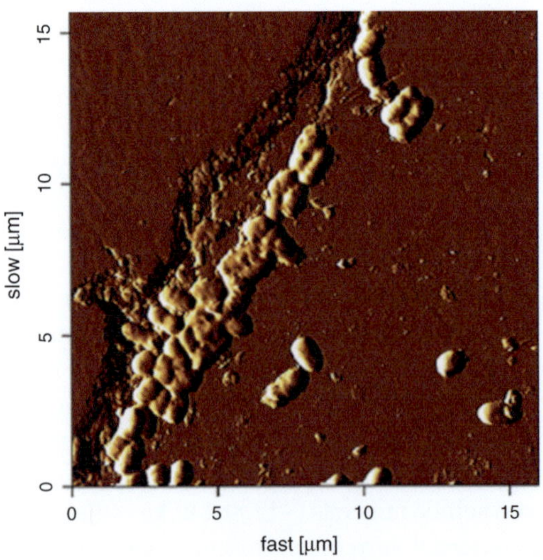

Fig. 6.1 Atomic force microscopy image of *Acidithiobacillus ferrooxidans* attached along a pyrite grain boundary (Noël et al. 2010)

6.3 Evolutionary and Revolutionary Developments in Commercial-Scale Biological Processing of Sulfide Ores and Concentrates

The stage was set in the mid- to late-1950s for the first commercial application of biological processing of sulfide ores, when technical papers were published about the newly discovered bacterium (Colmer and Hinkle 1947) *Thiobacillus ferrooxidans* (now *Acidithiobacillus ferrooxidans*) and its ability to facilitate oxidation of pyrite (Colmer et al. 1950; Temple and Colmer 1951) and other metal sulfides such as chalcocite (Cu_2S), covellite (CuS), chalcopyrite ($CuFeS_2$), and molybdenite (MoS_2) (Bryner et al. 1954; Bryner and Anderson 1957). The first commercial applications were dump (stockpile) bioleaching of submarginal grade, ROM (run-of-mine) copper sulfide ores. These applications led to further developments, which ushered in the contemporary era of biological processing. This section traces technical developments from the early years of commercial-scale biological processing through today's engineered heaps for treating sulfide ores and continuous stirred-tank reactors (CSTRs) for biological processing of concentrates.

6.3.1 Early Practices and Developments in Dump (Stockpile) Bioleaching of ROM Copper Ores

The first patent for commercial use of what became known as bioleaching issued in 1958 and was assigned to Kennecott Utah Copper (Zimmerley et al. 1958). Kennecott's Bingham

Mine near Salt Lake City, Utah (USA) employed the bioleach process at its ROM, copper dump leach operation. This biological process entailed stacking the ROM material in piles to depths of 100 m or more, applying dilute sulfuric acid over the ore piles, allowing the naturally occurring microorganisms to develop, and recovering the dissolved copper from the acidic solution that emerged from the bottom of the pile. The dissolved copper was recovered by cementation on iron, as described centuries earlier by King Liu-An and Paracelsus (Rossi 1990).

The dump leach operations at Bingham Canyon demonstrated to the copper industry that money was to be made by leaching submarginal grade ores. By the mid-1960s many copper operations were leaching submarginal grade, ROM ores in dump leach operations (Sheffer and Evans 1968). However, little was done initially with these ROM dump leach operations to enhance microbial activity. Most operations impounded acidic, ferrous iron-rich solutions from the cementation plant on the top surface of the dump where bacteria oxidized the ferrous iron to ferric iron. The ferric iron solutions in these impoundments flooded the dump material. This flood/leach method severely limited the amount of oxygen reaching the bacteria in the ore, because the solution filled the voids in the ore pile preventing air ventilation of the dump material. As a consequence, the oxygen-starved microbial populations were limited to the first few feet near the top of the dump (Bhappu et al. 1969) where some air was available.

Ironically, deliberate aeration of copper sulfide leach operations had been employed possibly as early at the eighteenth century at Rio Tinto in Spain (Jones 1905; Schlitt 2006). Ore was piled on top of horizontal stone flues that were connected to vertical chimneys surrounded by ore; this system ventilated the ore and increased copper leaching. In 1922, the Ohio Copper Company, which then operated Bingham Canyon, Utah, conducted underground (in-place) leaching of remnant block-caved, low-grade, copper sulfide ore. Rather than flooding the ore pile, solution was applied sparingly so only a thin film of solution flowed over the ore fragments, which prevented the voids in the ore pile to fill with solution. Large fans were placed in the galleries below the ore to ventilate the material, filling the void space in the ore pile with air (Anderson and Cameron 1926; Schlitt 2006). These measures increased the temperature in the ore pile as the microbial population produced ferric iron, which oxidized the sulfide minerals generating heat. Copper recovery also improved. However, the benefits of sparingly irrigating the ore and using aeration seemed to have been forgotten or ignored by dump leach operators and metallurgists when the ROM leaching process was reintroduced in the late 1950s and early 1960s.

By the 1970s, with increasing publication of research on the role of bacteria in dump leaching and the oxygen requirements of these microorganisms (Beck 1967), dump leach operators and metallurgists realized the importance of getting air into the dumps. The Anaconda Company experimented with "finger dumps" at Butte, Montana to increase exposure of the sides of the ore piles to the atmosphere (Robinson 1972). While effective in improving dump ventilation, finger dumps were somewhat impractical, because of the increased area needed for the ore piles. Pre-wetting of ROM dumps with acidified water, as the dumps were being built, was also initiated at Butte, Montana by the Anaconda Company (Schlitt 2006) and much later implemented during construction of the leach dumps at Toquepala in southern Peru (Gonzales and Nees 1996). At Butte, the acid-conditioned dumps were also aerated using horizontal and angled holes drilled into the face of the dump and cased with pipe through which air was blown. During the first leach cycle following pre-wetting and aeration, the copper tenor at Butte was twice that normally noted for similar dumps. The ultimate copper recovery was about 15 % greater than was typically obtained (Schlitt 2006). Other dump leach operations followed this practice of forced aeration with some variations including drilling and casing of vertical holes in assorted patterns. These efforts, which did improve copper recovery, have been summarized by Schlitt (2006). However, paradoxically none of these aeration tests performed in the 1970s led to widespread commercial use of forced aeration in dump leach operations over the ensuing 40 years.

The value of leach/rest cycles was also recognized early as a technique to promote improved copper extraction from ROM dumps. Rest cycles allow hold-up solutions in dumps to drain letting air fill the voids with subsequent oxidation of sulfide minerals by the microbially generated ferric iron; the sulfide oxidation heats the gases in the interior of the dump. The increased buoyancy of the heated gases induces convective air flow from dump faces. Leach cycles wash out dissolved copper and other soluble ions and add moisture to the ore bed to enhance microbial activity (Brimhall and Wadsworth 1973).

The development of solvent extraction technology for the recovery of copper from acidic leach solutions (pregnant leach solution, or PLS) coupled with electrowinning to produce cathode copper revolutionized the copper industry in the mid-1960s (Kordosky 2002). ROM dump leach plants converted from cementation to solvent extraction/electrowinning (SX/EW), which improved leaching practices by significantly reducing the amount of iron that was being added to the ore pile. Cementation had added large amounts of soluble iron to the leach solution

$$Cu^{2+}(aq) + Fe(s) \rightarrow Cu(s) + Fe^{2+}(aq) \quad (6.10)$$

and when this iron was oxidized by the bacteria, jarosite, a basic hydrous ferric sulfate compound, often precipitated in the impoundments on the top surfaces of the dumps that were initially used for the flooded leach process

$$3Fe_2(SO_4)_3 + 12H_2O \rightarrow 2HFe_3(SO_4)_2(OH)_6 \downarrow + 5H_2SO_4 \quad (6.11)$$

When new ore was placed, often by truck dumping, where the impoundments had been, the surface became compacted eventually resulting in impermeable layers within the dump where leach solutions would collect. This caused saturation of portions of the dump restricting aeration and also resulting in poor contact of the ore with the leach solution below the compacted zones. Dump leach operations, now generally referred to as "stockpile" leaching, usually remain active for many decades. Submarginal grade materials are continually being added to stockpiles as mining of huge open-pit operations continues creating massive volumes of material under leach. Consequently, some stockpiles under leach today still experience compaction and saturation problems resulting from poor practices carried out decades earlier.

6.3.2 Innovations in Heap Bioleaching/Biooxidation of Coarsely Crushed Sulfide Ores

6.3.2.1 Copper Sulfide Heap Bioleaching

The revolutionary innovation of solvent extraction/electrowinning (Kordosky 2002) in the mid-1960s allowed copper companies to leach copper oxide ores in ROM and coarsely crushed ore heaps and produce a high quality cathode copper product. The leaching of oxide ores is accomplished by irrigating with dilute sulfuric acid. Oxide ore minerals do not require an oxidation step and copper recovery is typically excellent. However, heap leaching of copper sulfide ores is another story, because an oxidation step is required.

The "thin layer" (TL) acid cure process, originally developed for copper oxide ores, was adapted for bacterial ferric iron leaching (termed "BTL") of mixed oxide/sulfide ores by Sociedad Minera Pudahuel (SMP), which applied the process at the company's Lo Aguirre mine in Chile (Bustos et al. 1991). The BTL process resulted in good copper recovery from both oxide and sulfide ores. The BTL process was soon implemented by Compañia Minera Quebrada Blanca and Cerro Colorado, both in Chile, to leach secondary copper sulfide ores, containing chalcocite and blaubleibender covellite, in the mid-1990s (Bustos et al. 1993). At Quebrada Blanca, ore is crushed in three stages to 100 % minus 9 mm, heated to increase the ore temperature, agglomerated with sulfuric acid and 85 °C water, and conveyed to the stacking area. The ore is stacked to form 6–6.5 m high lifts. After field trials at Quebrada Blanca in the mid-1990s demonstrated enhanced copper recovery with forced ventilation, aeration lines were placed under each lift to provide oxygen for the bacteria. Low-pressure fans were used to ventilate the heap. Shade cloth was placed on the top of the heap to reduce evaporative cooling. After the lift was effectively leached, the top surface was compacted, drain lines were placed on the compacted surface, and air lines were strategically placed above the drain lines in the ore to avoid solution from entering the air lines as a result of phreatic head. Another lift was stacked and leached (Schnell 1997). Similar practices were employed by Girilambone Copper Company at the company's copper heap leach operation in New South Wales, Australia (Schlitt 2006).

Heap bioleaching of crushed secondary copper (supergene) ores is regarded today as a conventional process given the number and size of these facilities and the overall contribution to global copper production (Brierley and Brierley 1999, 2013). Copper recovery for these coarse ore, secondary heap leaching plants is typically in the 80–90 % range of "recoverable copper", which is defined as the copper associated with the oxide and the supergene mineral assemblages (Scheffel 2006); "recoverable copper" does not include copper associated with the hypogene mineral assemblages. Secondary copper heap bioleaching is often described as a flexible and simple technology; however, this descriptor is misleading, because notwithstanding the general success of these plants, operating issues arise. Identifying and resolving these issues involve a multidisciplinary approach, because the problems are almost never the result of a singular factor (Brierley 2014).

Many problems encountered in secondary copper ore heap bioleaching are attributable to insufficient knowledge of the resource over life-of-mine (Ausburn and Baum 2014; Milczarek et al. 2014). The ore body may not have been adequately characterized from geological, mineralogical, chemical, metallurgical, and biological perspectives. Examples of insufficient characterization that result in later processing problems are

- Inadequate evaluation of the matrix rock for its acid consumption properties
- Lack of detailed mineralogical examination (Ausburn and Baum 2014) to identify ore minerals and expose potential locking of ore minerals within unreactive matrix rocks
- Insufficient chemical, metallurgical, and biological testing of representative samples to correlate ore mineralogy data with ore reactivity, quantify reagent consumption, assess leach kinetics of each copper sulfide mineral, estimate heat balance, confirm oxygen requirements, and assure microbial viability and activity in the ore and leach solution over time (du Plessis 2014)
- Failure to sufficiently evaluate the ore over life-of-mine and the water source for potentially toxic constituents (e.g., chloride, fluoride and nitrate) using appropriate column tests to assess dissolution of these constituents over time and their effect on the microorganisms (Brierley and Kuhn 2010)
- Failure to adequately test hydrologic characteristics and physical integrity of the ore that affect solution and air permeability and establish heap height, method of heap leaching (multi-lift, single lift with between lift liner, on-off pad), and aeration and irrigation strategies (Milczarek et al. 2014)

Once the heap bioleach is in operation, solving problems that may have been averted or at least diminished somewhat by rigorous upfront characterization is challenging and necessitates consideration of the chemical, microbiological, metallurgical, and hydrological factors. Some of the most frequently encountered problems in secondary copper heap leaching are inadequate permeability (Milczarek et al. 2014) of the ore bed to air and solution caused by excessive fine particles and/or clays in the ore when stacked, poor agglomeration of the ore, compaction by truck-stacking (when used), decrepitation of the ore by the leach solutions, and over-irrigation once the heap is stacked. These factors lead to poor wetting of the ore, preferential flow paths within the heap, impermeable layers, and solution saturation of the void spaces in the heap that displaces air. The result is inadequate distribution of microorganisms in the heap, poor microbial population development, and slow rates of microbial iron, and sulfur oxidation. The manifestation is declining redox potentials as ferric iron consumption outpaces the rate of microbial iron oxidation, decreasing copper tenor in the PLS, inability to leach the covellite reaction product of chalcocite leaching as the redox potential declines, and ultimately poor overall copper recovery (Brierley 2014).

Suboptimal temperatures in secondary copper sulfide heaps—particularly those located at high altitudes, such as the Andes Mountains, and the high deserts in temperate climates—are also a common problem. Heap temperatures can range from 10 °C or less in the winter to 20 °C or slightly higher in the summer. The microbial oxidation rate of iron slows as the temperature declines resulting in decreased dissolution of copper sulfide minerals with the consequence of less copper production. Sulfide oxidation is an exothermic reaction with the amount of heat generated dependent on which sulfide mineral is being oxidized, the amount of sulfide oxidized, and the rate at which it is oxidized. Although pyrite oxidation generates the greatest amount of heat of the sulfide minerals common to secondary copper sulfide deposits, pyrite content is frequently low in secondary copper sulfide ores, particularly in supergene ores, and the pyrite that is present often doesn't oxidize to any great extent. A probable reason for its non-reactivity is electrochemical; pyrite, unless it is amorphous or has substitutions in the crystal lattice that destabilize the structure, typically has a higher rest potential[4] than chalcocite and covellite and will not oxidize until the minerals with the lower rest potentials have oxidized. By the time the redox potential of the leach solution in the heap is sufficiently high to oxidize pyrite, the recoverable copper has been extracted. Some operations use thermal covers (Petersen and Dixon 2007) to minimize heat loss from the top surface of the heap due to evaporation and night-time radiation, to take advantage of solar heating, and to capture the heat that is generated from the oxidation of the copper sulfide minerals. This technique assists in raising the heap temperature a few degrees Celsius. Other operators have used waste heat from onsite power generation to warm agglomerated ore being placed on the heap (Schnell 1997) and submerged combustion units to heat raffinate. Benefits of ore and raffinate heating are difficult to quantify, however. Excessive irrigation is a contributing factor to heat loss in a heap (Dixon 2000). Over-irrigation causes heat loss to the PLS and can diminish the rate of oxidation of sulfide minerals, because of potential heap saturation that results in exclusion of air from void spaces within the ore bed. Excessive irrigation is particularly a problem in operations that have been acid leaching oxide ores and are transitioning to secondary copper sulfide ores. The dissolution of copper from oxides is dependent on delivering acid to the ore mineral, so the irrigation and solution application rates are important components of the process. Whereas, the leaching of copper sulfide minerals is little affected by the rate of irrigation or solution application rate and over application and irrigation will eventually decrease solution copper grades (Bartlett 1998). Slower application rates and implementing extended rest cycles, which allow heat to build in the heap, with controls to ensure that the heap does not dry out, aid in mitigating the issues associated with cool heap temperatures.

Aeration of secondary copper heaps was first evaluated in the mid-1990s (Schnell 1997; Walsh et al. 1997; James and Lancaster 1998). Today, many but not all, secondary copper heap leach operations employ forced aeration of the heaps (Scheffel 2006). The purpose of forced aeration is to provide oxygen and CO_2 for the microorganisms to maximize the population and enhance oxidation rates. The need to aerate is based on multiple variables across the multidisciplinary areas pertinent to heap bioleaching. Heap construction is a significant factor; for example, if the heap is a valley fill with only one sloping

[4] The rest potential of a mineral is that electro-potential at which the mineral will undergo corrosion. Minerals can be arranged in a galvanic (electro-potential) series for a specific electrolyte, such as acidic ferric sulfate. The more noble (resistant to corrosion/oxidation) the mineral, the higher it is in the galvanic series.

face open to the atmosphere, forced aeration may be a necessity. However, forced aeration may not be necessary, if the heap faces are well exposed to the atmosphere, the ore is quite permeable and the sulfide content of the ore is low, because convective air flow into the heap may be sufficient to sustain good microbial activity. The amount of time to oxidize the copper sulfide minerals may be longer than if forced aeration is used, but this may not be an important consideration, depending on pad space, heap design, and other factors (Scheffel 2006). Column testing with and without forced aeration may or may not predict whether forced aeration is required. Often sufficient air enters a typical metallurgical column test so the leach results from aerated column tests may be similar to those of non-aerated columns. Onsite pilot tests, which are not always completed for secondary copper sulfide ores, may also show forced aeration is unnecessary, because pilot tests are often of small tonnage and the heap may be exposed on the four sides to the atmosphere.

Despite some continuing challenges with commercial-scale leaching of secondary copper ores, the technology is the springboard to the more difficult undertaking of low-grade primary copper ore[5] bioleaching. It is imperative that technology is developed to economically and effectively leach low-grade primary copper ores, because 80 % of the world's copper resources is low-grade chalcopyrite; the grade of this resource is typically too low to concentrate and can't be economically processed other than by ROM stockpile or coarsely crushed heap leaching (Robertson et al. 2005). Meanwhile, stockpiles of ROM, chalcopyrite-bearing material mount at some mining operations as companies seek ways to retrofit existing stockpiles with technologies to enhance copper extraction and large deposits of low-grade primary copper sulfide ores have been discovered (du Plessis 2014).

[5] Chalcopyrite and enargite are "primary" copper minerals and are found in ore deposits of various geological origins. These primary minerals have not undergone alteration by weathering and are characterized as being refractory (difficult to leach).

6.3.2.2 ROM and Crushed Ore Heap Bioleaching of Low-Grade Primary Copper Ores—An Imperative Emerging Technology

It was readily apparent in the early 1960s that copper extraction from chalcopyrite-bearing ores was very poor, which led to active research programs on low-grade chalcopyrite leaching in the 1960s and 1970s (Dutrizac et al. 1969; Wadsworth 1972; Dutrizac and MacDonald 1974). Unfortunately research efforts precipitously declined in the late 1970s and early 1980s with the drop in world copper consumption and ensuing recession of the early 1980s (OTA 1988). With revival of the copper industry in the late 1980s, research efforts directed toward hydrometallurgical processing of copper sulfide minerals and chalcopyrite, in particular, gained momentum. However, much of the research was directed toward concentrates (Watling 2006). Research on bioleaching low-grade primary copper ores was seemingly a low priority. More than two decades have elapsed since the resurgence of the copper mining sector that stimulated research, but there is meager information in the published literature on bioleaching primary copper ores. As pointed out by Watling (2013), published studies on primary copper ores too often suffer from inadequate data on the mineralogy and chemistry of the samples used and the test work methodology to be of significant value.

What is known is that chalcopyrite ores tend to leach slowly and incompletely in an acidic ferric sulfate leaching solution at ambient temperature; about 10–20 % of the copper may be extracted before leaching ceases (Watling 2013). The sluggish nature of chalcopyrite bioleaching has been attributed to the mineral's crystalline structure and the changes it undergoes during acidic ferric sulfate leaching and/or passivation of the mineral in the potential range of approximately 640–840 mV (SHE) (Watling 2013). Three hypotheses have been proposed for the nature of the passivating layer: formation of a sulfur layer (Dutrizac 1989; Munoz et al. 1979); formation of copper polysulfide (Hackl et al. 1995; Tshilombo et al. 2002; Majuste et al. 2012); and precipitation of iron compounds (Parker et al. 2003). There is little question whether these products form during chalcopyrite bioleaching; the debate centers on whether the slowing and ceasing of chalcopyrite leaching are, in fact, the result of the formation of any of these over-layers (Crundwell 2013, 2014; Holmes and Crundwell 2013). Copper leaching is enhanced when pyrite is in direct contact with chalcopyrite due to the galvanic interaction between the two minerals (Nazari 2012) as chalcopyrite corrodes more rapidly than the pyrite. However, pyrite from different sources affects the rate of chalcopyrite leaching differently; some exert no influence while others, particularly pyrite containing silver, significantly enhance the rate of chalcopyrite leaching (Nazari 2012).

Research in the 1970s showed the thermophilic microorganisms, especially the extremely thermophilic archaea, improved the leaching of chalcopyrite ore (Brierley 1977) and the presence of these microorganisms accelerated galvanic leaching of the ore in the presence of pyrite (Berry et al. 1978). This research led to a large (6.4 t) long-term (2165 days) column leach test to evaluate bioleaching of a minus 15 cm, chalcopyrite-bearing ore (0.77 % Cu) starting at ambient temperature followed by 50 °C and finally 60 °C (Madsen and Groves 1983). Copper extraction rate increased substantially when the archaea were established in the column at 60 °C, however, the rate was not sustained past about 120 days. The decline in copper leach rate was attributed to the depletion of exposed chalcopyrite minerals in the ore, since the ore had been under leach for over 2000 days (Madsen and Groves 1983). Since 1983, research on ores and concentrates has confirmed chalcopyrite leaching is significantly enhanced using archaea at 60–82 °C (Watling 2013).

Many of the ROM stockpile bioleaching operations that have been operational for decades do contain chalcopyrite (Sheffer and Evans 1968; Gonzales and Nees 1996; Neira and Pearson 2012) and have in the past and continue to demonstrate profitable copper recovery. The percentage of copper leached from chalcopyrite in ROM stockpile operations has not been well documented over the decades often because the composition of the material placed on the early dumps was not carefully analyzed. Nevertheless, documented leaching of chalcopyrite occurs in some ROM stockpile operations, such as Toquepala in southern Peru, where about 80 % of the copper is present as chalcopyrite (Gonzales and Nees 1996) and Pinto Valley operations in Arizona (Neira and Pearson 2012). What characterizes ROM heap (dump) operations reporting chalcopyrite leaching? As early as the mid-1960s temperatures of 60–80 °C were documented in some low-grade dump leach operations with chalcopyrite ore and an increasingly rapid rate of copper solubilization at about 45 °C was noted in these operations (Beck 1967). Beck attributed this to chemical reactions rather than to bioleaching, because thermophilic bacteria and archaea were not known at the time and bacterial oxidation with the known mesophilic bacteria had been shown to be minimal at 40–50 °C. Increasing temperature in stockpiles is attributable to heat generation from pyrite oxidation, which will occur after leaching of sulfide minerals with lower rest potentials, have oxidized.

A pilot test at Kennecott's Bingham Canyon, Utah (now Rio Tinto Kennecott), initiated in 1995 and concluded in 1999, was a ROM heap with two lifts totaling 1.4 million t of ore grading 0.26 % Cu of which 0.17 % Cu (or about two-thirds of the copper) was associated with chalcopyrite. Details of the heap operation and performance have been published (Ream and Schlitt 1997; Schlitt and Ream 1997; Esdaile et al. 1999). Pyrite oxidation occurred, which maintained the temperature some 30 °C above ambient. Forced aeration was evaluated to see how it would affect oxygen levels in the heap; aeration was used on approximate 3 weeks cycles followed by a 2.5 week interlude. Overall, forced aeration didn't appear to significantly affect the copper leach rate. The elevated temperatures in the heap resulting from pyrite oxidation appeared to enhance natural convective airflow. While temperatures were independent of forced aeration, irrigation did influence the temperature. Modeling suggested that flushing water through an area of the heap would cool it down. It was also reported (Esdaile et al. 1999) that airflow rates increased during rest periods (no irrigation) and decreased when the area was under irrigation, suggesting that even a low irrigation rate causes some air-filled void spaces to flood. Copper recovery in the second lift approached 25 % in 13 months of leaching, although this recovery did include a contribution from the original lift.

At the Pinto Valley operation in Arizona (USA), low-grade (0.15 % Cu) chalcopyrite ore with a pyrite content of 0.7–1 % is bioleached in ROM stockpiles varying in depth from 15 to 182 m. Some 300 million tonnes of ore are under leach with an effective leach area of 144 ha. Copper recovery is 28.3 %. Three-dimensional modeling and a geophysical survey of the stockpiles have been used to identify areas of low permeability and solution saturated areas. Temperature and oxygen monitoring and a geotechnical site investigation have been conducted to better understand conditions within the stockpile that promote chalcopyrite leaching. Based on data from the characterization, improvements in irrigation have been made by skimming about 1 m off the top of the stockpiles to increase permeability and by acid injection to selected stockpiles where new ore has been added. Pyrite oxidation was found to be most effective near dump faces exposed to the atmosphere and this coincides with areas where the highest dump temperatures are recorded, as would be expected. Long rest periods and short leach cycles have also allowed temperatures in the dump to increase and the corresponding increase in oxygen with this regime has increased chalcopyrite leaching. The results of the characterization at Pinto Valley have the company looking to employ forced aeration of new dumps to improve performance (Neira and Pearson 2012).

Engineered heap technology was implemented in early 2006 at the Morenci, Arizona operations of Freeport McMoran Copper and Gold, Inc., with the objective of shortening the time scale required for chalcopyrite leaching by increasing the temperature in ROM stockpiles (Ekenes and Caro 2012). The engineered stockpiles, comprising an area of 0.5 million square meters, were equipped with aeration lines and monitoring wells and the stockpiles were drilled periodically to obtain solid samples for mineralogical and chemical analyses. The engineered stockpiles were initially inoculated with native bacteria via bio-augmented raffinate; bio-augmentation ceased after a period of time. The first measurable change observed was an increase in iron concentration in the PLS. The temperature of the engineered stockpiles increased over time and with placement of additional lifts the stockpile temperature was no longer adversely affected by seasonal temperature changes. The highest observed temperature in the stockpiles after about 2 years of operation and three lifts was in the 50–55 °C range. Pyrite oxidation was evident not only by increasing stockpile temperature and increasing PLS iron concentration but also by the amount of acid generated within the stockpiles. Copper recovery was not published, but it was noted that copper recovery outperformed the company's life-of-mine model by 25 % (approximately one million kg per month) (Ekenes and Caro 2012).

What has been learned from research, pilot plant operations, and ROM stockpiles containing chalcopyrite? Temperature is a critical factor in leaching low-grade primary copper ores. Getting temperatures in the heap into the thermophilic range of 60 °C and above is necessary to overcome the factors that cause leaching to slow and cease (Dew et al. 2011; Lee et al. 2011; Crundwell 2014). As pointed out earlier in this chapter, a high redox potential (>1000 mV SHE) is required to oxidize chalcopyrite at temperatures below 60 °C because of mineral passivation or the inherent mechanism of dissolution due to the semiconductor properties of chalcopyrite (see reaction 6.7). Such a high redox potential is very unlikely to be achieved in heap leaching. The most important consideration in heap leaching low-grade chalcopyrite is to increase the temperature in the heap to 60 °C and above through the oxidation of pyrite (Dew et al. 2011); pyrite oxidation requires a high solution redox potential (about 800–900 mV SHE) depending on the mineralogical nature of the pyrite.

Heap leaching primary copper minerals requires approaches that allow heat to build and be maintained in the heap. The amount of pyrite present in the ore and the rate at which it is leached are important considerations. Judicious control of solution application to avoid loss of heat in the PLS and to enhance convective air flow are critical; long rest/short leach cycles are an important component to this. Forced aeration must be considered on a case-by-case basis and its benefits are dependent on heap construction, permeability of the ore bed, the amount of pyrite present, the rate of oxidation of pyrite, heap temperature and other geotechnical, chemical, mineralogical, hydrological, and biological factors. Heap inoculation with the extremely thermophilic archaea will be necessary to operate the heap at temperatures above 55–60 °C (Logan et al. 2007), as there is little evidence the extremely thermophilic archaea will develop naturally in the heap or stockpile. In order to be successful, substantially more control of heap operating parameters is necessary for primary ore heap leaching than for secondary copper heap leaching.

6.3.2.3 Heap Bioleaching of Other Metal Sulfides

Heap and ROM stockpile bioleaching of other sulfide minerals, such as sphalerite, nickel sulfide minerals, and cobalt sulfide, is carried out similarly to that of heap bioleaching of copper sulfide ores. The same operational principles apply, however, the metals solubilized in the PLS are often recovered by methods other than solvent extraction. To date, only one major heap bioleach of a complex base metal sulfide ore has operated.

Bioleaching of a black schist deposit in Finland was initiated in 2008. The Talvivaara mine is one of the largest nickel deposits in the world. The sulfide content of the ore varies from 15 to 20 % and the sulfide minerals are pyrrhotite, pyrite, pentlandite, sphalerite, violarite, and chalcopyrite. The average grade of the ore is 0.23 % Ni, 0.50 % Zn, 0.13 % Cu, and 0.02 % Co (Saari and Riekkola-Vanhanen 2011). The complex mineralogy of Talvivaara and the project's location near the Arctic Circle make heap bioleaching challenging. Pyrrhotite oxidizes quickly and rapidly, consuming acid, requiring oxygen (2.25 kg O_2/kg S^{2-} as pyrrhotite), and generating heat. Initial temperatures in the heap are sufficiently high to support the extremely thermophilic archaea. However, as leaching progresses, temperature declines. The high acid consumption of pyrrhotite requires considerable initial acid addition, but acid is produced later in the leach cycle when pyrite oxidizes. Considerable attention to acid balance over the entire leach cycle is therefore required to operate such a complex circuit.

The Talvivaara heap leach has been managed as an on-off pad. After oxidation of the pyrrhotite and most of the sphalerite and pentlandite, the ore has been removed from the pad and placed in another pad area for secondary leaching. In the secondary bioleach pyrite and chalcopyrite leached along with cobalt, which is locked in the chalcopyrite matrix. Copper, nickel, cobalt, and zinc recoveries from PLS have been accomplished using gaseous hydrogen sulfide and pH control (Riekkola-Vanjanen and Palmu 2013).

6.3.2.4 Heap Biooxidation Pretreatment of Sulfide-Refractory Gold Ores

Heap biooxidation pretreatment entails the biooxidation of sulfide minerals, usually pyrite (FeS_2), arsenopyrite (FeAsS), or both, to expose micrometer-sized gold particles locked (embedded) within the sulfide matrix. The refractory-sulfidic ore is agglomerated with an acidic solution containing microorganisms. Because the ore heats during the biooxidation process as a result of pyrite and arsenopyrite oxidation, the ore is typically inoculated with mesophilic, moderately thermophilic, and extremely thermophilic microorganisms. When the heap biooxidation process is initiated, the microorganisms are cultured in a bioreactor. However, once the process is underway, fresh ore is conditioned and inoculated with effluent solution from the heap. The acid-conditioned, inoculated, and agglomerated ore is stacked on a HDPE lined pad on which is placed a layer of crushed rocks. Within the crushed rock layer is an array of perforated pipes for forced aeration of the ore pile. Low-pressure fans provide air for the ore heap (Logan et al. 2007). The heap is subjected to leach/rest cycles to conserve heat within the heap and to ensure drainage of the heap to allow void spaces to fill with air. Oxygen requirements can be high, depending on the amount of sulfide to be oxidized and the rate of oxidation.

Initially the stacked ore will be at ambient temperature, but as the sulfide minerals—particularly pyrite–oxidize, the ore bed heats. Initially the mesophilic microorganisms will be active, but as the temperature increases in the ore bed those organisms that are active at higher temperatures will be dominant. Eventually the extremely thermophilic archaea will dominate in those areas of the heap reaching temperature of more than 55–60 °C (Brierley 2003).

Once sufficient sulfide is oxidized to expose the gold, the heap is irrigated with fresh water to remove acid and soluble iron. This is an important step to avoid hydrogen cyanide production and to reduce cyanide consumption, because iron reacts with cyanide and can increase consumption to the point of making the process uneconomic. The oxidized and rinsed ore is then neutralized with lime. The neutralized ore can then be re-stacked and leached with a dilute cyanide solution to leach the gold. Alternatively the washed and oxidized ore can be lime-treated and milled in a CIL (carbon-in-leach) circuit to leach the gold with cyanide (Logan et al. 2007). Cyanide leaching requires a large pH change in the oxidized ore. Thiocyanate has been studied as an alternative to cyanide for extraction of gold from biooxidized residues, because thiocyanate can be used under acidic conditions (Li et al. 2012a, b, c, d, e). However, thiocyanate chemistry is complicated and consumption of the reagent can be unacceptably high, if conditions are not carefully controlled.

To date, only one industrial-scale plant for the biooxidation pretreatment of sulfidic-refractory gold ores has operated. Newmont Mining Company developed and employed the process to biooxidize a low-grade gold ore (about 2.64 g Au/t ore) at Gold Quarry, Nevada (USA), from 1999 to 2006 with more than 8.8 million t of ore biooxidized and about 12.2 t of gold recovered (Logan et al. 2007; Brierley and Brierley 2013).

Some gold ores are termed "double-refractory" because not only is the gold encapsulated in a sulfide mineral, but the ores contain carbonaceous matter, which adsorbs the gold-cyanide complex resulting in gold losses. This latter phenomenon is called "preg robbing". Mitigating preg robbing with microorganisms has been researched (Brierley and Kulpa 1992, 1993; Kulpa and Brierley 1993; Ofori-Sarpong et al. 2013), but a process has not been engineered or economically assessed.

6.3.3 Biological Processing of Sulfidic Ores in Vats

Although vats have been used extensively for leaching oxide ores and for cyanide leaching, vats have not been developed for biological processing of sulfidic ores. Nevertheless, some of the hydraulic and gas diffusion inefficiencies, inherent in many low-grade ore heap leaching operations, might be overcome using a saturated, continuous leaching reactor, designed for coarse ore particles in the 3–6 mm range (du Plessis et al. 2013; du Plessis 2014). Vat leaching could be considered a hybrid technology between conventional coarse heap leach technology and CSTR technology for sulfidic concentrates. The goal of vat biological processing is to have high-rate leaching kinetics by overcoming the diffusion-related limitations encountered in heap leaching. Vat leaching may be particularly useful for leaching chalcopyrite ores, because redox conditions could be controlled more effectively and diffusion limitations often inherent in leaching chalcopyrite might be overcome in a saturated reactor. The technology may also be useful in oxidizing ores that have lengthy leach times with high acid consumption; the shortened leach cycle of a vat with smaller particle sizes than a heap leach may be more efficient and cost effective by reducing acid consumption (du Plessis 2014). Vat technology for biological processing has not yet been developed for commercial use.

6.3.4 Stirred-Tank Biological Processing of Sulfide Concentrates

CSTR biological processing of sulfide concentrates was pioneered by Gencor Process Research, Johannesburg, with the first commercial plant for sulfidic-refractory gold concentrate processing commissioned in 1986 at the Fairview mine in South Africa. Commissioning of three BIOX® plants soon followed: Harbour Lights in 1992 and Wiluna in 1993, both in Australia; and Sansu, Ghana in 1994. In 1990, the Saõ Bento mine in Brazil installed a single BIOX® reactor in series with two pressure oxidation autoclaves. In total, 13 BIOX® plants have been commissioned. Six plants have subsequently been decommissioned (van Aswegen et al 2007; Biomin 2014; van Niekerk 2015). The largest BIOX® plant in operation is Kokpatas, Uzbekistan, commissioned in 2008, which processes 2138 t/day of sulfidic-refractory gold flotation concentrate. In 2013, Biomin South Africa (Pty) Limited acquired the BIOX® technology patents and Biomin continues development of the technology and is responsible for implementing the process at mining operations around the world.

Other CSTR biological processing technologies have been developed and implemented at commercial scale. BacTech commissioned three plants for biological processing of refractory-sulfidic gold concentrates: Youanmi, Western Australia, 1994; Beaconsfield, Tasmania, 1998; and Shandong, China, 2000. BIONORD®, a technology developed and owned by Polyus Gold, is used at the company's Olimpiada mine in Russia for sulfidic-refractory gold concentrate. BRGM, France, developed the BROGIM CSTR bioleach process for a polymetallic concentrate in Uganda (Morin and D'Hugues 2007). The Kasese Cobalt plant, commissioned in 1997, is the only currently operating stirred-tank biological process for base metals. The plant recovers cobalt, nickel, and copper.

All CSTR biological processing plants for pretreating sulfidic-refractory gold concentrates operate on basically the same principles and with the same process flow sheet (Fig. 6.2). The concentrate, which may be re-ground depending on gold recoveries in the mini-pilot plant run, is fed from the flotation circuit to a surge tank where

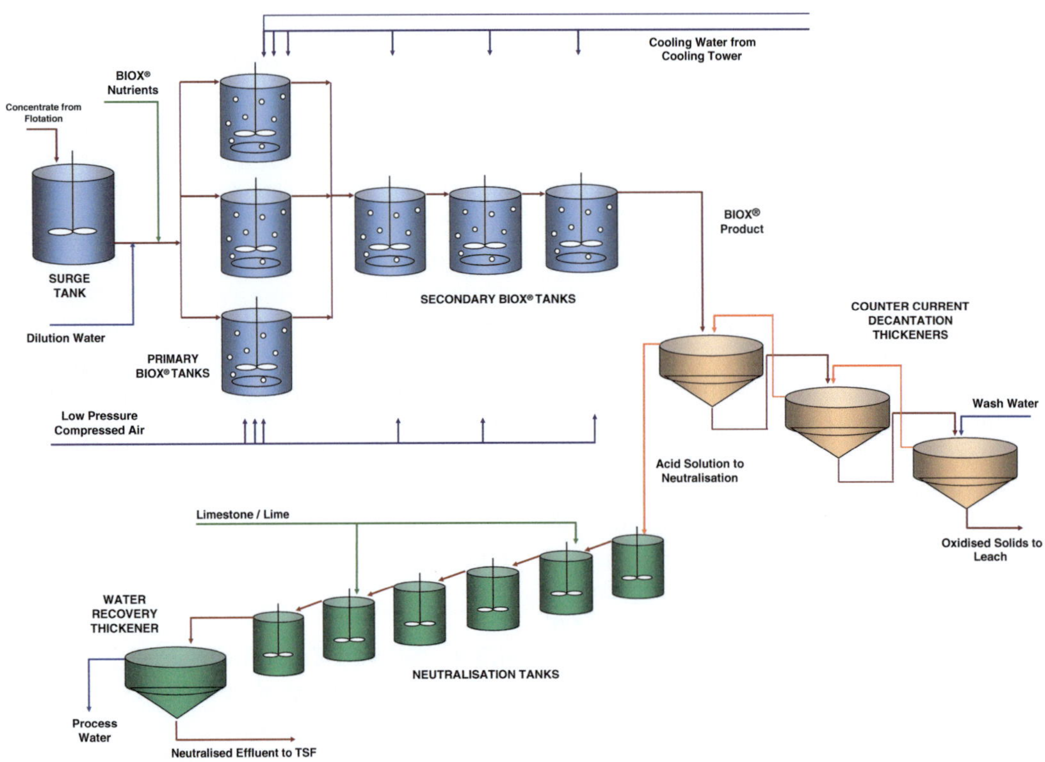

Fig. 6.2 Typical flow sheet for continuous stirred-tank biological processing of sulfidic-refractory gold concentrates (Courtesy of Biomin South Africa (Pty) Ltd.)

water is added. The biooxidation circuit is ordinarily configured with three primary reactors in parallel; this allows for a longer retention time of the solids to establish the microbial population and allows attachment of the microbes to the sulfide minerals. This prevents "wash-out" of the microorganisms from the circuit. The primary stage is followed by three (or more) secondary reactors in series.

Concentrate from the surge tank is typically diluted to 20 % solids before being fed along with nutrients (nitrogen, phosphorus and potassium salts) from a make-up tank by a feed splitter to the three primary reactors. The reactors are constructed of stainless steel to withstand the low pH and highly oxidizing conditions. Low pressure, compressed air is injected into the reactors through a diffuser ring at the bottom of the biooxidation reactor. Each reactor is equipped with a stainless steel impeller, which circulates slurry and breaks-up air bubbles from the sparge ring located directly below the impeller. The biooxidation reactors are equipped with internal cooling coils through which water is circulated from a cooling tower, because of the large amounts of heat generated from the oxidation of the sulfide minerals in the concentrate (Fig. 6.2).

The BIOX® technology underwent further development in 2013 and 2014 with the introduction of dual axial flow turbines for efficient dispersion of air in the biological reactors. This impeller design was installed at the Runruno plant in the Philippines. Development is underway for use of high-temperature microorganisms in at least one of the reactors to improve oxidation of reduced sulfur compounds produced from the ferric iron oxidation of sulfide minerals. Reduced sulfur compounds are cyanicides that increase cyanide consumption, which adds operating costs to the process. The thermophilic microorganisms, particularly the archaea, are more efficient at oxidizing these sulfur products, which should assist in reducing cyanide consumption.

Before plant commissioning, a microbial culture of mesophilic and thermo-tolerant microorganisms, adapted to the concentrate and used during laboratory piloting, is brought to the mine site and scaled-up in an onsite bioreactor. The industrial-scale reactors are started in batch using the microbial culture from the onsite bioreactor. The biooxidation circuit is then converted to continuous operation during commissioning (Fig. 6.3). CSTRs typically operate in the 40–45 °C range at a pH range of 1.1–1.5. pH is controlled in the reactors by adding limestone. The residence time of the mineral solids across a typical two-stage biooxidation circuit is 5 days. The percent of the sulfide to be oxidized for optimal gold recovery is determined in the lab in batch tests and during the mini-pilot plant phase. The plant is then designed for that percent sulfide oxidation, because the capital and operating costs of the plant are based largely on the amount of sulfide requiring oxidation. After the final stage of biooxidation, the contents of the reactor are subjected to solid/liquid separation. The solution is neutralized with limestone followed by lime before discharge to tailings. Any arsenic that leached during biooxidation is precipitated as stable ferric arsenate that meets environmental standards for disposal. The solids containing the gold are water washed in countercurrent decantation thickeners before being directed to the neutralization tanks for treatment with limestone and lime. The neutralized slurry is then thickened; the water is recovered for reuse and the thickened oxidized residue is cyanide leached to extract the gold (Fig. 6.2). Gold recoveries in CSTR plants (Fig. 6.3) are typically in the range of 95–98 % (Brierley and Briggs 2002; van Aswegen et al. 2007).

When base metal sulfide concentrates are bioleached in CSTRs the metal of value is in the solution from the last stage reactor. The base metals can be recovered from solution by standard technologies including solvent extraction/electrowinning for copper, sulfide precipitation, or other methods.

The BioCOP™ process, developed by Billiton in the late 1990s and early 2000s, was specific for treatment of refractory copper sulfide concentrates not suitable for commercial smelting due to the content of deleterious elements, such as arsenic. The process utilized extremely thermophilic archaea operating at temperatures up to 85 °C. The solubilized copper was subsequently recovered by conventional solvent extraction and

Fig. 6.3 CSTR biooxidation circuit at the Kokpatas, Uzbekistan BIOX® plant (photo courtesy of Biomin South Africa (Pty) Ltd.)

electrowinning, producing a high value copper metal product. After considerable pilot testing in South Africa a 20 ktpa (cathode copper) prototype unit to prove the extremely thermophilic BioCOP™ technology was constructed and began operation in 2003 at CODELCO's Chuquicamata mine in Chile. The prototype unit consisted of six 1260 cubic meter reactors of Stebbins construction—a ceramic design—capable of withstanding the highly corrosive conditions at 85 °C. The primary and secondary reactors had the world's largest Lightnin® A315 agitators with each having a diameter close to 5 m (Batty and Rorke 2006). The prototype unit operated until mid-2005. While the plant operated with minimal problems, the decision was made not to proceed with a commercial-scale unit. No public explanation has been made available; however, questions have been raised about the technology's economic competitiveness with other processes.

6.4 Reductive Mineral Dissolution by Biological Processing—An Emerging Process?

As discussed earlier in this chapter, biological processing involving oxidation of sulfide minerals is a conventional industrial practice. However, many metals of value occur in ores that are already oxidized, for example, nickel laterites, which are estimated to account for about 70 % of the world's nickel resources (Dalvi et al. 2004). In these lateritic ores, the nickel is largely associated with goethite, $\alpha\text{-Fe}^{III}O(OH)$, or limonite ore consisting of a mixture of hydrated iron(III) oxide-hydroxide minerals of varying composition.

The "Ferredox" processing concept, which was introduced in 2011 (du Plessis et al. 2011; Hallberg et al. 2011), uses the bacterium, *Acidithiobacillus ferrooxidans*, to couple the oxidation of elemental sulfur (S°) with the reduction of ferric iron contained in the mineral goethite (or limonite ore). The process is carried out under acidic and ambient temperature conditions in the absence of oxygen to facilitate the dissolution of metals. It has long been known that in the absence of oxygen, *Acidithiobacillus ferrooxidans* will oxidize elemental sulfur and use soluble ferric iron (Fe^{3+}) as an electron acceptor. The Ferredox concept is novel in that the ferric iron is associated with a solid mineral and, when the biologically catalyzed reduction occurs, the mineral undergoes dissolution releasing those metals associated with the mineral. The key reaction, when goethite is the ferric iron-containing mineral, is

$$S^{\circ} + 6FeO(OH) + 10H^{+} \rightarrow SO_4^{2-} + 6Fe^{2+} + 8H_2O \quad (6.12)$$

and this is carried out in the absence of oxygen. This bacterially catalyzed reductive dissolution reaction can also take place with the asbolane-lithiophorite group of minerals, (Mn, Co) (O, OH $_2$, $Ni(OH)_2$, releasing cobalt and manganese. In the case of this mineral group (du Plessis et al. 2011), the reaction may directly involve the oxidation of S° by *Acidithiobacillus ferrooxidans* in the absence of oxygen with electrons being accepted by manganese(IV)

$$Mn_3O_3(OH)_6 + S^{\circ} + 4H^{+} \rightarrow 3Mn^{2+} + 5H_2O + SO_4^{2-} \quad (6.13)$$

or by ferrous iron (Fe^{2+}) derived from the reductive dissolution of goethite

$$Mn_3O_3(OH)_6 + 6Fe^{2+} + 12H^{+} \rightarrow 3Mn^{2+} + 6Fe^{3+} + 9H_2O \quad (6.14)$$

There are currently no commercial plants that use Ferredox; however, a proposed flow sheet for the conceptual Ferredox process is shown in Fig. 6.4 (du Plessis et al. 2011). The key elements of the flow sheet are

- An acid consuming reductive leach of the limonite according to reaction 6.12
- Recovery of the metals of value (Ni, Cu, etc.) from the PLS
- Ferrous iron oxidation (reaction 6.1) to convert the iron to either soluble ferric iron or precipitated as jarosite or schwertmannite, an iron-oxyhydroxysulfate mineral. If precipitated as a ferric-oxysulfate, this can be used as a sulfuric acid generating step (reaction 6.11)
- Soluble ferric iron can be used as an oxidant for elemental sulfur oxidation and the resulting ferrous iron can be used in reductive leaching (reaction 6.14)

6.5 Motivations for Commercial Use of Biological Processes and Closing Considerations

Biological processing of sulfidic ores has been applied commercially for over half a century, when the discovery was made that microorganisms catalyze the oxidation of sulfide minerals. Little was done initially to enhance the microbial processes in what was called dump bioleaching of submarginal grade, ROM ores (now referred to more generally as "stockpile" leaching). However, when it became apparent through research that microorganisms required oxygen to catalyze sulfide oxidation, different approaches were taken to improve the ventilation of the stockpiles. By the mid-1980s and early 1990s CSTRs were in commercial use to biooxidize flotation concentrates in which gold was encapsulated in a sulfide mineral matrix. This biological pretreatment process greatly improves gold recovery from these refractory concentrates. During the same time period, crushed ore heaps were engineered for effective bioleaching of base metals and for biological pretreatment processing of low-grade, sulfidic-refractory gold ores. Today some 20 % of the world's mined copper and 3 % of the world's mined gold are now produced by biological processing. New microorganisms are being discovered, new engineered schemes for using microorganisms are being perfected and new approaches to engineering biological processes are being invented. These innovations are expected to increase the range of ores that can be processed by microbial methods to include highly refractory primary copper ores, such as chalcopyrite and enargite, and oxidized minerals, such as nickel laterites that are not amenable to conventional biological oxidation approach.

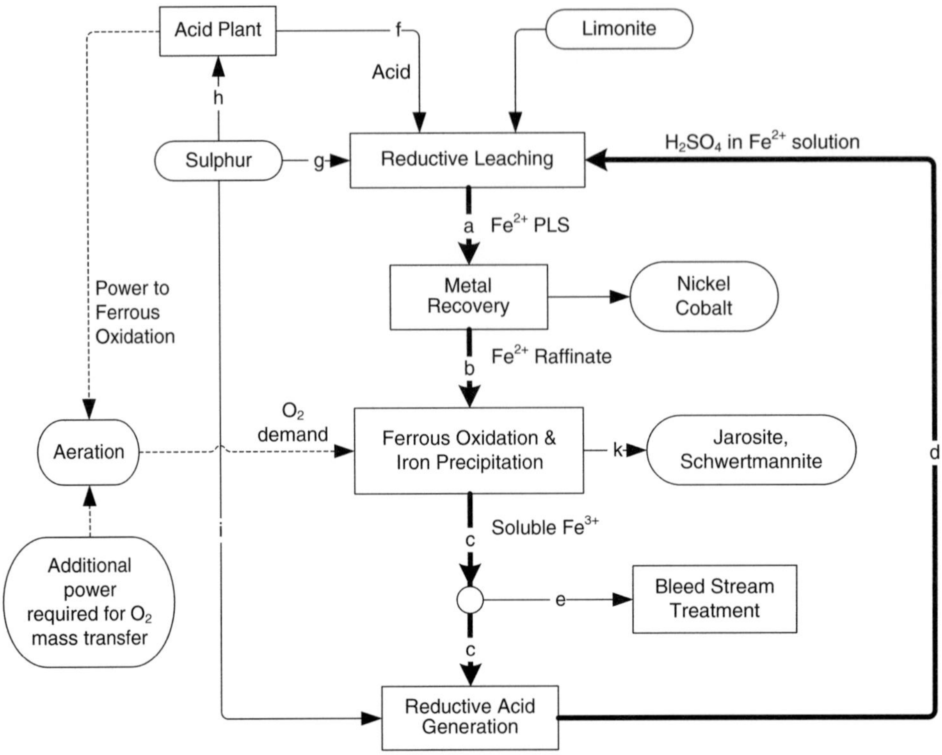

Fig. 6.4 Simplified Ferredox process flow sheet block diagram (du Plessis et al. 2011)

There are multiple reasons (Deloitte 2014) why biological processing of ores and concentrates is expected to grow in application and likely play an increasing role in future mineral processing:

Declining ore grades: Deposits of high-grade ores have largely been discovered and mined. Head grades have been steadily declining for many metals, including copper, nickel, and zinc. In some cases, copper grades are too low to support the cost of crushing, grinding, and flotation for processing by smelting. This leaves heap bioleaching of these low-grade ores as the viable option.

Easy-to-process-ore deposits are depleted: As existing mines become deeper and oxide and secondary copper ores are depleted, mining is taking place in the primary zone and newly discovered deposits have more complex ores. Hypogene ores, such as chalcopyrite and enargite, are typically low-grade and difficult to leach. Few cost-effective processing options exist, which is why a priority is being placed on developing and optimizing heap bioleaching of these ores. Complex ores are not only difficult to process by many conventional processing routes but also contain deleterious constituents making them unsuitable for some alterative processing. In biological processing of these materials, the deleterious constituents, such as arsenic, are first solubilized and then precipitated as a stable sludge that meets stringent environmental regulations and can be safely disposed in a tailings impoundment.

Increasing energy costs: As ore head grades decline, production costs increase. More energy is expended to recover the metal of value. In some countries where mining takes place (e.g., Chile and South Africa), energy costs are also increasing, because of greater demand for power by the public and private sectors and other bottlenecks in electrical supply. Biological processing, particularly heap leaching, requires less energy than some other alternative processes. Nevertheless, electrowinning of copper produced by heap bioleaching remains a significant consumer of electrical power.

Deeper mining and increasing population: As ore head grades decline and mines become deeper, open-pit and underground mining will no longer be profitable, because of increased cost of production.

The earth is currently home to 7.2 billion people and the current projection is that the earth will need to support a population of 9.6 billion by 2050 and 11 billion by 2100 (NRC 2014). Increasing population points to two matters of concern for mining:

- The amount of land throughout the world that is not populated by humans is decreasing. Consequently, mining activities are increasingly located near villages or larger population centers (Silver 2008)
- Most people envision a future with a greater standard of living than on average we currently have. This means a greater consumption of resources including metals (NRC 2014)

These two factors—deeper mines and increasing population—place significant challenges on the mining sector to reduce the footprint of mining, to cost effectively extract the metals necessary to sustain an increasing world population that expects a good quality of life, and to protect the environment. Practicable in situ (borehole) mining techniques must be developed to enable cost-effective mining of refractory and complex, low-grade metals at depth, protect groundwater, and reduce surface impacts. Commercially viable in situ leaching for metallic minerals in pristine formations entails making subsurface rocks permeable to solutions, containing solutions in underground mineral formations, controlling chemistry to selectively dissolve metals of value while leaving other subsurface material behind, and successfully returning the metal-bearing solutions to the surface for value recovery. Advancing in situ leaching for base and precious metal recoveries requires developments in: characterizing the ore body with minimal invasion; directional drilling of hard rock ore bodies; fracturing technology to enhance permeability; sensing technologies to monitor solution flow and chemistry; protecting groundwater; and robust extractive processes. Biological processes are expected to be an important technology for minimally invasive extraction technology.

References

Anderson, A. E., & Cameron, F. K. (1926). Recovery of copper by leaching, Ohio Copper Co. of Utah. *Transactions of AIME, 73*, 31–35.

Ausburn, K., & Baum, W. (2014, April). Maximise your copper extraction. *Mining Magazine* pp. 80–81.

Bartlett, R. W. (1998). *Solution mining—Leaching and fluid recovery of materials* (2nd ed.). Amsterdam: Gordon and Breach.

Basson, P. (2010). *The anodic dissolution of covellite in acidic, chloride solutions*. Thesis, University of Cape Town.

Batty, J. V., & Rorke, G. V. (2006). Development and commercial demonstration of the BioCOP™ thermophile process. *Hydrometallurgy, 83*, 83–89.

Beck, J. V. (1967). The role of bacteria in copper mining operations. *Biotechnology and Bioengineering, 9*, 487–497.

Bell, N., & Quan, L. (1997). The application of BacTech (Australia) Ltd technology for processing refractory gold ores at Youanmi Gold mine. In *Proceedings of the International Biohydrometallurgy Symposium IBS97-BIOMINE97*, Australian Mineral Foundation, Adelaide, Australia, Chapter M2.3.

Berry, V. K., Murr, L. E., & Hiskey, J. B. (1978). Galvanic interaction between chalcopyrite and pyrite during bacterial leaching of low-grade waste. *Hydrometallurgy, 3*, 309–326.

Bhappu, R. B., Johnson, P. H., Brierley, J. A., & Reynolds, D. H. (1969). Theoretical and practical studies on dump leaching. *Transactions of the Society of Mining Engineers AIME, 244*(9), 307–320.

Biology Online. (2014). Archaebacteria. Retrieved March 10, 2015, from http://www.biology-online.org/dictionary/Archaebacteria.

Biomin. (2014). Biomin BIOX® operations. Retrieved March 10, 2015, from http://www.biomin.co.za/.

Brierley, C. L. (1977). Thermophilic microorganisms in extraction of metals from ores. *Developments in Industrial Microbiology, 19*, 273–284.

Brierley, C. L. (2014). *Integrating biohydrometallurgy in contemporary mine production*. Paper presented at Biohydrometallurgy '14, Falmouth, Cornwall, UK, 9–11 June 2014.

Brierley, C. L., & Brierley, J. A. (1999). Bioheap processes—Operational requirements and techniques. In G. W. Jergensen (Ed.), *Copper leaching, solvent extraction and electrowinning technologies* (pp. 17–27). Littleton, CO: Society of Mining Engineers.

Brierley, C. L., & Brierley, J. A. (2013). Progress in bioleaching: Part B: Applications of microbial processes by the minerals industries. *Applied Microbiology and Biotechnology, 97*(17), 7543–7552.

Brierley, C. L., & Briggs, A. P. (2002). Selection and sizing of biooxidation equipment and circuit. In A. L. Mular, D. N. Halbe, & D. J. Barret (Eds.), *Mineral processing plant design, practice and control* (pp. 1540–1568). Littleton, CO: Society of Mining Engineers.

Brierley, J. A. (2003). Response of microbial systems to thermal stress in biooxidation-heap pretreatment of refractory gold ores. *Hydrometallurgy, 71*, 13–19.

Brierley, J. A., & Kuhn, M. C. (2010). Fluoride toxicity in a chalcocite bioleach heap process. *Hydrometallurgy, 104*(3–4), 410–413.

Brierley, J. A., & Kulpa, C. F. (1992). Microbial consortium treatment of refractory precious metal ores. US Patent 5,127,942, 7 July 1992.

Brierley, J. A, & Kulpa, C. F. (1993). Biometallurgical treatment of precious metal ores having refractory carbon content. US Patent 5,244,493, 14 September 1993.

Brimhall, D. B., & Wadsworth, M. E. (1973). Oxygen consumption in dump leaching. *Transactions of the Society of Mining Engineers AIME, 254*(3), 68–75.

Bryner, L. C., & Anderson, R. (1957). Microorganisms in leaching sulfide minerals. *Industrial and Engineering Chemistry, 49*, 1721–1724.

Bryner, L. C., Beck, J. V., Davis, D. B., & Wilson, D. G. (1954). Microorganisms in leaching sulfide minerals. *Industrial and Engineering Chemistry, 46*, 2587–2592.

Bustos, S., Castro, S., & Montealegre, R. (1991). The SMP bacterial TL—Leaching process at Lo Aguirre. In: J. C. Duarte & R. W. Lawrence (Eds.), *Proceedings IX International Symposium Biohydrometallurgy '91*, Troia, Portugal, 9–13 September 1991.

Bustos, S., Castro, S., & Montealegre, R. (1993). The Sociedad Minera Pudahuel bacterial thin-layer leaching process at Lo Aguirre. *FEMS Microbiology Reviews, 11*, 231–236.

Chávex, P., Obreque, J., Vera, J., Quiroga, D., & Castro, J. (2011). Microorganisms counting techniques in hydrometallurgy. In L. G. S. Sobral, D. M. de Oliveira, & C. E. G. de Souza (Eds.), *Biohydrometallurgical processes: A practical approach* (pp. 71–87). Rio de Janeiro: Centro de Tecnologia Mineral (CETEM).

Colmer, A. R., & Hinkle, M. E. (1947). The role of microorganisms in acid mine drainage: A preliminary report. *Science, 106*(2751), 253–256.

Colmer, A. R., Temple, K. L., & Hinkle, M. E. (1950). An iron-oxidizing bacterium from the acid drainage of some bituminous coal mines. *Journal of Bacteriology, 59*(3), 317–328.

Crundwell, F. K. (2013). The dissolution and leaching of minerals: Mechanisms, myths and misunderstandings. *Hydrometallurgy, 139*, 132–148.

Crundwell, F. K. (2014). The semiconductor mechanism of dissolution and the pseudo-passivation of chalcopyrite. In *Hydrometallurgy 2014* (Vol. 1, pp. 351–362). Montreal, Qubec: MetSoc.

Dalvi, A. D., Bacon, W. G., & Osborn, R. C. (2004). *The past and the future of nickel laterites*. Paper presented at PDAC 2004 international convention, Prospectors and Developers Association of Canada, Toronto, Canada, 7–10 March 2004. Retrieved March 10, 2015, from http://www.pdac.ca/pdac/publications/papers/2004/techprgm-dalvi-bacon.pdf.

Deloitte. (2014). Mining spotlight on sliding productivity and spiraling costs. Retrieved October 29, 2014, from http://www2.deloitte.com/content/dam/Deloitte/global/Documents/Energy-and-Resources/gx-er-sliding-productivity-and-spiraling-costs.pdf.

Dew, D. W., Rautenbach, G. F., van Hille, R. P., Davis-Belmar, C. S., Harvey, I. J., & Truelove, J. S. (2011). High temperature heap leaching of chalcopyrite: Method of evaluation and process model validation. In *Proceedings of international conference on percolation leaching: The status globally and in Southern Africa* (pp. 201–218). Misty Hills, Muldersdrift, South Africa, 8–9 Nov 2011. South African Institute of Mining and Metallurgy, Symposium Series S69, Johannesburg.

Dixon, D. G. (2000). Analysis of heat conservation during copper sulphide heap leaching. *Hydrometallurgy, 58*(1), 27–41.

du Plessis, C. (2014, April). Low-grade in the limelight. *Mining Magazine* pp. 83–84.

du Plessis, C. A., Mora-Huertas, N. E., Guimaraes, F. H., & Bowes, K. G. (2013). Sulfide ore leaching process. US Patent Application 2013/0291686 A1, 7 November 2013.

du Plessis, C. A., Stabbert, W., Hallberg, K. B., & Johnson, D. B. (2011). Ferrodox: A biohydrometallurgical processing concept for limonitic nickel laterites. *Hydrometallurgy, 109*, 221–229.

Dutrizac, J. E. (1989). Elemental sulfur formation during the ferric sulfate leaching of chalcopyrite. *Canadian Metallurgy Quarterly, 28*, 337–344.

Dutrizac, J. E., & MacDonald, R. J. C. (1974). Ferric ion as a leaching medium. *Minerals Science Engineering, 6*(2), 59–100.

Dutrizac, J. E., MacDonald, R. J. C., & Ingraham, T. R. (1969). The kinetics of dissolution of synthetic chalcopyrite in aqueous acidic ferric sulfate solutions. *Transactions of the Metallurgical Society AIME, 245*, 955–959.

Edelstein, D. L. (2013). 2011 Minerals yearbook—copper, [advance release, Jul 2013], US Geological Survey. Retrieved October 26, 2014, from http://minerals.usgs.gov/minerals/pubs/commodity/copper/myb1-2011-coppe.pdf.

Ekenes, J. M., & Caro, C. A. (2012). *Improving leaching recovery of copper from low-grade chalcopyrite ores.*

Paper presented at Society of Mining Engineers annual meeting, Seattle, Washington, 19–22 Feb 2012, Preprint 12-099.

Esdaile, L., Ream, B. P., & Schlitt, W. J. (1999). *The Bingham Canyon heap leach/SX-EW project: A 1999 update*. Paper presented at ALTA 1999 copper hydrometallurgy forum, Gold Coast, Queensland, Australia, 6–8 September 1999.

Gahan, C. S., Sundkvist, J. E., & Sandström, A. (2009). A study on the toxic effects of chloride on the biooxidation efficiency of pyrite. *Journal of Hazardous Materials, 172*(2–3), 1273–1281.

Gonzales, I., & Nees, M. R. (1996). Integrating the Toquepala and Cuajone SX/EW operations. *Minerals Engineering, 48*(10), 26–30.

Hackl, R. P., Dreisinger, D. B., Peters, E., & King, J. A. (1995). Passivation of chalcopyrite during oxidative leaching in sulfate media. *Hydrometallurgy, 39*, 25–48.

Hallberg, K. B., Grail, B. M., du Plessis, C. A., & Johnson, D. B. (2011). Reductive dissolution of ferric iron minerals: A new approach for bioprocessing. *Minerals Engineering, 24*, 620–624.

Holmes, P. R., & Crundwell, F. K. (2013). Polysulfides do not cause passivation: Results from the dissolution of pyrite and implications for other sulfide minerals. *Hydrometallurgy, 139*, 101–110.

ICSG. (2013). The world copper factbook 2013, International Copper Study Group. Retrieved October 26, 2014, from http://www.icsg.org/index.php/component/jdownloads/finish/170/1188.

James, B., & Lancaster, T. (1998). *Physical parameters of heap leaching at Girilambone Copper Company*. Paper presented at ALTA 1998 copper hydrometallurgy forum Brisbane, Queensland, Australia, 20–21 October 1998.

Johnson, D. B., & Hallberg, K. B. (2007). Techniques for detecting and identifying acidophilic mineral-oxidizing microorganisms. In D. E. Rawlings & D. B. Johnson (Eds.), *Biomining* (pp. 237–261). Berlin: Springer.

Johnson, D. B., & Roberto, F. F. (1997). Heterotrophic acidophiles and their roles in the bioleaching of sulfide minerals. In D. E. Rawlings (Ed.), *Biomining: Theory, microbes and industrial processes* (pp. 259–279). Berlin: Springer.

Jones, C. H. (1905). Wet methods of extracting copper at Rio Tinto, Spain. *Transactions of AIME, 35*, 3–17.

Kordosky, G. A. (2002). *Copper recovery using leach/solvent extraction/electrowinning technology: Forty years of innovation, 2.2 million tonnes of copper annually*. Paper presented at ISEC 2002 international solvent extraction conference, Cape Town, South Africa, 17–21 Mar 2002.

Kulpa, C. F., & Brierley, J. A. (1993). Microbial deactivation of preg-robbing carbon in gold ore. In A. E. Torma, J. E. Wey, & V. L. Lakshmanan (Eds.), *Biohydrometallurgical technologies, Vol. 1. The minerals* (pp. 427–435). Warrendale, PA: Metals & Materials Society.

Lacy, D. T., & Lawson, F. (1970). Kinetics of the liquid-phase oxidation of acid ferrous sulfate by the bacterium *Thiobacillus ferrooxidans*. *Biotechnology and Bioengineering, 12*, 29–50.

Lawson, E. N., Nicholas, C. J., & Pellat, H. (1995). The toxic effects of chloride ions on *Thiobacillus ferrooxidans*. In T. Vargas, C. A. Jerez, J. V. Wiertz, & H. Toledo (Eds.), *Biohydrometallurgical processing* (pp. 165–174). Santiago: University of Chile.

Lee, J., Acar, S., Doerr, D. L., & Brierley, J. A. (2011). Comparative bioleaching and mineralogy of composited sulfide ores containing enargite, covellite and chalcocite by mesophilic and thermophilic microorganisms. *Hydrometallurgy, 105*(3–4), 213–221.

Li, J., Safarzadeh, M. S., Moats, M. S., Miller, J. D., LeVier, K. M., Dietrich, M., & Wan, R. Y. (2012a). Thiocyanate hydrometallurgy for the recovery of gold. Part I: Chemical and thermodynamic considerations. *Hydrometallurgy, 113–114*, 1–9.

Li, J., Safarzadeh, M. S., Moats, M. S., Miller, J. D., LeVier, K. M., Dietrich, M., & Wan, R. Y. (2012b). Thiocyanate hydrometallurgy for the recovery of gold. Part II: The leaching kinetics. *Hydrometallurgy, 113–114*, 10–18.

Li, J., Safarzadeh, M. S., Moats, M. S., Miller, J. D., LeVier, K. M., Dietrich, M., & Wan, R. Y. (2012c). Thiocyanate hydrometallurgy for the recovery of gold. Part III: Thiocyanate stability. *Hydrometallurgy, 113–114*, 19–24.

Li, J., Safarzadeh, M. S., Moats, M. S., Miller, J. D., LeVier, K. M., Dietrich, M., & Wan, R. Y. (2012d). Thiocyanate hydrometallurgy for the recovery of gold. Part IV: Solvent extraction of gold with Alamine 336. *Hydrometallurgy, 113–114*, 25–30.

Li, J., Safarzadeh, M. S., Moats, M. S., Miller, J. D., LeVier, K. M., Dietrich, M., & Wan, R. Y. (2012e). Thiocyanate hydrometallurgy for the recovery of gold. Part V: Process alternatives for solution concentration and purification. *Hydrometallurgy, 113–114*, 31–38.

Logan, T. C., Seal, T., & Brierley, J. A. (2007). Whole-ore heap biooxidation of sulfidic gold-bearing ores. In D. E. Rawlings & D. B. Johnson (Eds.), *Biomining* (pp. 113–138). Berlin: Springer.

Madsen, B. W., & Groves, R. D. (1983). *Percolation leaching of a chalcopyrite-bearing ore at ambient and elevated temperatures with bacteria*. Bureau of Mines Report of Investigations 8827, United States Department of the Interior.

Majuste, D., Ciminelli, V. S. T., Osseo-Asare, K., Dantas, M. S. S., & Magalhães-Paniago, R. (2012). Electrochemical dissolution of chalcopyrite: Detection of bornite by synchrotron small angle X-ray diffraction and its correlation with the hindered dissolution process. *Hydrometallurgy, 111–112*, 114–123.

Majuste, D., Ciminelli, V. S. T., Eng, P. J., & Osseo-Asare, K. (2013). Applications of in-situ synchrotron XRD in hydrometallurgy: Literature review and investigation of chalcopyrite dissolution. *Hydrometallurgy, 131–132*, 54–66.

Milczarek, M., Keller, J., Yao, T. M., & Banerjee, M. (2014, April). Keeping a watchful eye. *Mining Magazine*, pp. 85–87.

Morin, D. H. R., & D'Hugues, P. (2007). Bioleaching of a cobalt-containing pyrite in stirred reactors: A case study from laboratory scale to industrial applications. In D. E. Rawlings & D. B. Johnson (Eds.), *Biomining* (pp. 35–55). Berlin: Springer.

Munoz, P. B., Miller, J. D., & Wadsworth, M. E. (1979). Reaction mechanisms for the acid ferric sulfate leaching of chalcopyrite. *Metallurgical and Materials Transactions B, 10*, 149–158.

Nazari, G. T. (2012). *Enhancing the kinetics of pyrite catalyzed leaching of chalcopyrite*. Dissertation, University of British Columbia.

Neira, M., & Pearson, T. (2012). Heap leaching of chalcopyrite ore at Pinto Valley operations. In *Proceedings of Hydroprocess 2012, 4th International Seminar on Process Hydrometallurgy* (pp. 404–413). Santiago: Gecamin.

Noël, N., Florean, B., & Sand, W. (2010). AFM & EFM study on attachment of acidophilic leaching organisms. *Hydrometallurgy, 104*(3–4), 370–375.

Norris, P. R. (2007). Acidophile diversity in mineral sulfide oxidation. In D. E. Rawlings (Ed.), *Biomining: Theory, microbes and industrial processes* (pp. 199–216). Berlin: Springer.

NRC. (2014). Can earth's and society's systems meet the needs of 10 billion people? National Research Council. Retrieved October 29, 2014, from http://www.nap.edu/catalog.php?record_id=18817.

Ofori-Sarpong, G., Osseo-Asare, K., & Tien, M. (2013). Mycohydrometallurgy: Biotransformation of double refractory gold ores by the fungus, *Phanerochaete chrysosporium. Hydrometallurgy, 137*, 38–44.

Olson, G. J., & Clark, T. R. (2004). Fundamentals of metal sulfide biooxidation. *Mining Engineering, 56*(8), 40–46.

OTA. (1988). Copper technology & competitiveness, OTA-E-367. Congress of the United States. Washington, DC: Office of Technology Assessment, US Government Printing Office.

Parker, A., Klauber, C., Kougianos, A., Watling, H. R., & Bronswijk, W. (2003). An x-ray photoelectron spectroscopy study of the mechanisms of oxidative dissolution of chalcopyrite. *Hydrometallurgy, 71*, 265–276.

Petersen, J., & Dixon, D. G. (2007). Principles, mechanisms and dynamics of chalcocite heap bioleaching. In E. R. Donati & W. Sand (Eds.), *Microbial processing of metal sulfides* (pp. 193–218). Dordrecht, The Netherlands: Springer.

Rawlings, D. E., Tributsch, H., & Hansford, G. S. (1999). Reasons why '*Leptospirillum*'-like species rather than *Thiobacillus ferrooxidans* are the dominant iron-oxidizing bacteria in many commercial processes for the biooxidation of pyrite and related ores. *Microbiology, 145*, 5–13.

Ream, B. P., & Schlitt, W. J. (1997). *Kennecott's Bingham Canyon heap leach program—Part 1: The test heap and SX-EW pilot plant*. Paper presented at ALTA 1997 copper hydrometallurgy forum, Brisbane, Queensland, Australia, 20–21 October 1997.

Riekkola-Vanjanen, N., & Palmu, L. (2013). Talvivaara nickel mine—From a project to a mine and beyond. In T. P. Battle, M. Moats, V. Cocalia, H. Oosterhof, S. Alam, A. Allanore, R. T. Jones, N. Stubina, C. Anderson, & S. Shijie Wang (Eds.), *Ni-Co 2013* (pp. 269–278). Hoboken, NJ: Wiley.

Robertson, W. E., Vercuil, A., van Staden, P. J., & Craven, P. (2005). A bacterial heap leaching approach for the treatment of low grade primary copper sulphide ore. In *Proceedings of the Third Southern African Conference on Base Metals* (pp. 471–484). The South African Institute of Mining and Metallurgy, Kitwe, Zambia, 26–29 June 2005.

Robinson, W. J. (1972). Finger dump preliminaries promise improved copper leaching at Butte. *Mining Engineering, 24*(9), 47–49.

Rossi, G. (1990). *Biohydrometallurgy*. Hamburg, Germany: McGraw-Hill.

Saari, P., & Riekkola-Vanhanen, M. (2011). Talvivaara bio-heap leaching process. In *Percolation leaching: The status globally and in Southern Africa* (pp. 53–66). Misty Hills, Muldersdrift, South Africa, 8–9 November 2011. The South African Institute of Mining and Metallurgy, Symposium Series S69, Johannesburg.

Sand, W., & Gehrke, T. (2006). Extracellular polymeric substances mediate bioleaching/biocorrosion via interfacial processes involving iron (III) and acidophilic bacteria. *Research in Microbiology, 157*, 49–56.

Sand, W., Gehrke, T., Jozsa, P. G., & Schippers, A. (1999). Direct versus indirect bioleaching. In R. Amils & A. Ballester (Eds.), *Biohydrometallurgy and the environment toward the mining of the 21st century* (pp. 27–49). Amsterdam: Elsevier.

Scheffel, R. E. (2006). *The rewards of patience*. Paper presented at the Society of Mining Engineers Annual Meeting, St. Louis, Missouri, 27–29 March 2006, Preprint 06-014.

Schippers, A. (2007). Microorganisms involved in bioleaching and nucleic acid-based molecular methods for their identification and quantification. In E. R. Donati & W. Sand (Eds.), *Microbial processing of metal sulfides* (pp. 3–33). Dordrecht, The Netherlands: Springer.

Schlitt, W. J. (2006). *The history of forced aeration in copper sulfide leaching*. Paper presented at the Society of Mining Engineers annual meeting, St. Louis, Missouri, 27–29 March 2006, Preprint 06-019.

Schlitt, W. J., & Ream, B. P. (1997). *Kennecott's Bingham Canyon heap leach program—Part 2. The column leach testwork*. Paper presented at ALTA 1997 copper hydrometallurgy forum, Brisbane, Queensland, Australia, 20–21 October 1997.

Schnell, H. A. (1997). Bioleaching of copper. In D. E. Rawlings (Ed.), *Biomining: Theory, microbes and industrial processes* (pp. 21–43). Berlin: Springer.

Sheffer, H. W., & Evans, L. G. (1968). Copper leaching practices in the western United States. U.S. Bureau of

Mines Information Circular 8341, United States Department of the Interior.

Silver, D. B. (2008). The mineral supercycle: Past, present and future. *Mining Engineering, 60*(6), 72–77.

Stott, M. B., Watling, H. R., Franzmann, P. D., & Sutton, D. C. (2001). The effect of solution chemistry on jarosite deposition during the leaching of chalcopyrite by the thermophilic archaeon, *Sufololobus metallicus. Process Metallurgy, 11*, 207–215.

Temple, K. L., & Colmer, A. R. (1951). The autotrophic oxidation of iron by a new bacterium *Thiobacillus ferrooxidans. Journal of Bacteriology, 62*(5), 605–611.

Tshilombo, A. F., Petersen, J., & Dixon, D. G. (2002). The influence of applied potential and temperature on the electrochemical response of chalcopyrite during bacterial leaching. *Minerals Engineering, 15*, 809–813.

USGS. (2014). Mineral Commodity Summary—Gold. U.S. Geological Survey. Retrieved March 10, 2015, from http://minerals.usgs.gov/minerals/pubs/commodity/gold/mcs-2014-gold.pdf.

van Aswegen, P. C., van Niekerk, J., & Olivier, W. (2007). The BIOX™ process for the treatment of refractory gold concentrates. In D. E. Rawlings & D. B. Johnson (Eds.), *Biomining* (pp. 1–31). Berlin: Springer.

van Niekerk, J. (2015). Personal communication, 13 February 2015.

Wadsworth, M. E. (1972). Advances in the leaching of sulphide minerals. *Minerals Science Engineering, 4*(4), 36–47.

Wadsworth, M. E. (1975). Physico-chemical aspects of solution mining. In *Proceedings of [anais do] III encontro nacional de tratamento de minérios: Beneficiamento, hidrometalurgia, aglomeraço* (pp. 408–454). Belo Horizonte, Brazil, 22–24 May 1975.

Walsh, D., Lancaster, T., & James, B. (1997). *Recent developments at the Girilambone heap leach-SX/EW operations.* Paper presented at ALTA 1997 copper hydrometallurgy forum, Brisbane, Queensland, Australia, 20–21 October 1997.

Warren, G. W., Wadsworth, M. E., & El-Raghy, S. M. (1982). Passive and transpassive anodic behavior of chalcopyrite in acid solutions. *Metallurgical and Materials Transactions B, 13*(4), 571–579.

Watling, H. R. (2006). The bioleaching of sulphide minerals with emphasis on copper sulphides—A review. *Hydrometallurgy, 84*, 81–108.

Watling, H. R. (2013). Chalcopyrite hydrometallurgy at atmospheric pressure: 1. Review of acidic sulfate, sulfate-chloride and sulfate-nitrate process options. *Hydrometallurgy, 140*, 163–180.

Zhang, R., Bellenberg, S., Castro, L., New, T. R., Sand, W., & Vera, M. (2014). Colonization and biofilm formation of the extremely acidophilic archaeon *Ferroplasma acidiphilum. Hydrometallurgy, 150*, 245–252.

Zimmerley, S. R., Wilson, D. G., & Prater, J. D. (1958). Cyclic leaching process employing iron oxidizing bacteria. US Patent 2,829,964, 8 April 1958.

7 Process Compression

V.I. Lakshmanan and Raja Roy

7.1 Introduction

Falling commodity prices along with rising production costs is putting a squeeze on the bottom line of the mining and metallurgical industry. In order to stay competitive, production costs need to be minimized without affecting the quality of the product. One of the innovative ways to achieve this goal is by process compression. Process compression is the elimination of a process step to make the process more economical. Many times technology available in other process industries can be applied to mining and metallurgical industry to eliminate certain process steps, if sustainable. Some examples of process compression are provided below.

7.2 Carbon-in-Pulp (CIP) Process

In CIP process, cyanide leach solution slurry containing gold is mixed with activated carbon in an agitation tank or in a counter current manner through a series of tanks. Gold cyanide complex is adsorbed on the activated carbon, which is separated from leach slurry by screening. CIP process minimizes costly solid/liquid separation steps.

Figure 7.1 shows the world gold production by recovery method in 2004 (Marsden 2006). CIP and CIL (carbon-in-leach) processes accounted for ~42 % of worldwide production in 2004. Separate estimates for CIP and CIL were not made as it had become difficult to make a clear distinction between these two processes as same plants had started using equipments for both processes.

Figure 7.2 shows the typical process flowsheet for CIP process (Hill 1986). The ore is crushed and ground to −100 mesh. It is then leached in cyanide at 45–50 % solids. Lime is added during leaching for maintaining the alkalinity of sodium cyanide. This is followed by counter-current carbon-pulp contact during which gold is adsorbed on activated carbon. Loaded carbon is then separated from pulp by screening. Gold is desorbed from activated carbon by stripping and recovered from preg strip liquor by electrowinning. Carbon is reactivated and recycled to the process.

7.3 Resin-in-Pulp (RIP) Process

RIP process is similar to the CIP process except that activated carbon is replaced by solid spherical polystyrene resin beads and has been

V.I. Lakshmanan (✉) • R. Roy
Process Research ORTECH Inc.,
2350 Sheridan Park Drive, Mississauga,
ON, Canada, L5K 2T4
e-mail: llakshmanan@processortech.com

V.I. Lakshmanan et al. (eds.), *Innovative Process Development in Metallurgical Industry*,
DOI 10.1007/978-3-319-21599-0_7

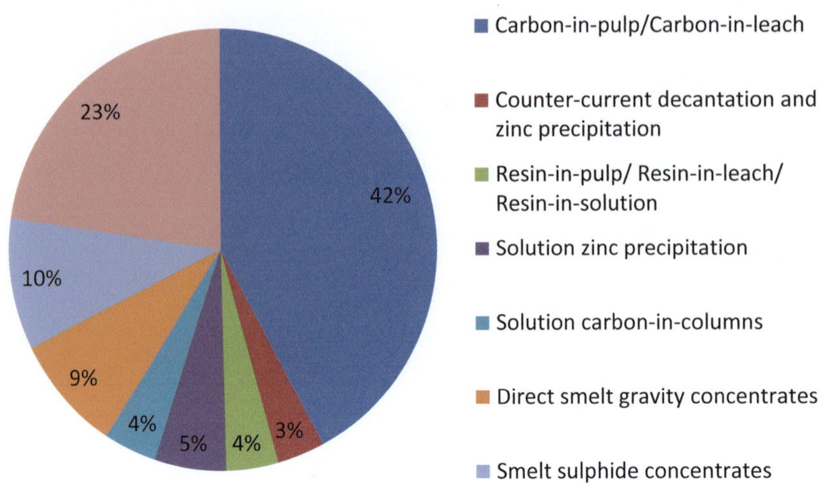

Fig. 7.1 World Gold production by recovery method in 2004, adapted from Marsden (2006)

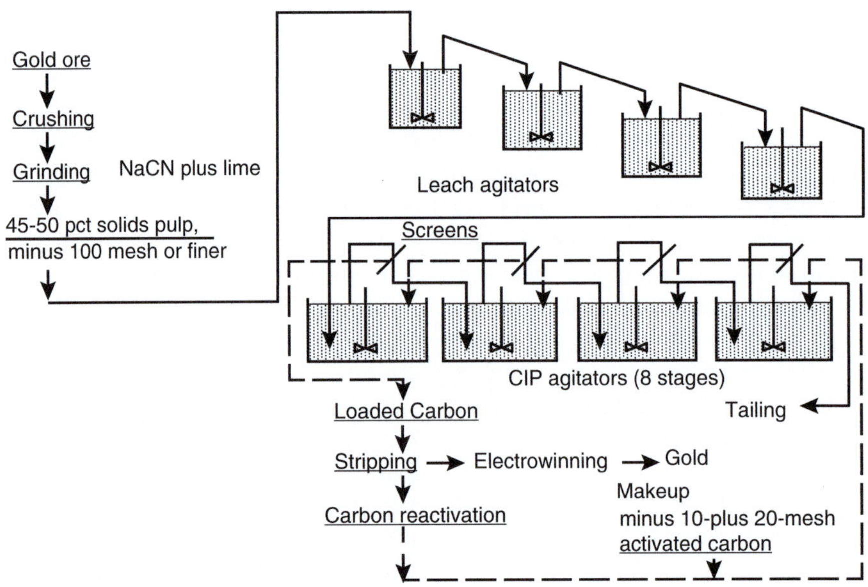

Fig. 7.2 Typical process flowsheet for CIP process (Hill 1986)

practiced in the uranium industry. Similar to the CIP process, Resin-in-Pulp (RIP) process minimizes costly solid/liquid separation steps. Traditionally, uranium ore is leached with sulfuric acid to produce uranium-rich pregnant leach liquor. The leach liquor is passed through ion-exchange resin columns, which remove uranium from leach liquor by forming a complex uranium anion. After resin is fully loaded, uranium is removed from the ion-exchange resins by passing an eluting liquid through the ion-exchange resin columns. Regenerated ion-exchange resin is then used to treat fresh batch of uranium-rich pregnant leach liquor. Due to low grade of uranium ore, process requires separation of pregnant leach liquor from large amount of residue by filtration, which adds significantly to the process cost. In addition, resin-packed columns also involve high capital cost. The RIP process eliminates the costly filtration step and reduces capital cost by

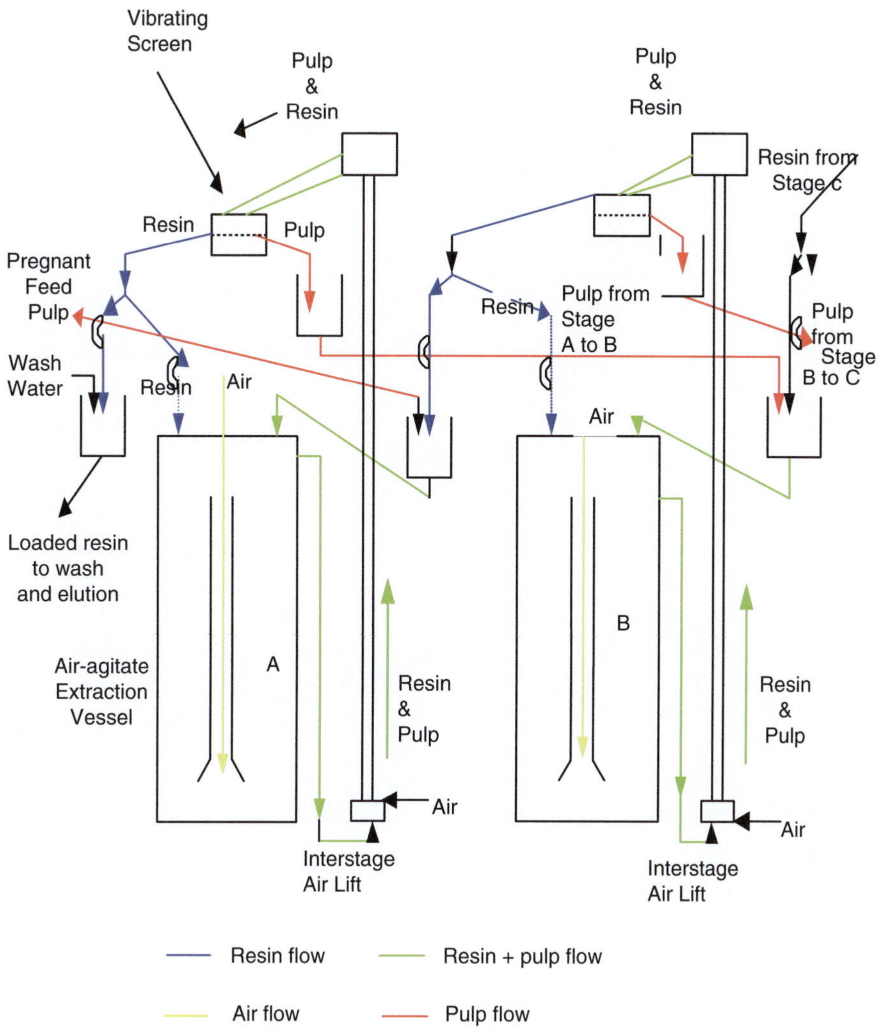

Fig. 7.3 Agitated vessel RIP process (Makhubela 2006)

eliminating the need for ion-exchange resin columns. Instead, moderately coarse resin beads are mixed with pulp and separated by screening when the ion-exchange resin is loaded with uranium. The RIP process is practiced in the USA, Russia, France, South Africa, China, and Canada for the processing of uranium ore (Mirjalili and Roshani 2007).

Figure 7.3 shows the schematic of a typical RIP process used in uranium industry (Makhubela 2006). Pulp and resin are mixed in an air-agitated extraction vessel and then pumped to a vibrating screen, which separates pulp from resin. Pulp flows to the next stage where it is again mixed with resin. Loaded resin passes through a resin splitter, which returns a fraction of loaded resin to the same tank to increase resin residence time. Rest of the loaded resin is then washed and eluted in elution columns.

7.4 Heap Leaching

Heap leaching is used for low grade ores. Ore is crushed and made into a heap on a pad. After leaching, leach liquor is pumped out for further processing. Heap leaching eliminates the need for grinding, which makes it an economical

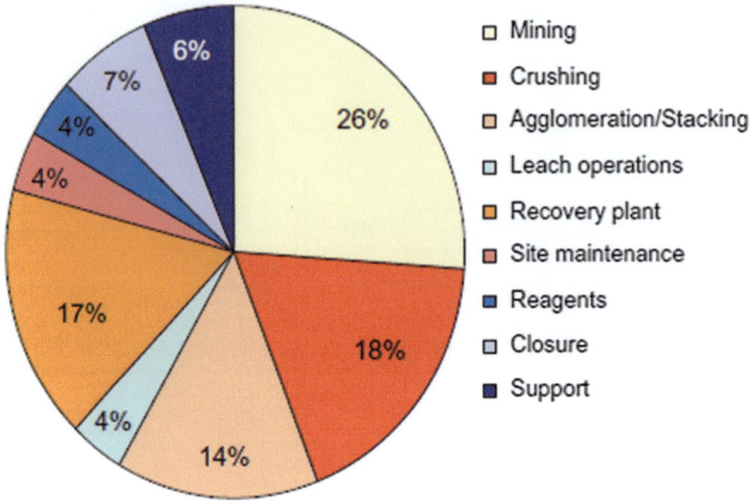

Fig. 7.4 Heap leach operating cost distribution (Dhawan et al. 2013)

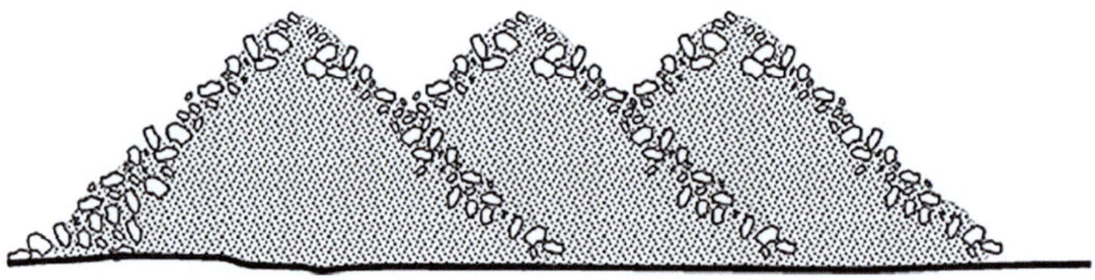

Fig. 7.5 Improper heap building resulting in segregation of coarse and fine particles (Dhawan et al. 2013)

process option for the treatment of complex ores. Heap leaching has been used for the treatment of copper, uranium, nickel, silver, and gold ores. Poor recovery due to non-uniform percolation has been faced by many heap leach operations, particularly the ones with significant clay content in the ore. To overcome this problem, modern heap leach operations include agglomeration/stacking as one of the process steps before leaching. As shown in Fig. 7.4, agglomeration/stacking cost is about 14 % of the total cost of heap leach operation (Dhawan et al. 2013). Improper heap building can result in segregation of coarse and fine particles as shown in Fig. 7.5, which will result in poor recovery due to non-uniform percolation of the lixiviant through the ore heaps (Dhawan et al. 2013). Agglomeration of ore particles results in agglomerates of similar size particles avoiding the problem of segregation. This results in uniform percolation of the lixiviant through the ore heaps as shown in Fig. 7.6 (Dhawan et al. 2013).

Figure 7.7 shows the flowsheet of a heap leach operation (Zanbak 2012). Run-of-mine ore is crushed, agglomerated, if necessary, and made into a heap on an impermeable lined pad. It is then heap leached by passing the lixiviant solution from the top. As the lixiviant solution percolates through the heap under gravity and atmospheric conditions, leaching process may take several weeks to few months for completion. Pregnant leach solution is collected in a pond and goes for metal recovery. Pregnant solution pond is lined with double layer of composite liners and fitted with leak detection pipes and pumps for the purpose of environmental protection. Barren solution is collected in a barren solution pond, which is recycled for leaching after making up for the lixiviant.

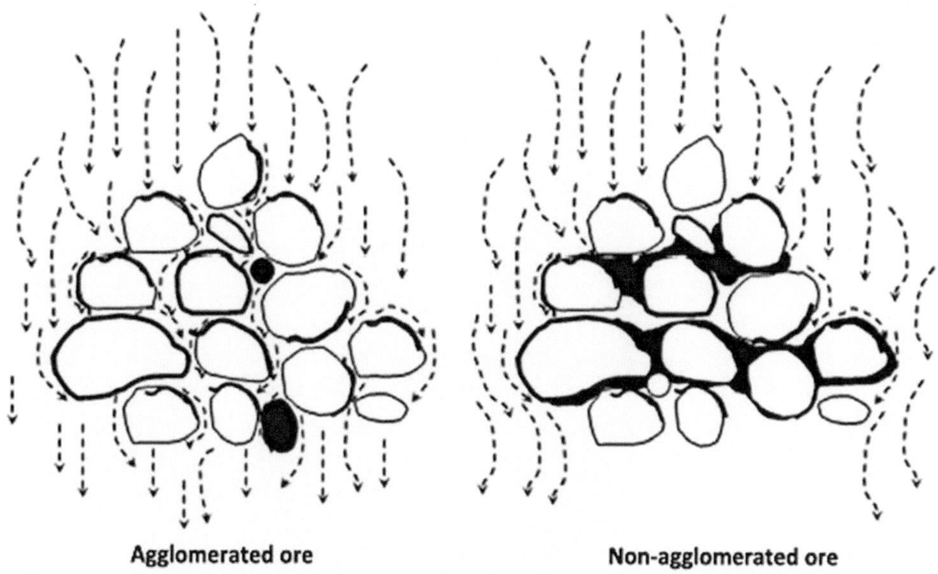

Fig. 7.6 Solution percolation in agglomerated vs. non-agglomerated ore (Dhawan et al. 2013)

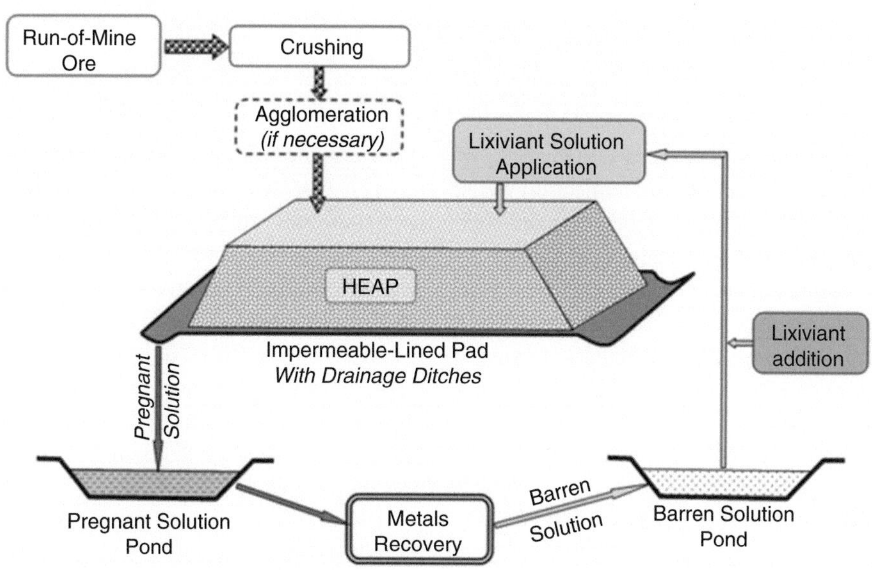

Fig. 7.7 Flowsheet of a heap leach operation (Zanbak 2012)

7.5 In-Situ Leaching

In in-situ leaching, the leaching solution is pumped through the ore body in-situ. In-situ leaching eliminates the need for crushing, grinding, and solid/liquid separation. In addition, in situ leaching minimizes issues related to tailings management.

Figure 7.8 shows the flowsheet of a heap leach operation for uranium ore (US Nuclear Regulatory Commission. Information Digest 2013–2014). Lixiviant, typically a solution of groundwater with sodium bicarbonate, hydrogen

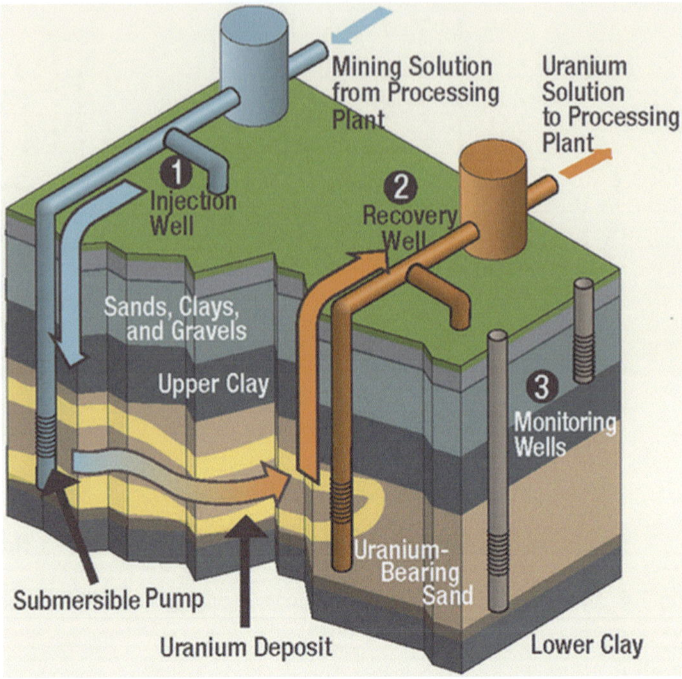

Fig. 7.8 Flowsheet of an in situ leach operation for uranium ore (US Nuclear Regulatory Commission. Information Digest 2013–2014)

peroxide, and oxygen, is pumped into the layer of earth containing uranium through injection wells. Uranium is leached in situ and the pregnant leach liquor is pumped back to the surface using recovery wells. Pregnant leach liquor is sent to the processing plant to make yellowcake. Samples from monitoring wells are checked regularly to ensure that uranium and chemicals are contained within the drilling area.

7.6 Mixer-Settler Equipment

Krebs mixer-settler equipment eliminates the need for multiple stages with conventional mixer/settlers. Figure 7.9 shows the schematic of an early prototype of a conventional mixer-settler (Knapp et al. 1958). The mixer settler is a rectangular box with multiple stages. Each stage is divided in two chambers: antechamber and a settling chamber. A mixing chamber is suspended in the antechamber using an overhead clamp. An impeller is provided in each mixing chamber for mixing the aqueous and organic phases. Aqueous and organic phases flow counter-currently through different stages. Aqueous phase from next stage and organic phase from previous stage enter the antechamber through underflow and overflow ports, respectively. Both phases flow into mixing chamber and are mixed by an impeller. Mixed phases flow into the settling chamber through an overflow arm. Aqueous and organic phases are separated in the settling chamber by gravity and flow to adjoining stages through appropriate ports.

Advantages of the conventional mixer-settler include relatively simple low cost design, stable operation, low maintenance, reliable scale-up, relatively good visibility of the process, and easy access for crud removal, while disadvantages include large settlers and footprint area, large organic inventory, and lengthy piping (Taylor 2007).

Krebs mixer-settler was developed by Krebs in France in 1970s. These mixer-settlers have been installed in many uranium solvent extraction plants and some copper solvent extraction plants. Figure 7.10 shows the schematic of a

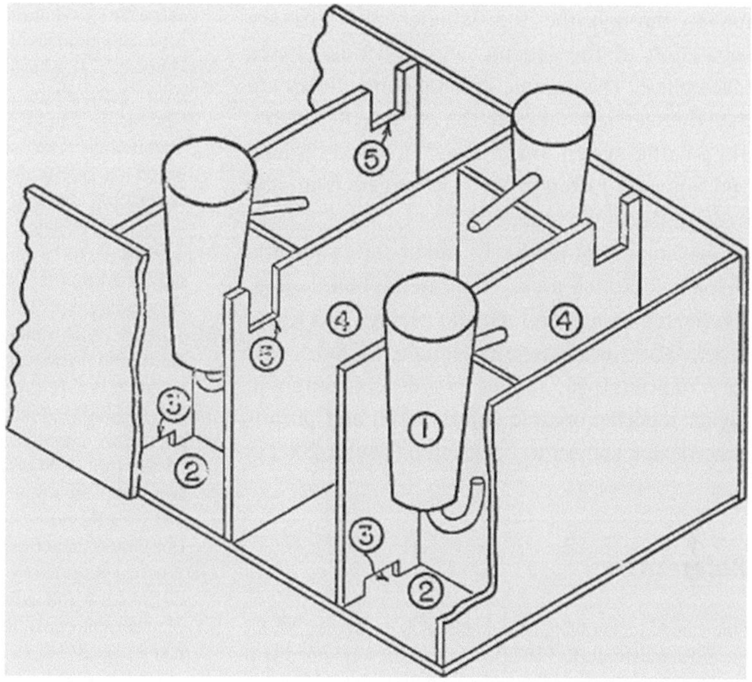

Fig. 7.9 Schematic of conventional mixer-settler (Knapp et al. 1958). (1) Mixing chamber, (2) antechamber, (3) heavy phase underflow port, (4) settling chamber, and (5) light phase overflow port

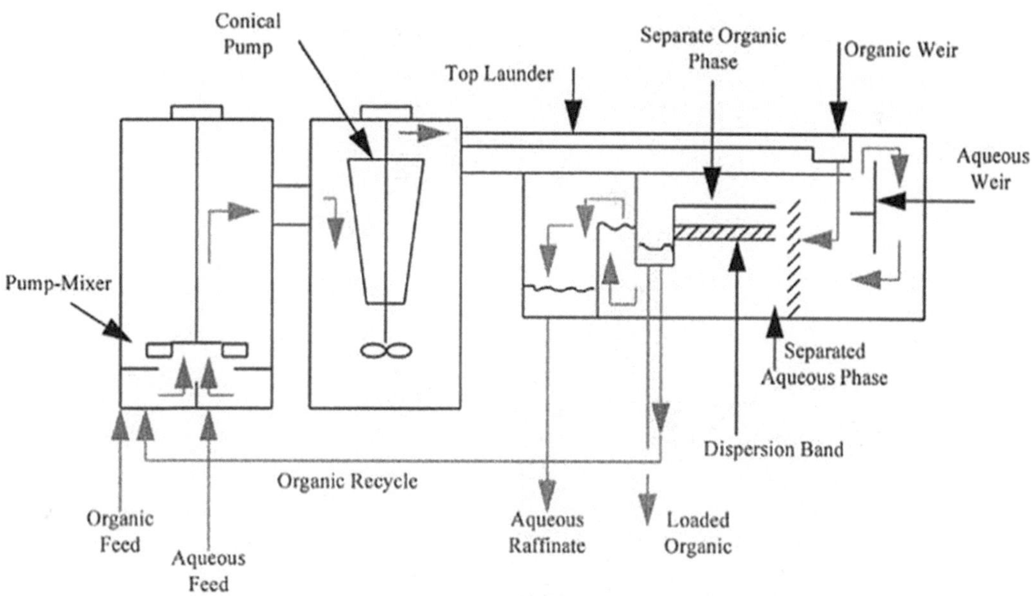

Fig. 7.10 Krebs mixer-settler elevation view (Taylor 2007)

Krebs mixer-settler (Taylor 2007). Organic and aqueous feeds pass through a pump mixer into an agitator, which is fitted with a low head, high volume, low shear conical pump. The pump consists of a static tulip-shaped stator and a six-bladed rotor. The conical pump is used to develop the head to lift the dispersion of organic and aqueous phases to the top launder. As the dispersion moves through the top launder, the primary separation of the organic and aqueous phases takes place. The organic and aqueous phases are then directed separately to the main settler below via a baffle system on one end. Loaded organic and aqueous raffinate are discharged from the main settler at the other end.

Advantages of the Krebs mixer-settler include smaller footprint area, simplified plant layout, shorter inter-stage and recycle piping, and lower capex, while disadvantages include limited access for crud removal, higher power consumption, higher reported organic entrainment, and proprietary design subject to license fee (Taylor 2007).

References

Dhawan, N., Safaradeh, M. S., Miller, J. D., Moats, M. S., & Rajamani, R. K. (2013). Crushed ore agglomeration and its control for heap leach operations. *Minerals Engineering, 41*, 53–70.

Hill, S. D. (1986). *The carbon-in-pulp process* (U.S. Bureau of Mines Information Circular No. 9059), pp. 40–43.

Knapp, L., Schoenherr, R., Barghusen, J., & Smutz, M. (1958, January). *A polyethylene box-type mixer-settler extractor*. United States Atomic Energy Commission. Research and Development Report. Unclassified.

Makhubela, T. R. (2006). *Base metal recovery by resin-in-pulp technology*. Master of technology thesis, Chemical Engineering, Tshwane University of Technology, Pretoria, South Africa, January 2006, based on Streat, M., & Naden, D. (1987). *Critical reports on applied chemistry*. (Vol. 19, pp. 1–50). New York: Wiley.

Marsden, J. O. (2006). Overview of Gold processing techniques around the world. *Minerals and Metallurgical Processing, 23*(3), 121–125.

Mirjalili, K., & Roshani, M. (2007). Resin-in-pulp method for uranium recovery from leached pulp of low grade uranium ore. *Hydrometallurgy, 85*, 103–109.

Taylor, A. (2007). *Review of mixer-settler types and other possible contactors for Copper SX*. ALTA Metallurgical Services. Retrieved August 30, 2014, from http://www.altamet.com.au/wp-content/uploads/2012/12/Review-of-Mixer-Settler-Types-and-Other-Possible-Contactors-for-Copper-SX.pdf.

US Nuclear Regulatory Commission. Information Digest. 2013–2014 (NUREG-1350, Volume 25). Section 4. Nuclear materials. Retrieved September 6, 2014, from http://en.wikipedia.org/wiki/In_situ_leach#mediaviewer/File:NRC_Uranium_In_Situ_Leach.png.

Zanbak, C. (2012, November). *Heap leaching technique in mining*. Retrieved May 29, 2015, from http://www.euromines.org/files/mining-europe/mining-techniques/batforheapleaching-feb2013-c.zanbak-euromines.pdf.

8 Process Selection

Shiv Vijayan and V.I. Lakshmanan

8.1 Introduction

The separation and removal of metals of interest from various matrices such as ores, or aqueous or organic solutions containing those metals can be achieved by a variety of metallurgical processing methods. The main goal in processing is to remove the metals of interest as selectively and economically as possible while permitting the final discharge of all the liquid, solid, and gaseous effluents generated from the processes to the environment. The effectiveness of a process is judged by the separation efficiency, that is, the ratio of metal concentration in the metal-depleted effluent to metal concentration in the feed matrix. Effectiveness is also judged by the reduction in volume achieved between the original material being processed and the final effluent requiring safe management as waste material. For example, metal separation efficiency in the range of 90–99 % and volume reduction factors in the range of 10–1000 are typical requirements in many applications involving solid and solution matrices.

In any metal separation process initiative, the end-deliverable is characterized by a set of performance aspects (PA) related to the implementation and operation of the overall process system. The common performance aspects are capital and operating costs (or life-cycle cost), and duration from concept identification to system start-up. In addition, there are other aspects, which impact cost and duration but are difficult to quantify up front in terms of these measures. These could be safety (measured through operating experience and consequences of system failure), ease of compliance with regulations, perceived environmental impact, utilization of existing resources (equipment and personnel), etc. A consensus on key aspects for such an initiative is generally arrived at among technology specialists, finance controllers, licensing experts or regulators and customer/stakeholders.

A structured approach for process selection from various options, which is discussed in this section, is based on a familiar, common sense approach to decision-making without bias. This simple approach can be considered as an effective and innovative tool for the process selection team. Such a simple but innovative approach would benefit the decision makers to arrive at a logical outcome in a consistent manner. Application of innovative approaches on all facets of the operation is essential in today's globalization of commodity markets, in particular the metal markets, in order to remain competitive and sustainable in the long term (e.g., Lakshmanan 1992).

S. Vijayan (✉) • V.I. Lakshmanan
Process Research ORTECH Inc.,
2350 Sheridan Park Drive, Mississauga, ON,
Canada, L5K 2T4
e-mail: shiv.vijayan@gmail.com

V.I. Lakshmanan et al. (eds.), *Innovative Process Development in Metallurgical Industry*,
DOI 10.1007/978-3-319-21599-0_8

8.2 Methodology

This section describes the first of a two-stage structured evaluation of process options to arrive at the best process for removing one or a group of metals of interest from a solid or a solution as the feed to the process. To arrive at the best process option, it is recognized that an integral logical analysis of the issues and their impact on performance is essential. The approach used here was evolved from the principles embedded in the Theory of Constraints Methodology (Goldratt 1999). The methodology involves a two-stage approach. In Stage 1, a four-step analysis is used (e.g., Tennankore and Vijayan 1999), namely, identification of options, identification of key performance aspects and associated measures, assessment of impact of the options on the performance aspects, and an overall assessment. A simple scoring method based on "1,4,9" ("low, medium, high" preference) is used to quantify the impact of the options on key performance aspects. The overall score is obtained through appropriate weighting of the score for individual performance aspects and is then used to short-list the preferred options for further detailed evaluation to arrive at the best option.

If weights for performance aspects are selected and fixed, a short list of options can be deduced for further detailed evaluation of impacts on the performance aspects to arrive at the best option.

8.3 Methodology Application

Several approaches essentially based on common sense performance factors have been and are being used to select the most promising option from a list of options. However, a structured approach discussed in this section is expected to provide an initial selection of a short list of options without bias, which can be used as the basis to perform sensitivity analysis of the effects of the performance factors, and to arrive at a consensus by the process selection team.

To illustrate the structured approach as an innovative and effective method to process selection, as an example, process options for vanadium extraction to produce marketable product(s) were considered. In order to evaluate the processes options for this example, the following key assumptions have been made.

- The need for vanadium extraction from flyash as the feedstock is justified on the basis of quantity, availability, and resource demand (and an opportunity to create value from industrial waste)
- Through a search of information in published literature, five process options were considered available and selected
- The recovered product and by-products were justified to have sufficient market
- There were no legacy or political issues applicable
- The goal for the methodology application and evaluation of the process options do not attest or recommend one or more processes. Any opinions and analysis may only be considered as an exercise for methodology considerations, and should not be interpreted as final answers to a vanadium recovery process for implementation.

Process options for the recovery of vanadium from flyash, and the selection of a short list of the most processing options by using the process selection methodology described in Table 8.1 are analyzed in the following section.

8.3.1 Methodology Application: Flyash and Its Utilization

The utilization of a waste material such as flyash as the secondary resource for vanadium is important not only for saving the raw material resource but also for closing the vanadium cycle, thus reducing the environmental impact (Xiao et al. 2010).

Flyash is generated as a waste effluent (e.g., thermal power plants) in the utilization of fossil fuels (e.g., coal and petroleum products). The value added and innovative utilization of the flyash would support economic benefits and sus-

Table 8.1 A summary of vanadium extraction process options

Process option	Description	Reference
Option #1	A hydrometallurgical conventional process involving concentrated H_2SO_4 leaching of flyash followed precipitation and purification to produce V_2O_5 as the product	US Patent 3,416,882 (Whigham 1968)
Option #2	A hydrometallurgical process that uses concentrated alkaline hydroxide leaching of flyash to recover vanadium selectively, followed by staged precipitation with lime, bicarbonate and CO_2 and CO_2-ammonia to produce a calcined V_2O_5 final product	US Patent 3,873,669 (Guillaud 1975)
Option #3	The carbonaceous feed material such as flyash is salt roasted at about 1000°C, leached with dilute alkali or water, followed by pure vanadium recovery as V_2O_5 product (and other value metals if desired) by elegant precipitation and drying steps	US Patent 4,539,186 (Schemel et al. 1985)
Option #4	A novel process for vanadium and zeolitic products recovery from flyash involving a preconditioning water wetting step followed by carbon removal by flotation, and pressure leaching with dilute NaOH, and leach liquor treatment by solvent extraction and precipitation steps	US Patent 4,798,709 (Lakshmanan et al. 1989)
Option #5	A pyrometallurgical process involving reduction of metal oxides in flyash in a gas fired smelter at about 1550°C to produce a ferrovanadium alloy as the final product	US Patent 5,685,244 (Goldfarb and Woodroffe 1997)

tainable environmental solutions (Singh and Gupta 2014). Flyash contains a large number of heavy metals and trace concentrations of specific heavy metals, and other substances that are known to be detrimental to health in sufficient quantities. Examples of such metals include arsenic, cadmium, chromium, molybdenum, selenium, thorium, uranium, vanadium, copper, zinc, and lead. These metal ions get released readily in aqueous environment, causing future threat to the environment. The unmanaged flyash may result in significant problems for the environment and ecology. Several utilization routes for flyash have been developed, considered for use, and some applied on large scale. Some of the application topics include (1) absorbent for cleaning sulfur compounds from flue gas; (2) adsorption of NO_x; (3) removal of mercury and boron; (4) removal of phosphate and fluoride; (5) removal of phenolic compounds and gaseous organics; (5) production of bricks as building material; (6) soil amendments associated with agricultural, wasteland reclamation and forestry sectors; and (7) recovery of value added metals.

It is the metal recovery aspect of the flyash utilization, in particular vanadium recovery, that has been selected in this section to demonstrate the process selection methodology. Previously, separation processes for recovering ferrosilicon alloy (e.g., Pickles et al. 1999), chromium (VI) (e.g., Dasmahaputra et al. 1998), gallium (e.g., Fang and Geaser 1996), and vanadium, (and nickel and magnesium) (e.g., Kuniaki et al. 1998) were reported.

8.4 Vanadium Recovery from Flyash: Process Options

Several approaches for the recovery of vanadium from fly ash have been developed (e.g., Ye 2006): A simplified flowchart in Fig. 8.1 illustrates the approaches and the process options considered to demonstrate the process selection methodology.

A summary of the selected five process options are given in Table 8.1 and illustrated

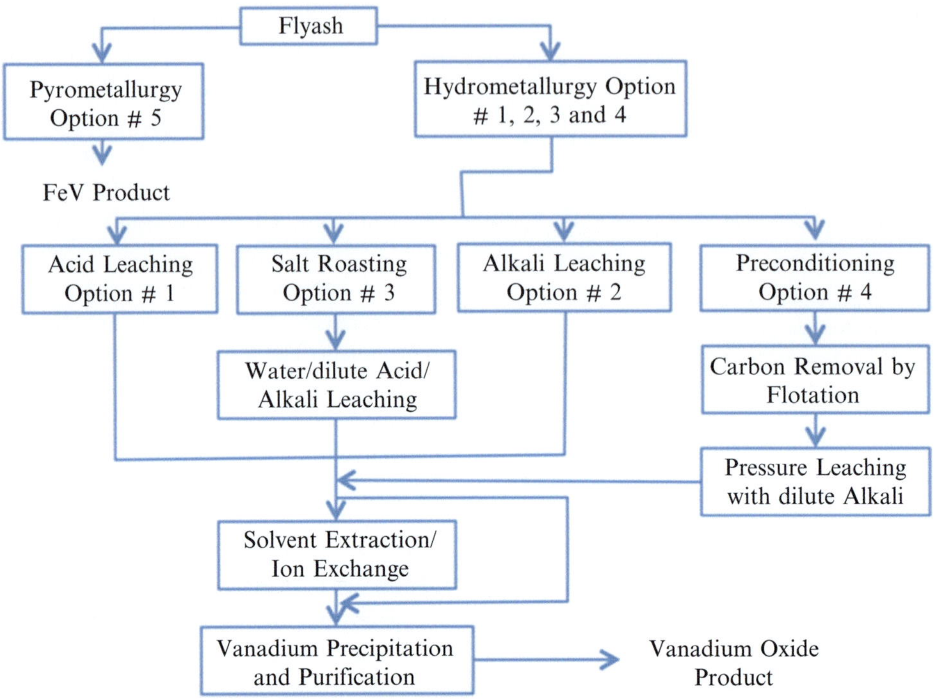

Fig. 8.1 A generalized flowchart of process options for vanadium extraction from flyash

through simplified process flowcharts in Figs. 8.2, 8.3, 8.4, 8.5, and 8.6.

The key performance aspects and associated performance measures are given in Table 8.2.

8.4.1 Description of Process Options

8.4.1.1 Process Option #1: Concentrated Sulfuric Acid Leaching

In Option #1 (Fig. 8.2), the flyash (or the vanadium concentrate) is leached with concentrated sulfuric acid (~98 %) to dissolve the vanadium and form a slurry. After settling/filtration, the vanadium in the clear solution is oxidized to its pentavalent state. The vanadium is precipitated with an agent such as ammonia and the hydrated precipitate is filtered, dried, fused, and formed into flakes. In this process, the leaching is carried out around 50°C and the oxidation and partial precipitation with ammonia are performed between 80 and 90°C in the pH range of 1.7–2.1. The separated precipitate, red cake is dried around 315°C and then fused in a fusion furnace at about 1000 °C in an oxidizing atmosphere to produce V_2O_5. The final flaked product contains typically 98 wt% V as V_2O_5, 1 % Fe as Fe_2O_3, 0.2 % Ni as NiO, 0.1 % Ti as TiO_2, and acid-insoluble materials of about 0.6 %.

Another earlier patent (Vezina and Gow 1968) has discussed concentrated sulfuric acid leaching to produce impure V_2O_5 and carbonate purification to produce ammonium metavanadate. A variation of the sulfuric acid leaching of flyash with precipitation using alkaline magnesium reagents (MgO, $Mg(OH)_2$, $MgCO_3$, or mixtures) followed by solvent extraction of the) leach liquor and vanadium purification steps has been reported in a patent by Pitts (1978).

8.4.1.2 Process Option #2: Concentrated Caustic Soda Leaching

In this process option (Fig. 8.3), an alkaline leach is used which selectively leaches the vanadium leaving iron and silicates in the residue. The

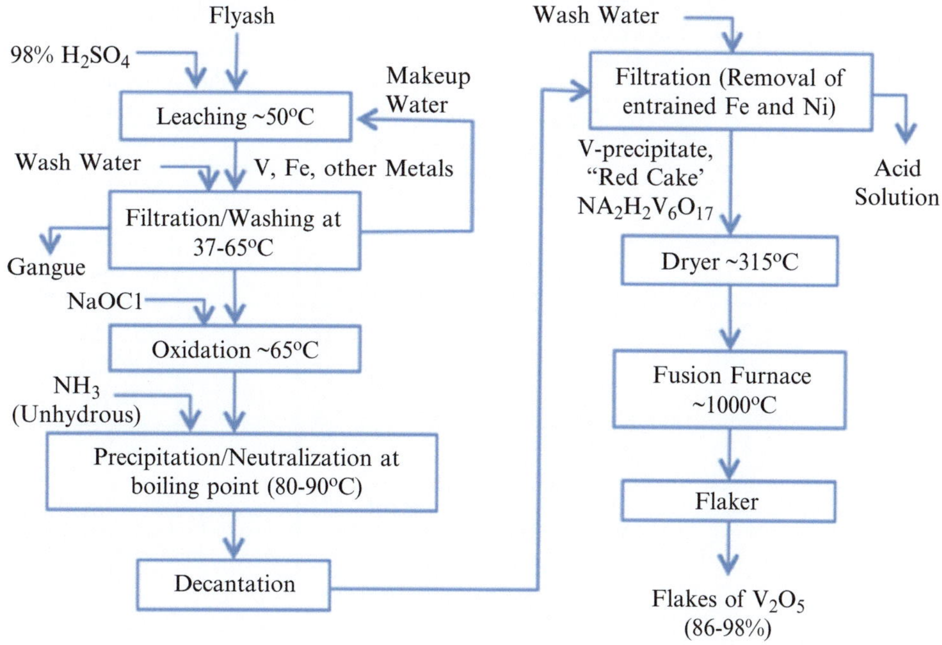

Fig. 8.2 Process flowsheet based on acid leaching of flyash (Whigham 1968)—Option #1

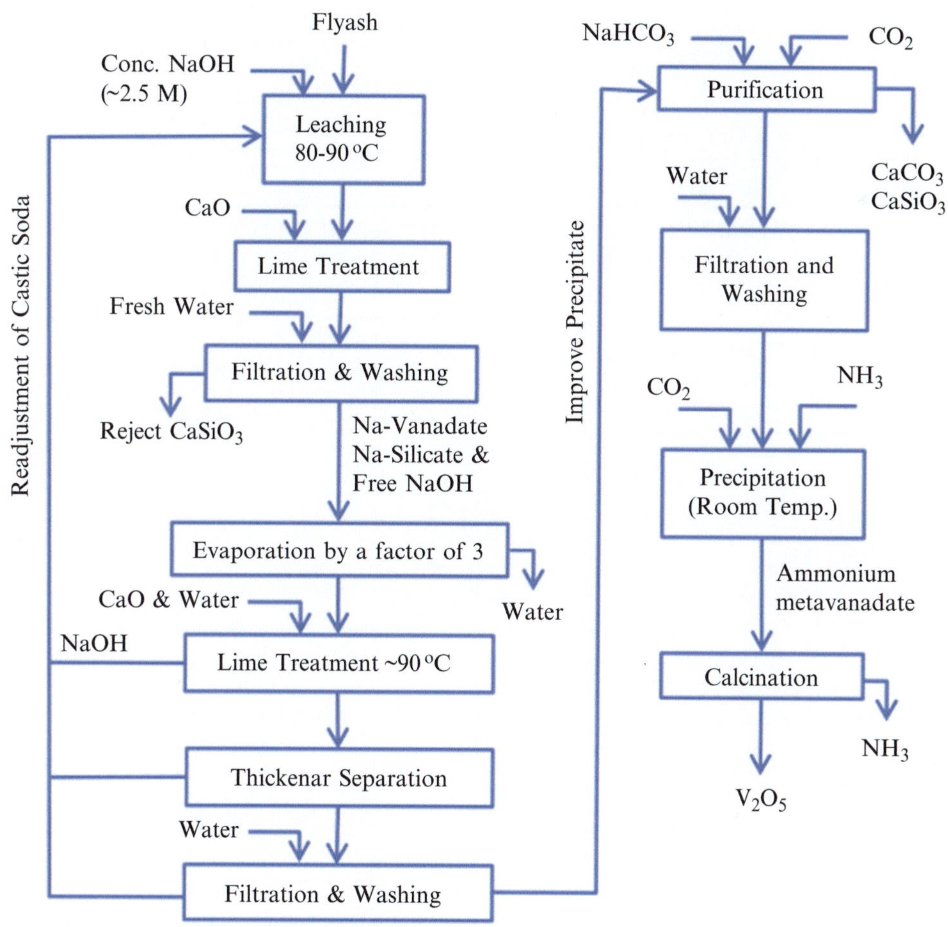

Fig. 8.3 Process flowsheet based on alkali leaching of flyash (Guillaud 1975)—Option #2

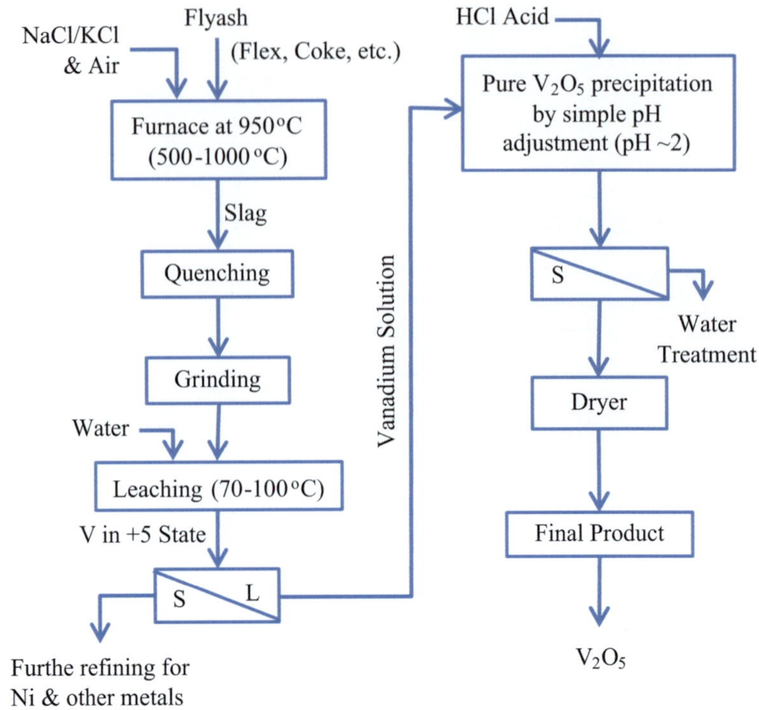

Fig. 8.4 A process flowsheet based on salt roasting and water leaching of flyash (Schemel et al. 1985)—Option #3

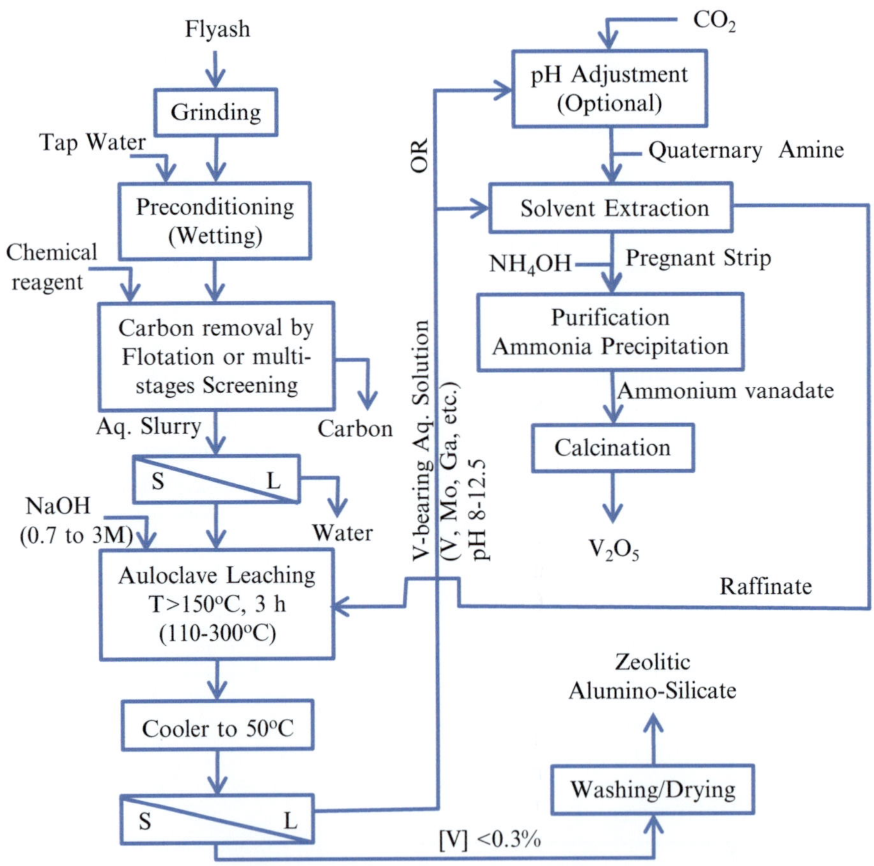

Fig. 8.5 Process flowsheet based on pre-conditioning and physical separation followed by alkali leaching of flyash (Lakshmanan et al. 1989)—Option #4

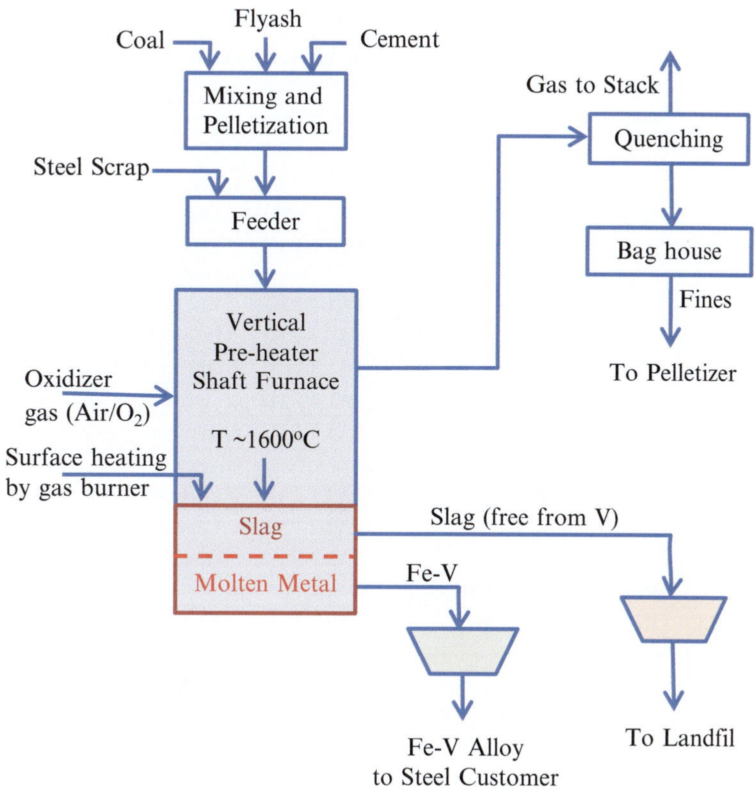

Fig. 8.6 Process flowsheet based on pyrometallurgical separation of vanadium from flyash (Goldfarb and Woodroffe 1997)—Option #5

process is suitable for flyash from different sources containing different amounts of free carbon and vanadium oxides. Here, the flyash undergoes selective leaching with concentrated sodium hydroxide (2–5 mol/L) in a temperature range of 80–90°C. The slurry containing V^{3+} (in solution) and calcium silicates and other solids is treated with CaO at near boiling temperature to soluble calcium vanadate ($V_2O_5 \cdot 4CaO$) and insoluble calcium silicate, and free NaOH. The solids are washed and the mixture filtered. The filtered solids contain all alumina, Fe, Ti, C, Mg, and alkaline salts. The solution containing over 80 wt% V and less than 1 % soluble silicate is volume reduced approximately by a factor of 3 in an evaporator. Excess lime is added to the concentrated solution to precipitate vanadium and calcium silicate. The slurry containing the precipitates is subjected to thickening, filtration, and washing. The separated calcium vanadate and silicate precipitates are treated extensively to produce pure V_2O_5 as the final product. The NaOH filtrates separated from various process steps are combined and concentration adjusted and recycled as reagent for vanadium leaching.

The purification of the mixture of calcium vanadate and silicate precipitates is performed by suspending the solids in an aqueous solution of $NaHCO_3$ and bubbling with CO_2 gas. The precipitated $CaCO_3$ and calcium silicate are rejected by filtration and washing. The filtrate containing vanadium is next reacted with ammonia and CO_2 at room temperature to precipitate ammonium vanadate, which is decomposed in a subsequent step to release free NH_3 and to produce pure V_2O_5. The filtrate containing NaOH and the NH_3 released from heat decomposition are recycled. The process refers to a pretreatment step to eliminate free carbon from the flyash, if necessary to avoid issues related to carbon residue in the process operation, by light roasting or flotation methods.

A variation of the alkaline leaching coupled with a second-stage sulfuric acid leaching is

Table 8.2 Key performance aspects and related measures for metallurgical process selection and evaluation

Performance aspect (PA)		Performance measure (PM)	
PA1	Technical		
PA1.1	Functionality/reliability	PA1.1-PM1	Key process and equipment performance
		PA1.1-PM2	Metal removal efficiency
		PA1.1-PM3	Process control and product quality
PA1.2	Technology maturity	PA1.2-PM1	Stage of process development
		PA1.2-PM2	Previous application of the process within the industry or other related industry
		PA1.2-PM3	Past experience
PA1.3	Design life	PA1.3-PM1	Life-time in years
PA1.4	Operational	PA1.4-PM1	Operator safety
		PA1.4-PM2	Operation monitoring, control and maintenance
		PA1.4-PM3	Waste management
PA2	Financial		
PA2.1	Payout period	PA2.1-PM1	Payback period in years
PA2.2	Life-cycle cost	PA2.2-PM2	Capital
		PA2.2-PM3	Operating (decommissioning and waste management)
PA3	Regulatory/health and safety		
PA3.1	Regulator acceptance	PA3.1-PM1	Environmental impact (human and nonhuman biota)
		PA3.1-PM2	Worker health and safety
		PA3.1-PM3	Ability for the process to meet more stringent regulatory requirements by changes to processing components/steps
PA4	Timelines		
PA4.1	Schedule	PA4.1-PM1	Process implementation for metal recovery
		PA4.1-PM2	Process plant refurbishment after the first design life
		PA4.1-PM3	Decommissioning after final design life

reported elsewhere for vanadium and molybdenum separation from flyash of heavy oil-fired power station (Stas et al. 2007; Stas et al. 2010).

8.4.1.3 Process Option #3: Salt Roasting Followed by Dilute Alkali or Water Leaching

In the published literature there are two patents (Schemel et al. 1985; Griffin and Etsell 1987) and some technical articles (e.g., Long et al. 2014) describe variations of the process involving salt roasting of flyash followed by alkali leaching for vanadium recovery.

This process option (Fig. 8.4) provides a method for removing vanadium and other metals from carbonaceous vanadium bearing materials. The feed material is introduced to the furnace lined with fused alumina refractory and maintained in the range of 750–1000°C. Air is introduced to the furnace at a rate controlled by the size of the furnace and the amount of carbon in the feedstock. The reaction results in the formation of a product which becomes a water-soluble salt. The product is quenched, and the slag is ground and subjected to leaching with water at a temperature of about 70–100°C. The water leaching brings vanadium in its pentavalent state to solution while other metals will be solids. After filtration and washing the solution is subjected to a unique precipitation step to form pure V_2O_5 by simple adjustment of pH by the addition of an acid such as HCl. The precipitate is separated by filtration and washing, and sent for drying, followed by calcination in a furnace at about 950°C for several hours in the presence of air to produce the final V_2O_5 product.

The objective of salt roasting is to render vanadium in water-soluble forms and is normally carried out in a rotary kiln. General retention times are up to 10 h (Ye 2006). The temperature of roasting depends on the type of salt being used. For example, when NaCl is used, the following reaction occurs:

$$2NaCl + H_2O(g) + V_2O_3 = 2NaVO_3 + 2HCl(g)$$
at 800–900°C.

The salt roasting is generally the cheapest and attacks vanadium selectively. The soda (Na_2CO_3) roasting is nonselective requiring relatively higher temperatures (900–1200°C) and is generally used to meet higher environmental requirements.

$$Na_2CO_3 + V_2O_3 = 2NaVO_3 + CO_2(g)$$

For roasting to be effective for vanadium oxidation, the process requirements are (1) free silica <3 %; (2) free lime <1 %; and (3) free oxygen >4 %. It should be noted that excess free lime produces insoluble metal vanadates and free silica forms a low melting oxide complex of Fe, Na, and Si ($Na_2O{\cdot}Fe_2O_3{\cdot}4SiO_2$).

8.4.1.4 Process Option #4: Preconditioning Followed by Carbon Removal and Pressure Leaching with Dilute NaOH

This process option (Fig. 8.5) allows the recovery of vanadium as V_2O_5 and zeolitic alumino-silicates as final products from flyash and related carbon bearing, heat-treated materials. The process steps include (1) separation of carbon by water addition and multistage screening or flotation; (2) pressure leaching of the carbon-depleted flyash with dilute alkali metal hydroxide solutions at elevated temperatures; (3) recovery of vanadium from the leach liquor by solvent extraction, and as an option removal of other value metals present in the leach liquor; (4) precipitation of vanadium from the strip liquor in solvent extraction; (5) drying and calcination of the separated precipitate to produce pure V_2O_5 as the final marketable product; and (6) treatment of the silicates in the leach residue separately to produce a zeolitic alumino-silicates as the second marketable product, adding economic value to the overall vanadium recovery process.

The process chemistry and operating conditions for the key process steps are as follows. The free-carbon separation may be achieved by grinding the flyash and subjecting the particles to multistage screening using conventional equipment or flotation using conventional reagents and equipment (e.g., Aunsholt 1984). The pressure leaching of the flyash with a dilute alkali such as NaOH (0.7 mol/L) or up to 3 mol/L is carried out at temperatures between 100 and 300°C. The leach liquor with a solution pH in the range of 8–12.5 (preferably 8.3–10) is typically cooled to about 50°C and subjected to solvent extraction using a quaternary amine and an oxine in kerosene diluent. Any middle phase between the aqueous and organic phases in solvent extraction is prevented by the addition of a suitable amount of modifiers such as isodecanol. The loaded vanadium in the organic phase is stripped with a sulfuric acid solution. The strip liquor is precipitated with ammonia, filtered, dried, and calcined using conventional process steps. The separated leach residue is washed and dried.

8.4.1.5 Process Option #5: Pyrometallurgical Process for Vanadium Recovery as Fe-V Alloy from Flyash

A pyrometallurgical process option is described in a patent by Goldfarb and Woodroffe (1997), and other variations of the process can also be found elsewhere (Ye 2006; Xiao et al. 2010). For Process Option #5 (Fig. 8.6), the patented process was chosen as the process. The process is a smelting-reduction process for reducing metal oxides, in particular vanadium oxides, present in oil and/or coal ash. The smelting furnace is fired by natural gas with oxygen or oxygen-enriched air. The molten metal layer is masked with a thick surface layer of slag to prevent re-oxidation of the metals.

This is a high-temperature process, which uses two heat sources to heat the ash, coal, and cement pellets fed to the furnace. The main source of energy is delivered by the top fired burners operating with natural gas or oil, and an oxidizer gas (air/oxygen). The secondary source of energy is derived from the combustion of carbon present in the ash or coal. Metal oxides in the mixture (oxides of V, Fe) are reduced to molten metals by the carbon in the slag layer and by other optional, stronger reducing agents such as aluminum. The molten metal gravitates to the bottom layer in the furnace. The overlying slag layer shields the molten metal against oxidation by the oxidation atmosphere of the furnace and the burner jets. The surface temperature of the slag layer is maintained at about 1600°C. The slag constitutes up to 90 wt% of molten materials. The slag layer (about 7–12 cm thick), free of vanadium, is withdrawn, quenched and packaged for landfill disposal. The off gas is cooled with water heat exchangers. The carry-over particulates in the cooled gases are separated in the hot bag house and recirculated in the pelletizer. The molten metal layer is tapped by tilting the smelter. The taped ferrovanadium, Fe-V, alloy is quenched and packaged in drums as product to steel customers.

The pyrometallurgical process is an environmentally friendly process. The slag generated as waste being a stable monolith, free of vanadium, can be readily disposed of without adverse effects to possible water leaching in the landfill sites.

The use of natural gas in the furnace instead of electrical energy (conventionally used) has been claimed to reduce significantly the cost of Fe-V production.

8.5 Process Option Analysis

General Considerations:

- The main marketable products from vanadium recovery are Ferrovanadium (Fe-V) alloy and vanadium pentoxide (V_2O_5). It is assumed that both products have comparable market demand.
- Hydrometallurgical processes for vanadium recovery from flyash are considered attractive due to the relatively lower operating (energy) costs (Mambote et al. 2008).
- The pyrometallurgical process for flyash treatment is an environmentally friendly process. The slag being a stable monolith, free of vanadium, can be readily disposed without adverse effects to possible water leaching in the landfill sites.
- It is important to recognize that the design of a low cost process for value-added product making with attributes such as closed-loop processing, which eliminates any harmful environmental footprints, would be an essential approach for the selection of the best process option for implementation.
- Alkaline leaching of flyash allows the treatment of silicates in the leach residue separately to produce a zeolitic alumino-silicates as the second marketable product, adding economic value to the overall vanadium recovery process.
- The processes selected are capable of producing marketable vanadium products such as regular grade FeV alloy 75–85 % V with high Al (1.5 % max) of low Al (0.5 % max), technical grade V_2O_5 (83–86 % pure), fused black oxide V_2O_5 (86–92 % purity), or 98–99 % or greater purity V_2O_5, depending on the purification route.

8.5.1 Assumptions and Scoring Scale for Performance Aspects

8.5.1.1 PA1: Technical

PA1.1: Functionality/Reliability

The performance measures selected for this performance aspect are (1) key process/equipment performance, (2) metal removal efficiency, and (3) process control and product quality.

The characteristics of acid leaching in Option #1 include (1) high acid consumption, (2) a requirement for acid-resistant equipment, (3) treatment of acidic wastewater effluent, (4) treatment of solid residues before discharge (Long et al. 2014), and (5) nearly all metals present in the source material are leached with vanadium, which would require extensive purification of the solution generated in the process downstream in order to obtain high purity vanadium pentoxide (Lakshmanan et al. 1989).

Alkaline leaching used in Option #2 results in (Long et al. 2014) (1) reduced impurities in the leach solution that is beneficial for subsequent processing, (2) low requirement for corrosion-resistant equipment, and (3) direct discharge of water and solid residues after simple treatment. High concentration alkaline leaching of flyash can recover about 84 % vanadium (Chmielewski et al. 1997). However, the residual ash would be left with about 15 % nickel, which may not be environmentally friendly as such metals can dissolve in water and reach the food chain.

Roasting of the source material such as flyash with sodium salts (e.g., NaCl) is generally preferred for high vanadium content ash. The chemistry can be complicated as a narrow temperature control would be necessary (Guillaud 1975).

The scores assigned to performance aspect, PA1.1, for Process Options #1–#5 are "1, 9, 4, 9, and 9," respectively (Table 8.3).

Table 8.3 Performance aspects and measures with scoring, weighting, and ranking for each vanadium recovery option from flyash

Performance aspect	Performance measure	Process options				
		#1	#2	#3	#4	#5
PA1—Technical						
PA1.1—Functionality/reliability	– Key process and equipment performance	1	9	4	9	9
	– Metal removal efficiency					
	– Process control and product quality					
PA1.2—Technology maturity	– Stage of process development	9	9	9	9	9
	– Previous application of the process within the industry or other related industry					
	– Past experience					
PA1.3—Design life	– Lifetime in years	4	9	4	9	4
PA1.4—Operational	– Operator safety	1	9	4	4	4
	– Operation monitoring, control and maintenance					
	– Waste management					
PA2—Finance						
PA2.1—Payout period	– Payback period in years	4	9	4	9	9
PA2.2—Life-cycle cost	– Capital	1	4	4	4	9
	– Operating (decommissioning and waste management)					
PA3—Regulatory/health						
PA3.1—Regulator acceptance	– Environmental Impact (human and nonhuman biota)	4	4	4	9	9
	– Worker health and safety					
	– Ability for the process to meet more stringent regulatory requirements by changes to processing components/steps					
PA4—Timelines						
PA4.1—Schedule	– Process implementation for metal recovery	4	4	4	4	9
	– Process plant refurbishment after the first design life					
	– Decommissioning after final design life					
Weighted score (equal weight[a])		3.6	5.9	4.3	6.8	8.4
Ranking		5	3	4	2	1
Weighted score (unequal weight[b]		3.7	6.4	4.5	7.9	8.0
Ranking		5	3	4	2	1

Notes: [a]Equal weighting of performance aspects (2.5, 2.5, 2.5, 2.5, respectively)
[b]unequal weighting of performance aspects (4, 1.5, 4, 0.5, respectively)

PA1.2: Technology Maturity

Technology maturity of the process options is to be judged and scored by considering key measures that include (1) stage of process development; (2) previous application of the process within the industry or other related industry; and (3) past experience with the process. In the present evaluation, the chemistry, operation, and equipment of all processes have been used in the industry in one form or the other for metal extractions from source minerals or waste materials. Some plant applications of the processes have occurred in the past for vanadium extraction from waste materials including from flyash. However, it appears that there are no plants producing vanadium from flyash. The assumption here is that all process options are well developed and are available for deployment when necessary. Thus, a score of "9" has been assigned to all five options.

PA1.3: Design Life

The single performance measure for the design life is lifetime in years. The design life of the hydrometallurgical process plants are expected to be relatively greater than the pyrometallurgical process plants. Thus, a high score of 9 is assigned to Process Options #2 and 4, and a score of "4" to Option #5. Similarly, for process complexity reasons, Options #1 and 3 are given each a score of "4".

PA1.4: Operational

Operator safety, and operation monitoring, control and maintenance are important performance measures. It is assumed that all process options have many common safety aspects built within the design and operating procedures. Aspects requiring more attention would be preventative maintenance and control and adequate operator training of autoclaves used for pressure leaching in Option #4, high-temperature salt roasting equipment in Option #3 and maintenance and control related to smelter/furnace, and flu gas treatment system in Option #5. The relatively large footprint of concentrated H_2SO_4 leaching plant (Option #1) is given a low score of "1" for operational aspect, for reasons of (1) large volume reagent usage, (2) relatively larger volume process and effluent stream handling with dissolved toxic metals present, and (3) corrosion issues associated with equipment and components in the process.

The solvent extraction or ion-exchange processing steps used in vanadium recovery from leach solutions, for example, in Options #1 and #4, can result in operational control issues due to third phase formation in solvent extraction operation, and hence loss of efficiency, and similarly a limitation on the availability of selective and high loading capacity resins in ion exchange.

Additional operational complexities of treating off gases from Options #3 and #5 and pressure leaching equipment in Option #4 have resulted in a medium score of "4" being assigned to each of these options (see Table 8.3).

The scores assigned to operational aspect, PA1.4, are "1, 9, 4, 4, and 4" for Process Options #1–#5, respectively (Table 8.3). A high score of "9" for Process Option #2 is mainly to reflect the attractive performance measures related to safety, maintenance and control and waste management.

8.5.1.2 PA2: Financial

The key performance measures are (1) payback period in years, (2) capital cost, and (3) operating costs that include the standard process plant operating cost plus the cost collected now for future decommissioning and waste management.

PA2.1: Payout Period

The payback period is defined as the time to recover the capital investment associated with the implementation of the process option. A relative score of 1 or 9 is assigned for payout period if the time is greater than or less than, respectively, the target payback period of 5 years.

PA2.2: Life-Cycle Cost

Life-cycle cost is an important factor that accounts for capital, operating, and all other costs such as waste management, plant refurbishment, and decommissioning. Determination of the life-cycle cost at the Stage 1 evaluation would be beneficial in the selection of the preferred option.

If sufficient data are not available to perform this cost calculation, it must be definitely done prior to detailed plant design in Stage 2 evaluation. The Option #1 is scored low ("1") on the basis of anticipated high cost of maintenance, high reagent cost (as nonselective reaction between H_2SO_4 and all metal oxides in the flyash), effluent management, equipment corrosion issues, and negative environmental impacts due to nonselective removal of toxic and other metals present in the flyash.

The Option #5 is given a high score of "9" because of the simplicity of equipment and processing steps, and relatively lower operating and waste disposal costs. Considering the attributes for low and high scores for Options #1 and #5, Options #2, #3, and #4 are assigned a medium score of "4"

8.5.1.3 PA3: Regulator/Health and Safety

PA3.1: Regulator Acceptance

The success of a process to meet regulatory acceptance may be evaluated as measured by environmental impact (human and nonhuman biota), worker health and safety, and ability for the process to meet more stringent regulatory requirements by changes to processing components/steps.

Handling of V_2O_5 in plant operation and loss of material to the environment can have negative impacts on humans and animals. Vanadium as V_2O_5 (CAS No. 1314-62-1) has been classified by IARC (IARC 2006) as possibly carcinogenic in humans, with inadequate evidence of carcinogenicity in humans and sufficient evidence in animals. In a study, inhalation exposure has resulted in increased incidence of alveolar/bronchiolar neoplasms in mice and male rats but no human carcinogenicity data has been reported. V_2O_5 is a respiratory irritant and at high doses can cause "boilermaker's bronchitis." Workers exposed to 0.1–0.3 mg/m^3 V_2O_5 for about 6 months have reported symptoms of ear, nose, and throat (ENT) irritations and exhibited signs of pharyngeal infection, etc. (OEHHA 1999).

Ferrovanadium, Fe-V, (CAS No. 12604-58-9) is a noncombustible solid but the dusts from the alloy can form explosive mixtures in air presenting fire and explosion hazard when exposed to heat or flame. The dust from Fe-V can be found in various sizes and it can enter the environment through industrial discharges or spills. Fe-V produces toxic vanadium oxide gas on combustion. It can react violently with strong oxidizers like chlorine (Vincoli 1997). Fe-V is generally stored at low temperature away from oxidizing agents. Health effects include irritation of ENT and affected organs are respiratory system and eyes. Exposure limits in general industry is 1 mg/m^3.

Thus, with regard to health and environmental aspects, Fe-V may be considered to be relatively superior to V_2O_5, but for this analysis both Fe-V and V_2O_5 products are assigned a medium score "4".

Chloride salt roasting (Option #3) can cause serious environmental pollution with relatively toxic gases and waster. In this context, the concentrated H_2SO_4 (Option 1) or high concentrations of caustic soda used in alkali leaching (Option #2) can produce large volumes of solid and liquid effluents, requiring comprehensive effluent management. As a result, a score of "4" has been assigned to Options #1, #2, and #3. The simplicity of processing steps that include the use of relatively less severe chemicals and low effluent generation in Options #4 and #5 has allowed a high score of "9". The details are summarized as follows.

It is expected that the regulator would view several characteristics of the selected process options favorably. However, certain features would be also looked upon not so favorably. For example:

- Option #1 nonselectively dissolves most metal oxides from flyash into solution, creating toxicity issues with regard to liquid effluent discharges. The solid residues require extensive treatments for safe management. Corrosion and spills related too large amounts of concentrated sulfuric acid would pose additional focus on safety of operating staff.

- Option #2 has some selectivity for vanadium leaching and use of caustic soda results in less corrosion and other operator safety issues. Proper treatment of the solid residue free of vanadium allows the production of another marketable product, a zeolitic material, which would minimize the amount of effluents to be managed. However, this aspect is not included in this process option.
- Option #3 involves oxidative roasting of the ash containing free carbon or no carbon at high temperatures, which can emit toxic gases requiring proper off gas treatment. But the leaching of the residue with water or dilute alkali ensures removal of vanadium easily. Equipment control and adequate maintenance are essential components of the reliability of the process.
- Option #4 has a room temperature physical separation step to remove free carbon, if present, and employs pressure leaching at moderate temperatures with dilute alkali. The use of established autoclave technology allows efficient extraction of vanadium. The marketable zeolitic alumino-silicates by-product produced by a simple physical treatment is an added feature that reduces environmental discharges and can offer favorable economics.
- Option #5 is considered to be a favorable process in most performance aspects except design life and operational aspects. The less number of processing steps and a small footprint for easy construction and decommissioning make this option very attractive. Effluent generation is insignificant and the stable slag produced in the furnace can be readily disposed of in landfills.

For reasons discussed above, the performance aspect of regulator acceptance is given a score of "4" with regard to Options #1, 2, and 3, and a "9" for Options #4 and #5.

8.5.1.4 PA4: Timelines

The schedule/timelines for (1) process implementation for metal recovery, (2) process plant refurbishment after the first design life, and (3) decommissioning after final design life are key measures that can have a significant impact on obtaining regulatory approvals, firming up financing arrangements and keeping costs under control. The relative differences for these factors among the various process options can also determine in concert with other performance aspects the selection of the preferred process. In the absence of adequate data to support differences among the five options, a score of "4" has been assigned to all five process options.

8.5.2 Overall Analysis: Ranking of Process Option

The scores are summarized in Table 8.3. If the weights are assigned to the performance aspects, then overall scores can be deduced to arrive at a ranking of the options. If equal weights are adopted for the main four performance aspects (2.5, 2.5, 2.5, 2.5), as shown in the bottom part of the table, Option #4, (a hydrometallurgical process involving a physical carbon removal step followed by pressure leaching with dilute alkali for extraction, and solvent extraction, precipitation and calcination for the purification of V_2O_5) and Option #5 (comprising a pyrometallurgical process to produce FeV as the final product) rank high compared to other process options. To assess the sensitivity of weighting on the performance factors, an arbitrary unequal weighting of the four performance aspects (4, 1.5, 4, 0.5) was chosen. If unequal weights are used and the technical and regulatory aspects are given significantly more importance ("4" each) than other performance aspects with finance as the second most important aspect, then again Options #4 and #5 appear to surface as the most attractive options.

The advantage of the proposed methodology is that the impact on all key performance aspects for the different process options are all considered and supported as much as possible either conceptually or based on past experience (and quantified if adequate data are readily available) and assembled in a single table to enable the effect of weighting to be easily assessed. Also, the effect of a change in any one score on the overall ranking can be easily assessed. This has the benefit of focusing actions to arrive at a short list of the best-preferred options. For example, if technical and life-cycle cost effects can be favorably brought to Options #4 and #5 through innovative developments, even Option #1 could jump to the short list of best options for detailed evaluations in Stage 2 analysis.

Thus, based on assigned weights and estimated scores for the impact on performance aspects, the Process Options #4 and #5 can be short-listed for consideration in the second-stage analysis.

8.6 Path Forward: Next Stage 2 Process Selection Analysis and Implementation

The short-listed options from the first-stage analysis will have to be further evaluated through a detailed calculation of the impacts to arrive at the best process option.

In the second-stage analysis, quantitative information for the various performance factors will be obtained for each of the short-listed process options. For example, sufficiently detailed analysis of the following items will be made. They include (1) performance data for the short-listed processes by tests at a suitable scale to obtain and verify design and operating information, (2) a detailed evaluation of the applicable environmental regulations and its impacts, (3) design of the short-listed process plants, and (4) estimation of the capital and operating costs of the plants including the life-cycle costs. The detailed data will be used to develop appropriate scoring scales for each of the performance measures.

The methodology used for the Stage 1 analysis will be repeated with the new data for the short-listed options. A parametric sensitivity analysis of all performance factors (performance measures and performance aspects) will be carried out with the revised scores in an iterative manner, as necessary.

At this stage, other performance factors should also be considered prior to taking the final decision for the implementation of the selected best process option. The other performance factors may include:

- Capital availability and return on investment tolerances
- Market process swings for the products produced by the selected process
- Market size ranking
- Emerging regulatory and sustainability issues

Some of the above aspects including technology maturity have been discussed recently (King 2014) in conjunction with a retrospective SWOT analysis for new processes for metal production. In addition, certain nontechnical questions that may be asked and accountable in the project implementation (King 2014) may include:

- Why are we doing the project? (Question at the project beginning and end)
- Is this a legacy project? (Determine if political or personal agenda overruling standard technical and engineering practices)
- Is the project repeating history? (Substantiate if the project is different from similar projects which have failed)

The results will be compared between the short-listed process options ,and the best preferred process option will be selected for implementation. It should be noted that the Stage 2 evaluation will require considerable efforts involving process plant design, cost estimation, and impact analysis pertaining to health, safety, and environment. Evidently, there is sufficient incentive to select a minimum number (e.g., 2 or 3) of the short-listed options from Stage 1 analysis. It is for this reason that adequate considerations of all available process options and performance factors should be identified and analyzed without bias or a priori judgement in the Stage 1 evaluation.

As stated earlier in Sect. 8.2, a detailed Stage 2 evaluation and demonstration of the methodology is beyond the scope of the current objectives and will not be pursued further.

8.7 Summary

- A structured first-stage evaluation of five process options for the recovery of vanadium from flyash as the source material to produce a marketable quality product, Fe-V or V_2O_5, was evaluated to select a short list of the preferred options. To achieve this goal a set of performance aspects and the related performance measures were used. On the basis of equal weighting factor for the different performance aspects, and scoring and ranking, two process options from among the five initial options considered were selected for the next-stage detailed analysis and to select the best option for implementation.
- If the weights for performance aspects are selected and fixed, a short list of options can be deduced for detailed evaluation of impacts on the performance aspects to arrive at the best option.
- The vanadium recovery example discussed to illustrate the process selection methodology should be viewed only as an exercise to illustrate the approach. The goal was not to recommend or endorse with special interests any one process.

References

Aunsholt, K. E. H. (1984). Process for the separation of coal particles from fly ash by flotation. U.S. Patent 4,426,282, Retrieved January 17.

Chmielewski, A. G., Urbanski, T. S., & Midgal, W. (1997). Separation technologies for metals recovery from industrial wastes. *Hydrometallurgy, 45*, 333–344.

Dasmahaputra, G. P., Pal, T. K., & Bhattacharya, B. (1998). Continuous separation of hexavalent chromium in a packed bed of fly ash pellets. *Chemical Engineering and Technology, 21*, 89–95.

Fang, Z., & Geaser, H. D. (1996). Recovery of gallium from coal flyash. *Hydrometallurgy, 41*, 187–200.

Goldfarb, V. M., & Woodroffe, J. (1997). Gas-fired smelting apparatus and process. U.S. Patent 5,685,244, Retrieved November 11.

Goldratt, E. (1999). *Theory of constraints*. New York: North River Press.

Griffin, P. J., & Etsell, T. H. (1987). Vanadium recovery from ash from oil sands. Canadian Patent CA 1,221,243 A1, Retrieved May 5.

Guillaud, P. (1975). Process for treatment of vanadium containing fly ash. U.S. Patent 3,873,669, Retrieved March 25.

IARC. (2006). Vanadium pentoxide. *IARC Monographs, 86*, 226–286.

King, M. (2014). The good, the bad and the ugly—Metallurgical project experiences and the lessons to be learned, Recycling of metals from industrial waste—A short course with emphasis on plant practice, June 24–26, sponsored by the Colorado School of Mines. Office of the Continuing Education.

Kuniaki, M., Nishikawa, K., Ozaki, T., Machida, K., Adachi, G., & Suda, T. (1998). Recovery of vanadium, nickel and magnesium from a fly ash of bitumen-in-water emulsion by chlorination and chemical transport. *Journal of Alloys and Compounds, 1–2*, 156–160.

Lakshmanan, V. I. (1992). Emerging technologies in Hydrometallurgy. *Mineral Processing and Extractive Metallurgy Review, 8*, 219–228.

Lakshmanan, V. I., Melnbardis, D., & Geisier, R. A. (1989). Process for treatment of fly ash. U.S. Patent 4,798,709, Retrieved January 17.

Long, S., Feng, Q., Zhang, G., & He, D. (2014). Recovery of vanadium from alkaline leach solution from roasted stone coal. *Science Asia, 40*, 69–72.

Mambote, C. R., Xiao, Y., & Schuiling, R. D. (2008). Vanadium recovery from fly ash—A review. In *Proceedings of XXIV International Minerals Engineering Congress—IMPC 2008* (pp. 3697–3707). Beijing, China: Science Press, Retrieved September 24–28.

OEHHA. (1999). Retrieved from www.oehha.org/air/acute-rels/.

Pickles, C. A., Mclean, A., Alcock, C. N., & Nikolic, Z. N. (1999). Plasma recovery of metal values from fly ash. *Canadian Metallurgical Quarterly, 29*, 193–200.

Pitts, F. (1978). Neutralization. U.S. Patent 4,126,663A, Retrieved November 21.

Schemel, R., Rodriguez, R., & Salazar, R. (1985). Method for leaching and recovering vanadium from vanadium bearing by-product materials. U.S. Patent 4,539,186, Retrieved September 3.

Singh, R. K., & Gupta, N. C. (2014). Value added utilization of fly ash—Prospective and sustainable solution. *International Journal of Applied Science and Engineering Research, 3*(1), 1–16.

Stas, J., Dahdouh, A., & Al-chayah, O. (2007). Recovery of vanadium, nickel and molybdenum from fly ash of heavy oil-fired electrical power station. *Chemical Engineering, 51*(2), 67–70.

Stas, J., Dahdouh, A., & Al-chayah, O. (2010). Extraction and separation of molybdenum and vanadium from alkaline leaching. *Chemical Engineering, 54*(2), 87–93.

Tennankore, K. N., & Vijayan, S. (1999). A structured approach for selecting the best option for active liquid waste treatment technology. In *Proceedings of International Symposium on Radiation Safety Management*, Taejon, Korea Republic, Retrieved November 287–298.

Vezina, J. E., & Gow, W. A. (1968). Method of recovering vanadium oxide. Canadian Patent 783,006, Retrieved April 16.

Vincoli, J. W. (1997). *Risk management of hazardous chemicals* (Vol. 1, pp. 1435–1439). Boca Raton, FL: CRC Press.

Whigham, W. (1968). Production of vanadium values from crude oil. U.S. Patent 3,416,882, Retrieved December 17.

Xiao, Y., Mambote, C. R., Jalkanen, H., Yank, Y., & Boom, R. (2010). Vanadium recovery as FeV from petroleum flyash. *The 12th International Ferroalloys Congress* (pp. 179–188). Helsinki, Finland, Retrieved June 6–9.

Ye, G. (2006). *Recovery of vanadium from LD slag, a state of the art report, Part 1—Facts and metallurgy of vanadium* (JK Report 88031, April 05). Stockholm, Sweden: Jernkontorets Forskning, Stalkretsoppet D81.

9 Metallurgical Processing Innovations: Intellectual Property Perspectives and Management

Corale L. Brierley and Peter D. Kondos

9.1 Introduction

Competition among mining companies focuses on cost-effectively out-producing the other. In the past competition has largely centered on accessing and developing the most favorable ore bodies. However, successfully competing is particularly challenging given the new realities of declining ore grades, increasingly complex ores, deeper ore bodies, lack of infrastructure and a trending sustainable development imperative to reduce water and energy usage, decrease waste production and minimize the carbon footprint. To remain competitive by overcoming these challenges necessitates an aggressive innovation approach (Hedges and Lloyd 2010; Stacy 2012; NRC 2002) toward metallurgical processing.

Metallurgical processing has constantly undergone changes as companies and organizations comprising this segment of the mining sector introduce uses for technologies developed for other applications (NRC 2002) and modify existing unit processes. These innovations are generally aimed at incremental or evolutionary, short-term (less than 3 years), operational improvements to increase productivity, reduce operating costs, and reduce capital expenditure (Stanway and Andrew 2014; Thompson 2014). Occasionally, however, a transformative, or revolutionary, discovery is made that fundamentally alters the way some aspect of metallurgical processing is done and which has a significant economic benefit.

Innovations in metallurgical processing emanate from many sources: mining companies; equipment manufacturers; chemical suppliers; engineering technology and services companies; government laboratories; metallurgical contract laboratories; universities; high technology firms; and independent consultants. Many evolutionary and revolutionary discoveries made by these entities are held by the inventors as intellectual property. In some cases, however, details of the innovations are published in the public domain, which diminishes the likelihood of others patenting the new process, machine or improvement.

This chapter discusses why innovation is vital to sustain a vibrant metallurgical processing industry; considers the research and development (R&D)/technology/innovation continuum; examines sources of innovations in metallurgical processing and strategies employed by various sectors of the industry to innovate and protect intellectual property; and looks at the challenges and barriers to develop and protect processing technology.

C.L. Brierley (✉)
Brierley Consultancy LLC, Highlands Ranch, CO, USA
e-mail: clbrierley@msn.com

P.D. Kondos
Barrick Gold Corporation, 161 Bay Street, Suite 3700, Toronto, ON, Canada, M5J 2S1

V.I. Lakshmanan et al. (eds.), *Innovative Process Development in Metallurgical Industry*,
DOI 10.1007/978-3-319-21599-0_9

9.2 Importance of Innovation to Metallurgical Processing

9.2.1 The Innovation Imperative

For mining to remain a vibrant industry in the face of intensifying challenges (Gleason 2015) innovation is not a luxury, but a necessity. Metallurgical processing imperatives the mining industry must accomplish are:

- *Effectively treat low ore grades*: It costs more to process low grade ores, because more material must be handled, requiring more energy and costs per unit of value. To cost-effectively respond to this challenge innovations are needed in materials handling equipment, sensors allowing more selectivity so less rock is moved, and metallurgical processing methods that are more productive (Thompson 2014).
- *Process mineralogically complex ores*: To successfully process complex ores requires new metallurgical processes including less energy intensive comminution methods, advancements in liberation technology, innovative physical separation technologies, novel processing methods that use nontoxic, efficient lixiviants, and improved solution purification and concentration technologies (Gentry 1998; NRC 2002).
- *Access deeper ore deposits*: These deposits can no longer be cost-effectively mined by conventional open-pit or underground mining techniques because of depth. In situ mining is an alternative and, although this process is principally used for extraction of water-soluble salts, brine, sulfur, and uranium (World Nuclear Association 2014), it has been demonstrated for copper oxide leaching (USBM 1989; NRC 2002). In situ mining is currently proposed for at least one copper oxide mine site. For in situ mining to be effectively employed for the deep deposits currently being identified as resources, the technology must be applicable not only to metal oxides but also to sulfide ores. This presents unique technological challenges that can only be met with innovation (Gentry 1998; NRC 2002): fracturing and rubblizing methods need to be perfected to employ at depth; techniques must be developed to increase the permeability of the fractured rock enabling effective contact of leach solutions and microorganisms with metal-bearing sulfide minerals; sensors for real-time monitoring of rock fracturing and solution flow are needed to monitor reagent effectiveness and solution flow at depth; and operational control technologies to protect the environment are required. Some of these advancements may be adapted from other applications such as hydraulic fracturing and horizontal drilling currently used by the oil and gas sector. Others may eventually be adopted as developments in advanced materials engineering, robotics and electronics emerge.
- *Operate in remote areas with a lack of infrastructure*: The remote location of deposits and lack of accessibility, power, and trained personnel impose limitations on the size and complexity of processing operations. Consequently, these remote locations have excessively high capital and operating costs. Innovations relating to materials transportation, waste rock removal at the source, and the start-up of small, modular processing plants are necessary to cost-effectively initiate and operate remote plants.
- *Diminish water use and treat saline/brackish water*: Hydrometallurgical processes require large quantities of water (Roberto et al. 2013). Most of the world's largest crushed ore heap leach and stockpile leach operations are located in high desert regions (Brierley and Brierley 2013; Schippers et al. 2014) where evaporative losses are high, water is scarce, and salinity of existing groundwater supplies is an issue. Water conservation measures, such as heap covers that reduce evaporative losses and desalination technology to treat poor-quality groundwater and seawater, are currently used. However, additional technology is needed to reduce water usage, enable greater recycling of process water, treat poor quality groundwater, improve and reduce the cost of

desalination and routinely use brackish or saline water in metallurgical processing.

- *Minimize energy use*: Metallurgical processes generally have high energy requirements (Roberto et al. 2013). Among the processing activities, comminution, pressure leaching, and electrowinning are the most energy intensive, but also consuming energy are pumping, agitation, drying, and material movement including air mass transfer. The biggest gains in energy reduction would likely come from innovations in comminution, such as selective blasting (Thompson 2014). Innovations in energy management, such as fuel substitution, renewable energy use and integration, electrical grid optimization, variable speed motor use, and plant energy use efficiency, must become standard features in modern operations.
- *Reduce waste*: Advances have been made in disposal of wastes produced from mining and metallurgical processing activities. Mine backfill of tailings is possible with developments in thickening. New uses have been discovered for some tailings (e.g., construction materials from clay-rich tailings and manganese tailings; raw material for glass, ceramics and bricks from bauxite red mud). Slag is used for road construction and in concrete and cement. Water treatment sludge that is high in iron is used in pigments. Sulfur emissions from smelters are converted to sulfuric acid for other industrial uses. However, new uses for metallurgical wastes could be discovered with research and development.
- Other approaches to reducing waste are: development of novel hybrid flow-sheets merging physical, chemical, and bioprocessing methods to avoid repetitive stages of the same process; discovery of selective leaching agents for dissolution of desired metals while minimizing the solubilization of impurities; and advances in separation technologies that avoid disposal of valuable or problematic constituents (Gentry 1998). Major progress in waste reduction will happen in mining with a changeover from open-pit/surface mining to underground mining, with a focus in mining all the ore and only the ore, minimizing dilution. This focus will result in smaller crushing and grinding plants, reduction of energy consumption and waste generation. In situ mining, when developed and perfected for comprehensive use with metal sulfide ores, will have a major impact on decreasing waste from metallurgical processing (NRC 2002).
- *Continue to improve health and safety*: Health and safety are already the highest priorities in mining and metallurgical operations and improvements are always being pursued. Innovations in automation would limit the number of workers in metallurgical processing thus improving safety. New nontoxic lixiviants would provide a healthier and safer work environment (NRC 2002; Thompson 2014). In situ mining and metallurgical processing in the subsurface would eliminate many health and safety hazards that currently exist in aboveground processing.
- *Address increasing global population uncertainties*: A growing global population, expected to reach 9.6 billion by 2050, poses an opportunity as well as a threat for mining (NRC 2014). The burgeoning population envisions a higher standard of living than they currently have, as more and more people join a global middle class. The demand for subsurface resources will escalate to keep pace with the manufacturing of goods and products for the expanding, global middle class. There is an expectation that the mining industry can meet this demand.
- Mining activities currently disturb about 1 % of the earth's surface (Gleason 2015), but as population mushrooms, it is expected that people will increasingly encroach on existing and future mine sites. This imposes greater social responsibility and environmental awareness on the part of mining enterprises (Gentry 1998) as the industry faces increased scrutiny exacerbated by the influences of social media (Gleason 2015). An important element of securing a social license to operate is to transform through technology. Innovations are necessary to appreciably minimize the footprint of mining operations. Moving metallurgical processing to the subsurface is one option to accomplish this.

- *Reduce carbon footprint*: One focus of the sustainable development imperative is reducing the carbon footprint of mining operations. Innovations that decrease energy use in metallurgical processing will have a substantial bearing on this aspect of sustainable development (Hedges and Lloyd 2010; Adey et al. 2011). Furthermore, the increased use of renewable fuels and energies, the ability to minimize the use of limestone by reducing the generation of acid throughout the process and innovative processing of high-carbonate ores will achieve a major reduction in greenhouse gases.

9.3 Sources of Innovation in Metallurgical Processing, R&D/Technology/Innovation Continuum and the Role of Intellectual Property

9.3.1 Sources of Innovation

Evolutionary innovations in metallurgical processing tend to come from within the mining industry itself. These are typically stepwise developments in existing unit processes (Batterham and Algie 1995; McNulty 1998) and also include the introduction of novel ways of combining unit operations. Most innovations in metallurgical processing are evolutionary.

Revolutionary or disruptive/transformative innovations, which are less common, are more likely to come from equipment manufacturers, chemical companies, engineering technology and services suppliers, government laboratories, metallurgical contract laboratories, universities, high technology companies often supported by venture capital, and occasionally independent consultants (McNulty 1998; Brierley 2008) or from industries other than mining. There are several reasons why innovations come from these entities: large equipment manufacturers and chemical and engineering technology suppliers typically have large R&D budgets—10–100 times those of mining companies on a percent of revenue basis (Stanway and Andrew 2014); for most of these businesses new products, new equipment and novel services are central to their business existence; and for government laboratories and universities R&D and innovation are fundamental mandates for existing. Operating mining companies, however, occasionally discover and implement transformative, innovative technologies. The introduction of revolutionary innovations into metallurgical processing is especially risky, because of the unproven nature of novel technology, the length of time required to substantiate the benefits of the innovation, the very large scale of most metallurgical operations, and the disruption to operations that introducing an innovation may cause (Marsden 2004; Thompson 2014).

9.3.2 R&D, Technology, and Innovation in Metallurgical Processing

"Innovation," "technology," and "R&D" are distinctive, complementary elements in the integrated process of improving metallurgical processing.

Innovation is the process of bringing a new idea, new product, new method, etc. to an application, adding significant value. An innovation is sometimes referred to as an invention, a narrower term which in US patent law is defined as a new, useful process, machine improvement, etc. (Dictionary 2014a). An innovation doesn't have to be new to the world only new to the enterprise or operation that applies it and it doesn't necessarily have to be a technology (Batterham and Algie 1995). Innovations in metallurgical processing can be described as discovering something new, which may be an entirely novel process, a new development within an established unit process, a new type of processing equipment or novel improvements in existing processing equipment (McNulty 1998). It may be simple as a small change in process conditions that leads to significant improvements in the process. In a broader context innovation can even involve a new business model that comprises commercial, organizational, and technological elements

(Stanway and Andrew 2014). Innovation is what gives a mining company a competitive advantage (Batterham and Algie 1995).

Technology, on the other hand, is the application of engineering or applied science knowledge for practical purposes (Google 2014a; Dictionary 2014b). Innovations that have been developed for metallurgical processing by applying engineering or applied science principles are usually referred to as "technical or technological innovations."

Research and development (R&D) in industry is work directed toward innovation and the introduction and improvement of products and processes (Google 2014b). R&D and innovation are integral parts of the same continuum, although innovation can take place in the absence of R&D.

Some metallurgical processing examples serve to clarify the definitions of innovation, technology, and R&D:

- *Flash-converting*: Development of Flash-Converting was carried out via a joint development effort, in which Outotec's (previously Outokumpu) flash smelting technology was combined with Kennecott Utah Copper's (KUC) solid matte-oxygen converting innovation (George and Richards 1983) to create the Kennecott-Outotec Flash-Converting technological innovation. This innovation set a new global benchmark for converting efficiency, energy effectiveness, environmental compliance, and operator safety. This technological innovation has become a global commercial success through a joint Kennecott-Outotec licensing agreement. Flash-converting is a technological innovation—an entirely new process—that emanated through R&D and technology contributions from two companies—a mining company and a technology company.
- *Pressure oxidation to leach copper sulfide concentrate*: Pressure oxidation technology that uses high temperature and high pressure to oxidize sulfide minerals to sulfate and sulfuric acid is not new because there have been industrial plants in operation in the USA since the 1950s. However, Phelps Dodge (now Freeport McMoran, Inc.) revived pressure oxidation technology in the early part of the 2000s at the company's Bagdad, Arizona operation to treat copper concentrates. The process stream from the autoclave is combined with effluent solution from stockpile leaching and directed to the solvent extraction/electrowinning (SX/EW) circuit for copper recovery. The acid generated from the autoclave process is recycled to the stockpile via the SX raffinate to derive the acid credit (Dreisinger 2006). This is an innovation because it represents new developments within established unit processes—the combining of pressure oxidation, stockpile leaching, and SX/EW—to add value.
- *Continuous stirred-tank biooxidation of base and precious metal sulfide concentrates*: In the late 1970s Gencor Process Research, Johannesburg, South Africa, researched and developed biological processing of gold concentrates in which the precious metal was embedded in a sulfide mineral matrix. The biooxidation process releases the gold from the matrix enabling significant improvement in precious metal recovery using cyanide. The process was scaled-up and pilot tested at the Fairview Mine in South Africa, leading to the development of the BIOX™ technology (van Aswegen et al. 2007). The technology was subsequently transferred to Billiton plc during a corporate reorganization and licensing of the BIOX™ innovation was initiated. R&D continued to develop the process for base metal sulfide concentrates resulting in the BioCOP™ and BioNIC™ technologies. During another reorganization the BIOX™ technology was transferred to Gold Fields plc where licensing continued. The BioCOP™ technology was transferred to BHP-Billiton where the Alliance Copper Joint Venture, established between BHP-Billiton and CODELCO, demonstrated the BioCOP™ technology near Chuquicamata, Chile, in a 20,000 tonne/day copper cathode plant (Dreisinger 2006). The BIOX™ technological innovation is now licensed to mining companies throughout the world by BIOMIN South Africa (Pty) Limited, the sole owner of the technology. BIOMIN also licenses base

metal sulfide concentrate bioprocessing. BIOX™, BioCOP™, and BioNIC™ illustrate a sustained research and development effort, are extensively patented and are trademarked. BIOX™, which has been in commercial use since 1986, is an engineered industrial process representing a new metallurgical processing method and is an innovation because it generates value in its application.

9.3.3 Intellectual Property in Metallurgical Processing

Common types of intellectual property (IP) rights include patents, copyrights, industrial design rights, trademarks, trade dress, and, in some jurisdictions, trade secrets. Industrial design rights, which protect the visual design of objects (Wikipedia 2014a), and trade dress, which refers to visual appearance characteristics of a product or its packaging or the design of a building (Wikipedia 2014b), are not common types of intellectual property in metallurgical processing. The metallurgical processing sector typically protects innovations with patents, copyrights, trademarks, and trade secrets/confidential information, as illustrated in Sect. 9.3.2. Trade secrets, sometimes referred to as confidential information, include an invented formula, practice, process, design, instrument, pattern, commercial method, or compilation of information which is not generally known or reasonably ascertainable by others and can provide an economic advantage over competitors (Lin 2012; Wikipedia 2014c). Trade secrets, however, do not protect against independent discovery by others or from inadvertent disclosure. Trade secrets do not provide a company with exclusive rights to a technology or innovation (Ferraloro et al. 2012).

Some mining companies choose to fully disclose innovations by publishing details of the invention in a widely disseminated, international technical journal. Reasons for publishing this information vary, but can be because the company wants to safeguard its rights to use the innovation without concern that it might be discovered and patented by a competitor or some other organization. The inventing company may not have the resources—financial or otherwise—to patent the innovation and protect the IP going forward. The principal risk in disclosing the innovation through publication is that some detail may have been omitted or the information that was published may give others ideas for patentable concepts or ways of circumventing the published invention.

9.3.4 Competitive Advantages of Innovation and Intellectual Property

Companies that successfully implement new developments to meet emerging challenges benefit by achieving a competitive advantage within the mining sector. They accomplish this by effecting certain strategies (Marsden 2004; Hedges and Lloyd 2010; Ferraloro et al. 2012; Stanway and Andrew 2014).

Protect the innovation from competitors. This is typically accomplished by patenting the innovation whether it is a processing technology, equipment, reagent, or software. Patenting protects the innovation for up to 20 years and provides a competitive edge. Typically not every detail of the innovation is included in patent applications and issued patents, as some know-how, operational expertise, and other information are retained as trade secrets and confidential information. However, in the mining industry it is difficult to retain trade secrets and proprietary information for long periods because of movement of people among companies, the generally open communication even among competitive companies, and in some cases alliances between competitors on specific projects and operations.

Apply the innovation more rapidly and better than competitors. A significant competitive advantage can be realized by being first to employ a new innovation, particularly one that offers increased productivity and reduced capital and/or operating costs. However, the potential downsides of being first are the issues often associated with implementing new technologies at scale, which may be avoided by being a "fast follower" when it comes to incorporating innovations (Marsden 2004). It is possible to mitigate some problems associated with implementing new technologies by completing a detailed evaluation of the mineral resource, having highly qualified technical people and a committed senior management, developing a realistically implementable flow sheet for processes, confirming the innovation does not infringe on other patents (freedom to operate), and assuring the innovation is environmentally compliant (Canterford 2011). Patenting technology does not give a right to use that technology, but rather gives a right to prevent others from using it, hence the need to ensure that there is freedom to operate technology being adopted.

Use innovation for a greater proportion of overall resource production than competitors. The more broadly the innovation is applied across the company's operations, the greater will be the advantages, be those increased production or greater cost savings (Marsden 2004).

Achieve more value than competitors. This may be attained by employing an innovation on a resource that doesn't respond well to existing technology, so more of the resource can be processed by the innovation or more metal can be produced. Greater value can also be attained by synergies between existing processes and new technologies. The by-product from the innovative process applied to a mineral resource may be beneficial for an adjacent resource or process (Marsden 2004). This is exemplified in the pressure oxidation/SX-EW/stockpile leach innovation described in Sect. 9.3.2.

Motivate people. Implementing innovations excites and inspires technical and operating employees, particularly when the technology is strongly supported by senior management of the mining company. Employees feel empowered and know they are creating value (Marsden 2004).

Provide value through business partnerships and increased share prices. The ownership of a novel technology can generate significant value in a company by creating a sizeable interest from the broader industry and providing opportunities for business partnerships. It is recognized in investment circles that there is a close relationship between innovative companies and high profitability; this is clearly reflected in the share price of innovative companies compared to peers.

9.4 Overcoming Barriers to Intellectual Property Development and Technology Implementation in Metallurgical Processing

The need for innovation in metallurgical processing is recognized as vital to the success of the mining industry (Stanway and Andrew 2014; VCI 2014; Gleason 2015). Despite this realization, there are numerous barriers or impediments associated with R&D, technological development, the introduction of innovation into metallurgical processing, recognizing innovation, and protecting inventions. These barriers impact not only mining enterprises but also companies and organizations involved in developing new technologies for metallurgical processing. There are, however, approaches that can be taken to overcome these barriers.

9.4.1 Barriers to Metallurgical Processing Innovation and Intellectual Property Development

There are various obstacles to innovating in metallurgical processes and protecting the inventions that emanate from technological developments. Some of these barriers are discussed below.

Introducing innovations into large scale operations. Introduction of novel processes into metallurgical operations that involve tens of thousands of tonnes of material daily is daunting. The risk to the mining company of introducing something new that has large infrastructure requirements and costs is often deemed too onerous to even be considered by operators (Stanway and Andrew 2014; Thompson 2014).

Cost of introducing an innovation. The capital invested in major operations is very large—hundreds of millions of dollars is commonplace—so introducing an innovation into such a capital-intensive business is often viewed as too risky (Thompson 2014). The introduction of revolutionary/disruptive/transformative innovations into metallurgical processing is particularly risky because of possible process upsets and disturbance to the overall mining operation. Risk can be managed, but this takes time, additional costs, and intellectual efforts (Marsden 2004).

Demanding physical condition. The metallurgical processing environment is demanding and the risk of failure of a new process in this environment is high. Consequently, operators are reluctant to take a chance on something that may fail (Thompson 2014).

Proving new technologies at scale is difficult. Some organizations, particularly universities, small high technology companies, and government laboratories involved in innovative metallurgical processing technologies, may have tested the new process at small scale on a limited number of ores or on nonrepresentative samples. Consequently, the innovation may be of questionable technical and commercial value. These organizations typically have no access to a mine site to pilot test or demonstrate a technology for metallurgical processing. They often have limited finances, technical expertise, and abilities to achieve commercial status of their technological development. Consequently, these organizations become dependent on selling know-how rather than metallurgical processes (Brierley 2008; Canterford 2011).

Collaborating with a mining company early on in the innovation process is prudent for universities, high technology companies, and government laboratories. The mine operator can provide representative samples, process water (when important to the innovation) and important input on flow sheets during the R&D phase. Because of the involvement, the operating company may have sufficient interest, confidence, and understanding of the value of the innovation to want to pilot test the process on-site. However, collaborating doesn't necessarily guarantee the operating mining company will evaluate the innovation in the field because doing so may disrupt production, be too costly, or, in the mining company's view, simply not worth the effort and risk (Brierley 2008; Canterford 2011; Thompson 2014).

Marginalizing innovations of small companies. Mining companies are more likely to turn to off-the-shelf technology rather than risk incorporating innovative products and services from small companies. The reticence largely stems from concern by mining companies that small companies lack the financial and technical resources to successfully roll out innovations in large mining operations (VCI 2014). By the same token the expectations of some small high technology companies may be too high (Brierley 2008). Lack of crucial insights of the metallurgical processing sector can blind high technology companies to the real value of their innovations (Stacy 2012). The innovation may be one unit process among several and placing a value on that one innovation can be difficult. High technology companies may have outside investors, who demand a rapid and high return on their investment. Consequently, the time frame to incorporate a new innovation into a metallurgical operation and the real value of the innovation may not be aligned with the expectations of the investors, particularly investors accustomed to the high returns of inventions introduced into the biotechnology and information systems sectors (Brierley 2008).

Resisting innovative approaches from outside the industry. The mining industry is not completely comfortable with or accustomed to seeking new technologies or ideas from outside of the industry (VCI 2014)—a concept known as "open innovation." Open innovation in the mining sector entails looking for external ideas well beyond the typical suppliers of metallurgical processing technology and services. Open innovation, however, is used by the parallel resource sector—petroleum—and by other industrial sector companies. Possible reasons why the mining industry is uncertain about open innovations may be related to questions about intellectual property, the value of the innovations, and the lack of understanding about how, when, and where to use the innovation (VCI 2014). There may also be a cultural barrier—that is, the mining industry, unlike many other industrial sectors, is simply not accustomed to looking beyond its own domain for novel ideas.

Quantifying the benefit of a new technological innovation can be difficult. Production data are erratic because of changes in ore grade and mineralogy. Consequently, it may be difficult to detect and quantify improvements achieved by a new technology (Napier-Munn 1998).

Innovating within an operating mining company presents a unique set of challenges. The boom and bust, cyclical nature of the mining industry is a serious detriment to pursuing technological development endeavors within a mining company. When metal prices decline, R&D and innovative pursuits are abandoned in cost-cutting measures and technical expertise is lost in layoffs (Gentry 1998; Stanway and Andrew 2014). Layoffs sometime mean that competitors will gain from the R&D, when the individuals involved in the technological development move to another company, which capitalizes on the knowledge.

Mining companies often don't align the R&D process to an overall strategy to increase profitability, don't phase the innovation process appropriately to capitalize on interim successes and don't always manage R&D outcomes well. If the innovation process doesn't have an in-house champion within the executive leadership ranks, preferably the company's CEO, the effort is destined for failure. The R&D effort will eventually lose alignment with the company's strategy and the effort will lack support at the top level of the company when there is a business downturn (Napier-Munn 1998; Ferraloro et al. 2012; Stanway and Andrew 2014; VCI 2014).

In some cases mining companies and other businesses involved in innovating for metallurgical processing don't recognize the value of their discoveries and consequently don't protect it adequately. Learning to identify and value an innovation are important skills that need to be acquired (Stacy 2012; Taberer 2012).

Other difficulties in implementing metallurgical innovations in the mining sector revolve around: *change in management, lack of support from management at the mine site, or both; technologies not ready for implementation; and lack of support from the innovator* (Stacy 2012).

9.4.2 Overcoming Intellectual Property and Innovation Roadblocks in Metallurgical Processing

"Transformational innovation" is broadly recognized by mining executives and others as key to the success of the industry to effectively operate in the future (Stanway and Andrew 2014; Thompson 2014; Gleason 2015). Successful introduction of innovations and management of intellectual property necessitate overcoming the barriers outlined above. Approaches to surmounting these impediments related to metallurgical processing innovation and intellectual property are achievable, but many approaches require important changes in the way the mining industry does business.

Organize and drive innovation from the top. The CEO of a mining company must be the commanding leader and champion of R&D for innovation to be successful within the company. The CEO and the executive leadership must link the development of innovations to the company's overall business strategy and a future vision. The executive management is responsible for protecting R&D efforts from business down cycles that often eliminate such programs and to spread the future vision by engaging and empowering all parts of the organization to implement new developments (Batterham and Algie 1995; Ferraloro et al. 2012; Stanway and Andrew 2014; VCI 2014).

Develop strategic partnerships. Collaborations with other mining companies and R&D organizations, such as universities, for complex metallurgical processing challenges allow for shared risk (McNulty 1998; Thompson 2014) and open the door for new developments. Such collaborations also assist the R&D organizations to better understand the metallurgical processing needs of mining companies and environments that innovations will have to perform within. Collaborations among mining companies, however, are only feasible in noncompetitive spaces.

R&D and innovation are central to the livelihood of many companies that supply products and services for metallurgical processing. These suppliers focus on the most important problems in metallurgical processing, such as reducing energy and water use, to retain their competitive position. Mining companies should be looking at strategic partnerships with these companies to innovate technologies for both the short- and long-term that respond to the most onerous challenges facing metallurgical processing.

Multi-partner collaborations require a greater management effort than in-house R&D efforts, so it is important for all partners involved to agree on a strategy and R&D approaches, carefully frame the objectives of the partnership, align the leadership (Stanway and Andrew 2014), and establish ownership rights and privileges of technological developments. Equally important is to design a plan capturing interim ideas to measure progress rather than awaiting some ultimate technological development that may or may not occur. R&D efforts occasionally result in a "black swan"—a surprise innovation or idea of significance, whose benefit may not be understood or appreciated until well after the discovery. Appropriately overseeing R&D collaborations to keep things on track and capitalize on unexpected developments is a priority.

Implement different approaches to deliver R&D results and acquire innovations. Capturing the benefits of innovation, whether it is conducted in-house by mining companies, through university-mining company consortia, or by strategic relationships with large or small technology and services companies, requires a variety of methods: reports accompanied by an opportunity to interact; workshops; simulators; new processes and hardware; technology demonstrations; training courses; consulting; research staff joining the company to transfer expertise; and interactions with third party suppliers, who profit from successful technology transfer (Napier-Munn 1998). To create value from these technology transfer methods the R&D has to be aligned with the overall business strategy of the mining company.

Capturing value from "open innovations"—new technologies and ideas from outside the industry—can be achieved by allowing a free-flow of ideas from employees and others and collaboration with suppliers (VCI 2014). The mandate of suppliers is to stay competitive by innovating, applying unconventional ideas to persistent and intractable problems and solving problems with solutions used by disparate industrial sectors.

Recognize and manage intellectual property. It is important to nurture an environment that recognizes and protects intellectual property. Nurturing includes training employees to identify legitimate inventions (Stacy 2012; Taberer 2012) especially focusing on processes, techniques, and methodologies when developing IP in metallurgical processing (Ferraloro et al. 2012). Engineers and scientists tend to downplay their ideas, believing solutions to be good engineering or good science and focusing on problem solving and not giving thought to the innovation they just developed (Ferraloro et al. 2012). Engineers and scientists may also overestimate the effort required to patent an innovation and underestimate the benefits of patent applications (Hedges and Lloyd 2010). In many cases the cost of protecting innovations by patenting is small in comparison to the revenue generated through application of the innovation (Ferraloro et al. 2012). Patenting should also be viewed by a mining company and its employees as a way to prevent competitors from patenting the process you developed and are using. Filing a patent provides a defense against potential infringement allegations (Hedges and Lloyd 2010) and creates a strategic domain, within which the company operates. As important as IP is to the mining industry, filing patent applications on flow sheets or processes that are not technically or commercially viable should be vigorously discouraged (Brierley 2008; Canterford 2011) and the legal hurdles of patenting in bioprocessing must be well understood, because there are unique IP considerations related to this sub-field of metallurgical processing (Taberer 2012).

Identify and implement approaches for pilot testing that limit disruption and risk. De-risking a metallurgical innovation is crucial to successful implementation. This can be accomplished through pilot plants, demonstration plants, and small-scale applications before implementation of the innovation into large-scale operations. Parallel studies in heat and mass balance and financial modeling are critical to identify fatal flaws and recognize opportunities for improvement.

Identifying champions within operations and providing incentives to individuals for taking risks may also offer additional opportunities to scale-up new technologies. Collaborative opportunities with other mining companies would share the risk associated with new technologies (Thompson 2014). Adopting "open innovations" from outside the mining industry may also offer practices and products that have been used in other industrial sectors; while these innovations would still require testing within mining operations, what has been learned could be leveraged to assess applicability and use in metallurgical processing thus reducing risk (Thompson 2014).

9.5 The Future of Metallurgical Processing: Innovation and Intellectual Property

The mining industry is indispensable in providing society with products to fuel global economic growth. In 2010, the mining industry represented 11.5 % of the world's GDP based on revenues and products sold. Considering services and support services, the direct contribution in 2010 to the global economy increased to 21 % and, when considering the value from industries dependent on mined products (e.g., materials for construction, transportation fuel, and fertilizers), the combined direct and indirect economic contribution is more than 45 % (Gleason 2015). Continuing to provide the world with products having such a substantial impact on the global economy is becoming increasingly difficult in the face of the challenges confronting the mining industry: effectively processing lower grade, more complex ores in deeper deposits; reducing energy and water use; reducing mine wastes; continuing improvements in health and safety; reducing the carbon footprint; and addressing community pressures at mine sites. Transformational innovation is essential to address these issues and for the mining industry to operate successfully in the future (Gleason 2015).

The mining industry must aggressively address its most intractable issues related to the innovation process and introduction of technological innovations in metallurgical processing to meet the material needs of a burgeoning global population of 9.5 billion by 2050. Innovation must be organized and driven by the mining company's CEO and executive management. Mining companies must partner with organizations skilled in R&D and with suppliers, whose corporate livelihood is dependent on developing innovative technologies and products. Different approaches must be implemented to capture the benefits of R&D and value from "open innovations"—new technologies and ideas from outside the industry. An environment must be created within mining companies whereby intellectual property is recognized and managed and new approaches and tools are identified and implemented for testing of metallurgical processing innovations to limit disruption to ongoing operations, minimize risk and navigate the complexities of the current environment in which mining companies operate.

The implementation of open innovation and virtual laboratory concepts for innovation in the mining industry has found limited support in the past, as exemplified by programs introduced by Barrick Gold Corporation (Strategic Technology

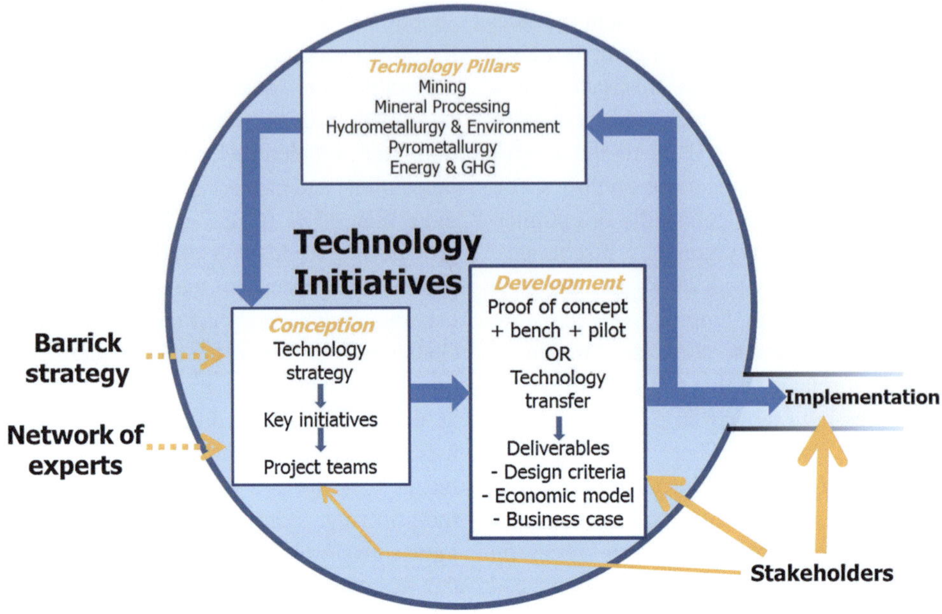

Fig. 9.1 Barrick's innovation model (Courtesy of Barrick Gold Corporation)

Solutions) and AngloGold Ashanti (Technology Innovation Consortium) (TIC 2014). In the case of Barrick Gold the working model (Fig. 9.1) involved a small technology team of subject matter experts working on the long-term technology vision of the company and having the flexibility to rapidly focus on significant challenges confronting the company.

However, the new realities the mining industry face require a more elaborate open innovation model—one that engages a diverse group of participants from a wider group of industries. A new innovation ecosystem explored by Barrick Gold in cooperation with Deloitte (Fig. 9.2) involves a combination of the corporate technology team (STS) with the technology center (AuTec) and the Intellectual Property Company (IPHP) operating as independent entities in close cooperation that facilitates partnerships and accelerates innovation implementation.

Perhaps novel approaches in seeking and implementing innovations will lead to metallurgical processing vastly different from what is done today. Imagine metallurgical processing carried out with emerging engineering developments (Hemingway and Loehr 2014): information technology (e.g., artificial intelligence); construction (e.g., 3-D printing); electronics (e.g., nano- and molecular sensors); robotics (e.g., molecular nanotechnology); synthetic biology (engineered microbes, molecules and biomolecular sensors); energy (e.g., fusion); and advanced materials (e.g., programmable matter). Some of these engineering innovations are commercially used now in medicine, manufacturing, and biotechnology. Novel engineering systems might be used to metallurgically process ore bodies in situ by identifying and characterizing deep ore bodies spatially, geologically and mineralogically; fragmenting ore bodies in place; selectively leaching and concentrating metals values; and monitoring and protecting the subsurface environment. Applying these emerging developments to metallurgical processing remains decades in the future, but now is the time for the mining industry to explore the possibilities.

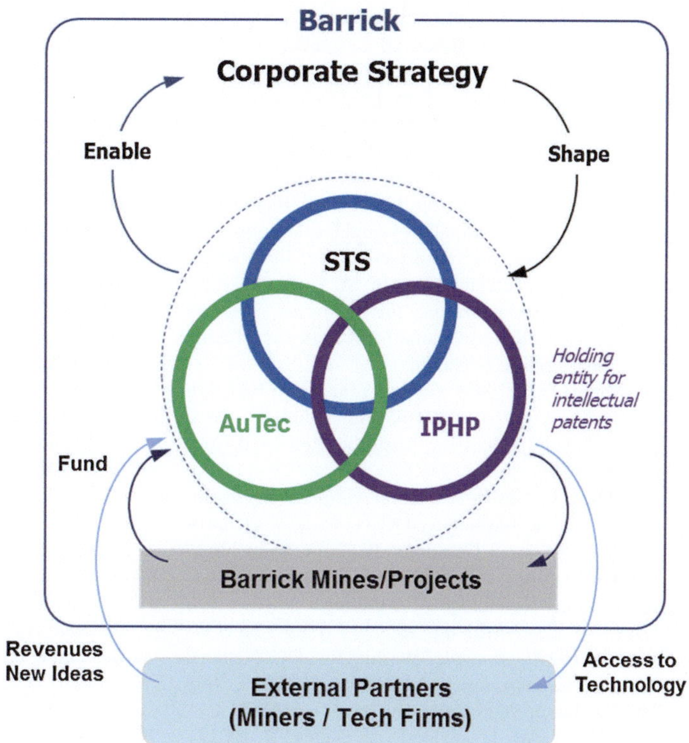

Fig. 9.2 Experimental innovation ecosystem (Courtesy of Barrick Gold Corporation)

References

Adey, E., Wall, F., Shail, R., Keech, J., Neal, W., Limprasert, R., Roba, C., & Delmore, C. (2011). Best practice for reducing the carbon footprint of the mining industry. *Impact monitoring of mineral resources exploitation*. from http://www.impactmin.eu/downloads/impactmin_d33.pdf. Retrieved December 18, 2014.

Batterham, R. J., & Algie, S. H. (1995). The role of technology in the minerals industry. In *Proceedings of the XIX International Mineral Processing Conference (IMPC)*. San Francisco, California, Retrieved October 20–25, 1995.

Brierley, C. L. (2008). How will biomining be applied in the future? *Transactions of the Nonferrous Metals Society of China, 18*(6), 1302–1310.

Brierley, C. L., & Brierley, J. A. (2013). Progress in bioleaching: Part B: Applications of microbial processes by the minerals industries. *Applied Microbiology and Biotechnology, 97*(17), 7543–7552.

Canterford, J. (2011). Keeping projects on the rails. In *Proceedings of Metplant 2011-Metallurgical Plant Design and Operating Strategies* (pp. 12–18). Perth, Australia, Retrieved August 8–9, 2011.

Dictionary. (2014a). Invention. Retrieved December 16, 2014, from http://dictionary.reference.com/browse/invention.

Dictionary. (2014b). Technology. Retrieved December 11, 2014, from http://dictionary.reference.com/browse/technology.

Dreisinger, D. (2006). Copper leaching from primary sulfides: Options for biological and chemical extraction of copper. *Hydrometallurgy, 83*(1–4), 10–20.

Ferraloro, A., Humphris, C., & Kaggwa, G. (2012). Sector missing out due to poor IP management. *Gold Mining Journal, 1*, 10–11. October–December 2012.

Gentry, D. W. (1998). Constrained prospects of mining's future. *Mining Engineering, 50*(6), 85–94.

George, D.B., & Richards, K.J. (1983). Solid matte-oxygen converting process. U.S. Patent 4,416,690, Retrieved November 22, 1983.

Gleason, W. (2015). Mining company of the future. *Mining Engineering, 67*(1), 26–25.

Google. (2014a). Technology. Retrieved December 11, 2014, from https://www.google.com/?gws_rd=ssl#q=define+technology.

Google. (2014b). Research and development. Retrieved December 11, 2014, from https://www.google.com/?gws_rd=ssl#q=define+research+and+development.

Hedges, M., & Lloyd, M. (2010). Claiming the green revolution – protecting investment in green mineral processing. In *Proceedings of the XXV International Mineral Processing Congress (IMPC)*. Brisbane, Australia, Retrieved September 6–10, 2010.

Hemingway, G., & Loehr, J. (2014). The living mine. *Mining Magazine*, p. 8, Retrieved December 2014.

Lin, T. C. W. (2012). Executive trade secrets. *Notre Dame Law Review 87*, 911. Retrieved December 10, 2014, from http://ssrn.com/abstract=2047462.

Marsden, J. O. (2004). Technology development and competitive advantages: Sustainable or short term? In E. C. Dowling & J. O. Marsden (Eds.), *Improving and optimizing operations; things that actually work! (plant operators' forum)* (pp. 127–137). Littleton, CO: Society for Mining, Metallurgy, and Exploration.

McNulty, T. (1998). Innovative technology: Its development and commercialization. In M. C. Kuhn (Ed.), *Managing innovation in the minerals industry* (pp. 1–14). Littleton, Colorado: Society for Mining, Metallurgy, and Exploration.

Napier-Munn, T. J. (1998). Getting value from R&D. In *Proceedings of AusIMM'98-The Mining Cycle* (pp. 117–121). Mount Isa, Australia, Retrieved April 19–23, 1998.

NRC. (2002). *Evolutionary and revolutionary technologies for mining*. Washington, DC: National Academy of Sciences. Retrieved December 10, 2014, from http://www.nap.edu/catalog/10318/evolutionary-and-revolutionary-technologies-for-mining.

NRC. (2014). *Can earth's and society's systems meet the needs of 10 billion people?* Washington, DC: National Research Council. Retrieved October 29, 2014, from http://www.nap.edu/catalog.php?record_id=18817.

Roberto, F., Snyder, J., Dodd, J., & Dunne, R. (2013). *Energy and water requirements in the treatment of refractory gold ores*. Paper presented at the World Gold 2013 Conference. Brisbane, Australia, Retrieved September 26–29, 2013, paper no. 86.00.

Schippers, A., Hedrich, S., Vasters, J., Drobe, M., Sand, W., & Willscher, S. (2014). Biomining: Metal recovery from ores with microorganisms. In S. Schippers, F. Glombitza, & W. Sand (Eds.), *Geobiotechnology I metal-related issues* (pp. 1–47). Heidelberg: Springer.

Stacy, C. (2012). *Recognizing and exploiting intellectual property*. Paper presented at the Society of Mining Engineers annual meeting. Seattle, Washington, Retrieved February 19–22 2012, Preprint 12–149.

Stanway, G., & Andrew, D. (2014). *Mining innovation: State of play 2013*. New York: Virtual Consulting International. Retrieved December 10, 2014, from http://govci.com/wp-content/uploads/2014/06/Mining-Innovation-State-of-Play-2013.pdf.

Taberer, R. F. (2012). Patenting in the field of percolation leaching. *Journal of the Southern African Institute of Mining and Metallurgy, 112*, 1031–1036.

Thompson, J. (2014, March). Nurturing technological growth. *MiningMagazine.com*. Retrieved December 10, 2014, from http://www.chrysalixevc.com/pdfs/MiningMagazine-March2014-ChrysalixEVC.pdf.

TIC. (2014). Anglo Gold Ashanti Technology Innovation Consortium. Retrieved March 27, 2014, from https://anglogoldashantitechnology.com/.

USBM. (1989). *In situ leach mining*. Information Circular 9216, U.S. Department of Interior. Retrieved December 17, 2014, from http://mines.az.gov/DigitalLibrary/usbm_ic/USBMIC9216InSituLeachMining.pdf.

van Aswegen, P. C., van Niekerk, J., & Olivier, W. (2007). The BIOX™ process for the treatment of refractory gold concentrates. In D. E. Rawlings & D. B. Johnson (Eds.), *Biomining* (pp. 1–31). Berlin, Germany: Springer.

VCI. (2014). Delivering mining innovation. *Mining Magazine*, p. 10–14, Retrieved December 2014.

Wikipedia. (2014a). Industrial design right. Retrieved December 10, 2014, from http://en.wikipedia.org/wiki/Industrial_design_right.

Wikipedia. (2014b). Trade dress. Retrieved December 16, 2014, from http://en.wikipedia.org/wiki/Trade_dress.

Wikipedia. (2014c). Trade secret. Retrieved December 10, 2014, from http://en.wikipedia.org/wiki/Trade_secret.

World Nuclear Association. (2014). In situ leach (ISL) mining of uranium. Retrieved January 13, 2014, from http://www.world-nuclear.org/info/Nuclear-Fuel-Cycle/Mining-of-Uranium/In-Situ-Leach-Mining-of-Uranium/.

Part II

Process Development

Conceptual Idea, Test Work, Design, Commissioning, and Troubleshooting

10

Ram Ramachandran and Alan Taylor

10.1 Conceptual Idea and Experiments

A conceptual idea can emerge unexpectedly or out of a deliberate search for either an improvement to an existing process or a completely new process.

10.1.1 Objectives and Examples

Objectives and incentives for process innovation vary widely. Some of the more common are presented below together with some representative examples from industry practice.

10.1.1.1 Enhancing Safety and Health of Operating or Maintenance Personnel

A classic example is the use of mercury for the recovery of gold and silver by amalgamation which was practiced for centuries. In recent times, amalgamation has been recognized as a major health hazard and has been largely replaced by gravity and cyanide leaching processes in commercial mining industry practice. (Unfortunately, it is still used in gold artisanal and illegal mining in some developing countries, where finding a safe easy to use alternative, suitable for small-scale application is more difficult.)

R. Ramachandran (✉)
Consulting Engineer, 9650 E. Peregrine Place, Scottsdale, AZ 85262, USA
e-mail: ramvasanti@aol.com

A. Taylor
ALTA Metallurgical Services, Level 13, 200 Queen Street, Melbourne, VIC 3000, Australia

10.1.1.2 Reducing Environmental Impact

Since about the 1970s, the pressure of the stricter regulation of sulfur dioxide and other noxious emissions has resulted in the desire to replace the smelting of copper sulfide ores and concentrates by the application of hydrometallurgy. Numerous processes have been proposed and tested with varying degrees of success. Most recently, pressure oxidation has emerged as a serious contender, having been commercially proven for gold and zinc production where it has been increasingly used instead of traditional roast-leach technology.

10.1.1.3 Improving Extraction Efficiencies

Up till the 1970s, zinc recovery by the roast-leach-EW process was significantly affected by the formation of zinc ferrite which provided the driving force for innovation. A major improvement was achieved through the development of the jarosite process in which most of the

V.I. Lakshmanan et al. (eds.), *Innovative Process Development in Metallurgical Industry*,
DOI 10.1007/978-3-319-21599-0_10

zinc is extracted from the ferrite and the iron is rejected as jarosite with low zinc content. Processes rejecting iron as goethite, which has an environmental stability advantage, and hematite, which offers the possibility of a saleable by-product, were also developed.

10.1.1.4 Reducing Capital and Operating Costs

Carbon in pulp was introduced on a commercial scale in the 1970s and rapidly replaced the traditional Merrill–Crowe process as the industry workhorse for the treatment of lower grade gold ores. Cost advantages accrued from the elimination of solid–liquid separation and the use of zinc dust for gold precipitation plus the reduction in soluble gold loss. The use of carbon was also extended to treating heap leach solutions.

10.1.1.5 Generating a New Product

Nickel pig iron (NPI) production was rapidly developed in China in 2006 to supply nickel to stainless steel producers due to high demand and escalating nickel prices at the time. NPI is essentially a low-grade ferronickel produced directly from nickel laterite ores which are too low grade for the traditional ferronickel smelting process. The remarkably short development period for NPI technology was made possible by the availability of numerous small-scale blast furnace and electric arc facilities in China formerly used for pig iron and other alloys.

10.1.1.6 Recovering a New By-Product

Examples of by-product recovery innovation include the development of processes for the recovery of uranium from nonconventional resources such as in the 1970s to take advantage of high demand and high uranium price. One example was the development of IX/SX technology for the extraction and recovery of uranium from copper leach solutions which can contain up to 40 ppm uranium. Large-scale commercial plants were operated at Anamax Twin Buttes in Arizona and Kennecott Bingham Canyon in Utah.

Another example was the development of processes for the extraction and recovery of uranium from phosphoric acid which typically contains 150–175 ppm U_3O_8. In the 1970s, various innovative two-stage SX processes were developed and applied at numerous commercial operations. More recently, the focus has been on IX technology which has the potential to reduce cost, avoid the phase disengagement problems in SX, avoid posttreatment of the acid, and improve environmental impact and safety.

10.1.1.7 Enhancing Product Purity

The successful introduction of solvent extraction in uranium ore processing in the 1960s led to the idea of developing selective copper SX extractants which made it possible to produce high purity electrowon cathode from copper leach solutions and resulted in the development of numerous new heap leaching projects. Before this, copper from leaching operations was produced as either relatively low-grade cathode or as cement copper by precipitation with scrap iron. Further details are presented as a case history at the end of this section.

10.1.2 Literature and Patent Search and Assessment

The first step in the development of an idea is to undertake a thorough literature and patent search in order to:

10.1.2.1 Determine Whether or Not the Idea Has Been Previously Proposed

If an idea has been previously proposed:

1. Use of published information and/or,
2. Contact with persons who were involved or worked on the idea

can either significantly assist with time, definition of scope, and cost of an initial study or provide sufficient information for abandoning the idea.

10.1.2.2 Locate and Assess Any Previous Patents or Patent Applications

The existence of patents or patent applications in some cases may be a roadblock to further development. However, in other cases it could open the door to either fruitful collaboration or an acceptable licensing arrangement which could yield significant savings in time and money.

10.1.2.3 Evaluate the Results of Any Previous Published Test Work

Previous test work can be a source of valuable information as to the potential viability of the idea and can provide a starting point for developing a test work program.

10.1.2.4 Evaluate Strengths, Weaknesses, Opportunities, and Threats for the Proposed Idea

The results of the literature and patent search provide input for an initial assessment of the proposed idea.

10.2 Scoping and Laboratory Scale Test Work

10.2.1 Scoping Test Work and Desk Top Study

The next step is to undertake exploratory laboratory test work followed by "desk top" technical and economic studies. The test work is aimed primarily at the key steps in the process in order to demonstrate viability at a preliminary level and generate data for the "desk top" studies. The test work is generally carried out at bench scale in batch mode, though continuous mode may be needed in some instances.

The components of the study typically include:

- Technical feasibility
- Possible applications
- Advantages and disadvantages
- Economic potential
- Environmental implications
- Chances for success

Assuming a positive outcome, a budget and timescale is developed for a "proof-of-concept" laboratory test work program.

10.2.1.1 Case History: Development of Selective Copper SX Extractants (Kordoski 2002)

The recovery of copper from sulfuric acid leach solutions by solvent extraction/electrowinning was proposed by General Mills in the USA in 1960, inspired by the successful development of uranium SX in the 1950s using their Alamine 336 extractant. Until then, copper leach solutions were treated by cementation with scrap iron or by direct electrowinning. General Mills postulated that the introduction of a solvent extraction step ahead of electrowinning would improve electrowinning efficiency and enable high-grade cathode copper to be produced. Unlike uranium, this would require the development of an entirely new SX extractant.

An initial market survey was disappointing as it showed little interest and even skepticism. Nevertheless, a small group of three enthusiasts within General Mills under the leadership of Joe House worked on the development of a new reagent in between their main duties and in their own time. This resulted in late 1962 in the formulation of LIX 63, a hydroxyoxime.

However, LIX 63 had a significant drawback in that it was not effective below pH 3 and therefore not applicable to typical acidic copper leach solutions. Technically, this problem could be solved by partial neutralization or by switching to ammonia leaching. Both of these solutions were not regarded as economical.

The group was not satisfied and further work resulted in the development of LIX 65, a ketoxime, which, with the addition of a portion of LIX 63, formed LIX 64. After successful laboratory test work, LIX 64 was trialed in a number of SX/EW pilot plants resulting in its first commercial application at Ranchers Exploration and Development Corporation in Arizona in 1968, treating solution from oxide ore heap leaching. In 1969 the LIX 65 component was replaced with LIX 65N to form LIX 64N which had greater extractive strength, faster kinetics, faster phase separation, lower entrainment, increased

Fig. 10.1 Ranchers Bluebird, Arizona: first commercial copper SX/EW operation (Reprinted with permission from The Journal of the South African Institute of Mining and Metallurgy)

copper/iron selectivity, and lower viscosity. LIX 64N became the workhorse of copper SX for many years. Extractants were also developed by Ashland Chemicals, Shell International Chemicals, and ICI Acorga in the late 1960s and early 1970s.

The success of Ranchers and the subsequent Bagdad operation, also in Arizona, led to a growing acceptance of copper SX/EW as a commercially proven technology and ultimately to the transformation of copper hydrometallurgical processing. A photograph of the first commercial copper SX/EW operation at Ranchers Blue Bird, AZ is shown in Fig. 10.1.

The above case history illustrates how the dedication of three chemists resulted in an innovation—the development of a solvent for the extraction of copper from dilute leach solutions. Based on this innovation, currently around 23–25 % of cathode copper is produced by Leach-SX-EW process enabling the metallurgical industry to exploit oxide copper ores and low-grade sulfide ores. However, the authors do not have a firsthand knowledge of how this innovation happened except from published papers.

Hence we have chosen a process from the zinc industry wherein one of the authors had a hands-on experience in the development of process improvement—not necessarily an innovative one—to illustrate how a concept is taken from laboratory scale to commercial production. The steps involved are described in some detail in Sects. 10.2–10.5. This was the first time that this process improvement was done in the zinc industry.

The process improvement consisted of combining the leaching of zinc calcine and fume into one continuous leach step. Until then, the calcine and fume were leached separately in a batch mode due to the nature of impurities in them. For the same reason, the leach solutions were purified and electrowon separately. The process improvement also involved developing a continuous purification step—again done in a batch mode—to produce a single purified solution for electrowinning.

10.2.2 Laboratory Scale Test Work

In the previous section, details of exploratory laboratory test work to test the process steps are outlined. This is followed by a discussion of "desk top" studies to assess the potential advantages of the process. Based on these studies:

1. Time and expenses are committed for systematic process development test work.

2. A budget and time scale is developed for a scoping level "proof-of-concept" laboratory test work program.

Once the above two items are in place, a process schematic containing the various process steps with inputs and outputs—with as much available data as possible—should be drawn up. The next step is to design "proof-of-principle experiments." These experiments are in general laboratory scale experiments in which the new concept is proven in some way. This identifies the information that needs to be collected in laboratory scale testing.

10.2.2.1 Steps in Process Development

Process development involves four steps as listed below:

1. Laboratory testing
2. Pilot plant testing
3. Demonstration plant testing
4. Testing in a new/existing plant

Each development step serves different purposes in the Total Development Work Process. The choice of the number of steps from the above list depends on the character of the development work. In many applications, simulation has replaced a part and/or whole of the test work and is expected to play a major role in process innovation. Simulation can be used in all and/or any steps shown in Fig. 10.2 and acts as a supplementary activity in testing at any stage in the process development work. Simulation or simulation programs have made great advances in the recent past; hence it is advisable to install proper equipment and programs for simulation prior to the start-up of any major new installation of process technology. Well-developed simulation programs can be used for training operators before and after the start-up of new process technologies and new production plants.

10.2.2.2 Laboratory Scale Test Work

Laboratory testing is often considered as the starting point for process development; this includes experimental work in the laboratory in a batch mode and is invariably followed by continuous mode of testing. The three areas of process metallurgy are: (a) pyrometallurgy, (b) hydrometallurgy, and (c) electrometallurgy. Of these three areas, it is somewhat easy to do laboratory scale test work for hydrometallurgical and electrometallurgical processes and get meaningful data. However, it is not easy to do laboratory tests for pyrometallurgical processes due to the small scale of operation and get useful data. For such operations, it is suggested that a small pilot plant scale testing would give (a) better process control and (b) reliable process parameters.

10.2.2.3 Technical and Economic Study of a New Process

The outline of laboratory testing described above can also be used for any process that has been conceived and developed from scratch. Once the process concept design has been established, a simple block diagram showing inputs and outputs should be established. This diagram defines the information needed from the tests and provides input for the planning of the research stage.

The data collected from preliminary laboratory testing should also provide a scoping level technical and economic study of the process. For simplicity, a hydrometallurgical process involving leaching and solution purification laboratory test work approach will be illustrated by the following example:

An existing electrolytic zinc refinery consists of the following unit operations:

1. Two separate circuits for leaching of the raw materials, viz., calcine and fume.
2. Two separate circuits for the purification of the respective leach solutions, viz., calcine leach solution and fume leach solution.
3. Two separate electrowinning units for the respective purified leach solutions.

The leach and purification steps were run in a batch mode, while the electrowinning steps were run in a continuous mode independently of the leach and purification circuits.

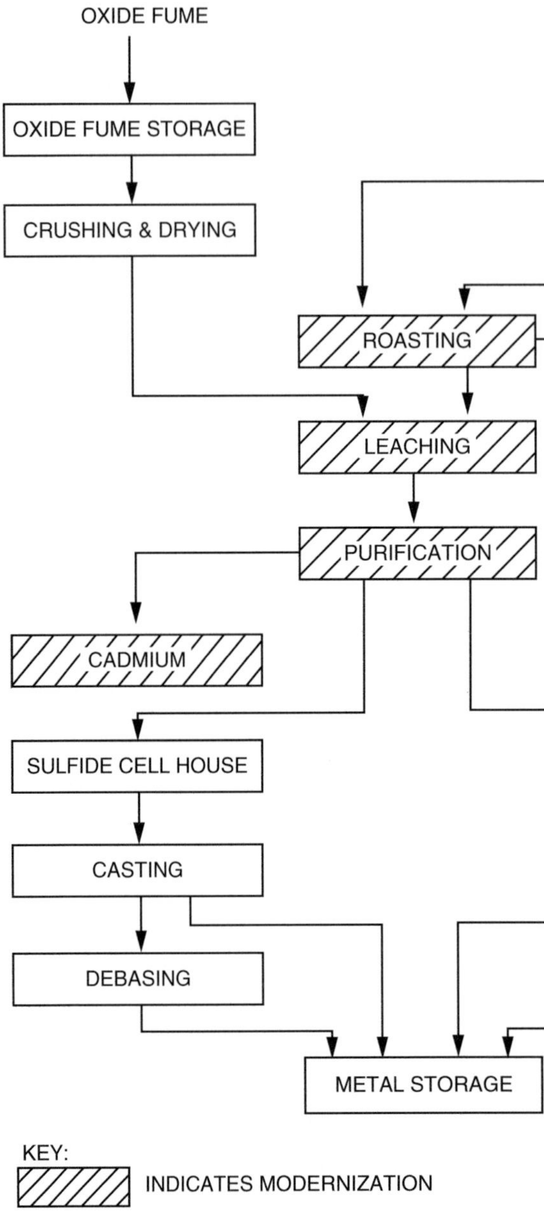

Fig. 10.2 Schematic of the modernized plant. (from Ramachandran and Cardenas 1983). Copyright 1983. Reprinted with permission of The Minerals, Metals and Materials Society

The goal of the project was:

1. To combine the leaching of the calcine and fume into one step and run the same in a continuous mode.
2. To run purification tests of the above single leach solution in a continuous mode. This stage would also involve optimizing the number of purification steps needed to remove "all" the impurities to produce a purified solution suitable for electrowinning.
3. To run the electrowinning operations in two electrolytic units but with the same purified leach solution.

Historically, calcine and fume were leached and purified separately due to the different nature of impurities present in them. The leaching of the calcine and fume in a single step was quite an innovative approach. This would also warrant a unique approach to the purification of the leach solution as it would have to address the removal of "all" the combined impurities from calcine and fume in one single step.

The project was divided into four steps. They were:

1. Calcine/fume leach tests—batch
2. Calcine/fume leach tests—continuous mode
3. Leach solution purification tests—batch
4. Leach solution purification tests—continuous

10.2.2.4 Batch Tests

The objective of the batch tests was to establish optimum operating parameters of the leach step to obtain maximum zinc recovery. The process variables were:

(a) Temperature
(b) Retention time
(c) Calcine/fume ratio
(d) Final pH

Based on the plant data that was already available on calcine leach and fume leach separately, tests were run using a preliminary range of operating variables. The calcine/fume ratio was established based on the weight ratio of calcine to fume that is available for processing in the modernized plant. For batch tests, the following parameters for the variables were chosen:

1. Calcine/fume ratio:	Variable based on availability of calcine and fume (65/35
	was optimum based on minimization of deleterious
	impurities.)
2. Final pH:	~3.2–3.5 at temperature (~4.8–5.2 at room temperature);
3. Temperature:	70, 80, and 90 °C (three levels)
4. Retention time:	3, 4, and 5 h at each temperature

Based on the calcine/fume ratio used, the amount of leachant (spent electrolyte) was varied to achieve maximum zinc extraction. With this was approach, the final pH reached around 3.2–3.5.

Data generated from these batch tests were used to establish the operating parameters—and their ranges—for continuous leach tests. In addition to maximizing zinc recovery, extraction of impurities such as copper, cadmium, nickel, and cobalt was also established. The above test work discussed was for converting an "existing" batch leach process for two different types of raw materials—treated separately—to a "continuous" leach process wherein the two raw materials were treated in one step.

10.2.2.5 Continuous Leach Tests

In these tests, three leach tanks (all of same size) were used in series. In addition to the four variables that were tested in batch tests, two additional variables were evaluated in these tests. They were:

1. Distribution of calcine/fume mixture between leach tanks.
2. Distribution of leachant between leach tanks.

Preliminary screening tests showed that calcine/fume mixture and the leachant be added in the first two tanks and the third tank be used only to provide retention time for maximum zinc extraction. The retention time was varied by varying the flow rate of the leachant. Also, based on the results of the batch tests (zinc extraction), only two levels of temperature, viz., 70° and 80 °C and two levels of retention time, viz., 3 and 4 h were chosen.

All continuous tests were done based on a statistical design of experiments. Using a factorial design, 16 experiments (2^4—four variables at two levels) were done to establish the optimum parameters for maximum zinc extraction. Details of the designed experiments are not described here as they can be found in any textbook on Statistical Design of Experiments.

Based on test results, the following parameters were finalized for maximum zinc extraction (~95 %):

1.	Temperature:	80 °C
2.	Retention time:	3 h
	(Three tanks in series):	1 h in (each tank)
3.	Distribution of calcine + fume mixture:	
	Tank One:	90 % by weight (all of calcine)
		Balance fume
	Tank Two:	10 % by weight (fume only)
	Tank Three:	Nil
4.	Distribution of leachant:	Tank One: 93–95 % by volume
		Tank Two: 5–7 %
		Tank Three: Nil

Additional tests were done to study the effect of grinding the calcine to a much finer size (~98 % minus 325 mesh from 74 % minus 325 mesh). This did not appreciably improve zinc extraction and hence was eliminated as a variable.

As a follow-up, tests were done to collect preliminary data on:

1. Settling of leach slurry in a simulated laboratory thickener.
2. Amount of suspended solids in the thickener overflow.
3. Use of flocculants for settling leach slurry.
4. Pulp density of the thickener underflow.
5. Washing and filtering of thickener u/flow to establish water-soluble zinc in the leach residue.

All the leach solutions—produced from batch and continuous tests—were saved for planned batch and purification tests.

10.2.2.6 Batch and Continuous Laboratory Purification Tests

Based on an approach similar to the leach tests, batch and continuous purification laboratory tests were performed to gather sufficient data for running a pilot plant. Detailed description of the purification tests—both batch and continuous tests—is not provided here as the procedure is quite similar to the leach tests.

However, due to the nature of the impurities present in the impure leach solution, purification was done in a number of stages—either two or three—dependent on the final quality of the electrolyte desired for electrowinning. The process variables for the *continuous* tests are summarized below:

1. Number of purification stages.
2. Temperature.
3. pH at temperature.
4. Zinc dust added for purification.
5. Particle size of zinc dust.
6. Zinc dust distribution between reaction tanks.
7. Type of activator in each stage.
8. Activator distribution between reaction tanks.
9. Total retention time.

Variables 2 and 4 through 9 were varied for each stage. Variables 1 and 3 were kept constant based on batch tests.

In summary, an approach for a systematic study of doing laboratory tests—batch and/or continuous runs—has been illustrated using a hydrometallurgical process involving leach and purification operations, i.e., two process steps run in series.

The same approach can be used in evaluating a "new" process that has been conceived for (a) production of a new product or (b) making a radical change to an existing process to reduce operating cost and improve environmental issues. Assuming that the "new" process has four independent process steps (aka unit operations), tests would be conducted as follows:

1. Batch testing of each of the four process steps.
2. Testing the four process steps in a continuous mode but run independently.
3. Combining all the four process steps in one "continuous mode." In this run, the following variables would also be included for evaluation:
   (a) Effect of Recycle Streams—if any.
   (b) Establish steady-state operation, run for a few days to observe changes—if any.

(c) Establish bleed streams with volume and composition—if any.
(d) Develop process options for the treatment of bleed streams.

10.2.2.7 Process Simulation Model

It is always a good practice to develop a preliminary process simulation model to:

(a) Investigate the effect of linking the process steps
(b) Provide support for the scoping and laboratory study and
(c) Assess the need for further proof of concept test work—if any

A typical process simulation model that can be used is the commercially available METSIM model. This model essentially provides a good mass and heat balance for the complete process and provides answer to the questions raised above. Based on the issues arising from this simulation model and scoping study, additional laboratory scale test work should be done until all process parameters for "all" the steps of the new process (for all the unit operations) are well established.

10.2.2.8 Data for Patent Application

A new innovative process invariably has a potential for obtaining a "patent' coverage to

(a) Protect the proprietary nature of the new idea and
(b) Possibly market the process to potential customers for royalty revenues

The scoping and laboratory scale test work that has been described in this section was for a hydrometallurgical process. A similar approach can be had for evaluating an electrometallurgical process. However, there are possible limitations in developing "good" process parameters for a pyrometallurgical process at the laboratory level. In such a case, a small-scale pilot plant operation—that provides good process control—may be needed to develop process parameters.

In summary, assuming that the scoping and laboratory study confirms the viability and advantages of the process, the next step would be to prepare a budget and schedule the pilot plant scale test work program.

10.3 Pilot Scale Test Work

Prior to doing any pilot scale test work, it is prudent to examine (a) whether a pilot plant is absolutely necessary and if so, what is the purpose of the pilot plant? Some of the guidelines to answer the first question are:

An integrated pilot plant with all the process steps and recycle flows is needed if (Harmsen 2013):

1. More than one new process is involved.
2. The process contains one new process step and a complex recycle flow.
3. The process contains a novel solids handling step.
4. The feed material details are not completely known.
5. Feed material may contain trace components which may affect operation of the new process unit.
6. Formation of trace components in one unit may affect the operation of another unit.
7. Recycle streams increase the buildup of trace components with possible consequences of fouling and corrosion.

A process step is considered "new" if (a) new chemistry is involved, (b) a new piece of equipment is involved or both. Complex recycle flow is defined as a recycle flow over at least two process units. Sometimes, if only one "new" process step is involved, a dedicated mini-test may be sufficient to validate the process design and process model.

Second, the purpose of any pilot plant scale test work is:

1. Validate commercial scale process "concept" design.
2. Confirm laboratory scale test work and preliminary process flow sheet.

3. Establish range of operation for all the process variables.
4. Proving the operation of the process on a continuous basis.
5. Generate additional data for improved process simulation program.
6. Testing a range of feed raw materials.
7. Evaluate materials of construction for corrosion, etc. This can be done by building a construction material test section with coupon testing so that corrosion rates are determined for various process stream compositions.
8. Produce product samples in sufficient quantities for customers for evaluation of product performance.
9. Providing data for environmental studies.

Based on the type and complexity of the process, there are a variety of pilot plant concepts. They are:

1. Piloting parts of the process only.
2. Fully integrated mini plant.
3. Fully integrated large-scale pilot plant to test equipment that would be used in a commercial plant.
4. Large-scale testing of parts of the process—either at the plant location or by a technology and/or equipment supplier.

The choice of a particular concept/s depends on what data needs to be generated for a final design of the full-scale plant.

It is a good idea to develop a pilot plant design team consisting at least of (a) an experienced process engineer from the engineering group, (b) a process engineer from the operations, (c) a safety engineer from operations, and possibly (d) an employee from a company with experience in pilot plant design and construction. The role of the safety engineer is to continually advise the team about safety issues related to the process steps and develop an ongoing safety manual for use during the commissioning of the plant.

In summary, the pilot plant should be a scaled-down version of the commercial plant, containing all process steps and recycle streams. In addition, the final outcome of the pilot plant testing is: (a) to collect data for the design of a full-scale plant and (b) to conduct a detailed technical and economic feasibility study.

10.3.1 Pilot Plant Testing: Continuous Leach and Purification: (Ramachandran and Cardenas 1983)

In Sect. 10.2, laboratory scale batch and continuous scoping tests—for both leaching and purification—were described in some detail. Based on acceptable test results for continuous laboratory tests and since the final design was for a continuous process, it was decided to run only "continuous" leach and purification tests on the pilot plant scale. Statistically designed pilot plant tests using a factorial design were done to confirm the laboratory data.

10.3.2 Results of Continuous Pilot Plant Leach Tests

The results from the continuous pilot plant leach tests confirmed the data that was developed in the laboratory scale and scoping tests. In addition, to the leach operating parameters, data for a downstream operating unit such as thickener for neutral leach residue were developed to assist in the design of the full-scale plant.

Based on prior operating experience and in the interest of time, only laboratory scale tests (not pilot plants) were carried out:

1. For washing and settling of leach residues
2. For vacuum filtration of the washed and settled residues

These data were used in the development of design data for a full-scale plant.

10.3.3 Additional Leach Operating Data

1. Average pH at leach temperature in:
   - Tank One: 1.5–1.7
   - Tank Two: 3.0–3.3
   - Tank Three: 3.8–4.0
2. Free Acid in Tank One: 4–8 g/L

10.3.4 Neutral Leach Slurry Thickener Data

Thickener loading in gal/min/ft^2 and thickener underflow density in grams/liter compared well with preliminary design data based on similar plant operations. The type and the amount of flocculant required for optimum settling rates were established.

10.3.5 Settling and Filtration of Washed Leach Residue

Variables optimized in these tests were:

1. Degree of washing needed to optimize low levels of water-soluble zinc in leach residue.
2. Settling rate of washed leach residue.
3. Filtration rate of settled washed leach residue.

10.3.6 Results of Continuous Purification Pilot Plant Tests

All the operating variables established—and somewhat optimized—in the laboratory continuous purification tests were confirmed at the pilot plant level. The data collected from these tests were used for the design of the full-scale continuous purification plant.

Prior to the start-up of these continuous purification tests, a review of the impurities in the neutral solution showed that by careful design of number of purification stages, it would be possible to produce separate concentrated impurity products. These products could then be treated at either the same plant or other sister plants for recovery of valuable products such as copper and cadmium. Based on this premise, a four-stage continuous purification system was tested at the pilot plant scale. They were:

1. Stage One—Partial Copper Cementation
2. Stage Two—Balance Copper plus Cadmium Cementation
3. Stage Three—Cobalt, Nickel, and Germanium Cementation
4. Final Polishing Stage

Pilot plant results confirmed the parameters for all the purification stages. Additional data collected at the pilot plant level were:

1. In order to get cobalt consistently to the desired critical level of 0.1 mg/L in the final purified solution, it was decided to run Stage Three of the purification step as a "batch" process instead of a "continuous" process. This change would give the plant better control over the removal of cobalt to acceptable levels. This decision emphasized the need to run pilot plant tests so that there were no surprises at the full-scale plant level to get the desired results.
2. The need for a final—also optional—polishing stage was established to address the possibility of the changing levels of the impurities in the neutral leach solution due to variations in the feed material, both calcine and fume, to the leach step.
3. Plate and frame filter presses and/or pressure filters were chosen as solid/liquid separation options for all stages of purification.

In summary, data collected at the pilot plant scale tests formed the basis for

1. Designing a full-scale plant.
2. Conducting a detailed technical and economic study.

Finally, Sects. 10.2 and 10.3 give a systematic description of laboratory and pilot plant testing—batch and continuous—of modifications proposed to an existing hydrometallurgical batch leach and purification processes. Using these process parameters information, Sect. 10.4 will describe additional data collection needed for the design of a full-scale plant.

A corollary of running these pilot plants, viz., leach and purification tests was that additional pilot plants were run to develop supporting information for the design of a full-scale plant. They were:

1. Settling tests on the neutral leach slurry.
2. Settling and filtration tests on washed neutral leach slurry.
3. Settling tests on purification slurries (unsuccessful); hence filtration tests for all four purification stages.
4. Pilot plant tests for the process development of a flow sheet for treating the second stage copper–cadmium purification cake for cadmium recovery. In these tests, cementation of cadmium using zinc dust was evaluated as an alternate to electrolytic recovery of cadmium.

A similar approach can be adopted for developing process data for: (a) any new process or (b) modification of an existing process. Based on the type of process, viz., pyrometallurgy, hydrometallurgy, electrometallurgy, and/or a combination of any of these process steps, process parameters should be defined—with appropriate ranges—for evaluation. This would be followed by systematic testing of the parameters by conducting appropriate laboratory scale and pilot plant tests as described in this section.

10.3.7 Chemical Analytical Support for Pilot Plant Operations

Any pilot plant campaign should be supported by quality analytical support for samples generated in the pilot plant. The turnaround time for the analyses should be quick to enable the pilot plant operators to make decisions for additional tests, if needed, for generating all necessary data. This can be best provided either by an on-site facility and/or a special arrangement with an outside laboratory or a combination of both. Setting up of an on-site analytical facility or to get analytical results from an outside laboratory will be a function of the location of the pilot plant, viz., distance from an existing plant with analytical facilities and/or distance from urban areas where outside laboratories are available.

10.4 Data Collection for Design of Full-Scale Plant

In Sects. 10.2 and 10.3, details of laboratory scale test work and pilot plant scale test work were described in some detail. In this section, information collected at the laboratory and pilot plant scale will be utilized to put together data for the design of a full-scale plant. This information is collected in the final stages of the laboratory test work when the flow sheet is somewhat fixed and continued through the pilot plant program. In some cases, an engineering firm is involved even at this early stage to specify test work program and design of the pilot plant.

The design of a commercial plant requires:

1. Data collected by the process development team from laboratory and pilot plant tests.
2. A detailed flow sheet showing the inputs and outputs in each unit operation step.
3. Involvement of an experienced engineering firm to ensure that the required data is generated and collected.
4. In some cases, appropriate involvement of suppliers of specialized equipment items.

The role of an engineering company—critical and essential—covers a number of aspects of data collection program for the design of the full-scale plant. They are described below:

10.4.1 Specification of Test Program

The specification normally covers: (a) all the feed materials that are likely to be processed, (b) ratios of the different feed materials to be used covering a wide range based on their availability, (c) variations in seasonal climatic changes—if any. The specifications may also include all the process variables for evaluation.

10.4.2 Design of Pilot Plant

Based on the data obtained from the detailed laboratory test work, a pilot plant must be built to generate all the necessary information for the design of a full-scale plant. Details of laboratory and pilot plant operations for a specific project, viz., continuous leaching of zinc calcine and fume and purification of the neutral leach solution were described in Sects. 10.2 and 10.3, respectively.

Based on the climatic conditions in which the full-scale plant would be run, it may be necessary to house the pilot plant in a climate-controlled enclosure. In some cases, the pilot plant may not be capable of generating all of the necessary data and supplemental tests may be required. At this point, data that may be already available in literature and other similar operating plants should be gathered and incorporated into the design process.

It is strongly suggested that the operating staff of the projected commercial plant be included as part of the design team in the pilot plant phase of the project. This step provides operators' point of view in the design of the pilot plant. It also enables them to get valuable experience in the commissioning and subsequent running of the full-scale plant.

10.4.3 Collection and Monitoring of the In-Process Conditions

During the running of the pilot plant, data on the in-process conditions such as temperature, pressure, densities and viscosities of solutions and slurries, pH, Eh, etc., should be collected. These data will be helpful in the selection of:

1. Proper mechanical equipment such as pumps, etc.
2. Electrical equipment with their power requirements.
3. Process control equipment for process automation.

10.4.4 Sampling and Analyses of Process Streams

All process streams should be sampled at appropriate sampling stations built into the pilot plant design and analyzed for all the relevant constituents in an on-site laboratory. The data feedback from the ongoing analysis will help in fine tuning the operating conditions and optimize them, for example, maximum zinc recovery in the leach step and maximum removal of impurities in the purification step.

In addition, all inputs and outputs from the pilot plant should be recorded routinely. This data, in combination with the analytical information, will be the basis for a complete mass balance. Simultaneously, heat and electrical energy requirements should be collected for an energy balance.

10.4.5 Process Model

Prior to the design of a full-scale plant, it is advisable to develop a process model using all the data collected during the pilot plant run. Sometimes, additional data—not developed in the pilot run—may be needed for the development of the process model. These data should be researched and obtained from published literature and technical handbooks.

One of the most common process models used in the process metallurgical industry is METSIM™. It is available commercially for a fee. This model provides a comprehensive mass and heat balance for the complete process. It should be used as a blueprint for examining "what if" scenarios of changing variables on the potential outcome of the process. Some of this hypothetical testing may or may not reveal information that necessitates additional testing. In that case, those tests must be done and the new data used to revise the process model until it predicts the anticipated behavior of the process under development. The process model always plays a major role in predicting an outcome of the flow sheet under development—especially when the feed materials or the ratio of different feed materials need to be changed due to unforeseen circumstances.

10.4.6 Identification and Assessment of Environmental Impact

This aspect of process development is always done in consultation with an experienced environmental specialist who has a good understanding of the proposed flow sheet. It consists of:

1. Identifying the streams and their chemical analysis—specifically effluent streams.
2. Defining the volume and chemical analysis of effluent streams that may need treatment prior to discharge.
3. Defining the quantity and chemical analysis of residual solids that may be produced in the process for safe disposal.
4. Establishing procedures to determine whether the solids:
   (a) Are a hazardous waste or not
   (b) Pass the Toxicity Characteristic Leach Procedure (TCLP) to establish whether it is a "characteristic" waste or not
5. Holding technical discussions with the environmental authorities (local, state, and federal) about the regulations that would be proposed for the safe disposal of the waste streams—both liquid and solid. It is always a good idea to be proactive in discussions with the environmental authorities about the current and proposed regulations so that all process steps needed to address the treatment of waste streams can be incorporated in the proposed process flow sheet right from the start.
6. At this point in the design of the full-scale plant, considerations should be given as to whether any zero discharge options for the liquid stream should be considered. Again, the stringent and changing regulations for discharge of liquid streams and the potential for monetary fines and law suits for possible violations will be a driving force to consider this option.
7. Zero discharge options can be achieved in the following ways:
   (a) The effluent stream is treated in a conventional waste water treatment plant for impurities' removal and the treated water evaporated in a thermal unit. All solids generated by these steps are disposed of in an environmentally acceptable manner.
   (b) In case the effluent stream has a low level of impurities, it can be sent to a small tailings pond without any chemical treatment. The tailings pond may already be available close to the plant near a mine site or can be newly constructed. The overflow from the tailings pond can be recycled back to the process. The volume of recycled stream back to the main process plant should be varied to study (a) the buildup of impurities' in the main plant and its effects on the process and (b) its effect on the water balance. Once the recycle volume and its effect on the water balance is determined, the number of recycles necessary to establish whether a steady state, with respect to the level of impurities, is attained should then be determined.
   (c) In case the results of these recycle tests reveal a water balance or impurities' levels issue due to not attaining a steady-state operation, then it becomes imperative to bleed a certain volume of the effluent stream for waste water treatment.

10.4.7 Corrosion and Erosion Studies for Selection of Materials of Construction

Corrosion and erosion studies should be done at the pilot plant level. Prior to running the pilot plant, specifications of procedures for monitoring corrosion and erosion should be written up in consultation with a corrosion specialist. Testing of various materials—in the form of "test" coupons—should be done during the running of the pilot plant. The data so collected should be shared with the engineering company that will be doing the full-scale design of the plant.

10.4.8 Additional Data Collection for Design of a Full-Scale Plant

In Sects. 10.2 and 10.3, laboratory scale and pilot plant scale test work, respectively, were described in some detail. The outcome was the optimization of all the operating parameters for the continuous leaching of zinc calcine and fume and the continuous purification of the neutral leach solution.

The following variables were tested at the laboratory and pilot plant levels to gather additional data for the design of the full-scale plant.

10.4.9 Variables for the Continuous Leach Tests

1. Zinc calcines produced from various zinc concentrates based on their projected availability.
2. Increased levels of zinc fume in the calcine/fume mixture. This was done because large amounts of zinc fume were available for treatment at a low cost.
3. However, this attempt brought in increased levels of deleterious impurities such as arsenic, antimony, and germanium which provided considerable challenges in the purification step.
4. This attempt also forced the use of increased levels of soluble iron in the leach step—with addition as an iron salt—to remove the above deleterious impurities by precipitation and/or adsorption.
5. In turn, this affected the settling characteristics of the leach residue due to the gelatinous nature of the ferric hydroxide in the residue. Consequently, extensive testing on the settling rates of the leach residue was needed for optimizing the thickener design.
6. Pilot plant scale thickener tests were done to establish thickener settling area requirements.
7. Stainless steel (316 L grade) was used for the leach tanks to examine it as an alternate to the conventional mild steel with lead-lined acid resistant bricks. However, this idea was rejected due to the dissolution of nickel—an undesirable impurity—at levels not acceptable for the purification step. The fall-back position was mild steel, lead-lined acid resistant brick.

10.4.10 Variables for the Continuous Purification Tests

1. In spite of considerable testing in a continuous mode, it was not possible to remove cobalt consistently in the second stage of purification to the desired level of 0.1 mg/L. Supplemental testing for 2 weeks on a scale smaller than the pilot plant scale did not give results encouraging enough that would have enabled the engineering team to recommend a "continuous" mode of operation for the second stage purification for cobalt removal. Since this was a critical variable that would affect the current efficiency in the zinc electrowinning step, it was decided to keep this purification step as a "batch" operation.
2. This decision required a few extra storage tanks to be included in the final design for holding the solution after the continuous first stage purification.

10.4.11 Summary

In this section, generic procedures for collecting data for the design of a full-scale plant have been outlined in a few steps. The main thrust is how all the necessary information collected during the pilot plant is put together by a design team involving pilot plant operators, operating personnel, and a reputable outside engineering company for the full-scale design. Environmental and corrosion specialists should be part of the team on an as-needed basis.

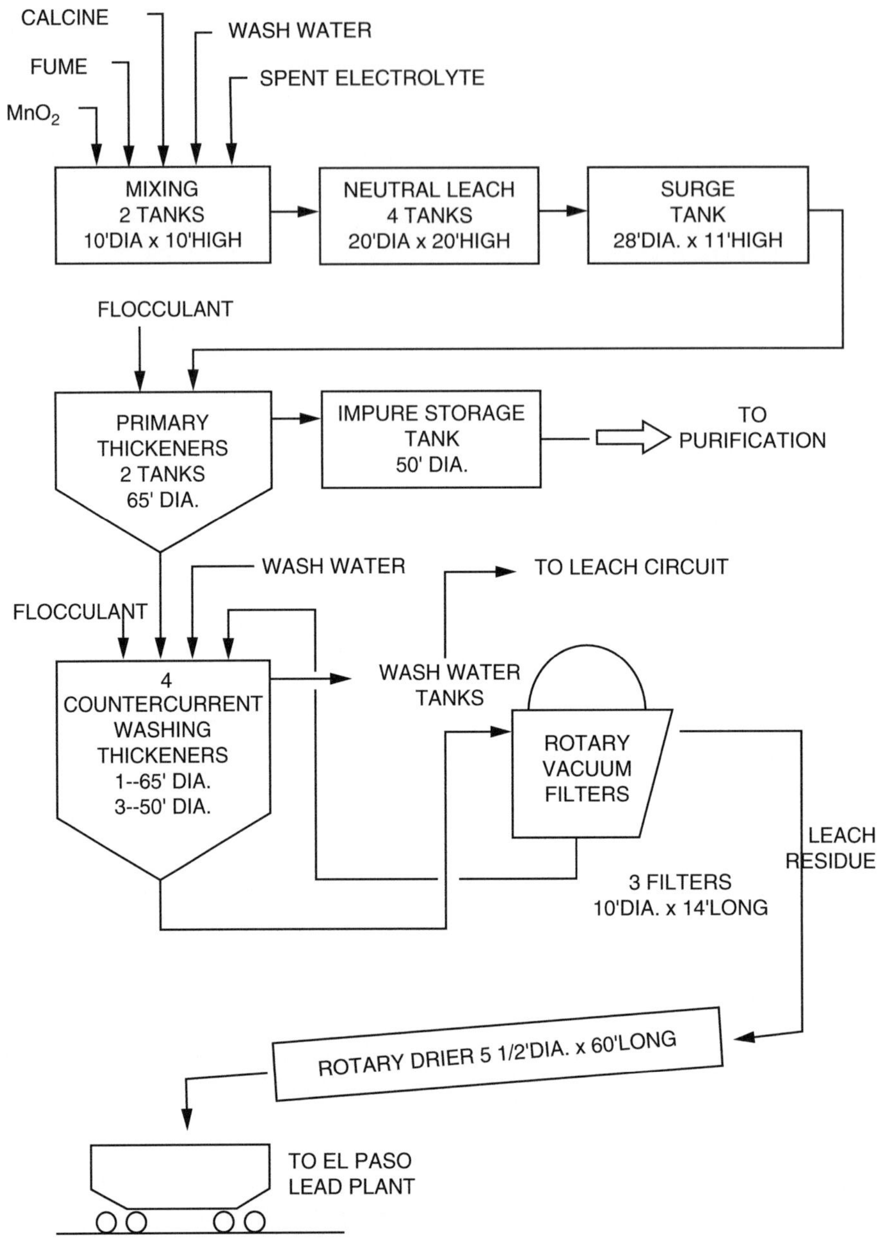

Fig. 10.3 Schematic of continuous leach circuit with plant scale equipment detail (from Ramachandran and Cardenas 1983). Copyright 1983. Reprinted with permission of The Minerals, Metals and Materials Society

The outcome of the detailed pilot plant testing resulted in the development of the following flow sheets:

Figure 10.2: Schematic of the modernized plant.

Figure 10.3: Schematic of the continuous leach circuit with plant scale equipment details.

Figure 10.4: Schematic of the continuous purification circuit with plant scale equipment details.

A brief description of these procedures as applied to the laboratory and pilot plant data collected—described in Sects. 10.2 and 10.3—for the

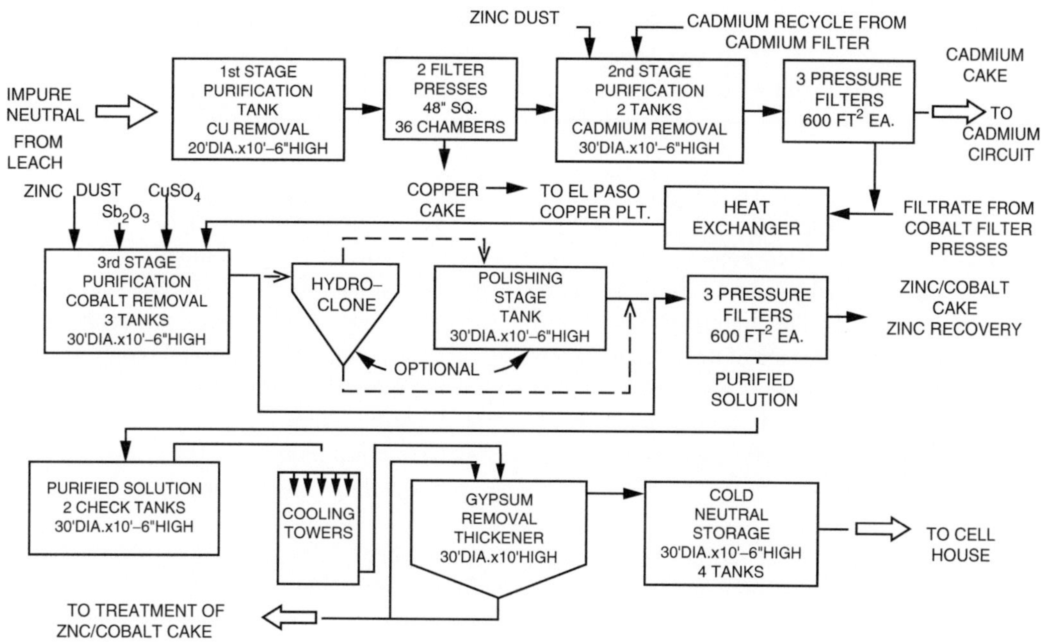

Fig. 10.4 Schematic of continuous purification showing plant scale equipment details (from Ramachandran and Cardenas 1983). Copyright 1983. Reprinted with permission of The Minerals, Metals and Materials Society

process development of an existing electrolytic zinc refinery is included in this section. Data collected from additional pilot plant runs—described in Sect. 10.3—were also used for the full-scale design of (a) settling and washing thickeners, (b) filtration of thickened slurries, and (c) treatment of the copper–cadmium cake for cadmium recovery using cementation as the main process step.

10.5 Commissioning and Trouble Shooting

10.5.1 Start-Up Organization (Harmsen 2013)

Prior to the commissioning of any plant, a start-up organization should be put together for the success of the project. There are various models for the start-up organization. Based on the scale of the start-ups, a model where the final production staff does the precommissioning, commissioning, and start-up works the best. It is also a good idea to have the researcher and the process engineer involved in the design, a process control specialist, and an analytical chemist as part of the start-up team. The start-up team should interact with the final operating staff regarding the complete process for a few months prior to the start-up date.

The start-up team should be headed by an experienced start-up leader with prior experience in major start-ups. He/she should be assisted by a qualified assistant to provide 24/7 leadership during the start-up.

10.5.2 Start-Up Preparation

The complete start-up team should go through a "potential" problem analysis and come up with answers ready to be implemented, if needed. This exercise will be helpful in developing a step-by-step procedure culminating in a complete start-up plan. This plan should be documented and distributed to all personnel involved in the start-up (Harmsen 2013). The next step is the start of the actual step-wise commissioning phase. Various steps of the commissioning phase are described below.

10.5.2.1 Commissioning

Commissioning is a staged process which starts when mechanical and electrical work is completed by the construction contractor. The terminology, scope, and sequence of the various stages can vary with the type and complexity of the plant and with the preferences and practices of the engineering group, technology supplier, and the operating company.

Generally, engineering is carried out by an outside firm, whereas the technology supplier can be part of the operating company as well as a separate organization.

A typical commissioning process comprises:

1. Precommissioning
2. Dry Commissioning
3. Wet Commissioning
4. Hot Commissioning
5. Performance Testing

Typical activities may include:

1. Precommissioning
   - Checking for conformity to design against drawings and specifications
   - Checking commissioning and operating procedures
   - Checking the status of all process equipment and associated electrical, controls, and instrumentation
   - Checking the status of all ancillary facilities including safety and fire protection/fighting; rescue and first aid; water; air; fuel; power supply/generation; steam generation; reagent storage and supply systems; tailings disposal systems; environmental control systems; laboratory facilities; and associated electrical, controls, and instrumentation
   - Flushing, cleaning, and drying
2. Dry Commissioning
   - Planning and scheduling
   - Start-up and shut-down protocol
   - Checking piping and instrumentation against P&IDs
   - Calibration of instrumentation
   - Verification of control logic
   - Checking direction of motors
   - Functional checks of mechanical equipment and control systems, power and steam generation, environmental controls, and fire protection/fighting systems
   - Pressure testing
   - Operator training

   Note: Dry commissioning of packaged facilities is best carried out in the supplier's factory
3. Wet Commissioning
   - Testing ore handling systems under load
   - Testing the operation of solution handling systems with water
   - Checking and rectifying leakages in equipment, tankage, piping, and ponds
   - Testing power and steam generation facilities
4. Hot Commissioning
   - Introduction of feed materials
   - Operation of all facilities at low capacity under continuous closed loop conditions and increasing to design conditions of temperature and pressure
   - Monitoring for leakages in equipment, tankage, piping, and ponds, rectifying as required
   - Checking of instrumentation and controls, recalibrating as required
   - Implementation of sampling and analytical support activities
   - Commencing verification and updating of the process model with operating and analytical data
   - Implementation of environmental monitoring, controls, and reporting activities
   - Progressive increase of plant throughput to design conditions
5. Performance Testing
   - Continuous operation for a specified period under design conditions

- Verification of process and equipment efficiencies
- Sampling and data gathering to verify throughput, production rate, product quality, quantity and composition of effluents, consumption of reagents, water, steam, power, and fuel
- Compliance with process and equipment guarantees

Commissioning Team

The commissioning team is typically multidisciplinary and may include:

1. A commissioning manager
2. Process specialists
3. Instrumentation/controls and electrical technicians
4. Maintenance personnel
5. Technology supplier's representatives
6. Equipment supplier's representatives
7. Operating personnel

The composition of the team typically varies as the program progresses through the various stages. Additional temporary personnel should be included during commissioning and the initial operating period to collect data, collect additional samples, and monitor the process while the regular operating crew is occupied in the mechanics of start-up.

Completion and Hand-Over

The point of completion of commissioning and hand-over from the engineering firm to the operating company varies and can occur at the end of wet commissioning, hot commissioning, or performance testing.

10.5.2.2 Trouble Shooting

Trouble shooting is carried out throughout the commissioning period. In each phase, punch lists (with priority) drawn up and rectification is undertaken before moving on to the next phase.

- During precommissioning and dry commissioning, initially the problems will likely be related to functional issues with equipment, electrics and controls, and interfacing with supplier packages.
- Wet commissioning issues will be more related to leakages and utility supply systems.
- During hot commissioning, problems typically relate to equipment capacity, process design, product quality, environmental and safety issues, malfunction and calibration of instrumentation and control systems, additional leakages, achieving and controlling operating conditions (such as temperature and pressure), and early indication of failure of materials of construction program of testing alternatives.
- Performance testing will reveal any capacity bottlenecks and provide a review of the problems rectified during hot commissioning and further information on materials of construction.

Some issues may require:

- Additional design and construction work by the engineering firm
- Further test work and process development by the technology supplier
- Replacement of faulty or inadequate equipment items
- Changing to alternative equipment designs
- Additional environmental control systems
- Modifications to ancillary facilities
- Longer term materials of construction testing

In some cases, it may be necessary to engage specialist consultants to provide additional expertise.

10.5.3 Post Start-up Report

It is important to document—such as a daily log done 24/7—all the activities of a start-up of a plant. This should be done until a steady-state operation is attained over an accepted length of time and the operating personnel have a degree of comfort in running the plant. The report will have the following information:

1. Start-up time.
2. Deviations from design performance such as product quality, production rate, and utility needs.
3. Deviations from design specifications and conditions.
4. Critical analysis of the relation between deviations from expected performance.

The data so collected will be useful as:

1. Reference document for the operating personnel—current and future.
2. Errors and learning points can be used for making improvements in the design of the plant.
3. On a long-term scenario, continual quality control improvements learnt from the start-up can be implemented on a company wide basis.

10.5.4 Commissioning and Trouble Shooting of Modernization of an Existing Electrolytic Zinc Plant

In Sects. 10.2–10.4, laboratory scale, pilot plant scale testing and data collection for design of a full-scale plant, respectively, were described in some detail. Based on the generic description of commissioning and troubleshooting in this section, the continuous leach and purification were brought on stream using the various commissioning steps outlined above.

The commissioning team consisted of:

1. Start-up manager.
2. Process specialist from the engineering company that designed the modernization plant.
3. Research engineer and pilot plant manager involved in laboratory and pilot plant testing.
4. Instrument/controls and electrical engineering staff from the existing plant.
5. Maintenance engineering personnel from the existing plant.
6. Equipment suppliers' representatives.
7. Operating personnel.

Preliminary meetings were held among the commissioning team to establish a line of action for start-up of the plant. The commissioning process steps listed at the start of this chapter were followed in a sequential order.

After completion of pre- and dry commissioning, wet commissioning was done as follows:

1. Testing of solids handling systems under load for a short period of time. Since the leach tanks were yet to be tested, these solids were temporarily diverted to a storage system for later reuse. Load cell and other related instrument calibrations were confirmed.
2. All solution handling systems such as leach tanks, purification tanks, storage tanks, pumps, and pipes were tested for leakage and fixed.
3. All utilities, viz., power and steam generation facilities were tested for smooth and continuous operation.

Hot commissioning followed wet commissioning. This involved:

1. Gradual introduction of feed materials, viz., zinc calcine and fume, pyrolusite ore (MnO_2) for iron oxidation, spent electrolyte from electrowinning (EW) operations as the leachant, water, if needed, for pulp density adjustment, etc.
2. Running of the leach section at reduced capacity to set up recycle streams—if it was part of the design and study effects of the same.
3. Continued checking and fixing of leaks in the flow system.
4. Implementation of all other aspects of hot commissioning outlined in this chapter.
5. Progressive increase of plant throughput to attain design conditions.
6. This operation took about 3 weeks and resulted in steady-state operations running at full design capacity.
7. During this time, pre- and dry commissioning of the continuous purification was completed

and the purification process run at low design rates to check the process chemistry for the removal of impurities.

8. This was done by running the purified solution through a small test section of the electrowinning (EW) plant. The parameters evaluated included current efficiency, cathode quality, etc.
9. Based on the feedback from the EW tests, the operating parameters for the purification steps were fine-tuned to get the desired impurities' removal.
10. The electrowinning testing helped in being able to run the purification steps close to design parameters in a relatively short time.
11. Additional spare storage tanks and thickeners which were readily available were put to use for temporarily storing leach and purified solutions during the commissioning process. While doing so, it was decided to use two large 40,000 gal thickeners to store purified solutions, test the solution for impurities, and then proceed to the use of the solutions in electrowinning. This safety step played a major role in preventing any upsets in the EW step. Also, these thickeners helped to precipitate and remove gypsum from the purified solution. This step was useful in preventing major gypsum precipitation in the EW piping and cells thereby reducing cleaning and maintenance costs.
12. During the commissioning steps, environmental monitoring, sampling procedures, and controls were set up and the reporting protocol established for use by the operating personnel.
13. Also, sampling procedures and analytical support needs were established as per protocol.
14. In summary, a continuous leaching and purification plant for combined treatment of calcine and fume in the electrolytic zinc industry for the first time was commissioned. Also, troubleshooting with its attendant solutions was completed in a timely manner.

10.5.5 Summary

In summary, this chapter describes the steps that were taken to successfully commission a new project from start to finish. This case study of the modernization of an existing plant was chosen as an example for its simplicity. A similar approach can be applied for any process whether or not developed from scratch through an innovative idea.

Lastly, this chapter describes how an innovative idea, new or improvement of an existing one, can be taken from concept to a successful operating process.

References

Harmsen, J. (2013). *Industrial scale-up: A practical innovation guide from idea to commercial implementation* (pp. 45–51). Oxford, England: Elsevier. ISBN 978-0-444-62726-1.

Kordoski, G. A. (2002). Copper recovery using leach/solvent extraction/electrowinning technology: Forty years of innovation, 2.2 million tonnes of copper annually. *The Journal of the South African Institute of Mining and Metallurgy, 102*, 445–450. November/December 2002.

Ramachandran, V., & Cardenas, R. I. (1983). Modernization of ASARCO's Corpus Christi electrolytic zinc plant. In *Proceedings of the 3rd International Symposium on Hydrometallurgy* (pp. 971–984). Atlanta, GA: Publication of the Metallurgical Society of AIME. Retrieved March 1983.

Part III

Process Optimization

An Integrated Mining and Metallurgical Enterprise Enabling Continuous Process Optimization

11

Ananth Seshan and B.K. Gorain

11.1 Introduction

The last few decades have seen the advancement in information technology as no other period in history. The same period has also a seen dramatic increase in computational power and a correspondingly significant decrease in the cost of electronics. In addition, internet and telecommunication technologies (especially wireless) have exploded in the first decade of the twenty-first century. The relative advancements in these areas (collectively termed as digital technologies) have resulted in several synergies to deliver innovation in diverse applications. The mining industry has been relatively slow in adopting these technologies compared to the other industries such as chemical or oil and gas.

11.1.1 Challenges

Mining companies globally are going through a series of challenges such as declining ore grades with complex metallurgy, increasing depth of ore bodies, high capital and operating costs, lower productivity, shortage of technical resources, significant pressure from stakeholders to deliver along with growing environmental and corporate social responsibility issues. In the case of the iron and steel industry, there is overcapacity, and as a result, erosion of profitability due to the recent downturn that has brought about a huge gap between supply and demand. Improvement in profitability would need an improvement in operational performance. In cement manufacturing, there is a need for energy cost reduction, increased productivity, the need to accommodate alternate fuels, and the need to maintain consistent clinker quality.

Consequently, over the last few years, there has been a major focus amongst some metallurgical industry leaders on operational performance and excellence. This has been only possible because of the replacement of a silos-based approach by an integrated view of the organization. Such an integrated view is now possible via the latest state-of-the-art technologies. Even so, such a transformation calls for a change in culture, which, for most organization is not trivial. Attempts in the past to integrate functions such as through Mine-to-Mill initiatives in the mining industry have demonstrated significant benefits to many operations (McKee 2013). Still, sustaining these benefits has clearly been a challenge due to manpower turn over and lack of proper systems and structures that integrate information

A. Seshan (✉)
5G Automatika Ltd., 203-203 Colonnade Road, Ottawa, ON, Canada K2E7K3
e-mail: aseshan@5gautomatika.com

B.K. Gorain
Barrick Gold Corporation, 161 Bay Street, Suite 3700, Toronto, ON, Canada M5J 2S1

V.I. Lakshmanan et al. (eds.), *Innovative Process Development in Metallurgical Industry*,
DOI 10.1007/978-3-319-21599-0_11

between business functions within an operation and also between operations and the corporate.

11.1.2 Opportunities

The advent of digital technologies have opened-up opportunities to integrate interdepartmental information for a unified understanding of, and across, an entire organization. The advances in instrumentation, equipment control systems, supervisory control systems (SCADA/Historian), and industrial networks have helped in the direction of enterprise wide integration. The availability of real-time information across an enterprise has given rise to numerous possibilities to improve organization performance. Some of the potential opportunities are as follows.

11.1.2.1 Informed and Timely Decision Making

The current state-of-the-art allows for production performance to be linked to planning, design, and the supply chain. Integrated software tools have allowed for an organization to share not just raw process data but intelligence that is abstracted from such data to drive well-informed decisions (called "actionable intelligence"). Therefore, changes in performance happening in any part of the production/mining value chain (ecosystem) can have an instantaneous impact on the other parts—making the interactions between the various stakeholders of the ecosystem dynamic and real time. This provides a better capability to all stakeholders in the ecosystem to predict outcomes that are relevant to their respective scope.

11.1.2.2 Cost Optimization

In the cement industry, today, it is possible to optimize the lowest cost fuel mix for kilns based on process and business constraints and taking into account the up-to-date market information. Similarly, in the mining industry, the cost of operations could be optimized in a dynamic manner through real-time integration of information across the entire mining value chain. An example of a useful synergy in the mining industry is the one between production and maintenance, allowing companies to innovatively maximize asset availability at minimal maintenance cost. In other words, of late, maintenance has increasingly ceased to be a cost center and has become an "opportunistic" driver to improve asset, and therefore, production performance.

11.1.2.3 Asset Optimization

Asset availability has been one of the key performance metrics for the mining and the metallurgical industry. Availability typically reduces due to inadequate and timely information on the condition of the critical assets to perform proactive maintenance and asset health care. Predictive intelligence of assets is an evolving method to maximize asset performance and to achieve better return on physical assets. Increasingly, equipped with predictive intelligence, proactive maintenance of assets is taking over reactive maintenance in the mining and metallurgical industries. Proactive is herein referred to as a combination of preventive and predictive maintenance. Critical assets in the mining and metallurgical industries have increasingly been subject to usage-based preventive maintenance and appropriate predictive maintenance to avoid unplanned downtime. The monitoring of usage is done by the use of advanced software tools and open standards available for enterprise application integration. The usage of critical machines is tracked by a real-time software and when they cross a threshold for a preventive maintenance call, a maintenance work order is automatically generated by the real-time software to maintain the health of the assets in good condition at all times Enterprise Gateway (2010). These software tools deliver the capability of optimizing the maintenance functions in such a manner to achieve maximum availability of production at a minimum cost of maintenance.

11.1.2.4 Energy Savings

Related to the operational performance of machines is energy efficiency. Mining and metallurgical processes such as haulage, comminution, pressure oxidation and electrowinning consume a significant amount of energy. The energy intensity of a metallurgical process is always a critical factor to an organization from multiple points of

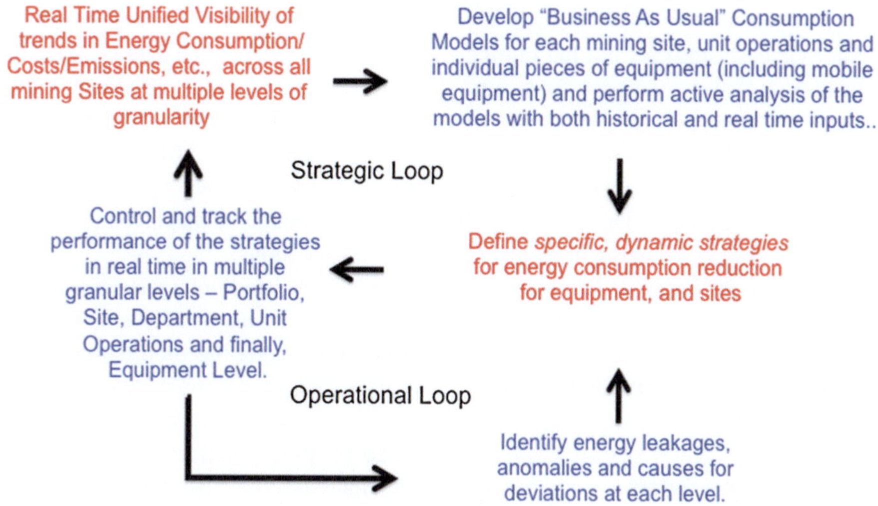

Fig. 11.1 The integration (and traceability) of strategic and operational control of energy consumption in a multisite facility

view—viz., profitability, greenhouse gas emissions, and system/process degradation. Many metallurgical operations are starting to have an advanced and intelligent energy management system in place. The purpose of energy management varies from implementing operational control of energy consumption in real time in a local facility to achieving a strategic goal of the enterprise to reduce energy consumption across the portfolio. These are normally performed in two parallel "loops" (refer Fig. 11.1)—the strategic loop that allows for an aggregated and comparative evaluation of the energy performance of various sites to arrive at useful strategies and the lower level control loop that implements the strategies via an existing control systems infrastructure.

The state-of-the-art solutions for achieving improvement in energy management allow for the *dynamic* generation of energy saving strategies in response to evolving operational conditions (as opposed to a shrink-wrapped set of pre-programmed strategies), and furthermore, provide improved traceability (integration and visualization) of the downstream implementation of such strategies via the distributed control of multiple, individual pieces of equipment.

11.1.3 The Way Forward

It is now possible for the mining and metallurgical companies to leverage recent advancements in information and telecommunication technologies as part of their operational excellence drive.

Some mining companies are already working in this direction by integrating their operating data, scattered across multiple locations, into centralized database with dashboards and analytics capability with an aim to make a step change in improving productivity and reducing operating costs through better collaboration between different business units.

Companies like Rio Tinto and BHP-Billiton in Australia, Codelco (in partnership with Honeywell called Kairos mining) in Chile, and some other companies and operations have taken steps in the direction of a partially or fully integrated operations strategy (BHP-Billiton 2013; Sherring 2012; Zamora et al. 2010).

The following sections of this chapter will discuss in details as to how these evolving technologies could be utilized. This chapter will elaborate on the concept of an integrated mining and metallurgical enterprise that comprises an architecture involving five logical layers. These sections will also illustrate, via case studies from the mining, cement and iron and steel industries, how automation systems and concepts are involved in each of these layers. In a subsequent section, the technology enablers for the five-layer integrated enterprise shall be explained. The chapter will conclude by providing a stepwise implementation methodology for the integrated enterprise.

11.2 The Integrated Metallurgical Enterprise

11.2.1 Optimization Goals

The Integrated Metallurgical Enterprise integrates the various functional elements of a metallurgical organization from strategic, tactical, and operational points of view. *It allows for a simultaneous focus on both production and cost control.* This initiative provides the industry with an ability to deal with frequently changing business dynamics and is therefore strategic to many mining companies.

For instance, in the mining industry, the target for tangible improvements through this initiative could be 5–10 % increase in productivity and a 15–20 % reduction in operating costs as the first step. These targets are realistic and are based on improvements that have been demonstrated by some mining companies pursuing even through a simpler strategy such as *Mine-to-Mill.*

Another simple strategy is to integrate Geology, Mining, and Processing with a unified visualization of the entire enterprise—focusing on a common target such as mining product quality (including grade and fragmentation) that delivers the best value in processing instead of just focusing on mining tonnage. This tactical goal in turn can result in the acquisition and tracking of relevant operational data—in the example above, the data collected and tracked are for instance, ore dilution, ore loss in waste, blast fragmentation, top size and hardness of rocks, mineral liberation during milling, metal recovery in process plant along with energy and material consumption for the mining and processing steps involved. The use of a web-based integrated dashboard to share such data between the different stakeholders or functions along with the ability to interact and interrogate the database will allow personnel in different functions to cooperate as they all have one common goal as this goal is continuously displayed and monitored in strategic locations.

This integrated enterprise must have appropriate value drivers and KPIs. As an example, if a mining manager has indices that measure only tonnage and production cost per ton mined, then there is no incentive to change a well-established blasting practice because such a change may increase the blasting cost per ton of ore mined. On the other hand, an integrated enterprise approach is defined holistically by a "hierarchy" of metrics at various levels of abstraction and the organization strives to optimize across the hierarchy.

Having an integrated database with estimates of key value drivers and KPIs as mentioned above allows benchmarking of an operation in terms of its unique set of KPIs. This database could receive information from a number of sources such as:

- The mine dispatch system
- The plant process data historian (PI system)
- The Laboratory Information Management System (LIMS)

An integrated enterprise strategy will require systems to measure, monitor, and control the raw data that result in the estimation of the key value drivers and KPIs across the mining value chain with an emphasis on optimization of the interfaces between business units to produce the desired outcomes. The key advantages of such an approach in the mining industry are availability of more accurate, reliable, and quantified information to assist with the following applications:

(a) Mine-to-Mill reconciliation and optimization
(b) Ore stockpile management
(c) Geometallurgical modeling
(d) Dynamic mine block modeling and planning
(e) Dilution control

(f) Grade Engineering (a new concept from CRC ORE mentioned in McKee (2013))
(g) Mill feed stabilization

11.2.2 Optimization Philosophy: Optimization of the Parts vs. Optimization of the Whole

As discussed earlier, one of the key differentiating philosophical insights seen in the industry now as a step change from traditional approaches has been to strive for optimization across the entire organization instead of only in parts. This goal is achieved via real-time integration of the enterprise. This philosophy has proven to be successful in many industries including oil and gas, manufacturing, and recently in mining by bringing many fragmented solutions together.

The "optimization of the parts," no doubt, allows improving efficiency of individual units and is still an important part of the mining or any metallurgical business. But this results in potential economic benefits to be missed as the focus is on constant execution to meet the targets of the individual business units, and the improvement opportunities associated with a systemic approach come only as an afterthought (Sherring 2012). As stated earlier, focusing on parts alone is tantamount to working in various business unit silos resulting in limited conformance to life-of-mine (LOM) plan in the case of the mining industry and looking beyond the plan is often challenging.

In order to address these challenges and to improve profitability, it is becoming imperative to look at a mining enterprise holistically requiring a strong interaction of professionals in various business disciplines involving Geology, Mining, Processing, Maintenance, Environment, Health and Safety, Finance, Supply Chain and Logistics, IT, HR, Community, and Public Relations. All of these professionals need to interact with a common automation platform that will allow for multiple applications to share data between them.

There are many examples that suggest that integrating different business disciplines such as Mine-to-Mill involving true integration of geology, mining, and processing functions have led to significant productivity gains and cost savings along with intangible benefits including a more cohesive, satisfied, and performance-oriented workforce transcending the traditional boundaries (McKee 2013).

The Rio Tinto experience, as shared by Sherring (2012), suggests that an integrated operations strategy also allows a balanced focus on planning, execution, and improvements with an enabling organizational and working environment.

The Kairos Mining's Collaboration Centre (a joint venture of Codelco and Honeywell) provides real-time monitoring of plant operating data and automated control systems for multiple sites with a focus on knowledge management and long-term process performance improvements (Zamora et al. 2010).

11.2.3 Five-Layer Architecture for an Integrated Mining Enterprise

Typically, an integrated Mining enterprise could have five logical layers as illustrated in Fig. 11.2. The first and the lowest logical layer involves raw data acquisition and monitoring of various unit processes, machines, sensors, and transducers in the production environment.

The data acquired from the lowest layer is used by the control and the data analysis layer. This forms the second layer. The controllers in turn could be organized in a hierarchy (within this layer) based on the sophistication required. The lower level controllers are normally regulators—meaning they are mandated to maintain a control variable within reasonable bounds of a reference set point.

The higher level controllers comprise supervisory and intelligent controllers. These controllers will be able to respond to the dynamics and actively alter the set points or cope with the nonlinearity as they present themselves using various

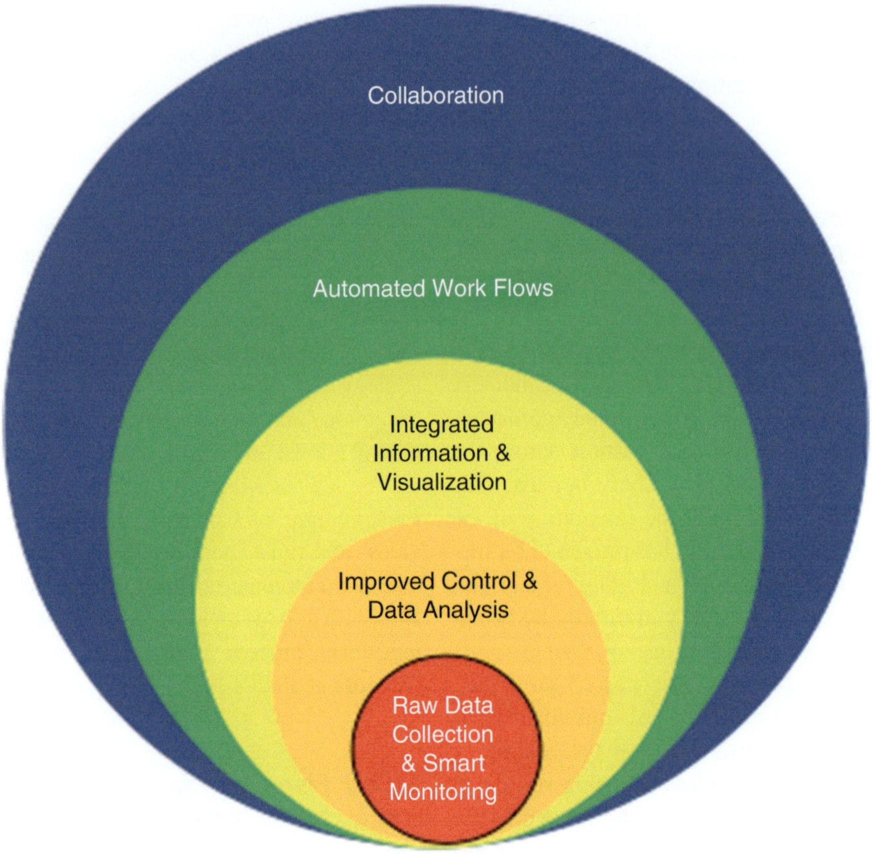

Fig. 11.2 Five layers of an Integrated Operations Enterprise (based on Sinclair 2012)

Artificial Intelligence based or mathematical modeling techniques. Such controllers are very domain specific and are constructed to meet the challenges of a specific process dynamic. Also this layer houses "Historians" for recording historical data and "Human/Machine Interfaces" for understanding the status of a process/machine visually for an operator. All of the subsystems in this layer are amenable to integration with the other layers in the architecture. Each low level controller, historian, and supervisory controller acts as modular component in the architecture that can exchange data with the other elements through standard conventions across a standard networks.

The third incremental layer is a plant-wide abstraction—one that integrates and visualizes the data across the individual business units—this layer comprises a web and mobile platform for unified visualization. This layer houses Manufacturing Execution Systems (MES) and Manufacturing Operations Management Systems (MOM).

The fourth logical layer is the action layer—one that integrates workflows between different business units across the enterprise, not restricted only to the plant or production environments. In other words, this layer comprises systems that can enable integrated and automated work flows between the different business units.

And the final layer enables "collaboration" between multiple stakeholders across different functional units within the enterprise. In other words, the collaboration layer allows for optimization across multiple departments—often, one whose objective is to achieve "continuous improvement" such as minimization of costs, downtimes, variation, and maximization of the return on assets under multiple constraints spread across the different departments. The availability of the integrated enterprise architecture not only enables seamless synergy between the multiple stakeholders to effectively and actively contribute, but also allows for the collective multidisciplinary team to generate predictive actions to effect control and optimization in a future time period based on the current state.

11.2.3.1 Logical Layer 1: Raw Data Collection and Smart Monitoring (Examples)

Using Radio and Smart Tags for Ore Tracking

In the mining industry, ore tracking is becoming a necessity as mining operations are becoming immensely complex. This tracking enables effective integration between the mine and the mill. In other words, the *tracking allows for information flow to be synchronized with material flow throughout the cycle.* Such tracking improves efficiency and the effectiveness of the process—especially when certain dynamic factors are at play, such as change in ore characteristics (Isokangas et al. 2012). Using RFID tags, a system tracks ore types from the mine to the mill and through the process plants. Physical RFID tags are now available from a number of suppliers, with some tags specially designed for mining applications. They can be passive or active RFID tags. Active tags require a battery power source and therefore may not be appropriate for long-term stockpile applications.

This system helps to identify the origin of the ore and its behavior throughout the mining process. These tags travel through a mine and process plant in a series of steps. Initially, the tag and insertion location is logged using a handheld computer or PDA, and then it is inserted into the rock mass in the same holes where blasting explosives are placed. The tag travels with the ore through digging, transport, and processing before being detected by sensors that are positioned for recording the time and the tag at various points. The RFID tag data is then loaded into a centralized database and analyzed as required.

The use of a tag system allows development of relationships between ore characteristics (available in the mine block models) and operating parameters in the mine and processing plant (such as ore dilution, fragmentation, stockpile residence times, segregation, energy consumption, ore grade) along with actual mine production and metal recovery. With this knowledge, ***operating parameters can be optimized in real time*** to respond rapidly to changes in ore characteristics, thus resulting in control of operating costs and improving productivity.

Fragmentation Analysis

Optical sizing technology for measuring size distribution of fragmented rocks after blasting has been used for many years now. The WipFrag fragmentation sizing system was originally designed using a roving camera and operator-assisted analysis. Using optical sizing technology, mining blast professionals could evaluate, reassess and redesign their blasts, while understanding the effect of their design on their final product. In addition, they could begin to quantitatively evaluate the effect of geological structure of their blasts. Although the accuracy of this method is low, but still useful as an alternative method of screening large masses of rocks.

The new generation of the Split-Online digital image analysis has been applied for accurate, continuous, and rapid measurement of rock fragmentation. Cameras are installed along key stages of crush, convey, and milling processes to determine rock fragmentation size. The critical data for mining operations, including particle size, shape, color, and texture, are calculated by the advanced split algorithms and can be reported to a centralized database to enable real-time evaluation by mine operators and management. Installation of the Split-Online camera systems in various stages of the comminution process at the Morila gold mine in West Africa resulted in a 10 % mill throughput improvement (Gillot 2006). Split-Online cameras along with ore tracking system have significant potential for operations and can provide information on the less understood interface between mine and mill, which is a key for successful implementation of an integrated operations strategy.

Using Piezo Electric Sensor Array for Slurry Flow Meters and Density Gauges

A new innovative technology for noninvasive flow measurement using piezo-electric sensor array has recently achieved wide acceptability in the mining industry with some significant benefits compared to the traditional flow meter technologies such as Electromagnetic, Ultrasonic Doppler, Differential Pressure, or Coriolis (Markoja 2011). These flow meters have recently been used for better quantification of recirculating loads in a grinding circuit and also allow metal balancing and reconciliation in a process plant for better Mine-to-Mill reconciliation.

Using Light and Chemometrics for Online Monitoring of Mineralogy and Assays

Online mineralogy and assays measurements provide opportunity to optimize process plant in real time. This will allow a tighter control of tailings losses, which otherwise is difficult to achieve in an offline-based mineralogy and assays measurements that are used presently. Online mineralogy measurement from BlueCube™ technology is a major breakthrough and has recently been successfully implemented in many precious and base metals industries (Mumbi 2012). This technology is based on diffused reflective spectroscopy combined with propriety chemometric techniques.

11.2.3.2 Logical Layer 2: Improved Control and Data Analysis

Supervisory Control Hierarchy

The control layer is responsible for local control of the process. The sophistication of controllers depends on the complexity of the process—simpler PID controllers are sufficient for linear control assumptions but more sophisticated approaches are necessary for processes that exhibit nonlinear dynamics. Control of such processes requires the development of a mathematical model of the process or plant dynamics. The parameters defined in the mathematical model are then dynamically calibrated before they are implemented as part of the controller of the process or plant. The controller is designed to cope with the dynamics of the plant/process for various set point and output conditions.

The problem arises when a critical process variable which is an input to the controller cannot be measured or the dynamic parameters of the model are imprecise or if there are external disturbances to the process that are not known or are not modeled—which are also the realities of the mining and metallurgical industry. The controllers degrade rapidly and the need for active refinement of the controller is necessary—which introduces the need for a supervisory layer. In these cases, the set point itself has to be changed to cope with the changing dynamics of a process, and/or, to compensate for the errors or the incompleteness in the model of the process. In such cases, a hierarchical structure of controllers will be used to control and optimize the process at various levels of abstraction.

The low level controller normally drives the system towards a local set point but does not guarantee global convergence especially in the wake of disturbances and influencing factors that are outside the scope of the local controller. In order to compensate for such factors, a supervisory control loop will be necessary. For instance, supervisory control is needed when the set point itself has to be modified actively and dynamically. Figure 11.3 shows a supervisory control hierarchy proposed by Bergh et al. (2007) for a copper solvent extraction pilot plant.

Solvent extraction is an important operation in hydrometallurgy—this involves a process of transfer of soluble metal compounds occurring between aqueous and organic phase. Extraction is a chemical reaction between the metal ion in the aqueous phase and the extractant from the organic phase. The control objective requires a trade off between the concentration and flow of the Pregnant Leach Solution, the flow and concentration in the organic phase of copper, the degree of entrainment of organic in the aqueous, the aqueous carryover in the organic, etc.

In such a process, the control is shown to be organized in three levels by Bergh et al.

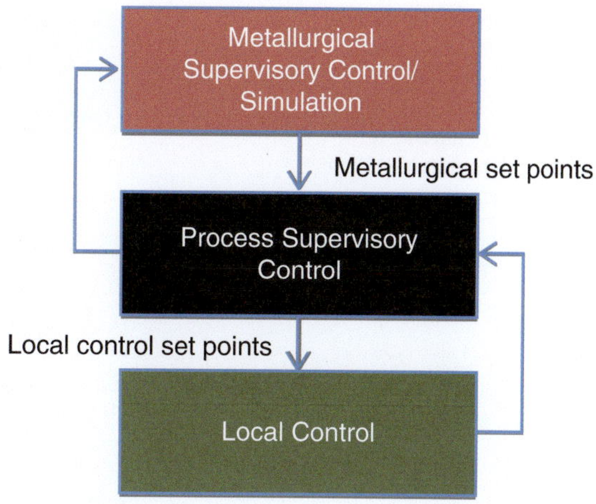

Fig. 11.3 An example of a supervisory control hierarchy (from Bergh et al. 2007, © 2007 Canadian Institute of Mining, Metallurgy and Petroleum)

(2007)—(a) local control loop as specified earlier; (b) hydrodynamic supervisory control loop; and (c) metallurgical supervisory control loop. In their example, field measurements included flow, level, and conductivity—these are provided as data inputs to a Programmable Logic Controller (PLC). The output of the PLC is the set point for the local level controller.

The hydrodynamic supervisory control loop acts as a process supervisory controller whose mandate is to control the flow rates between different process units and the stock solution levels in a coordinated manner. This would enable every process unit to be operated in *different* internal, external, or overall organic/aqueous ratios.

The metallurgical control loop can then provide the set points for local and the overall organic/aqueous ratios in order to modify the metallurgical targets of the copper concentration in different streams. The inherent problem here is the uncertainty in the measurement of the concentration of each stream. In order to achieve reasonable values for the concentration two major approaches have been reported in the literature—a prediction approach that relies on the use of statistical or Artificial Intelligence techniques; or fitting a dynamic model by means of experimental data. In both cases, the values obtained for the concentration of copper in the various streams have been used for developing a metallurgical strategy—such as the best operating conditions (set points as mentioned above) for the solvent extraction plant given the dynamic inputs.

Another example of metallurgical supervisory control driving lower level controllers is found in iron and steel making—especially in the characterization of slags. It is well known that the composition of the slags have an impact on specific physical and thermodynamic properties of steel, such as viscosity, density, activity, sulfide capacity, and so on. A system was recently developed based on the theoretical and experimental findings of Seetharaman (Thermoslag Ver. 2.0 2010) to provide a reverse optimization of slag composition, given certain desired properties of steel which are in turn based on a specific application. That is, the metallurgical supervisory controller in this case was capable of providing the appropriate ranges for the slag composition for achieving the lowest viscosity, or highest sulfide capacity, for a given set of components. These set points are then usable by the low level controllers to achieve the desired result.

One of the main innovations in the control layer is the integration of the control design technology with Artificial Intelligence techniques. The basic idea is to represent the equipment or plant under consideration as a *Mixed Logical Dynamic System*, i.e., systems evolving according to continuous dynamics, discrete dynamics, and logical rules. In particular, fuzzy control,

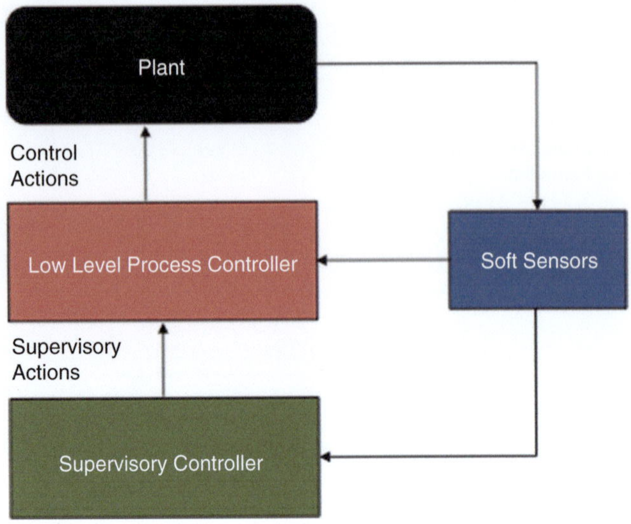

Fig. 11.4 Soft sensors serving as inputs to low level and supervisory controllers—As an alternative to unreliable or difficult to measure physical sensors

genetic algorithms, and neural networks have been extensively used to provide inputs, referred to in the industry as "soft sensors," to the lower level controllers (Fortuna et al. 2007). The values for the soft sensors are derived from process-dependent rules, which are in turn, designed or "trained" for each specific application. The soft sensors provide a major advantage—they act as "signals" of variables that are difficult or impossible to measure at high sampling rates. In other words, this technique is useful when it is impossible or difficult to physically measure a process variable that is important for the control and optimization of the process. For instance, in mineral processing, an obvious example to be used in grinding is predicting particle size average as a function of the mill state and history. Another case would be the construction of "free lime soft sensors" for cement and lime kiln control. Yet another example is assessing the temperature distribution of a kiln or furnace in a continuous manner.

Also such soft sensors are used to provide a backup for critical process measurement devices (Fig. 11.4). In the case of a failure of critical process measurements, a soft sensor can provide the control strategy with a usable "estimate" of the missing measurement. This allows the controller to continue to work to its objectives while the failed device is repaired.

The lower level controllers are normally implemented using Programmable Logic Controllers (PLCs) or Distributed Control Systems (DCS). The tactical level controllers are implemented using either SCADAs or PCs. The supervisory metallurgical control loop is normally executed offline using a PC. These three hierarchical levels are normally found in any sophisticated control system controlling a complex metallurgical process.

PCs and DCS systems are generally needed for implementing dynamic controllers. A commonly implemented dynamic controller in the metallurgical industry is Model Predictive Control (MPC). MPC is based on the "receding horizon" principle that allows for future optimal control actions to be computed only for a short future time horizon: $[t, t+T]$, where t is the current time and T is the prediction horizon length. The most important innovation in this type of controller is that only the first term of the sequence is implemented. Subsequently, a new sequence that replaces the previous one is computed when a fresh set of updated measurements are available. And once again the first term in this sequence is only implemented and this cycle goes on. In other words, every sequence is computed in an incremental manner.

MPC thus involves extensive mathematical modeling of the process in question, and the selection/design of a suitable objective function. MPC is often used for control and optimization of kilns, furnaces, mills, etc. (Alvarez 2005).

Kiln Alternative Fuels Optimization: Case Study

With increased focus on reducing the cost of operation of cement plants, organizations have started adopting alternate fuels for kilns. This has introduced some challenges—for instance, one has to cope with the different characteristics of the different alternative fuels.

ABB developed a kiln (cement) control strategy (Alvarez 2005) that achieved optimal kiln operation. These control strategies were based on neural networks and fuzzy control. This system incorporated an Alternative Fuels Optimization Module based on Neuro-Fuzzy controller integrated to an MPC. This advanced control application has been reported to have achieved optimized use of alternate fuels, reduction of waste, and strict satisfaction of environmental, contractual, and technical constraints.

The main idea here was to use the data gathered by the data acquisition layer and additional data from the market and LIMS to calculate the lowest cost fuel mix that satisfied the process and business constraints. The basic element of this algorithm as reported in the paper was a dedicated kiln model that was used for Model Predictive Control. The mathematical model estimated cooler, flame, burning zone, back end and preheater temperatures, kiln energy requirements, emission and volatiles levels, etc. The model parameters were tuned using a combination of neural networks and Kalman filtering techniques.

Process Historian

At the turn of this century, it was envisaged that process data needed to be acquired in a real-time basis with accurate time stamps to observe the global trend of the system. This was much beyond the need of just collecting data for tracking a local controller's set point. This resulted in the development of the Process Historian. Process Historian have come a long way in acquiring plant management information about production status, performance monitoring, quality assurance, tracking and genealogy, and product delivery with enhanced data capture, data compression, and data presentation capabilities. The historians allow for archival of time-based process data that can be used in a future period for elaborate analysis.

11.2.3.3 Logical Layer 3: Integrated Information

Mining Example: Mine-to-Mill Integration: Optimization of Blasting Costs vs. Milling Costs

The use of RFID tags to synchronize information flow with material flow was discussed earlier. The emphasis therein was on the data acquisition technology. In this section, a further elaboration of Mine-to-Mill strategy is provided from the point of view of illustrating how the integrated information (which is the subject matter of this chapter) can enhance the performance of mining and downstream processing activities.

One of the key objectives of the Mine-to-Mill integration is to maximize the profitability of operations through a holistic approach to the optimization of ore fragmentation. Generation of fines by blasting can have a significant impact on the following:

- SAG mill throughput increase (10–30 %)
- Reduced overall energy costs (up to 30 % decrease in kWh/t)
- Higher excavator productivity
- Higher truck loading
- Better primary crusher productivity
- Better heap leach permeability

The key task for Mine-to-Mill optimization is to identify the optimum feed size distribution for the crusher or the mill. The optimum feed size distribution is typically generated through blasting (ROM size distribution), crusher settling along with stockpile and feeder management.

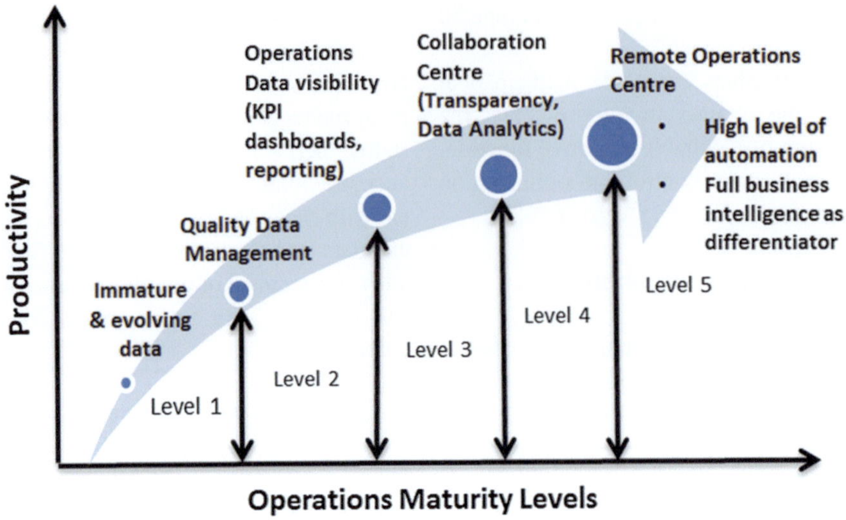

Fig. 11.5 Operations maturity levels required for improved productivity

Mine-to-Mill now is a proven methodology and has been applied at many large open pit operations around the world (McKee 2013; Renner et al. 2006). This involves rock characterization, benchmarking, and process modeling. Many of the projects have achieved 10–30 % increase in mill throughput. The issues facing success with this approach are not just technical, but involve cultural change and sustained implementation.

These Mine-to-Mill challenges are similar to that faced by business improvement opportunities such as in Lean and Six Sigma implementations and could be addressed through a dedicated and systematic effort. The concepts involved in Mine-to-Mill could be readily applied to Mine-to-Metal or for Ore-to-Profit applications. The key principle is to make changes upstream to improve quality of feed stream to reduce costs and improve efficiency through a holistic approach involving the entire mining value chain.

Integrated Operations Support and Quality Production Reporting

A recent development in the mining industry has been the establishment of operations centers, both on site and remote, where a substantial part of the total operation is monitored and controlled. Such centers are designed to oversee multiple aspects of an operation and this directly assists in establishing the desirable integrated thinking. While operations centers are not essential to an integrated thinking approach, there appears to be little doubt that they will help facilitate this goal as evident from recent successes from many companies. Figure 11.5 shows the various levels that an operation must achieve to obtain improved productivity.

The key target for any organization that is embarking on an integrated enterprise is to focus on obtaining quality and reliable operating data. This step is critical because some operations may not be willing to present their operating data on visible dashboards if day-to-day operating data obtained from various instrumentation, historians and analytical tools in operations are not reliable. Most operations rely on month-end inventory to report gold production as an example. Daily accurate reporting of gold production is challenging because of limitations in reliable metal balancing and accounting processes—due to sometimes poor reconciliation between mine and mill and also

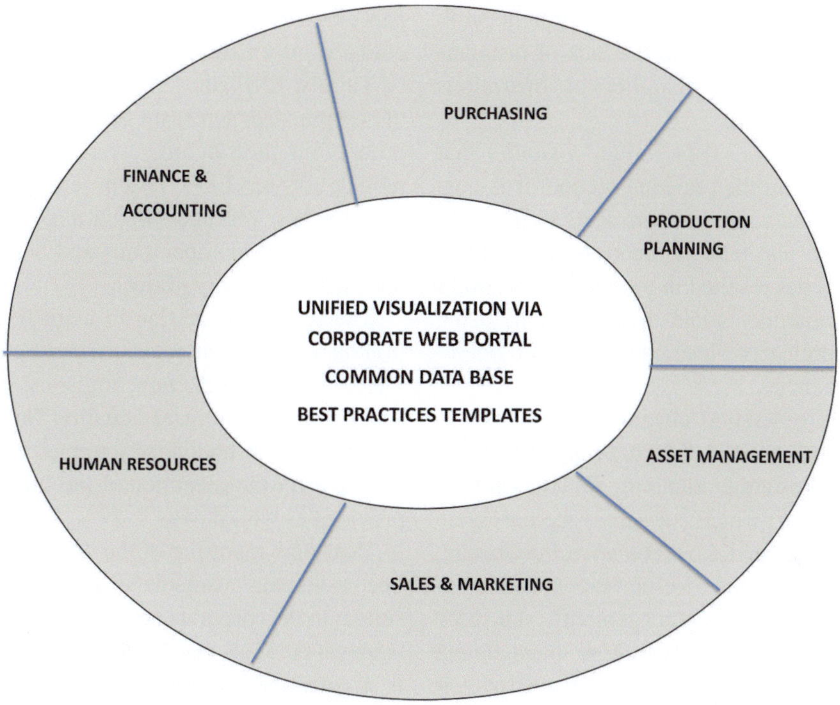

Fig. 11.6 Integrating enterprise operations using ERP software

within the different processing steps in an operation. This is not the fault of operations as embracing new ideas and relevant technologies require a joint effort between different functions in operations and also the corporate requiring an integrated platform to maximize value within the mining value chain.

To achieve this, typically, a comprehensive mobile and web dashboard displaying up to date values for the relevant KPIs are available to all important stakeholders across the enterprise to monitor and track performance and extract useful reports from a common database. This replaces the conventional silo-based individual reports generated by disparate systems in different departments that are based on distinct and separate databases. Due to the disparity in the master data stored in the different systems, reports generated from these systems do not reflect one truth.

11.2.3.4 Layer 4: Automated Workflows

Plant to Enterprise Integrated Workflows

The corporate operations in the metallurgical industry as in the case of any manufacturing organization involves finance, purchasing, sales and marketing, human resources, asset management, and production planning actions whereas the production operations are mainly related to the execution of production processes which will involve several functions such as scheduling, inventory, logistics, and quality. Figure 11.6 below shows the different functions in the execution process and how they are interrelated to each other and the enterprise level planning processes. Until recently, the workflow between different processes were human centric and paper based. The operations of the production environment and that of the corporate (enterprise) were distinct and separate and were fraught with delays and latencies due to the lack of instantaneous and real-time visibility of operations across the enterprise.

With the advent of the unified visibility that was explained in the previous section, it is now possible for the various departments to start taking action in real time in response to "current" events. This has resulted in integrated workflows between departments and their functions using software applications leading to increased operational efficiencies.

The first generation integration in the workflows was achieved in the corporate side using Enterprise Resource Planning (ERP) applications (Fig. 11.6) which cut across the corporate functions mentioned above (Finance, Purchasing, Marketing, Human Resources, Production Planning, and Asset Management). An ERP solution with integrated workflow management provided three immediate advantages—firstly, it allowed for instantaneous and unified tracking of "operational states" of business processes across all related corporate functions as opposed to only getting periodic manual updates extracted from individual silos of information. Also, the dynamic occurrences of events in any division or department (within the corporate environment) were immediately notified to the other departments via alerts or information in a common corporate portal.

Secondly, ERP used a common database across all functions as opposed to maintaining individual silos of data. The common database allowed for automation applications to share data between the different departments mentioned above. As a result, interdepartmental workflows were possible. For instance, in the cement or iron and steel industries, finance could trigger (integrate) the purchasing workflows for raw materials (order processes) more effectively based on the current and actual production plans that in turn could be actively formulated based on the actual current demand. The availability of real time or near real-time information allowed for the individual departments to take informed decisions and thereby reduce waste and costs significantly. This was not possible in manual, paper-based workflows that relied on historical or outdated information.

Thirdly, ERP allowed for the implementation of best practices across the enterprise and reduced process variance. In other words, each functional module of an ERP system encapsulated the industry best practices which allowed for standardization of the operations and better compliance with regulatory norms.

The second generation in work flow integration happened in the plant environment with the integration of production, logistics, production scheduling, inventory, and quality. The individual functions were subdivided into modules and best practices for the execution of the functions were provided as workflows.

Real-time reporting of the "state" of the plant operations was available in a plant portal very similar to the corporate portal. These applications were called MES. An MES system by definition is a software application that can provide the real-time information of the "state" of the plant floor equipment, processes, etc. so that plant personnel can optimize the plant performance in real time. As in the case of an ERP system, it is a resource orchestration and execution system in the plant floor.

The advantages of an MES system are multifold but the following are probably the most important ones:

(a) Improvement in the overall equipment effectiveness (OEE)
(b) Reduction in paper-based and manual processes
(c) Reduction in inventory

Of the above, the first item encapsulates three important aspects of improvement of plant performance—improvement in availability, productivity, and quality. The real-time calculation of OEE has now become a standard best practice in the industry for tracking production performance. Most MES systems have built in alarms and thresholds that are set to alert appropriate personnel if the OEE values are reducing beyond a desirable level to immediately take appropriate corrective action (Fig. 11.7).

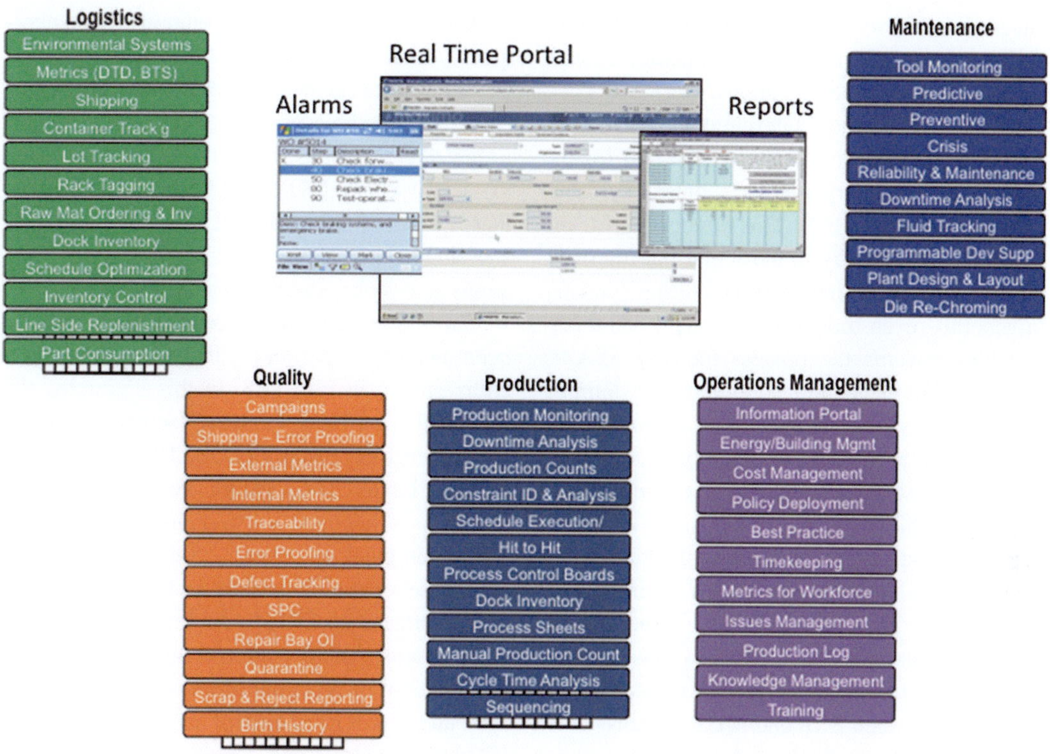

Fig. 11.7 Integrated plant operations using MES

Apart from improving production performance, MES systems help in the integration of multiple functions as stated above based on a common database very similar to the ERP applications. Since the various functions like logistics, maintenance, and production can share data, it is possible to optimally schedule production taking into consideration raw material input, availability of equipment, etc. Furthermore, MES applications are used to track optimality of a process—for instance, MES can track the efficiency of a process or equipment for that matter such as a boiler or a pump or a cooling tower and respond with alarms whenever the values go off limits. MES systems are also employed for tracing the complete history of the operations as they happen in the plant floor based on batches, operations within batches, equipment condition, and personnel operating the equipment. Such traceability is important in the metallurgical industry for meeting regulatory norms.

In summary, MES implementation has allowed the industry to eliminate waste, monitor, and alert suboptimal performance, integrate and automate workflows between different functions, track the execution of processes, improve standardization and quality, and because of all of the above, reduce the cost of operations.

Even though the ERP and MES applications have contributed to integrate workflows within the corporate and the plant floor environments respectively, they were considered as two distinct islands of automation and were not integrated to each other until the middle of the last decade. The need for integration of the workflows beyond the plant or the corporate was precipitated by the need for better efficiencies, and to remove the latencies, in the day-to-day transactions between these two environments within a manufacturing organization. This has resulted in the birth of the third generation in workflow integration, which is widely called, Plant to Enterprise (P2E) Integration. P2E integration allows for real-time integration of business processes in the plant operations with those in the corporate operations.

In parallel, the International Standards Association (ISA) introduced a standard for the integration of enterprise and control systems, ISA-95 (Gehman 2013). This is not only for the metallurgical industry in particular but the overall process and discrete manufacturing industry in general—even though the standard has a certain bias towards the process industry. ISA-95 consists of models and terminology for defining the workflows between the control systems in the plant and other software application in the enterprise. These standards are used in the industry to determine which information has to be exchanged between systems (software applications) for sales, finance, logistics production, maintenance, and quality. This information is structured in Unified Markup Language (UML) models, which are the basis for the development of the standard interfaces between ERP and MES systems.

The ISA-95 standard can be used for several purposes, for example, as a guide for the definition of user requirements, for the selection of MES suppliers, and as a basis for the development of MES systems and databases and to integrate specific departments within the enterprise regardless of whether they are part of the plant floor or the enterprise. In that sense, ISA 95 has provided a guideline to break the artificial walls that existed between the plant and the enterprise and to allow for "data sharing" between these two environments between applications that reside in each of these environments. As a result, it is possible to implement a business process workflow that transcends multiple applications today. By doing so the manufacturing industry is not held hostage by the limitations of the software packages and solutions—the individual software solutions collaborate and integrate to implement business process of the organization in an automated and efficient manner. Today, ISA 95 is one of the most powerful standards for integration of plant to the enterprise and vice versa.

These Plant-to-Enterprise solutions were subsequently identified as being part of "Manufacturing Operation Management (MOM)" solutions which encompassed a broader scope of connecting not just a plant to the enterprise but multiple plants and even the supply chain. MOM is different from MES in that it allowed for organizations to integrate outsourcing to internal operations, to standardize operations across plants and supply chain partners.

The above-mentioned wider scope has recently been included as part of the ISA-95 systems hierarchy shown in Fig. 11.8. In this hierarchy, the low level controllers, Human Machine Interfaces and Supervisory Controllers and Data Acquisition occupy Levels 0–2, MOM solutions/software platforms, scheduling software, inventory control, and other plant operations occupy Level 3 and enterprise operations (such as those controlled and automated by ERP, EAM, etc.) occupy Level 4. In general, MOM platforms aggregate the vast quantities of data coming from controls, automation, and supervisory control and data acquisition (SCADA) systems and convert them into useful information about the production operation. In that sense, MOM platforms facilitate a more comprehensive, real-time view of all the plants and the supply chain as an integrated unit.

Typically, an MOM solution comprises one or more of the following attributes:

- A configurable solution as opposed to being hard coded to a particular application or domain
- Standard integration to Enterprise Systems such as ERP and EAM systems
- Standard integration to Industrial Automation and Plant Control Systems
- Capability to represent manufacturing data and equipment in a standard model
- Capability to perform business process modeling and integrate workflows automatically
- Capability to visualize the current status of the plant and the enterprise in multiple media
- Capability to aggregate, analyze, and respond to real-time manufacturing events

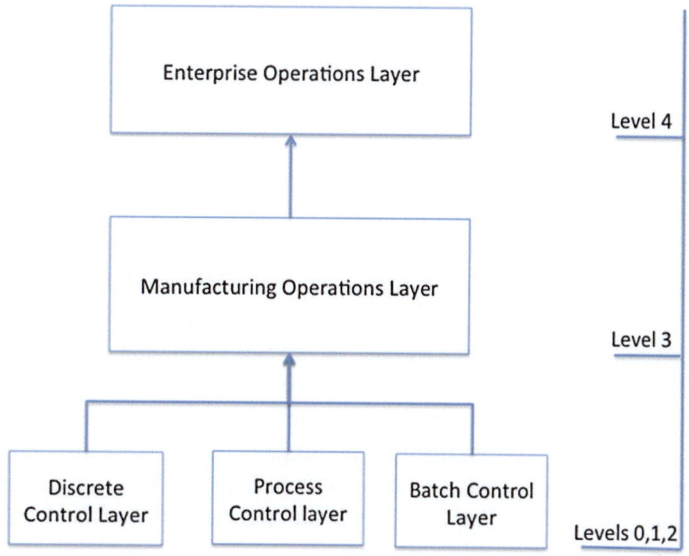

Fig. 11.8 The manufacturing value chain—ISA 95 Standard

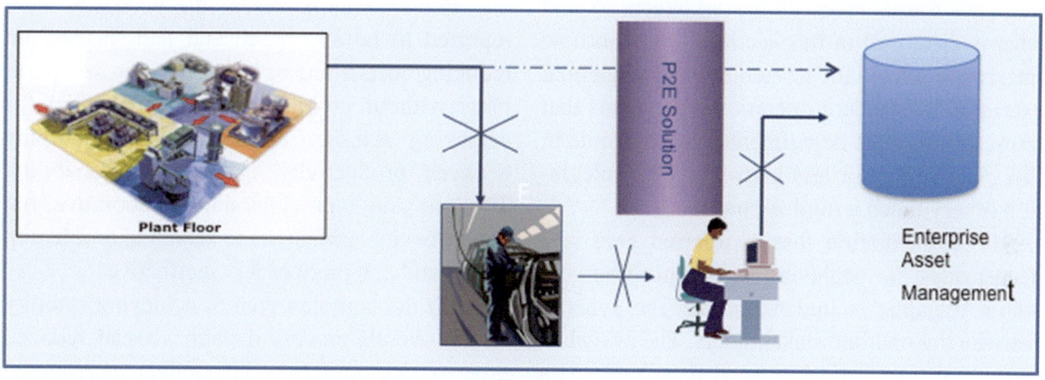

Fig. 11.9 An MOM solution—Real-time production to asset management integration example showing the replacement of the human loop by an automated P2E solution (Enterprise Gateway 2010)

The example below illustrated in Fig. 11.9 is an MOM use case that integrates production and maintenance to achieve better usage-based active Preventive Maintenance. Traditionally, Preventive Maintenance in the metallurgical industry has been time based—that is based on an elapsed duration of time—regardless of the usage of the machine during the elapsed time. Therefore, a machine could be either over maintained or under maintained based on how much the machine has been used during the said time period.

With the advent of the MOM platform, a use case for integration of actual usage of a machine to a Preventive Maintenance schedule is possible. The figure shows the MOM solution collecting the usage (run time, number of cycles, etc.) from the machines in the shop floor and automatically updating the Enterprise Asset Management (EAM) system on the corporate side in real time. The EAM system, on the other hand, triggers a workflow for generating a PM work order automatically once a threshold of usage is crossed for any machine. By doing so, the EAM system that was an isolated system previously has now become integrated to the plant dynamics. The wastage due to over maintenance or under maintenance of machines is avoided.

11.2.3.5 Logical Layer 5: Collaboration Layer

The discussion on the four layers thus far explained how automation in data access, communication, control, visualization, and workflow integration can help in building an integrated enterprise. This infrastructure can be helpful to an organization to "collaboratively" achieve optimized solutions to problems that impact multiple departments and stake holders in a holistic manner. This collaboration function constitutes the fifth and the final layer of the five-layer architecture.

There is a difference between the type of collaboration that is achieved via workflow integration (the fourth layer of the architecture) and what is discussed in this section. The workflow integration allows for a seamless and sequential execution of standard operational processes that transcend multiple departments. The dynamic in this case is more or less known and the integration is very much a routine one.

The collaboration that is referred here is a deeper quest for achieving an optimization goal whose dynamic is understood via the synergy between the multiple stakeholders. The availability of real-time data, historical data, meta data (information), integrated work flows (wherever possible), and unified visibility only helps in achieving this collaboration. The collaboration is enabled by automation (the integrated architecture) but involves the human loop—supervisors, managers, domain experts, and analysts to collectively solve the problem in hand using the data, meta-data, and knowledge generated from the said automation. The following section outlines an example of collaboration using a case study in the cement industry—for removing process variation. The first part of the discussion illustrates how the cement company collaborated to solve a problem in hand. The second part explains one way to perform collaboration to continuously *prevent* problems.

Collaboration: Cement Industry Case Study

A key challenge for any process manufacturing industry is to maximize the utilization of its critical assets, while maintaining customer satisfaction, cost, safety standards, and product quality. Cement manufacturing processes is impacted adversely like any other industry when the equipment is unreliable, when the processes are unstable, when raw material quality is not consistent and when excess variation exists in how the plant is operated. A case study has been reported in the literature (Arora 2007) that discusses a collaborative effort at Adelaide Brighton Cement Ltd. for achieving a common goal of removing *downtime*. In this particular case, there was a need for predictability in (a) the volumes of cement, clinker or lime produced, (b) the quality of the product(s), and (c) the cost of manufacturing. Previous initiatives of the company were reported to be silo based and had focused on reducing costs alone or trying to improve operations without probing into a deeper need for achieving stability in the dynamic interplay between productivity, quality, and reliability. The company's previous non-collaborative, but departmental approaches to solve the problem in parts had been more or less ineffective.

With the common goal of achieving stability in the overall process dynamics (read reduced process variation), the collaborative team started probing into the causes of variation in the process and their occurrence patterns. This required collection of data. However, the team had to collect data manually which was in itself affecting plant performance as the process engineers were spending up to 3 days per month simply generating reporting data and not focusing on improving plant operations. This lack of timeliness for data analysis meant that it would be days or weeks to understand the plant dynamics. The Process and Engineering teams also manually gathered and recorded downtime data using a number of sources such as Control Room log sheets, MS Access databases, and MS Excel spreadsheets. As a result, there were different points of view and no single version of the truth. This motivated the company to use appropriate tools for productivity, conformance (quality), and reliability.

The company embarked on using some automated data collection and analysis tools (the first two layers of the integrated architecture). The online data immediately started providing trends that pointed to why the process is varying. A downtime reduction software was used as part of the Reliability Improvement Plan to assist the site to record and act on emerging failures in "real time," before the stoppages manifest into larger, more costly downtime events. This also improved the productivity.

The plant also introduced a production system that provided the following real-time data: Metrics such as, fuel efficiency GJ/t, power consumption—kWh/t, production data—sacks/h, tph equipment reliability—run hours, % utilization, % reliability, % quality, % performance factor (% of MDR [Maximum Demonstrated Rate]), wastage—rejects tons, etc. were tracked (Level 3 of the integrated architecture). With the basic three-layer infrastructure in place, the collaborative team was able to address the common problem of process variation.

The team identified and categorized the sources of process variation to be as follows: (a) variation in raw material attributes and quality; (b) variation in operator performance and training; (c) variations due to seasonality and demand patterns; (d) variations in instrumentation—calibration and accuracy; (e) variation in maintenance checks—frequency and quality; (f) variation in the work instructions, KPIs.

With the automated systems and technology, the company had access to relevant and timely information and therefore the ability to monitor many sections of the plant via centralized control rooms and effect useful actions. They identified certain trends such as differences in trip rate across shift groups. This was identified therefore as a management problem and one of the standardizations across the various groups.

The synergy between the groups provided for the understanding that the stabilization of the variation in the process has to be done in an incremental, evolutionary, and continuous manner taking into consideration all of the above-mentioned factors that contribute to variation. This data driven, continuous improvement collaborative initiative has been useful in significantly improving raw mill control, kiln, and raw mill stoppages. The company concluded that the inclusive methodology adopted with all concerned employees and their collaboration and commitment with the focus on stabilizing the process continuously has delivered the improvement in kiln operation by significantly reducing stoppages and downtime by 50 %.

Proactive Collaboration: Asset Performance Management Example

The previous case study demonstrated a real situation of solving an existing problem of process variation via data-driven collaboration between multidisciplinary teams—namely, production, quality, finance, and maintenance. It also demonstrated the value of real-time data and information in addressing process optimization that involve multiple stakeholders in the company. In this section, the concept of collaboration is taken to the next step—to perform proactive interference in the dynamics of a plant in order to introduce desirable effects and eliminate undesirable performance in a future time period.

In asset intensive industries (such as the metallurgical industry), one of the important strategic goals is to achieve a better return on assets as they constitute a significant portion of the investment. In order to not lose track of such a strategic goal, there is a need for synergy between multiple stakeholders of the business to foresee all possible conditions that can lead to the goal not being tracked in a future time period and avoid them beforehand. The reader must note the difference between this approach and the conventional approach of identifying the reasons for a problem after it has occurred.

Asset Performance Management (APM) is an evolving methodology that allows for *collaboratively* and proactively achieving a desired performance in the assets that will identify a good majority of failures before they happen as opposed to addressing all of the problems after the failures have occurred.

APM is not a onetime initiative but a continuous improvement process. It departs from the conventional view of looking at an operational piece of equipment as a depreciating asset—but rather as a driver of business performance and sustainability. Traditionally, there have been different primary drivers for an APM collaborative initiative—most of them fundamentally aim at either maximizing profit, avoiding risks, getting better return on production assets, minimizing costs, minimizing variation, etc.

The emerging trend is one that looks at APM evolving in an organization due to a synergistic combination of all of the above drivers. In other words, no longer is a mill only interested in regulatory compliance as a driver for APM—but also wants to put in place a multipronged strategy of maximizing productivity and quality as well to achieve better profitability. Slowly mining companies are transforming from defining multiple initiatives that address asset performance (as a part of those initiatives) to an Asset Performance Initiative that spans multiple disciplines and stakeholders within an organization to track superior performance in a future time period.

Typically, in the mining industry for example, the processing units are mandated to maximize productivity—from an asset performance standpoint that means every asset (excavators, conveyors, crushers, grinding mills, etc.) must maximize their effectiveness and efficiency for a given set of dynamic conditions. The corporate governance/compliance department would like to continuously reduce the risks and exposure—from an asset performance point of view, this means the assets in question must have some redundancy built in their critical systems in addition to having a record of failures, the impact of these failures (penalties, loss), and the probability of occurrence of these failures in any given future time period. The quality department would like to reduce the Cost of Quality—this means (from an asset performance point of view) the preventive and the appraisal costs of quality of a product output from the assets should be increased to reduce quality problems in a future period. The maintenance department would like to reduce its annual maintenance costs as a percentage of the overall operational costs. The corporate executive on the other hand would like to maximize the return on the assets in order to provide maximum returns to the shareholders. This would mean the increase in net profit without an increase in new capital expenditure. Therefore, an APM process by definition would have multiple, individual departmental goals that need to be simultaneously achieved—in other words, the strategies employed within the APM process for the achievement of a particular departmental goal in a future time period cannot violate the achievement of the stated goal of another department.

Table 11.1 Optimization of multiple objectives

Department	Goal	To not violate
Production	Maximize operational profitability	Maintenance
Maintenance	Decrease maintenance costs	Production, Quality
Quality	Reduce cost of quality	Maintenance
Corporate governance	Reduce risk via redundancy and prediction of failures	Production and maintenance
Executive	Return on production assets	Corporate governance production, maintenance, and quality

The example in Table 11.1 below shows how in an APM process (that spans multiple departments) every department strives for optimality in the achievement of the goals across multiple related departments.

Let us take the first row as an example—the production department's goal of maximizing profitability from its operational assets in this case is achieved without violating the goal of the maintenance department, which is, to keep the maintenance costs of these assets at a minimum. The same is true with the quality department—which has to decrease the cost of quality without increasing the cost of maintenance (which is the maintenance department's goal), and so on. The multidisciplinary APM team therefore has to work out a set of "common minimum strategies" that collectively achieve the goals of every department without violating the goals of the other related departments.

Typically, such strategies are defined directly for implementation at the operational levels—by not only tracking certain operational metrics which are "loosely coupled" to the strategies, but to also predict the future values of these metrics and use the workflow integration layer to generate proactive actions.

Active Criticality Analysis for Assessment of Future Risks

Active Criticality Analysis (ACA) is a collaborative practice under an overarching APM program that is adopted to choose the "right assets" to monitor at any given time to achieve the best optimization in the future. ACA is a dynamic method that is extensively used in any risk-based asset management strategy. Simply put, ACA provides a Pareto ranking of the critical assets, at any given time, based on the consequences of the failure of these assets to the business. ACA does not result in a static selection of assets—it is an active time varying dynamic. Therefore, a regular and timely assessment of criticality is required depending upon changes in the operational dynamics.

An organization can use qualitative or more detailed quantitative techniques to arrive at asset criticality—both approaches have merit. For instance, the calculation of the likelihood of a failure of an asset (or a component of an asset) could be considered as a probability between 0 and 1, or can be considered qualitatively as belonging to fuzzy intervals such as "less likely", "more or less likely" and "most likely". The likelihoods increase or decrease dynamically based on the current status of the asset ("symptoms" of failure) and the past history of failures. A real-time measurement system can monitor for the symptoms and update the criticality map dynamically (of the "current" estimation of the likelihood of failure in a future time period). A typical ACA normally identifies a relationship between such likelihoods of failure of an asset to the business consequences of such failure to the organization. Once again, the consequences can be quantitatively determined or qualitatively established to fall within certain intervals such as "catastrophic," "major," and "minor". Based on the current inputs from the real-time measurement system if there is a change in the critical ranking of the asset, then, such changes are reported to the asset management system, which will in turn, implement the appropriate maintenance strategy in response to the changes.

The outcome of ACA is normally the decider of the type of maintenance an asset would need—say, preventive, corrective, and so on—in order to effectively manage the overall risks of failure in a future time period. In many instances, a previously planned maintenance regime may have to be refined in the wake of some new risks identified by ACA. The ACA, therefore, forces the maintenance organization to always (and actively) concentrate on the most important assets at any time (from a business point of view)—and therefore renders the maintenance actions to be more effective. Figure 11.10 shows the relationship between the real-time identification of the symptoms of failure and the dynamic ranking of the criticality.

Metrics Hierarchy

Every collaborative program needs a set of metrics, and it is obviously important that the metrics are easily measurable and the measurement relates to the achievement of the goals—in other words, if the goal is the reduction of maintenance costs at a strategic level in a given future time period, then there has to be one or more measurements that relate directly to the goal of reducing costs and so on. An illustration is provided in Fig. 11.10 for a typical set of measurements that can be made (and tracked) from strategic, tactical, and operational points of view (this illustration provides the measurement framework and examples of some metrics—the metrics could be different from the ones presented below based on the specific industry/goals) using the various layers of the integrated enterprise architecture. The figure also illustrates how the measurements are at multiple levels of abstraction and are related to each other.

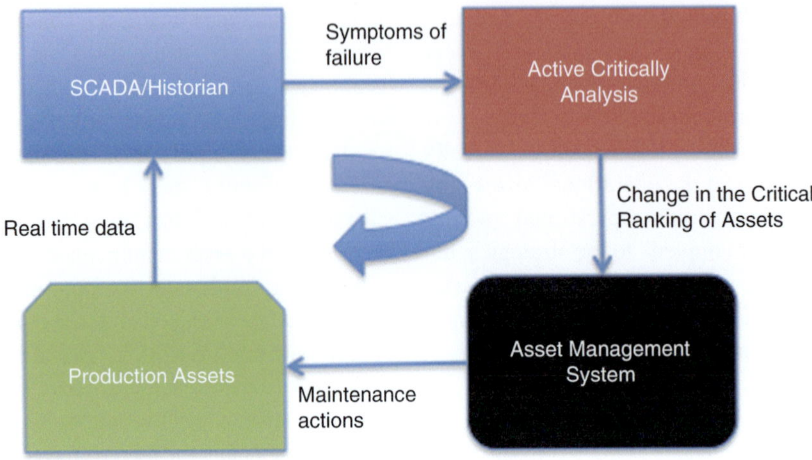

Fig. 11.10 The continuous cycle of active criticality analysis of assets

The example shows three measurements at the strategic level corresponding to the strategic (and collaborative) goals of reduction of maintenance costs, reduction in risks and increase in the net return on production assets during a given future time period. The reduction of maintenance costs in this case stems from two measurements—maintenance costs as a percentage of Replacement Asset Value (RAV) and maintenance costs as a percentage of Regulatory Failures. Even though some of the costs elements that shall be accounted for in the calculation of these two metrics may overlap, the metrics independently maintain two distinct measurement points for understanding/observing the progress of achieving the collaborative goal of reducing maintenance costs. The second collaborative goal, namely, reduction of risks, is tracked (once again) by the maintenance costs as a percentage of Regulatory Failures. The third collaborative goal, achieving a better return on assets, is tracked by the NROPA (Net Return on Production Assets) measurement—which is a formula that requires net profit and total asset value as inputs. The optimization of all of these goals in a given future time period will be the objective of the "progress function." The trends (rates of change) in the progress functions (which are the gradient of the progress function) can be computed at any given time to track the velocity of progress.

The tactical level metrics (in Fig. 11.11) act as an intermediary between the physical and strategic metrics. Continuous improvement cells/programs are normally involved in tracking the progress of the tactics to improve asset performance by measuring the tactical metrics. It is obvious that by focusing too much on the operational metrics one loses sight of the big picture (from a collaboration stand point) and on the other hand by defining the strategic metrics without a relation to their practical implementation and measurement process one loses the relevance of the strategic metric. The tactical layer measurements bridge these two worlds.

The operational level measurements are the fundamental layers that collect fine granular operational data that in turn feed into the tactical and strategic layers. While the tactical and strategic layer measurements are abstracted logical measurements, the operational measurements are physical measurements.

Optimization of Reactive and Proactive Interventions

An important outcome in the collaborative pursuit of optimizing the cross-disciplinary goals to maximize asset performance is to have a well thought out plan for future maintenance interventions. Too much of intervention is costly—too little is again costly. In order to implement the strategic requirements of reducing the cost of maintenance and/or increasing

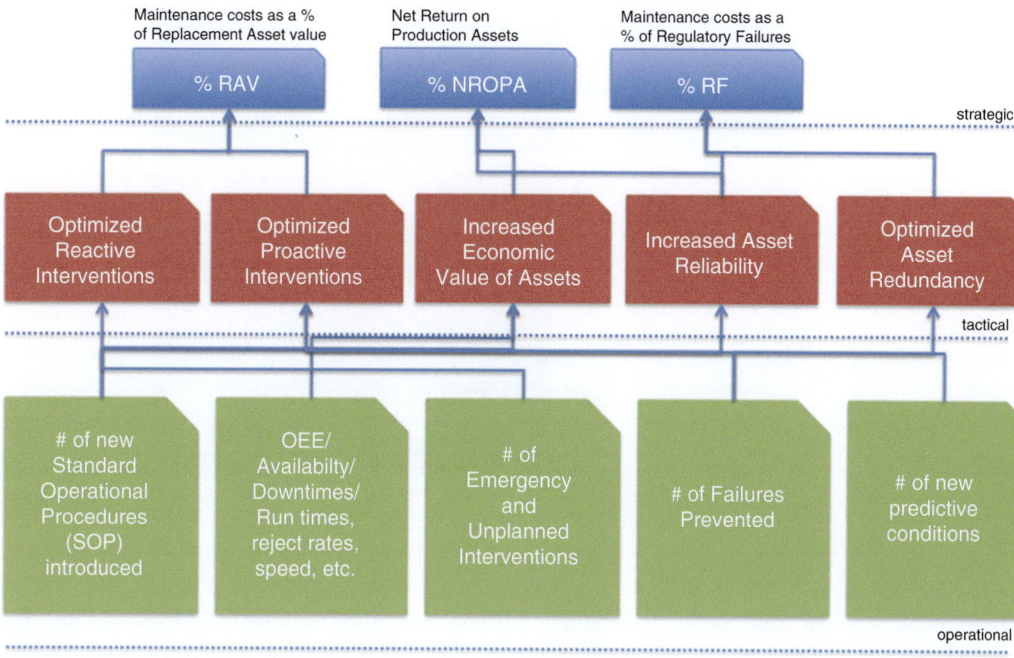

Fig. 11.11 APM metrics hierarchy—Example

the net return on assets, one has to look at arriving at an optimal balance of proactive and reactive interventions.

Optimality can be achieved by understanding the elements of both proactive and reactive maintenance activities. Proactive maintenance activities include any activity that is performed for maintaining an asset before they fail. Reactive activities therefore are those that are performed after the failure. Proactive maintenance comprises two types of activities—Preventive Maintenance and Predictive Maintenance. Reactive also comprises two types—Emergency Maintenance and Corrective Maintenance. Achieving optimality would mean increasing the cost of Proactive Maintenance appropriately in order to reduce the cost of Reactive Maintenance substantially.

As stated earlier, Reactive Maintenance comprises emergency and corrective maintenance. Of these, the emergency maintenance activities are the costliest. Even if it is not possible to eliminate them, they have to be reduced substantially for the organization to reduce its maintenance costs. Emergency Maintenance results from two main reasons: (a) lack of a proper maintenance policy and (b) lack of reliability.

The lack of proper maintenance policy herein means the lack of clarity on what constitutes "emergency" within the organization. The exact meaning of emergency should be clarified in the "maintenance policy" and a common method to categorize any maintenance job as emergency or non-emergency should be available as a "standard operating procedure" (SOP). The adequacy of the maintenance policy is determined at the strategic level, and the standard operating procedures are performed at the operational level. But collaboration at the tactical level is needed to monitor the performance of the operational level as regards to emergency maintenance—by (a) analyzing the emergency activities in the past; (b) use the data in (a) to predict possible emergency situations in the future (true positives); (c) predict emergency situations that will be "hidden" until it is too late (false negatives); and (d) predict those situations that will emerge as not emergency ones (false positives).

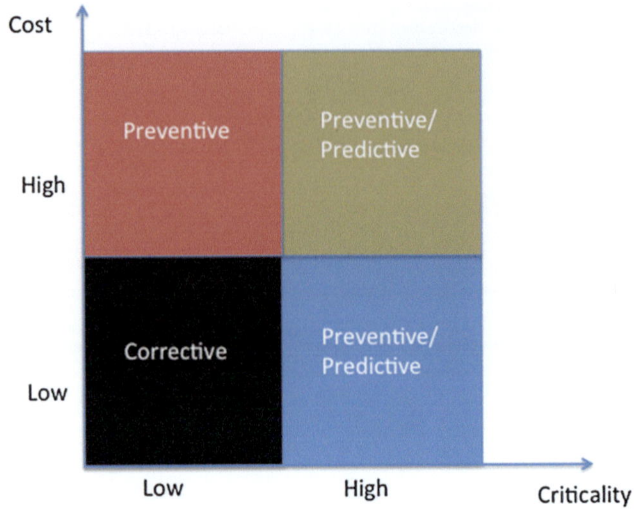

Fig. 11.12 Optimal maintenance planning based on criticality and cost

True positives are the right "emergency responses" as per policy—that is, the responses of the maintenance organization are exactly aligned with the policy and are performed as per the standard operating procedures. False positives are emergency responses that result in a "false alarm"—maintenance calls wherein the team upon reaching the site realizes that they are not actually emergency in nature. Therefore, these are wasted emergency interventions. Finally, there are false negatives—emergency situations that were not identified which consequently results in a catastrophic failure.

Such real-time monitoring of the type of maintenance responses, validation of the responses against policy, adherence to the SOP and predicting false positives, false negatives, and true positives will help in the understanding of the "effectiveness" of the emergency maintenance practices within the organization. Appropriate actions can be taken based on the monitored data—such as, if there are emerging symptoms of a variance in the adoption of the SOP, then, more training on the SOP is initiated before the variance becomes widespread. If there are higher number of false positives (rate of change in the number of false positives in a given period), then a proactive investigation is required into why the responsible owners for making the call are suddenly unable to do their job correctly. There can also be simple hidden reasons for ordering emergency maintenance such as the rates for maintenance work being higher under emergency maintenance than in non-emergencies. It is up to the collaborative team to understand these reasons and attempt to continuously drive the emergency maintenance costs down by improving the effectiveness of emergency responses.

The second aspect that affects emergency maintenance costs (for that matter even the corrective maintenance costs) is the lack of reliability. Reliability in simple terms is inversely proportional to "failure rate." Every asset will have a history of failures and the frequency of such failures—the inverse of the frequency of failures is the failure rate. If the failure rate is low, then the asset is supposed to be more reliable. Failure rates and failure histories of assets are maintained in the EAM systems. Based on the failure history maintained in the EAM systems P-F Graphs (Potential Failure Graphs) can be generated for critical assets. The P-F curves will predict the possible future failure states for a piece of equipment which is at a certain current state. This input can drive Preventive Maintenance of these assets.

While all of the above-mentioned discussed corrective and emergency responses that can be avoided, there are instances when Corrective

Maintenance is the right one to adopt. Corrective Maintenance is good if the criticality of the asset and the cost of the maintenance activity are both low. Figure 11.12 below illustrates the "calculus of interactions" between these two variables—asset criticality and the unit cost of maintenance for the activity.

It can also be noted that three fourths of the maintenance activities fall under the proactive maintenance (either predictive or preventive or a combination of both).

Finally, as a result of the collaboration at the operational and tactical levels the Net Return on Production Assets (NROPA) can be monitored and continuously improved. The NROPA establishes how the entire asset base of the organization returns to the shareholders—a metric that is vital to track at the strategic level of an integrated metallurgical enterprise.

11.3 Technology Enablers

11.3.1 Introduction

Monolithic control solutions were common until the turn of the century to implement process control and automation. They were point-to-point and one-off solutions that would be solved with (often very expensive) custom programming. Over the last decade and a half, the trend has dramatically changed. Many plants use several different automation controller families from different generations and different vendors. As stated above highly integrated metallurgical enterprise requires that all of these disparate systems are seamlessly integrated using standards rather than bespoke methods. This means that the underlying technology infrastructure, that comprises the information, communications, and software engineering components, must allow for integration standards that cope with the heterogeneous multi-vendor system landscape. This section outlines some of the important software engineering and information technology advancements that have enabled easier and standard implementation of the five-layer integrated enterprise architecture. The discussions provide an overview on how standardization has happened in each of the layers to improve interoperability and quality.

11.3.2 Unified Object Model

The metallurgical industry can have many thousands of equipment, transducers, sensors, and actuators. These physical objects need to be modeled in a software application for control and optimization. The model should not only capture the behavior of the physical objects but also capture the interaction between them in the physical world. Therefore, there is a need for mapping the physical objects to corresponding "software objects."

Software objects need to exhibit two properties—modularity and reusability. For instance, when a centrifugal pump is defined as a software object, it should be possible to reuse the same definition for every other centrifugal pump within the organization. Furthermore, it should be possible to abstract the properties and attributes of a "pump class" and define a centrifugal pump as an instance of the pump class and thereby use the same definition of the pump class to define all types of pumps within the organization. The same is true with every other physical asset.

Based on the attributes of a physical asset, a software object would have identifiers, events, and incidents associated with them. An identifier is a unique reference to the physical object within the software program. There are multiple conventions in practice to define the identity of an object. The most common and standard convention is the Common Object Model that considers the software object as a container that holds different "aspects" of the object. Typical aspects of an object are its control program, operator faceplate, trend configuration, function specification, and so on.

In the Common Object Model, all interactions with an object in the system are carried out with the help of its "aspects." Software components are built and integrated in a seamless way with the help of a "loose coupling" between objects and aspects. In other words, the aspects and the objects are distinct and separate elements and can be defined as hierarchical concepts individually within the model. They are associated only contextually by the control component of the software program during its execution.

In the metallurgical industry in particular, relevant data is scattered around in MES, CMMS, LIMS (Laboratory Information Management Systems), and ERP systems. The Common Object Model allows for addressing objects in a unified way, across systems, no matter where they are physically located. As a result, applications can be written to be more abstract and reusable in different contexts.

Without the use of such a Common Object Model, each of these systems would need to be addressed individually and separately. Also, such point-to-point addressing implies that the server where the data resides must be explicitly specified every time. This reduces the flexibility and scalability of the implementation because servers cannot easily be changed. Control solutions that are developed for a particular process within a plant cannot be reused or transported to a similar process elsewhere in the organization. Also by connecting directly to the underlying systems, the end applications would need to contain a larger portion of access logic, making the applications less transferable to other contexts, where different underlying systems might be used.

An example of this is seen in the point-to-point condition monitoring systems versus an integrated APM discussed in the previous section. In a point-to-point condition monitoring system, the asset in question becomes an island, and the condition of the asset is monitored normally by a proprietary, monolithic control program whose actions (business logic) are not shared with the rest of the stakeholders within the organization. The business logic of such programs resides in a separate server that is directly addressed by the application. On the other hand, in a Common Object Model implementation, the business logic that will encompass, the definition of the asset class, the asset, its condition, root causes for the condition, the action taken, the after effect of the action, the financial ramifications, the spare part inventory status, etc. are all commonly shared by multiple applications at an abstracted layer. In the former case, even though the condition monitoring is effective for that particular asset for a limited time until the asset is retired, there is no abstracted base of knowledge created in the organization that can be reused, compared, benchmarked, etc. for effective monitoring of the condition of every other related or similar asset or asset condition in any other part of the enterprise.

11.3.3 Field Device Configuration Technologies

The metallurgical plants have several thousands of digital field devices distributed remotely. Typically, analog field devices required access to the devices for calibration and measurement—which presented a challenge. Digital field devices, on the other hand, could be accessed remotely. However, because there are very large number of devices and multiple vendors supplying these devices, the management of these devices in real time still remains a challenge. The following standards and technologies allow for effective management and maintenance of field devices in the process and metallurgical industry.

11.3.3.1 FDT/DTM

Field Device Tool (FDT) technology enables a standard communication interface between field devices and systems. The key feature is its independence from the communication protocol and the software environment of either the device or the host system (see www.fdt-jig.org). In other words, any device can be configured, operated, and maintained—regardless of the supplier, type, or communication protocol.

FDT allows a device to be managed by a user interface called "Device Type Manager (DTM)." DTMs are of two types based on the functions they perform—(a) Device DTMs act as a user interface to the field devices and (b) Communication DTMs which act as connectors to the software communication components.

The major drawback of FDT is that data is trapped inside the DTM and does not allow for external applications to get access—which is where the EDDL technology described below scores better.

11.3.3.2 EDDL

Electronic Device Description Language (EDDL) is a text file used to describe properties of, and precise declaration of data from digital devices (see www.eddl.org). Since it is not an executing software, it is easy to manage and maintain. It is agnostic to any specific communication protocol. Therefore, it is possible to integrate data from multiple protocols such as HART and Fieldbus. Also adding new devices to an EDDL host is quite easy—like copying a file. Enhanced EDDL offers sophisticated graphical features such as trend charts and tables beyond a simple text declaration.

EDDL uses OPC-UA or OPC-DA (to be described later in section 11.3.5) to make available the information to external applications. A common use of EDDL is in parameter reconciliation. For instance, a user would be able to compare the data from a digital device against data stored in an external database without the need for a software program. Another widespread application of EDDL in the mining and metallurgical industry is in the analysis of valve positioners and variable speed drives. More specifically, step responses and valve signatures can be plotted using EDDL.

11.3.3.3 FDI

Field Device Integration (FDI) is the latest state of the art in this area and will use a subset of the OPC UA technology within a client–server architecture. FDI will combine the advantages of FDT and EDDL technologies. FDI technology will provide a common solution for managing intelligent field devices during all phases of the lifecycle of the device from installation through calibration and diagnostics to retirement.

The FDI technology is based on client–server architecture that helps control systems manufacturers to simplify the use of device data in distributed control system environments (Level 2 solutions). Furthermore, such data could also be made available to higher level MES systems (Level 3 systems and solutions). Data is managed centrally and therefore inconsistencies are avoided.

11.3.4 Field Device Networks

11.3.4.1 Fieldbus

A fieldbus is a plant network protocol that is used in the context of *distributed control* of plants. Such systems include devices such as transducers, actuators, and controllers. Several network topologies are supported by Fieldbus such as, daisy chain, star, ring, and branch. In the iron and steel, cement or mining industries, the market is dominated by Foundation Fieldbus and Profibus.

As a general rule, applications that are controlled and monitored by PLCs (programmable logic controllers) tend towards PROFIBUS, and applications that are controlled and monitored by a DCS (digital/distributed control system) tend towards FOUNDATION Fieldbus. Zuruwski (2005) provides a comprehensive account of Fieldbus and its use in the process industry.

FOUNDATION Fieldbus

FOUNDATION Fieldbus provides a communications protocol for control in which each device has its own "intelligence" and communicates via a digital, serial, two-way communications system. FOUNDATION High-Speed Ethernet (HSE) is ideally suited for use as a control backbone. The technology is designed to run at 100 Mbps, and is applied for device, subsystem, and enterprise integration (see www.fieldbus.org).

HART

Highway Addressable Remote Transducer (HART) is a master/slave field communications protocol. A good majority of analog field devices installed in plants worldwide are based on HART technology. The unique feature of the HART protocol is that it makes use of a principle called—Frequency-Shift Keying (FSK)—to superimpose digital communication signals at a low level *on top of* the 4–20 mA analog signal.

Thus, this is a good technology to implement in "brownfield" situations when there are existing analog systems that need to be converted into digital. The HART protocol communicates at 1200 bps without interrupting the 4–20 mA signal and allows a host application (master) to get two or more digital updates per second from a field device. As the digital FSK signal is phase continuous, there is no interference with the 4–20 mA signal (see www.hartcomm2.org). Many metallurgical and mining plants utilize this technology as a transition mechanism from analog to digital or to use both digital and analog within the same infrastructure.

WirelessHART is an open wireless standard established to address the needs of the process industry for simple, reliable, and secure wireless communication in industrial plant applications. Close on the heels of WirelessHART is the ISA 100 standard (Bassett 2013) that provides interoperability between different protocols including the HART protocol. The important aspect is that it supports IPv6-based technology (the latest Internet protocol for industrial applications and one that can support Industrial Internet of Things).

11.3.4.2 PROFIBUS

Even though PROFIBUS is similar to Fieldbus and uses the same physical layer, it is not the same. PROFIBUS got its name from "Process Field Bus" and it has application/industry-specific protocols such as PROFIBUS DP, PROFIBUS PA, PROFINET, and Profisafe. PROFIBUS DP uses RS 485 multi-drop serial interface to connect remote IO systems such as motor controllers to a Programmable Logic Controller. PROFIBUS PA is based on IEC 61158 standard (see www.profibus.com) that is commonly used in the process industry.

11.3.5 Field Data Access Standard: OPC

During the turn of the millennium, the first OPC standard (Object-Linking and Embedding for Process Control) was developed by several automation suppliers together with Microsoft (Iwanitz and Lange 2006). This standard is based on Microsoft Common Object Model and describes how applications should exchange real-time data on a Microsoft platform. As OPC was a good solution to an urgent problem, the standard was rapidly supported by more and more automation system and product vendors. The OPC Foundation (www.opcfoundation.org) currently has more than 400 members worldwide.

The OPC convention has eliminated the need for specific drivers and protocols for every application in the manufacturing industry—metallurgical industry included. Consequently, there is no need for the age old bespoke or custom solutions that are error prone and expensive. Normally, the OPC server is provided by the vendors along with their supplied device or equipment. This was a significant transformation in the industry (refer Fig. 11.13)—to also supply a standard means of communication of the data (from their device or equipment) as part of the supply of the equipment/machines.

Since OPC servers are software programs developed on a standard convention for the exchange of data between the sensors, and/or controllers of instruments connected to machines, it is possible for any third-party vendor to develop an OPC server for a machine if it is not already supplied by the OEM with the machine or equipment. While the original idea behind OPC was to provide standardized communication between hardware devices and applications, it is now also being used as an open interface between applications. OPC is therefore a software interface that can run on a hardware server.

11.3.5.1 OPC DA

OPC Data Access is a standard for exchanging data between various devices and pieces of equipment. An OPC DA server contains a structured list of "tags" or I/O points. Each tag represents a data item. A data item contains a timestamp and a flag for data quality. OPC clients "subscribe" to data items in the OPC server. The subscription to OPC items could be event based (such as "upon change") or time based (a certain sampling interval).

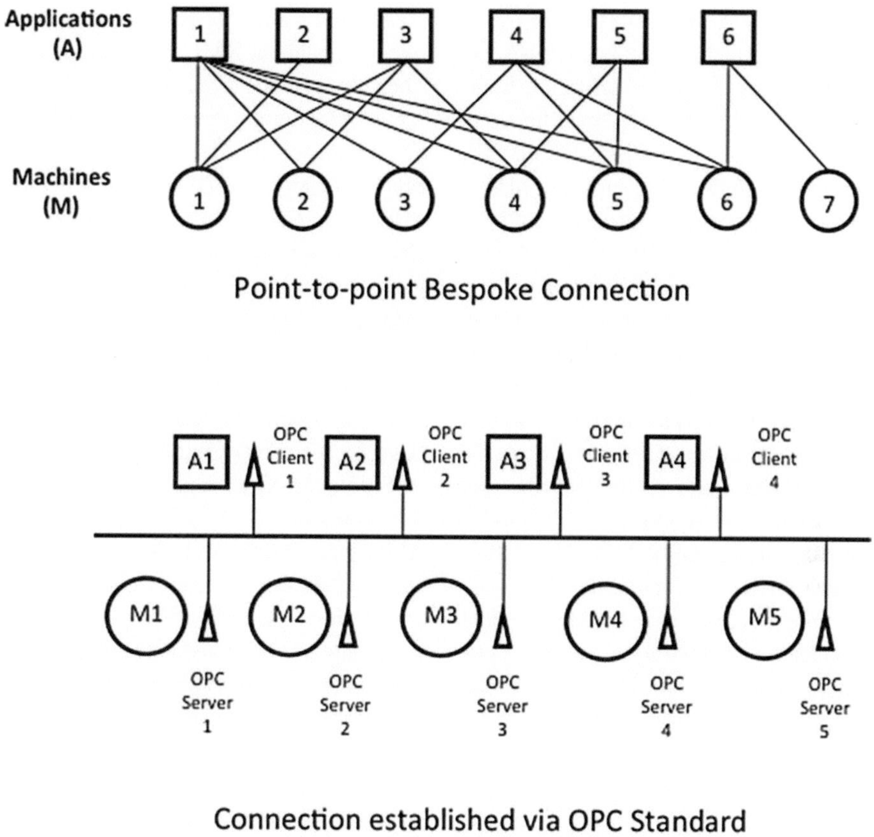

Fig. 11.13 Traditional vendor-specific bespoke solution replaced by standard OPC integration

11.3.5.2 OPC AE

Another commonly used OPC specification is the OPC Alarms & Events. This specification is responsible for handling multiple levels of alarms and events occurring in a mine, mill, clinker, kiln, furnace, etc.—as produced by alarm and event engines.

11.3.5.3 OPC HDA

The third specification of OPC is the OPC Historical Data Access or OPC HDA. This expands the capabilities of OPC DA, which is focused on current data, to historical data. OPC HDA provides the data for any historical time line that enables software solutions in the higher level layers of the integrated enterprise architecture to perform analysis of trends and predict future events based on history.

11.3.5.4 OPC UA

The standard OPC specifications such as OPC DA for data access, OPC A&E for alarms and events, and OPC HDA for historical data access are quite commonly used in the industry to exchange data between multiple equipment and multiple layers in the integrated five-layer architecture. There are several thousand OPC products on the market and all major automation vendors provide OPC solutions (Burke 2008).

OPC unified Architecture (OPC UA) is a next generation OPC specification that unifies the different specifications of all the three types of specifications provided above, thus enabling a single point of entry into a system for current data and alarms and events, together with the history of both. These are provided by a simple, generic set of services.

OPC UA provides a rich information model using object-oriented techniques—which is superior to a rather simple tag-based definition found in the standard OPC DA and HDA specifications. In OPC UA, it is not only possible to offer a measured value and its engineering unit, but also to indicate that it was measured by a specific type of sensor. As a result, quite a bit of meta-data is encapsulated in the data acquisition stage (Level 1) which provides rich information to the higher level analysis and visualization applications. Information modeling is a big improvement of OPC UA compared to the standard OPC and provides a lot of opportunities. OPC UA allows for incorporation of information models that are application-specific or vendor-specific models without any loss in generality. OPC UA defines two main functions of interoperability; the communication infrastructure and the meta model. The communication infrastructure defines how information is exchanged, and the meta-model specifies what information is exchanged.

OPC UA also provides for services that can run on different communication infrastructures. It can use the meta model as a basis for defining appropriate parameters for the services. Also, vendor-specific or standard information models can be built on top of the basic model. In other words, OPC UA defines several standard information models for data access, alarms and conditions, programs, historical data, and aggregate functions. Multiple information models can be supported in a single server. The information about the models can be read by the services and thus clients only knowing the services can be provided access to the information.

11.3.6 Interoperability Standard—XML

Extensible Markup Language (XML) is a simple and flexible text format that allows for interoperability between different systems and control platforms (Skonnard and Gudgin 2001). The W3C (World Wide Web Consortium) is the authority that maintains the XML specification (w3c.org/xml).

An XML Schema is used to express a set of rules to which an XML document must conform. The rules provide a means for defining the structure, content, and semantics of XML documents. For instance, the Business to Manufacturing Mark up Language (B2MML) schema provides structured and standard transactional information templates for the transactions that typically happen between the enterprise and the plant floor in batch processes. This is an XML representation of the ISA 95 standard explained earlier. The B2MML schema allows for the ISA 95 standard to be incorporated into several MOM software (Level 3 software) solutions.

Similarly, XML representation has helped in the confluence of OPC-DA. OPC HDA and OPC Alarms & Events and the capability of MOM applications to use these for interoperability.

Another neutral XML format is the AutomationML (www.automationml.org). This schema is used for exchanging plant engineering information. For instance, in the mining industry, the data that is common to Geology and Processing can be shared using the AutomationML standard. The major contribution of this standard is the enabling of *interdisciplinary engineering methods* that need to be employed across *heterogeneous* tools. The representation of the objects is based on the Common Object Model explained earlier. AutomationML is based on IEC 62714 standard. The standard eliminates the need for writing custom programs for importing and exporting data from one tool to another in order to perform engineering analysis.

11.3.7 Web and Next Generation Internet Technologies

11.3.7.1 HTML/HTTP

The HyperText Markup Language (HTML) is based on ISO/IEC 15445 standard. HTML allows users to specify web-based user interfaces analogous to how EDDL provides a text standard for describing parameters of digital devices. HTML is used in the configuration of smart devices from a web interface. A HTML web interface, apart from being convenient to a user to get access from anywhere, is an Internet connection that also helps in the reduction of cost and footprint in local devices. This is because attaching a dedicated display device to every sensor would be too expensive. Also even if there is a display device, it is only locally accessible. On the contrary, a device that can implement a small web server and offer the user interface via HTML over the network would be less costly and more versatile providing access to many users across geographies.

The HyperText Transfer Protocol (HTTP) has been jointly developed by IETF (Internet Engineering Task Force) and W3C (see http://tools.ietf.org/html/rfc2616).

11.3.7.2 Web Services

Web services are higher level programmatic interfaces that enable application-to-application communication using standard web techniques (see www.w3.org/2002/ws). Most large mining and metallurgical organizations use multiple software systems. When data from one software system needs to be communicated to another, web services are used as standard. The system that requests the data is called the service requester and the one that provides the data is called the service provider.

The key aspect of web services is that it is agnostic to any programming language or platform. Thus, they use XML files for data exchange. Many web services are based on the SOAP standard (see www.w3c.org/TR/soap). SOAP (originally, Simple Object Access Protocol) is a lightweight protocol intended for exchanging structured information in a decentralized, distributed environment. An alternate method of implementing web services was proposed by Fielding (Fielding 2000).

11.3.7.3 The Industrial Internet of Things

The traditional manufacturing industry is undergoing a digital transformation. Such a transformation is expected to bring about a fusion between the real and the virtual worlds of manufacturing. Intelligent Robots, autonomous drones, 3D printing, intelligent sensors, smart grids, intelligent communication networks, etc. are expected to be playing a role in both the real and the virtual worlds in order to convert the current factories into smart factories. In other words, every smart factory will turn into a "cyber-physical" production system. This revolution is called by various names—*Industrial Internet of Things (IIOT), Industry 4.0 (to denote the fourth industrial revolution)*, or *Industrial Internet*. In this scenario, the distinction between products and services will diminish.

More specifically, IIOT is said to bring about four important transformations (Schlaepfer and Koch 2014):

(a) Vertical Networking of smart factories, smart logistics, and smart services including personalized (or customized) manufacturing;
(b) Horizontal Integration that will include integration of business partners and customers leading to new models of cooperation and collaboration between them across different geographies;
(c) Life Cycle Engineering that integrates manufacturing process, product engineering, and after support and maintenance;
(d) Technology Acceleration that will include the adoption of new disruptive technologies and services that are easily scalable, in turn, driving the need for a cultural change in the organization.

IIOT will also enable machine-to-machine (M2M) interaction, and consequently, will enable machines to respond intelligently based on such real-time inputs. It is expected to enable deep analysis, derived from the plethora of data that will be available via the peer-to-peer machine networks, via distributed computer processing. Analysts are forecasting billions of connected smart IP devices in a wide range of applications.

Leading oil and gas companies such as Shell have already embarked on building an IIOT infrastructure (called Smart Fields) wherein sensors, advanced analytics, automation, and data management are used to unlock production, reduce operating costs, and optimize assets. In the Smart Field initiative, Shell has introduced the concepts of Smart Wells and 4D Seismic Acquisition. A Smart Field Consortium at Standford University is exploring advancements in optimization techniques in hydrocarbon production systems. The program integrates strategic planning, production, and back end modeling, and data storage. In other words, the program is aimed at data collection, simulation, model updation, and optimal control (Cameron and Durlofsky 2014; Rajaieyamchee and Bratvold 2009; Sarma and Chen 2008).

Another example of IIOT is the capability of performing Big Data Analytics of the data collected and stored in a distributed environment. For instance, in the oil and gas industry mentioned above, thousands of Electrical Submersible Pumps (ESPs) pull oil from reservoirs worldwide, onshore and offshore—but many ESPs fail unpredictably, resulting in missed production targets. Prescriptive Analytics predicts ESP failures by modeling the dynamics of pumps, production, and subsurface characteristics. The software can prescribe actions based on the modeled data to reduce pump failures and mitigate production losses.

What is missing from the IoT and IIoT marketplace today is a common framework for enabling multi-vendor interoperability across multiple applications, supporting different communication media such as wireless, power line, and wired. In order for IIoT to be truly effective and widely adopted, the connectivity and interoperability need to be enabled between diverse products and applications. Without this interoperable framework, the industry will be going back in time and re-adopting proprietary platforms, severely limiting widespread market acceptance. In order to meet this challenge, General Electric is working on a standard software platform called Predix (Chauhan 2013) which allows for the standard way to run industrial scale analytics.

Fortunately, the semiconductor industry is also delivering dramatic advancements in low-cost and reliable communication transport technologies. Because of the unforgiving environments in which these industrial devices exist, including harsh physical conditions and mission critical processes, IIoT solutions must meet the challenging requirements of industrial-strength reliability, hardened security, wired and wireless connectivity, and backwards compatibility with large installations of legacy devices. By steering industry towards open systems technology, companies can deliver exciting new applications, while better access to data opens up new horizons in the control systems marketplace.

11.3.7.4 Control Networks and ISO/IEC 14908-Based Systems

Increasingly, industrial systems today are becoming a cluster of distributed intelligent devices that communicate over a control network. The state-of-the-art network technology allows for intelligent devices to communicate with other devices. This peer-to-peer communication enables direct M2M interactions without the need for a centralized control or a human intervention (when unnecessary).

Such intelligent devices need to be optimized for their specific application domain as well as their network transport method. Control networks often support multiple physical layer options, such as twisted pair, power line, radio, and Ethernet. Network management and device configuration services need to be standardized to simplify the configuration and installation process and provide a common platform for multiple manufacturers' configuration tools. Finally, the application layer needs to be standardized to provide interoperability between devices from multiple manufacturers.

A network standard that is emerging in the area of IIOT and one that provides for an interoperable framework is ISO/IEC 14908-1 (http://www.iso.org/iso/catalogue_detail.htm?csnumber=60203). The IEC 14908 is a protocol for network control and allows for both peer-to-peer and master/slave configurations. The master/slave configuration allows for the aggregation and/or accumulation of information across multiple devices as necessary to generate meta-data or further abstracted intelligence. This suite adheres to the seven-layer network protocol and is optimized to the needs of control systems. It also has an application layer that provides the much needed interoperability.

The protocol is optimized for networks of devices sending small control messages (typically less than 277 bytes) reliably to tens or thousands of devices. The management of the complexity (frequency and usefulness of the messages) due to the chain reaction that can be generated with each subsequent receiver of the message in turn responding and/or propagating a new message to the network is an important research area. Regardless, it is safe to assume that there will be several hundreds of thousands of messages that will be sent back and forth and therefore collision detection and avoidance becomes an important aspect in network control. IEC 14908 standard supports p-Persistent CSMA with Collision Avoidance and optional Collision Detection to optimize performance in networks with "short, bursty" messages. The ISO 14908 standards suite provides multiple options for the interface between Medium Access and Physical Layers, such as ISO/IEC 14908-2 for twisted pair communication, ISO/IEC 14908-3 for power line communications, and ISO/IEC 14908-4 for IP communication. The application layer standard ISO/IEC 14908-5 ISO/IEC 14908-6 includes a common framework for device profile definitions, common data type definitions, and standard configuration properties.

11.4 Implementing an Integrated Enterprise Strategy

11.4.1 Five Key Steps to Integrated Operations

The key steps involved in developing and implementing integrated enterprise are shown in Fig. 11.14. The example emphasizes on the operational integration aspects of the integrated enterprise taking the mining industry as an example.

11.4.1.1 Step 1: Identify Key Value Drivers and KPIs

The first key step is to identify the key value drivers along with KPIs for each business unit. In the case of mining, during the initial stages of this project, it is proposed that only a few key drivers from Mine-to-Mill be used for simplicity. Some examples of the key value drivers from Mine-to-Mill are mine ore fragmentation and hardness, grinding specific energy, power split between mills and mill product size.

Key drivers and KPIs to quantify planned vs. actual ore mined, variations in mining plan, grade control model, reserve model vs. mine production vs. metal production, mining dilution along with metal recovery and losses in process plant will be included. The mine dispatching system

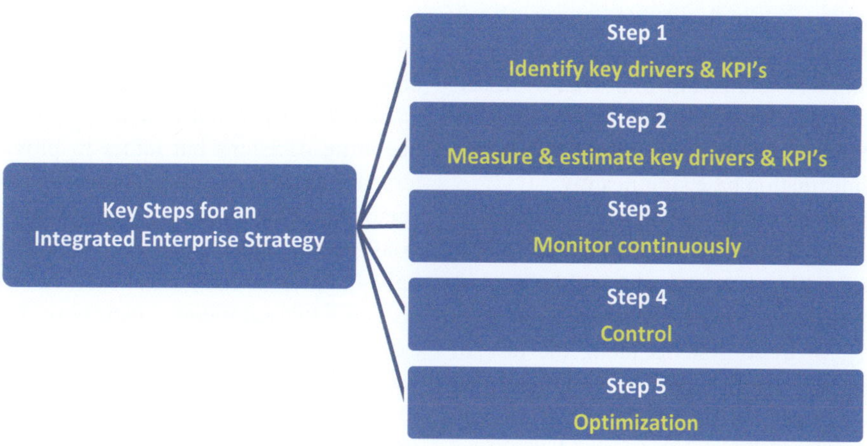

Fig. 11.14 Five key steps involved in development of an integrated enterprise strategy

will be enhanced by including data contained in the block model and possibly using the GPS coordinates to locate detailed information on the mined ore. This text file will be automatically retrieved and the values entered into a central database along with all other key value drivers. Each operation will have its own unique value drivers and will need to be identified in discussion with site personnel.

It is important to note that these value drivers are not necessarily the KPIs measured in some operations. This central database will, however, include all the existing KPIs (both technical and financial) measured by the different business units in operations using software such as plant data historian (PI), Laboratory information management system (LIMS), Operations reporting system (ORS) along with budgeting tools such as Xeras and Hyperion. The main aim in the initial stages of the pilot program will be to have a simplified system with a small number of value drivers and KPIs displayed in the integrated dashboard relevant to the needs of the various business units in an operation, regional, and corporate offices.

11.4.1.2 Step 2: Measure the Key Value Drivers

The next key step is to be able to measure or estimate these key value drivers and KPIs as accurately as possible. There are two important aspects to quality measurements, one is to use "best practices" and the other is "training and mentoring" to ensure these "best practices" are sustained. The existing measurement in the operations should be retained or improved as necessary. Additional sensors or measurement devices should be included only if this improves the existing information base significantly.

It is highly recommended that a core technical group, highly trained in measuring these value drivers and KPIs be created who will be responsible for assisting our operations in high level monitoring of these systems along the task of promoting best practices and providing training needs as required by the operations. This core technical group will consist of individuals based in corporate and Regional offices with assistance from external consultants with high level of expertise. This group will be able to assist in standardization of these measurement techniques and will also help to bridge gaps in understanding the key drivers to better define the interfaces between different business units. This model has been shown to be successful for Codelco in Chile with Kairos Mining (a joint venture between Codelco and Honeywell) based in Santiago providing this core technical expertise to all their operations in remote locations (Zamora et al. 2010).

Enabling technologies such as SmartTag Ore Tracking system along with a Split-Online Fragmentation analyzer are important to this project as they will provide robust measurements of ore movement and characteristics such as size distribution, hardness, and other attributes available in the mine block model and could be easily transferred to the central database. These are important tools that can assist in tracking movements of specific materials during trials and in monitoring ore dilution by placing the tags along the boundary of ore and waste in addition to confirming the residence time and constitution of stockpiles for calibration of stockpile models.

11.4.1.3 Step 3: Monitoring of the Value Drivers

Continuous monitoring of these value drivers and KPIs of different business units on an integrated platform or dashboard is an important part of this Integrated Operations strategy. This monitoring will be carried out by operators and professionals in different business units in an operation along with the core technical group based in corporate or a regional office for example. The role of core technical group is not to interfere in day-to-day operating decisions but rather to provide long-term trends and recommendations to operating personnel, which is difficult to pursue by operations due to other priorities. Various business units in an operation control only the value drivers and KPIs pertaining to their area of work and expertise and do not interfere with the operating decisions of other business units. Any deviation from the normal range of value drivers and KPIs (Red Flags) are communicated automatically (also through mobile devices) to the relevant business units for rectifying the situation based on the control logic (empirical models) developed through leveraging the centralized database with a historical record of operating efficiency.

This centralized database is a connection between resource/reserve model, mine planning, mine dispatch system, truck movements to the crushers or stockpiles, plant data historian, equipment monitoring status, production information along with laboratory assays of plant feed and products. Other items that could be eventually linked with the centralized database are energy, water, and environmental monitoring systems along with health and safety. Codelco's integrated dashboards based in Santiago include both technical and financial information and have been instrumental in providing an effective communication between various operational personnel in different operations and corporate personnel.

The concept of using an integrated dashboard is to promote a collaborative environment, data transparency, and a whole system visibility. These attributes are prerequisites for identifying opportunities that allow a step change in improved productivity and reduce costs.

The question of how often the monitoring system updates itself is very important. A system which updates itself once a day is easier to maintain and reconcile but lacks fidelity to detect ore changes that affect mill performance for example. It is recommended that such a centralized database be updated with ore movements at least once a day, preferably every shift (every 8 or 12 h). For a weekly comparison against benchmarking expectations, this would be 14 or 21 data points. Having a systematic update very frequently generates a large database that soon becomes unmanageable as well as sensitive to short-term stockpile volumes, like the coarse ore pile ahead of the grinding circuit.

11.4.1.4 Step 4: Control

Once the key drivers are measured and monitored on an integrated dashboard through a centralized database, the next step is to be able to use this information for controlling the processes and produce the desired outcome. An important consideration here is to quantify the optimum range of these key value drivers and KPIs based on models or logics developed through the historical record of operating efficiency in the database. This is where interaction between the different business units is critical as it is important to note that the optimal range from a holistic whole-system perspective could be very different from the optimal range from a particular business unit perspective. For example, the optimal powder factor in blasting could be much lower from a mining perspective only but when considered holistically from a Mine-to-Mill perspective, the optimum blasting factor could be almost double with higher blasting costs for optimum mill throughput and overall energy savings.

Once the optimum range of key value drivers are identified based on a holistic perspective, then this becomes a powerful tool from a control viewpoint. This information can be visually presented in the integrated dashboard for easy viewing using different color codes for values within the operating range and for those outside the operating range. These indicators will prompt an operator in a business unit to take corrective action immediately. The wider visibility of this information in an integrated dashboard across the operation value chain will prompt quick action to rectify the situation though a quick root cause analysis resulting is reduced productivity losses.

The control could be done either manually or automatically depending on the situation. When process models and automation tools are available, it is best to utilize them to improve control. Development of models and control logic is an area that should be encouraged as a strategic activity by the core technical team in cooperation with the operating team using the historical record in the centralized database. This will allow maximum utilization of operating data, which unfortunately is not fully utilized when business units operate in silos. With high turnovers in operation with operating knowledge lost over time along with unavailability of resources at certain time, it is very important to maximize utilization of operating data and convert them into models for automatic control purposes as much as possible.

11.4.1.5 Step 5: Optimization

One of the advantages of sharing information by different business units on an integrated platform through the centralized database brings a different dimension to operations optimization, which otherwise would be difficult to conceive in a conventional operations scenario. Because the integrated enterprise promotes a strong collaborating environment, there is a constant need for each business unit to share knowledge and maintain their optimal range of key value drivers. **Transparency drives behavior** and the visual display of dashboards makes it imperative to react immediately, as other business units are relying on this corrective action. This promotion of a collaborative environment is an important attribute of the integrated enterprise.

11.5 Concluding Remarks

Mining and metallurgical companies are facing various challenges at several fronts. It is becoming immensely important to be able to obtain a wide range of information involving various disciplines in the mining and metallurgical value chain, involving different geographical locations, cultural groups, governments, and a wide range of stakeholders, for making critical decisions. Survival of many mining companies depends on the capability to make well-informed decisions in complex situations.

Recent advancements in information technologies, sensors and measurement tools, process control, and automation provide an opportunity for the mining companies to achieve operational excellence across an *Integrated Enterprise*. The advent of the Industrial Internet of Things shall further allow for collaboration and synergies—not only between departments, groups, and processes, but also between machines.

Various case studies discussed in this chapter suggest that mining and metallurgical industries at large can adopt some of the initiatives taken by some of the industry leaders. Technology change is happening rapidly at this current time, but such a change needs to be accompanied by a cultural transformation within the organization. Needless to say, many of the digital technologies that are expected to be implemented in the mining and the metallurgical industries in the next few years are going to be disruptive. It is safe to say that the best strategy is to take a phased approach with smaller steps, depending on the immediate and long-term goals of the respective organizations.

References

Alvarez, E. (2005). Advanced process control to meet the needs of the metallurgical industry. *World of Metallurgy: ERZMETALL, 58*(3), 123.

Arora, V. (2007). *Improving plant operations*. Retrieved from http://www.schneider-electric.ca/documents/solutions/success_story/136-adelaide-broghton-cement-in-australia.pdf.

Bassett, R. (2013, May). Understanding wireless technology, Automation World.

Bergh, L. G., Lucic, E. M., & Zuleta, F. M. (2007). Supervisory control project of copper solvent extraction pilot plant. In R. L. Stephen & G. A. Eltringham (Eds.), *Process control, optimization and six sigma* (Vol. 7). Montreal, QC, Canada: The Canadian Institute of Mining, Metallurgy and Petroleum.

BHP-Billiton. (2013). Retrieved from http://www.stepchangeglobal.com/wp-content/uploads/2013/09/IROC-Brochure-201305031.pdf.

Burke, T. (2008). OPC and Intro OPC UA, Presentation at OPC UA DevCon. Munich.

Cameron, D. A., & Durlofsky, L. J. (2014). *Optimization and data assimilation for geological carbon storage*. Taylor & Francis Group/CRC Press.

Chauhan, N. (2013). Modernizing machine to machine interaction—A platform for igniting the next industrial revolution. GE Software White paper, Retrieved September, 2013.

Enterprise gateway (2010)—A MOM solution for integrating production and maintenance in real time (5G Automatika Internal Report 240310).

Fielding, R. T. (2000). *Architectural styles and the design of network-based software architectures*. Irvine, CA: University of California.

Fortuna, L., Graziani, S., Rizzo, A., & Xibilia, M. G. (2007). *Soft sensors for monitoring and control of industrial processes*. London: Springer.

Gehman, D. (2013, May). ISA-95: Integrating manufacturing's future, Automation World.

Gillot, P. (2006). Pit-to-plant optimization at morila gold mines. In *Proceedings of the SAG 2006 conference*, Vancouver.

Isokangas, E., Sonmez, B., Wortley, M., & Valery, W. (2012). Using smarttag to track ore in process integration and optimization projects: Some case studies in a variety of applications. In *Platinum 2012, SAIMM conference*.

Iwanitz, F., & Lange, J. (2006). *OPC: Fundamentals, implementation, and application*. Heidelberg, Germany: Hüthing.

Markoja, B. (2011, December). *Cidra SONARtrac flowmeters: An alternative flow measurement technology*. Paper presented at ISTOG Winter Conference, Florida, USA.

McKee, D. (2013). Understanding mine to mill, A monograph published by the Co-operative Research Centre for Optimizing Resource Extraction (CRC ORE). Brisbane, Australia.

Mumbi, M. (2012). Process optimization of mixed copper ores through real-time mineralogical analysis. Presentation from Quantum Minerals, Zambia.

Rajaieyamchee, M. A., & Bratvold, R. B. (2009). *Real time decision support in drilling operations using Bayesian decision networks*. SPE 124247, 2009 SPE EUROPEC/EAGE Annual Conference and Exhibition held in New Orleans, Lousiana, USA.

Renner, D., La Rosa, D., DeKlerk, W., Valery, W., Sampson, P., Noi, S. et al. (2006). *AngloGold Ashanti Iduapriem mining and milling process integration and optimization, Vol. 1*. SAG 2006 Conference Proceedings, Vancouver.

Schlaepfer, R., & Koch, M. (2014). Industry 4.0—Challenges and solutions for the digital transformations and use of exponential technologies, Deloitte.

Sherring, A. (2012). *Automation and remote collaboration: Will it change the future of mining?* Presented at the Honeywell Users Group Symposium, extract written in the PACE magazine.

Sinclair. M. (2012). *Remote Collaboration Centres (Honeywell)*. Retrieved from https://www.posccaesar.org/svn/pub/PCA/MemberMeeting/201210/Day2/RemoteCollaborationCenters-MikeSinclair-Oct2012.pdf.

Skonnard, A., & Gudgin, M. (2001). *Essential XML quick reference*. Upper Saddle River, NJ: Addison-Wesley.

Thermoslag, ver. 2. (2010). Swedish Steel Producers Association.

Zamora, C., Schifferli, G., Castelli, L., & Bonomelli, A. (2010). *Kairos mining: A comprehensive approach to sustained long term benefits in copper processing plant automation*. Paper presented at the 42nd Annual Canadian Mineral Processors Conference, Ottawa, Retrieved June 19–21, 2010.

Zurawski, R. (Ed.). (2005). *Industrial communication technology handbook* (Industrial technology series, Vol. 1, pp. 7–10). Boca Raton, FL: CRC Press. Retrieved 4 Feb 2013. ISBN 0849330777. LCCN 2004057922.

11.5.1 Further Reading

Alsmeyer, F. (2006). Automatic adjustment of data compression in process information management systems. In W. Marquardt & C. Pantelides (Ed.), *Proceedings of the 16th European Symposium on Computer Aided Process Engineering*. Garmisch-Partenkirchen, Germany. Retrieved May 9, 2009, from http://bit.ly/tWouV.

American National Standards Institute/International Society of Automation, ANSI/ISA-99.00.01-2007. (2007). *Security for industrial automation and control systems: Concepts, terminology and models*. Research Triangle Park, NC: ISA.

Anderson, B. D. O., & Moore, J. B. (1979). *Optimal filtering*. Upper Saddle River, NJ: Prentice-Hall.

Anderson, R. J. (2008). *Security engineering: A guide to building dependable distributed systems*. Hoboken, NJ: Wiley. ISBN 978-0-470-06852-6.

ANSI/ISA-95.00.01-2000. (2000). *Enterprise-control system integration Part 1: Models and terminology*. Research Triangle Park, NC: ISA.

ANSI/ISA-95.00.02-2001. (2001). *Enterprise-control system integration Part 2: Object model attributes*. Research Triangle Park, NC: ISA.

ANSI/ISA-95.00.03-2005. (2005). *Enterprise-control system integration, Part 3: Models of manufacturing operations management*. Research Triangle Park, NC: ISA.

ANSI/ISA-95.00.05-2007. (2007). *Integration, Part 5: Business-to-manufacturing transactions*. Research Triangle Park, NC: ISA.

ANSI/ISA-99.00.01-2007. (2007). *Security for industrial automation and control systems: Concepts, terminology and models*. Research Triangle Park, NC: ISA.

Åström, K. J., & Hägglund, T. (2006). *Advanced PID control*. Research Triangle Park, NC: ISA.

Bemporad, A., & Morari, M. (1999). Control of systems integrating logic, dynamics, and constraints. *Automatica, 35*(3), 407–427.

Biehl, M., Prater, E., & McIntyre, J. (2004). Remote repair, diagnostics, and maintenance. *Communications of the ACM, 47*(11), 100.

Bondi, A. (2000). Characteristics of scalability and their impact on performance. In: *ACM Proceedings of the 2nd International Workshop on Software and Performance* (pp. 195–203). Ottawa, Canada, ISBN: 1-58113-195-X.

Brandl, D. (2006). *Design patterns for flexible manufacturing*. Research Triangle Park, NC: ISA.

Brändle, M., & Naedele, M. (2008, November/December). Security for process control systems: An overview. *IEEE Security & Privacy, 6*(6), 24–29.

Bristol, E. H. (1990). Swinging door trending: adaptive trend recording? In *ISA National Conference Proceedings*. New Orleans, LA.

Byres, E. J. (2002, September).The Myth of obscurity. *InTech* (p. 76). Research Triangle Park, NC: ISA.

Chappell, D. A. (2003). *Using S88 batch techniques to manage and control continuous processes*. Woodcliff Lake, NJ: World Batch Forum.

Cheever, G., & Schroeder, J. (2007). *Remote service*. ABB Review Special Report Automation Systems. Retrieved from May 01, 2009, from http://bit.ly/Lybsh.

Clark, D. (2008). Optimum process scheduling. *International Mining*, Retrieved January 2008.

Cutler, C. R., & Ramaker, B. L. (1980). Dynamic matrix control—A computer control algorithm. In *Proceedings Joint American Control Conference*, San Francisco.

DeMarco, T. (1982). *Controlling software projects: Management, measurement & estimation*. New York: Yourdon Press.

Ebel, M., Drath, R., & Sauer, O. (2008). Automatische Projektierung eines Produktionsleitsystems der Fertigungstechnik mit Hilfe des Datenaustauschformats CAEX' In atp—Automatisierungstechnische Praxis 50, Heft 5, Seiten 84–92. München: Oldenburg Industrieverlag.

Electric Power Research Institute Advanced Control Room Alarm System. (2005). *Requirements and implementation guidance* (ERPI report 1010076). Palo Alto.

Farb, D., & Gordon, B. (2005). *Agent GXP FDA Part 11guidebook*. Los Angeles, CA: University of HealthCare.

Franke, R., & Vogelbacher, L. (2006). Nonlinear model predictive control for cost optimal startup of steam power plants. *Automatisierungstechnik, 54*(12), 630.

Fritzson, P. (2004). *Object-oriented modeling and simulation with Modelica 2.1*. Piscataway, NJ: IEEE Press.

Früh, K. F., & Maier, U. (2004). *Handbuch der Prozessautomatisierung* (3rd ed.). München, Germany: Oldenbourg Industrie.

Ganz, C., & Layes, M. (1998). *Modular turbine control software: A software architecture for the ABB gas turbine family control system*. Second international workshop on development and evolution of software architectures for product families, Las Palmas.

Garcia, R. G., & Gelle, E. (2006). Applying and adapting the IEC 61346 standard to industrial automation applications. *IEEE Transaction on Industrial Informatics, 2*(3), 185–191.

Garpinger, O. (2009). *Design of Robust PID Controllers with Constrained Control Signal Activity* (Licentiate Thesis). Lund University, Sweden.

Grosspietsch, K. -E., & Kirrmann, H. (2002). Fehlertolerante Steuerungs- und Regelungssyteme. Automatisierungtechnik at, *50*(8).

Gutermuth, G., & Hausmanns, Ch. (2007). Kostenstruktur und Untergliederung von Automatisierungsprojekten. In atp—Automatisierungstechnische Praxis 49, Heft 11, Sieten 40–47. München: Oldenburg Industrieverlag.

Hägglund, T. (1999). Automatic detection of sluggish control loops. *Control Engineering Practice, 7*, 1505–1511.

Hall, K. H., Staron, R. J., & Zoitl, A. (2009). Challenges to industry adoption of the IEC 61499 standard on event-based function blocks, downloaded from IEEE Xplore on March 25, 2009, 1-4244-0865-2/07.

Harris, T. (1989). Assessment of control loop performance. *The Canadian Journal of Chemical Engineering, 67*, 856–861.

Hölscher, H., & Rader, J. (1986). *Microcomputers in safety technique, An aid to orientation for developer and manufacturer*. Rheinland, Germany: TÜV.

Horch, A. (1999). A simple method for detection of stiction in process control loops. *Control Engineering Practice, 7*(10), 1221–1231.

IBM Globalization. (2009). www.ibm.com/software/globalization (Last retrieved May 22, 2009).

IEC 61508-3. (1998). *Functional safety of electrical/electronic/programmable electronic safety-related systems—Software requirements*. Geneva, Switzerland: IEC.

IEC 61511. (2003). *Functional safety—Safety instrumented systems for the process industry sector*. Geneva, Switzerland: IEC.

IEC 61511-1. (2008). *Functional safety—Safety instrumented systems for the process industry sector -Part 1: Framework, definitions, system, hardware and software requirements*. Geneva, Switzerland: IEC.

IEC 61513. (2001). *Nuclear power plants—Instrumentation and control for systems important to safety*. Geneva, Switzerland: IEC.

IEC 62061. (2005). *Safety of machinery—Functional safety of safety-related electrical, electronic and programmable electronic control systems*. Geneva, Switzerland: IEC.

International Electrochemical Commission, IEC 61346. (1996). *Industrial systems, installations and equipment, and industrial products—Structuring principle and reference designations. Parts 1–4*.

International Electrotechnical Commission. (2003). *IEC 61131-3: Programmable controllers—Part 3: programming languages Edition 2.0.*. Geneva, Switzerland: IEC. ISBN 2-8318-6653-7.

International Electrotechnical Commission (IEC 61346). (1996). *Industrial systems, installations and equipment and industrial products—Structuring principles and reference designations*.

International Electrotechnical Commission (IEC 61508–1). (1998). *Functional safety of electrical/electronic/programmable electronic safety-related systems—General requirements*. Geneva, Switzerland: IEC.

International Electrotechnical Commission (IEC 61508–2). (2000). *Functional safety of electrical/electronic/programmable electronic safety-related systems—Requirements for electrical/electronic programmable electronic safety-related systems*. Geneva, Switzerland: IEC.

International Electrotechnical Commission. (2006). *IEC 61508: Functional safety of electrical/electronic/programmable electronic safety-related systems, Parts 1 to 4*. Geneva, Switzerland.

ISA-88.00.05. (2010). *Batch control Part 5: Implementation models and terminology for modular equipment control (working draft standard)*. Research Triangle Park, NC: ISA.

ISA-88.01. (1995). *Batch control Part 1: Models and terminology* (R2006). Research Triangle Park, NC: ISA. See also IEC 61512.

Isaksson, A. J., & Graebe, S. F. (1999). Analytical PID parameter expressions for higher order systems. *Automatica, 35*(6), 1121–1130.

Isaksson, A. J., & Graebe, S. F. (2002). Derivative filter is an integral part of PID design. *IEE Proceedings - Control Theory and Applications, 149*(1), 41–45.

Jaikumar, R. (1993). 200 years to CIM. *IEEE Spectrum, 30*, 26–27.

Johannesson, G. (1994). *Object-oriented process Automation with SattLine*. Lund, Sweden: Studentlitteratur.

John, K. H., & Tiegelkamp, M. (2001). *IEC 61131-3: Programming industrial automation systems, English edition*. Heidelberg, Germnay: Springer. ISBN 3-540-67752-6.

Kristiansson, B., & Lennartson, B. (2006). Robust tuning of PI and PID controllers. *IEEE Control Systems Magazine, 26*(1), 55–69.

Laubner, R., & Gohner, P. (1999). *Prozessautomatisierung 2*. New York: Springer. ISBN 3-540-65319-8.

Li, Y., Ang, K. H., & Chong, G. C. Y. (2006). Patents, software and hardware for PID control. *IEEE Control Systems Magazine, 26*(1), 42–54.

Liefeldt, A., Gutermuth, G., Beer, P., Basenach, S., & Alznauer, R. (2005). Effizientes Engineering—Begleitende Fortschrittskontrolle grober Projekte der Automatisierungstechnik. In atp—Automatisierungstechnische Praxis 47, Heft 7, Seiten 60–64.

Maciejowski, J. M. (2002). *Predictive control with constraints*. Upper Saddle River, NJ: Prentice-Hall.

Mahnke, W., Leitner, S. H., & Damm, M. (2009). *OPC unified architecture*. New York: Springer.

Meyer, B. (2000). *Object-oriented software construction*. Upper Saddle River, NJ: Prentice Hall.

Microsoft Go Global Developer Center. (2009). msdn.microsoft.com/goglobal (Last retrieved May 22, 2009).

Müller, J., Gote, M., & DeLeeuw, V. (2007, January). NAMUR Umfrage plant asset management, *Automatisierungstechnische Praxis Atp*.

NAMUR NA 35. (2003). *Abwicklung von PLT-Projekten/Handling PCT Projects*, NAMUR Worksheet NA35, Version of March 24, 2003. Available from www.namur.de.

NAMUR NE 100. (2007). *Use of lists of properties in process control engineering workflows*. Namur Recommendation NE100, Version 3.1. of November 2, 2007. Available from www.namur.de.

Narasimhan, S., & Jordache, C. (1999). *Data reconciliation and gross error detection*. Houston, TX: Gulf Professional.

O'Brien, L. (2002). *Total automation business for the process industries world wide outlook—Market analysis and forecast through 2006*. Boston, MA: ARC Advisory Group.

O'Dwyer, A. (2006). *Handbook of PI and PID controller tuning rules* (2nd ed.). London: Imperial College Press.

OPC foundation. (2002). *Alarms and events custom interface standard V1.1*.

OPC Foundation. (2002). *Alarms and events custom interface standard version 1.10*.

OPC Foundation. (2008, December). *Analyzer devices* (Draft version 0.30.00).

OPC Foundation. (2003). *OPC foundation data access custom interface standard V3.00*.

OPC. (2009, February). *OPC historical data access specification version 1.2* OPC *UA Spec. Part 1—Concepts* (version 1.01); OPC. (2009, February). *UA Spec. Part 2—Security model* (version 1.01); OPC. (2009, February). *UA Spec. Part 3—Address space model* (version 1.01); OPC. (2009, February). *UA Spec. Part 4—Services* (version 1.01); OPC. (2009, February). *UA Spec. Part 5—Information model* (version 1.01); OPC. (2009, February). *UA Spec. Part 6—Mappings* (version 1.0); OPC. (2009, February). *UA Spec. Part 7—Profiles* (version 1.0); OPC. (2009, February). *UA Spec. Part 8—Data access* (version 1.01); OPC (2007, November). *UA Spec. Part 9—Alarms and conditions* (Draft version 0.93q); OPC. (2007, January). *UA Spec. Part 10—Programs* (version 1.00); OPC. (2007, January). *UA Spec. Part 11—Historical access* (version 1.00); OPC. (2007, November). *UA Spec. Part 12—Discovery* (Draft version 1.00.03); OPC. (2008, July). *UA Spec. Part 13—Aggregates* (RC version 1.0); OPC. (2008, December). *UA devices* (Draft Version 0.75).

Pettersson, J., Ledung, L., & Zhang, X. (2006). Decision support for Pulp Mill operations based on large-scale on-line optimization. In *Preprints of Control Systems 2006*, Tampere, Finland, Retrieved 6–8 June.

Rao, C. V. (2000). *Moving horizon strategies for the constrained monitoring and control of nonlinear discrete-time systems*. Ph.D. Thesis, University of Wisconsin.

Reierson, B. (2006, April). Myths and realities of asset optimization. *Plant Engineering* Retrieved from www.plantengineering.com/article/CA6308016.html.

Richalet, J., Rault, A., Testud, J. L., & Papon, J. (1978). Model predictive heuristic control: Applications to industrial processes. *Automatica, 14*, 413–428.

Rivera, D. E., & Morari, M. (1987). Control relevant model reduction problems for SISO H_2, Hinf, and u-Controller synthesis. *International Journal of Control, 46*(2), 505–527.

Rode, M., Franke, R., & Krüger, M. (2003, March). Model predictive control for boiler start-up. *ABB Review*. Available from http://www.abb.com/abbreview.

Sarma, P., Chen, W. H. (2008). *Applications of optimal control theory for efficient production optimization of realistic reservoirs*. IPTC 12480, Paper presented at the International Petroleum Technology Conference help in Kuala Lumpur, Malaysia, Retrieved December 3–5, 2008.

Samad, T., McLaughlin, P., & Lu, J. (2007). System architecture for process automation: Review and trends. *Journal of Process Control, 17*, 191–201.

Sauter, T. (2007). The continuing evolution of integration in manufacturing automation. *IEEE Industrial Engineering Electronics Magazine, 1*, 10–19.

Scholten, B. (2007). *The road to integration: A guide to applying the ISA-95 standard in manufacturing*. Research Triangle Park, NC: ISA.

Seborg, D. E., Edgar, T. F., & Mellichamp, D. A. (2003). *Process dynamics and control* (2nd ed.). Hoboken, NJ: Wiley.

Shobrys, D. E., & White, D. C. (2000). Planning, scheduling and control systems: Why can they not work together. *Computers and Chemical Engineering, 24*, 163–173.

Siewiorek, D. P., & Swarz, R. S. (1982). *The theory and practice of reliable system design*. Bedford, England: Digital Press. ISBN 0-932376-13-4.

Skogestad, S. (2003). Simple analytic rules for model reduction and PID controller tuning. *Journal of Process Control, 13*, 291–309.

Song, F. J. (1996). Next generation of structural engineering automation systems. In *Computing in civil engineering* (pp. 494–500). New York.

Thomasson, F. Y. (1995). Controller tuning methods. In N. J. Sell (Ed.), *Process control fundamentals for the pulp and paper industry*. Norcross, GA: TAPPI Press.

Thornhill, N. F. (2005). Finding the source of nonlinearity in a process with plant-wide oscillation. *IEEE Transactions on Control System Technology, 13*, 434–443.

Thornhill, N. F., & Horch, A. (2007). Advances and new directions in plant-wide disturbance detection and diagnosis. *Control Engineering Practice, 15*, 1196–1206.

Thornhill, N. F., Shouk, M. A. A., & Shah, S. L. (2004). The impact of compression on data-driven process analyses. *Journal of Process Control, 14*, 389–398.

U.S. Food and Drug Administration. *FDA 21 CFR Part 11*. Available from http://www.fda.gov/cder/gmp/index.htm.

Unicode Consortium. (2006). *The Unicode standard, version 5.0*. Upper Saddle River, NJ: Addison-Wesley.

Valdez, G., Sandberg, D. G., Immonen, P., & Matsko, T. (2008). Coordinated control and optimization of a complex industrial power plant. *Power Engineering Magazine*, pp.124–134. Retrieved November.

VanDoren, V. (2008). Advances in control loop optimization. *Control Engineering*. Retrieved January 5, 2008, from http://www.controleng.com/article/CA6559117.html.

VDI/VDE. (2008). Plant asset management (PAM) in the process industry—Definition, model, task, benefit. *VDI/VDE Guideline* No. 2651.

VDI/VDE 2182. (2007, August). *VDI guideline IT security for industrial automation—General model, Sheet 1; Draft for comments. Gründruck VDI*. Berlin, Germany: Beuth.

Watson, M. J., Liakopoulos, A., Brzakovic, D., & Georgakis, C. (1995). Wavelet techniques in the compression of process data. In Proceedings of the American Control Conference, Seattle, WA.

Whitt, M. D. (2004). *Successful instrumentation and control systems design*. Research Triangle Park, NC: ISA. ISBN 978-1-5617-992-1.

Woods, D. D. (1995). The alarm problem and directed attention in dynamic fault management. *Ergonomics, 38*(11), 2371.

Part IV

Equipment

Equipment Development, Design, and Optimization

12

Mark F. Vancas and Ram Ramachandran

12.1 Challenging the Status Quo

There is no such thing as the status quo. If one is in the business of producing anything, chances are someone else also produces the same product, is thinking about doing so, or producing the same product at a lower cost. This sets up, like it or not, a competitive situation whereby each producer tries to increase their profits by garnering a larger share of the marketplace or decreasing their production costs.

One may be the industry leader in one's field and may have the lowest operating costs of the industry, but if you sit on your hands, the competitors will eventually catch up. Consequently, if you continue to rest on your laurels and the competitors advance, you, in effect, are falling behind. Hence, there is no such thing as the status quo.

M.F. Vancas (✉)
MFV Engineering Inc., 5310 E. Golder Ranch Dr., Tucson, AZ 85739, USA
e-mail: bsxtucson@aol.com

R. Ramachandran
Consulting Engineer, 9650 E. Peregrine Place, Scottsdale, AZ 85262, USA

12.2 Staying Ahead

To stay ahead of the game, a company must advance, must decrease costs, increase productivity, or both, and, to do this, the company must be innovative. The upper management personnel of a company may have achieved their position through their own innovative abilities but most likely, in well-established companies, they have achieved their position due to their management abilities. The definition of management is "Getting things done through other people." So the upper management people do not need to actually come up with all the innovations, but they do need to be able to recognize an innovation when they see one.

Innovation cannot be summoned, coerced, dictated, or forced. Innovation "just happens." Innovation can be encouraged by creating an atmosphere in which innovation is welcomed. That said, no company will allow changes that could potentially hurt production and/or profitability.

12.3 Degrees of Innovation

Improving existing equipment is one of the easier paths to take toward being innovative. The development of innovative equipment involves several steps, depending on the degree of variation of the design from existing practice. The degrees of innovation include: improving the design of

V.I. Lakshmanan et al. (eds.), *Innovative Process Development in Metallurgical Industry*,
DOI 10.1007/978-3-319-21599-0_12

existing equipment, extending the size of the largest current equipment, and/or developing a radically different equipment design.

In any processing plant, many pieces of equipment are utilized to produce the desired end product. These various pieces of equipment have most likely been around for a long time and have proven to be reliable and robust enough to withstand the abuses of production on a continuous basis. One might think that the tried and true equipment is the best available because of just that tried and true reputation and the, oh so common, “if it ain’t broke, don’t fix it” philosophy. In many cases, this may be entirely true but there are cases where times have changed and different materials of construction or better instrumentation have become available that make changes possible. Graphite composite materials, for example, can be made to be as strong as, or stronger than, metallic components and possibly lighter weight as well. Use of variable speed motors makes some operations more energy efficient thus reducing operating costs. Because of the ongoing effect, operating cost reductions do not need to be large to add up to a big contribution over time.

Economy of scale can be used effectively, particularly in older operations. In older operations, chances are that certain equipment sizes were used because they were the largest available at the time. Well, times change, maybe now the equipment manufacturer, by being innovative, has found a way to make their equipment even larger than before. This allows the end user to use fewer pieces to achieve the same result. Fewer motors usually mean lower capital cost and reduction in operating cost. Mixing is a good example where new designs of impeller blades have enabled designers to mix larger tanks with the same or higher efficiency resulting in either a savings of power or an increase in mass transfer.

Making radical changes to existing equipment is probably one of the most difficult innovations to be approved by management. These types of changes must be supported by an extensive amount of test work and irrefutable evidence. Often times the best evidence is actually the successful adoption of similar or identical practice or piece of equipment in another industry. The innovation part comes from recognizing that various industries have common principles and some of those principles may be applicable elsewhere. This is why keeping track of what is going on in other, totally unrelated industries, often times leads to “breakthroughs” in one’s own industry.

12.4 Developmental Steps

How does a company that actively promotes and encourages innovation go about being innovative? The steps involved in the development can include: desktop design, mock-ups, small-scale demonstration, and/or full-size demonstration.

Obviously, someone must first envision the innovation and write down their thoughts and/or sketch up their idea. This, initially, is probably a simple statement of concept to promote a healthy discussion among colleagues, see their reaction to the new idea and establish whether the idea has any merit or not. Also, somewhere in the collective experience of your colleagues there may be some specific experience that says that your idea has been tried before and failed.

After sharing such thoughts with colleagues and others, maybe a miniature version will be built. In today’s world of design, this first “miniature” may well be an electronic 3D model. With the software available these days, there is little need to actually make a scale model unless one wishes to get a feel for how the manufacturing and assembly will actually be done.

In case, the scale model (electronic or physical) does not show any “fatal flaws,” the time may come to build a working model that can be used to demonstrate the idea in a laboratory or in a potential customer’s office/plant. A major problem with scale models is the scale factor. Too small and the idea may not be clearly demonstrated, too large and there may be problems with securing enough feed stock and/or disposing of the waste materials or products, particularly if the products are high value or the chemicals/compounds are considered hazardous. A real key aspect of scale models is identifying which variables are critical, viz., the real determining variables.

Obviously if a scale model prototype is warranted, then a full-scale demonstration version may also be needed. In this case, the developers need to either have their own production facility or have a client that will help with building a full-scale model based on the results of the small-scale model.

12.5 Sources of Innovation

All of the degrees of innovation require someone or a team to come up with an idea that can offer a better solution to a problem than what is experienced with the existing equipment. Seldom are new developments created in a vacuum.

When talking about equipment development, the developers of innovative equipment can come from various sources such as: Manufacturers and/or their suppliers or operators trying to make their job easier. Manufacturers are always looking to improve their products and reduce costs. They do a very good job, for the most part, of advancing their technology at the same time. Some manufacturers/suppliers provide chemicals to their industry. The new chemical formulations that they come up with have the same effect of improving the end users' operations. This can be demonstrated by the use of existing equipment in different applications or with different chemicals. Reagent suppliers often modify the chemistry of their product so as to make the reagent more selective or more acceptable to the environment.

Radical departure from the "status quo" is probably one of, if not the most, difficult types of innovation to get accepted by industry. Reasons for the opposition can be anything from "If it ain't broke, don't fix it," "not on my watch," to "we want to be the first to be second." Many times the management is perfectly happy to let someone else test the waters with new innovations and only if the new innovation is proven elsewhere, will management then jump on the bandwagon and allow testing in his/her facility.

Not all innovations come from someone on the inside thinking "outside the box." Sometimes innovations come from someone on the outside of the box looking in. This is where "cross pollination" between industries can be particularly beneficial.

12.6 Relax

Innovative ideas crop up at odd times and in odd places and for odd reasons. For any innovative idea to move forward and finally become useful to society, there must be a system/procedure/method to record those ideas and share them with colleagues when appropriate.

Operating problems often times lead to innovations that have no resistance when the "the management" says: "We do not care what it takes, get this running!" When production is in jeopardy, management is most often willing to try any reasonable remedy. This is the innovator's dream situation which leads to many new improvements. Once again a key factor is to let one's brain relax and let the answer "come to you."

12.7 Persistence

Innovative processes often times take a long time to fully develop and get tested. Often times these developing innovations take more time that what is really required. That is because the project/people get "paralyzed by perfection." This happens because some feel that a failed test is bad. Some, however, do not believe that a failed test is bad but rather believe the quickest way to work out the bugs is to "get it wrong the first time." Failure is an expected part of development. The key to success is to be persistent and carry on. Burt Rutan, the legendary avionics innovator believed that to be true. He also believed that to be a true innovator one needs to be able to go out to a shop and build a prototype with his/her own hands. Rutan firmly believed that in building the first prototype, the innovator will get it wrong the first time and immediately see ways to improve the design. This finally leads to optimizing the design.

12.8 Optimization

In a production facility there is a constant push to optimize the use of the equipment. Optimizing can be simply: (a) fine tuning the controls on a burner to get more complete combustion, (b) improving the maintenance of the equipment so that unplanned shutdowns occur less frequently, (c) matching the size and capacity of the equipment to the standard work week so as to avoid having to work overtime at premium labor rates. Regardless of the method, optimizing every aspect of the operation will lead to better efficiencies and lower costs both of which help the company advance in comparison to the competition.

12.9 Intellectual Property (IP)

Quite often an innovation produces unexpected results that must be investigated further and optimized. These unexpected results could result in additional Intellectual Property that may or may not need to be protected. Intellectual Property is always a "sticky wicket." If IP Intellectual property (IP) protection is justified, should the protection be in the form of patents or copyrights.

Some companies actively promote and pay for "brainstorming" which actually results in innovations that in turn benefit the company. For example, 3M allows employees to spend 10 % of their time as "think" time to come up with any idea that may benefit the company.

In general, all companies should encourage innovation in their employees that would help the company develop processes/products and benefit financially. The company should also have a suitable program to reward their employees—in addition to their salaries and benefits—based on their contribution and benefit to the company. In short, all companies should promote innovation from within and reward the employee in proportion to the value realized by the company.

12.10 Recognize Individuals

For a company to really be innovative, the management philosophy must promote, encourage, and reward innovative thinking of the individual employees. Such an atmosphere will lead to improvements in design of equipment and processes and to optimization of existing technology. The key to all this is in the brain power of the employees. One must always remember that the collective knowledge of a company walks out the door with the employees every day and they do not have to walk back in the next day. The power of a company lies with the employees and their ingrained desire to innovate.

The preceding is a generic description of the various factors involved in the design and development of process equipment in the process industry. Once developed, the equipment is optimized over a period of time in service to give best performance.

12.11 Case Studies in Equipment Design and Development

Examples of equipment developed for the metallurgical process industry are given below. It covers mineral processing and hydrometallurgy areas. Equipments developed in the pyrometallurgy area are discussed in Chap. 16.

12.11.1 Mineral Processing

Three pieces of widely used equipment—manufactured by FLSmidth—in the mineral processing are shown below (Figs. 12.1, 12.2, 12.3). Figure 12.4 shows a flotation cell Manufactured by Outotec USA. Figure 12.5 shows a mixer for agitated tanks used in leaching operations.

A brief description of the mixer's operating details is given here.

Many mineral processing plants incorporate large tank cells to effect conditioning of the ore prior to the extraction process. More notably, the Gold and Nickel industry routinely require significant retention times, sometimes measured in days, thus the need for large tank agitation. At the heart of these large tanks is a robust mixer mechanism to maintain adequate suspension over extended time periods. Selecting the proper materials of construction, agitation speed, and blade design are critical for these mixer drives as well as providing longevity of operation.

Fig. 12.1 Cutaway view of FLS-Dorr Oliver 300 Cell Flotation. FLSmidth also manufactures WEMCO flotation cells with varying capacities. Reproduced with permission from FLSmidth

Fig. 12.2 General view of FLS-Dual Drive Thickener. Reproduced with permission from FLSmidth

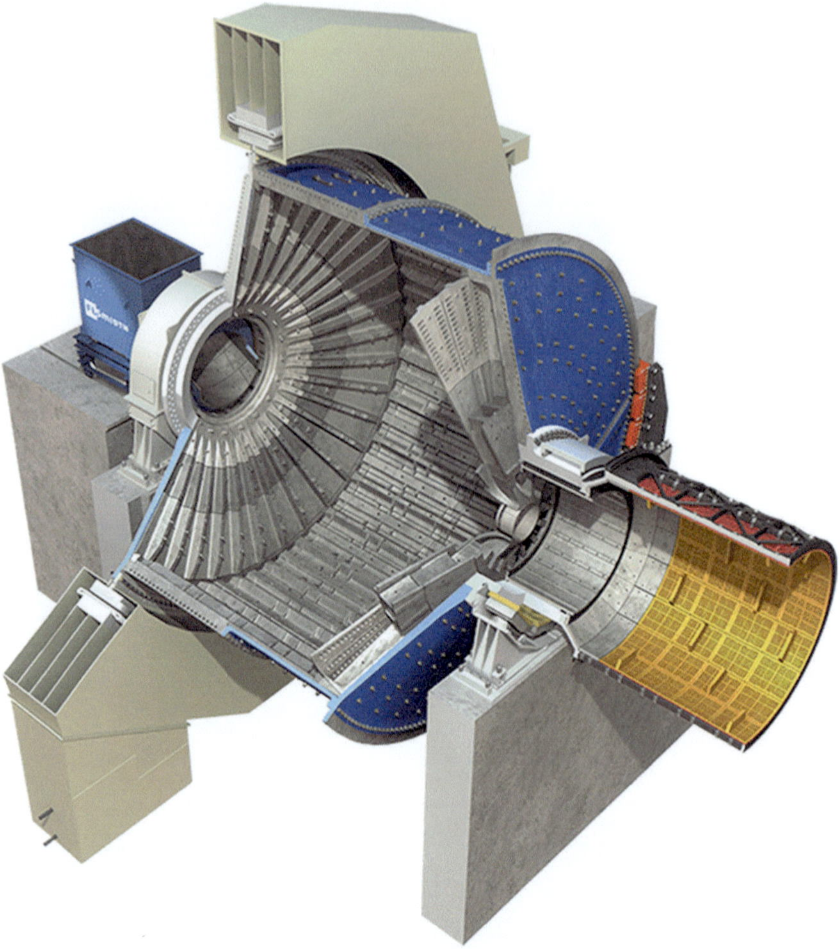

Fig. 12.3 Cutaway view of FLS SAG Mill. Reproduced with permission from FLSmidth

The photo represents industry-proven technology for these large mixer drives used in tank leaching CIL and CIP gold and nickel, zinc, and other processing circuits.

Outotec Oyj in Finland has developed the Vertical Smooth Flow (VSF™) mixers for use in SX-EW plants. The VST™ maintains smooth agitation throughout the solvent extraction plant to avoid oxidation of organic and development of overly small droplet size in dispersion.

12.11.2 Hydrometallurgy

The autoclave has been the workhorse of many leaching processes in hydrometallurgical operations. It is widely used in the pressure leaching of base metal sulfides, refractory gold ores, and nickel concentrates and mattes. A cutaway view of an autoclave used in pressure leaching technology is shown in Fig. 12.6.

The following two examples illustrate: (a) a feel for equipment development and (b) application of technology across various applications.

Contributed by: Dr.Adrian Deneys
Business Development Manager
PRAXAIR Inc.
E-mail: Adrian_Deneys@Praxair.com

Two examples transferring combustion and gas injection metallurgical process technology from steelmaking to copper smelting are pre-

Fig. 12.4 Flotation Cell-Outotec (USA). Reproduced with permission from Outotec (USA) Inc.

Fig. 12.5 Mixer for agitated tanks. Reproduced with permission from Outotec (USA) Inc.

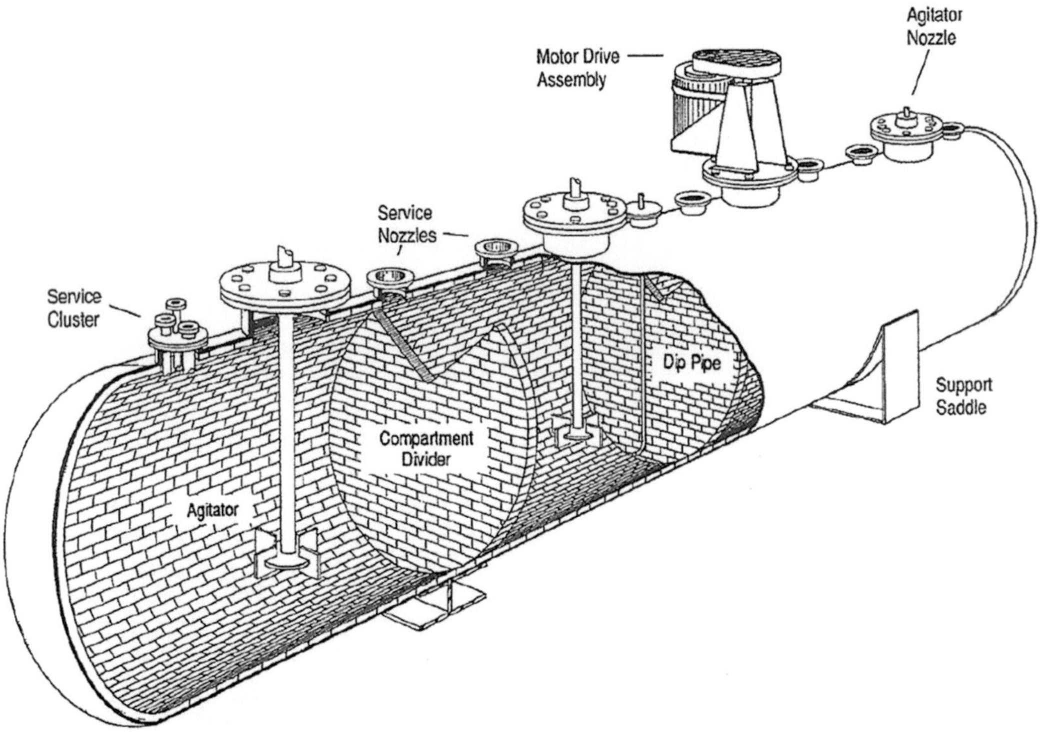

Fig. 12.6 A four compartment *brick* and *lead-lined horizontal* autoclave (Berezowsky 1991). Copyright 1991. Reprinted with permission from the Minerals, Metals and Materials Society

Table 12.1 Summary of Praxair Coherent Jet technology (CoJet®) system installations (see Fig. 12.7) (Deneys et al. 2013)

Process	No. of furnaces	First installation or trial
Electric arc furnace (steelmaking)	144	1996
Copper converter (MK reactor)	1	2004
Copper anode furnaces	2	2007/2009

sented. Coherent jet technology combines gas injection and combustion technology in a single process. The technology was developed for steelmaking in 1996 and transferred to copper converting and fire refining from 2003 to 2009 (Table 12.1).

Annual process licenses across stainless steelmaking (AOD process), electric arc furnace steelmaking, and two nonferrous metallurgical processes (Outotec flash smelting and Sirosmelt bath smelting) are shown in the figure above. The variations in the annual licenses granted (given by the slopes of the lines) are attributed to many complex factors. Some factors may include relative capital investment, productivity enhancements, market sizes, communications within the markets, and channels to markets.

The Vale Inco MK Reactor Process (Liu et al. 2005) applied Praxair's coherent jet (CoJet®) technology in 2004. The MK Reactor process is described in detail elsewhere (Warner et al. 2003). Vale Inco pioneered the process in 1993 with improvements in feeding, reductant addition, and refractories. In 2000 Vale Inco tested sonic to supersonic velocity oxygen lances and achieved some improvements. However, the oxygen lances formed long tip accretions. Cleaning the oxygen lances limited furnace availability and the rapid decrease in oxygen efficiency hindered productivity. Praxair developed coherent jet technology at low oxygen supply pressure

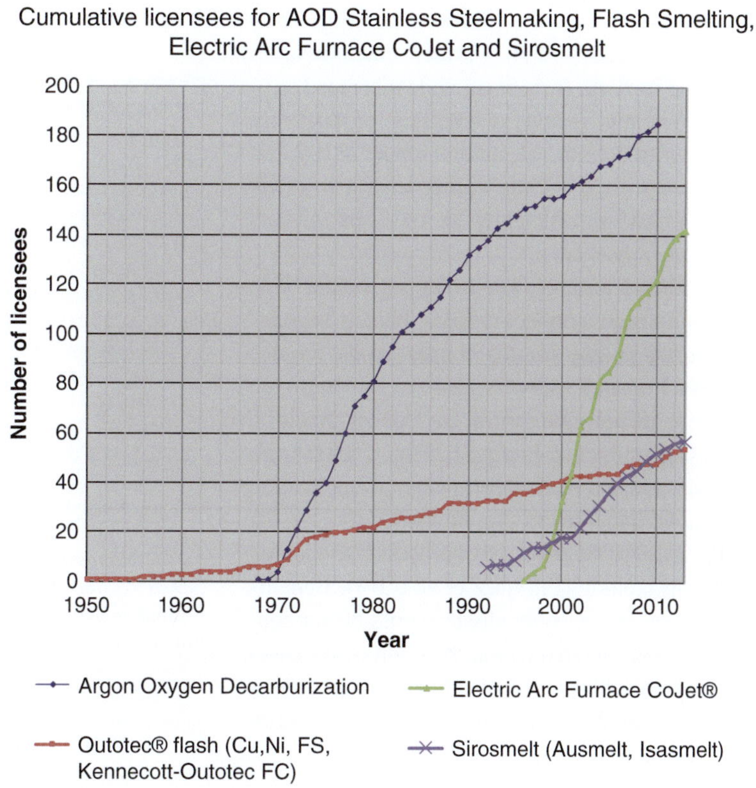

Fig. 12.7 Licensees for Argon Oxygen Decarburization for stainless steelmaking, Electric Arc Furnace CoJet®, Outotec flash smelting, and Sirosmelt bath smelting. References: Praxair internal data (AOD and Electric Arc Furnace CoJet), www.outotec.com, www.xstrata.com, data from October 2013 (Deneys et al. 2013)

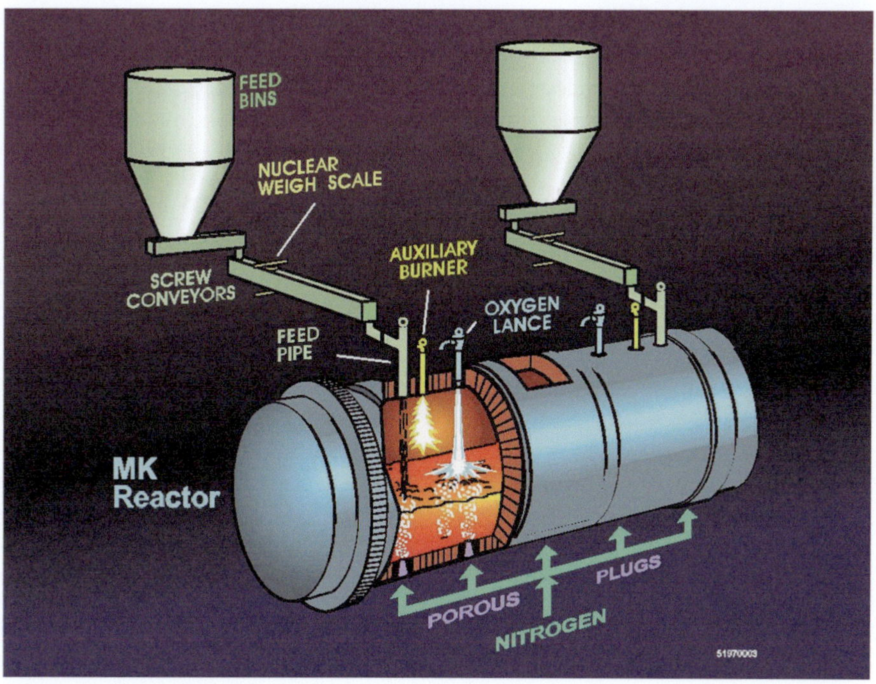

Fig. 12.8 Vale Inco MK Reactor CoJet® system schematic (Liu et al. 2005)

(around 20psig) for the MK Reactor (Fig. 12.8) process in 2003–2004. Further customization occurred in 2010. Benefits include increased oxygen efficiency, increased smelting productivity (record production, Vale annual report 2010), and increased turndown (the ability to run the MK Reactor process efficiently at lower production rates, when needed).

In 2005 Kennecott Utah Copper visited Vale Inco and began a project to apply Praxair's coherent jet technology for the copper anode furnace process. Praxair developed mixed oxygen/nitrogen coherent jet technology for the copper smelting process (Deneys and Enriquez 2009). The technology was adopted in 2009 following the application to remove large hearth accretions (Enriquez et al. 2010). In 2010 the coherent jet process was adopted for scrap melting (Enriquez et al. 2013). Benefits include the ability to rapidly heat cold blister copper, increased scrap melting, accretion control, and to desulfurize high levels of sulfur in blister copper (Deneys et al. 2013). Top blown coherent jet refining process developments continue.

12.12 Conclusion

This chapter summarizes the driving forces that are involved in the design and development of equipment for the metallurgical process industry. Optimization of existing equipment to obtain best results is always an ongoing process. The chapter concludes with an example of (a) equipment development and (b) adaptation of a developed technology across various disciplines of the metallurgical industry.

References

Berezowsky, R. M. G. S., et al. (1991). The commercial status of pressure leaching technology. *Journal of Metals, 43*, 9–15. February, 1991.

Deneys, A., & Enriquez, A. (2009). Scrap melting in the anode furnace and the development of coherent jet technology in copper refining. In Joël Kapusta & Tony Warner (Eds.), *International Peirce-Smith Converting Centennial* (pp. 321–338), TMS (The Minerals, Metals & Materials Society).

Deneys, A., Mahoney, W., Mathur, P., Masterson, I., Chan, A., & Manley, S. (2013). Twenty Years of Innovation in Praxair's Coherent Jet (CoJetTM) Gas Injection Technology. In *Proceedings of the Ralph Lloyd Harris Memorial Symposium* (pp. 319–341), Metallurgical Society of CIM MS&T Meeting, Montreal.

Enriquez, A., Nexhip, C., Poplar, G., Deneys, A. (2010 June). An update on praxair coherent jet technology in anode refining at Kennecott Utah Copper. In *Proceedings of Copper 2010* (pp. 2327–2339), GDMB Society for Mining, Metallurgy, Resource, and Environmental Technology: Hamburg, Germany.

Enriquez, A., Nexhip, C., Krippner, D., Wright, K., Deneys, A. C., Ma, Y., Manley, S. (2013). An update on Kennecott Utah Copper's use of Praxair's coherent jet technology in anode fire refining-developments since 2010. In *Proceedings of Copper/Cobre 2013 (Herman Schwartz Pyrometallurgy Symposium)* (pp. 279–296), IIMCh: Santiago, Chile.

Liu, J., Warner, A. E. M., McCann, D., Hall, D. E., Mallette, D., Bradley, J. A., Mackenzie, E., Mahoney, W. J., & Deneys, A. (2005). Improving Inco's MK reactor performance through application of Praxair's Coherent Jet (CoJet) Technology. In A. Ross, T. Warner, & K. Scholey (Eds.), *Converting and fire refining practices* (pp. 61–75), Warrendale, PA: The Minerals, Metals, and Materials Society, TMS.

Vale annual report 2010, http://www.vale.com/canada/en/investors/annual-reports/20f/pages/default.aspx.

Warner, A. E. M., Osborne, G. E., Liu, J., Bichel, B., Hall, D. E., & Mallette, D. (2003). Inco's oxygen top-blowing, nitrogen bottom-stirring (TBBS) technology for converting chalcocite to "semiblister", Copper 2003 (pp. 371–384), Santiago.

Part V

Sustainable Development and Environmental Management

Sustainability Considerations in Innovative Process Development

13

Krishna Parameswaran

13.1 Introduction

The mining industry comprising extraction (mining), mineral processing (also referred to as beneficiation, milling, or concentration), and metallurgical processing (smelting and refining) is capital, water, and energy intensive with potential for significant environmental impacts, if the environmental aspects are not properly managed. In order to be sustainable, industry needs to carefully address all of these aspects.

The scope of this book is "mineral and metallurgical" processing, involving both physical and chemical processing and high- and low-temperature processing—namely, pyrometallurgical and hydrometallurgical processes. The industry needs to be innovative in how it goes about developing mineral and metallurgical processes. Due consideration consequently needs to be given to addressing capital intensity, water and energy use, environmental impacts, and industrial hygiene and safety considerations. How processes are developed today has a bearing on whether the industry is, and will remain, sustainable in the future.

In mineral and metallurgical process development, addressing these aspects starts at the conceptual stage and continues through bench-scale testing, pilot plant testing, commercial demonstration, and ultimately commercial operation. Finally, how a new process is integrated into other unit operations at a facility requires careful deliberation, calling for a holistic approach through all aspects of mining, mineral and metallurgical processing, looking at both upstream (mining and exploration) as well as downstream (fabrication and production of end-use product) operations, product use, and ultimate end-of-use recycling or disposal.

The examples used to illustrate some of the essential considerations of sustainable development are drawn from the primary copper industry because of the author's experience with that sector. There are two routes for primary copper production: (1) hydrometallurgical—leach, solvent extraction, and electrowinning and (2) pyrometallurgical—milling, froth flotation, smelting, and refining. For a more detailed understanding of these processes, the reader should refer to a standard textbook such as Schlesinger et al. (2011).

In this section, we discuss the importance of sustainable development and environmental management considerations in the mining and metallurgical industry and their implications for innovative process development.

K. Parameswaran (✉)
President, tfgMM (Trusteeship for Future Generations Mining and Metallurgy) Strategic Consulting, 34365 North 96th Way, Scottsdale, AZ 85262, USA
e-mail: kparameswaran0346@gmail.com

V.I. Lakshmanan et al. (eds.), *Innovative Process Development in Metallurgical Industry*,
DOI 10.1007/978-3-319-21599-0_13

13.1.1 Is Mining Sustainable and What Are Sustainable Mining Practices?

We start by examining whether mining and mineral development is sustainable (Rajaram and Parameswaran 2005) because at first blush it appears it is not. The reason for such a view is that no mine lasts forever because mineral resources are finite and nonrenewable. The other reason is that certain historic mining practices, which are no longer acceptable, have had significant negative impacts on the environment. However, mining is important for the economic development of nations. In addition, because every item in modern commerce is made from something that is either mined or grown, it is impossible to contemplate sustainable development, or for that matter, any development that is not to some extent based on mineral development.

The 1987 World Commission on Environment and Development report "Our Common Future," also referred to as the "Bruntland Commission Report" defines sustainable development as:

> Development that meets the needs of the present without compromising the ability of future generations to meet their own needs. (WCED 1987)

It is true that mines run out of ore and, therefore, a mine at some point has to cease operations. This makes consideration of sustainability in mining even more important because:

> Sustainability encompasses many more values than the continuing availability of the resource being developed. Indeed, it is the very fact that mineral development will end some day that makes integration of those other sustainability considerations into the mining process highly appropriate. (Pring 1998)

With respect to being a nonrenewable and finite resource, there are several examples of copper mines with long lives. One good example in the United States is the Bingham Canyon copper mine, 25 miles southwest of Salt Lake City, Utah, operated by Kennecott Utah Copper Corp. and owned by Rio Tinto has been producing copper since 1906. Over that period, the mine has produced around 19.5 million tons of copper, 27 million troy ounces of gold, 240 million troy ounces of silver, and around 1.1 billion pounds of molybdenum, and the Cornerstone Project is expected to increase mine life well beyond 2019 and there is significant mineralization beyond the Cornerstone Project. It is estimated that the remaining mineralization equals what has been mined to date.

Examples of other copper mines, mills, and smelters with long lives in Arizona include ASARCO LLC (Asarco) Ray Mine and the Hayden Concentrator, which celebrated their centennial in 2011 and the Hayden Smelter which celebrated its centennial in 2012. Freeport-McMoRan Inc. (FMI) Morenci mine started underground mine operations in 1881 and transitioned to open-pit mining in 1937. FMI's Bagdad mine started underground mine operations in 1928 and transitioned to open-pit mining in 1945 and its Chino mine started open-pit operations in 1910 and concentrator in 1911. Asarco's Mission mine celebrated 50 years of operation in 2011 and FMI's Sierrita mine started open-pit operations in 1957. Asarco's Silver Bell mine commenced open-pit operations in the mid-1950s and operated through 1984. It has operated its Leach, SX-EW facilities since 1997.

Very few manufacturing facilities can rival that record of longevity. However, there are other mines that may have ore sufficient for 10 years of operation, particularly in gold mining. We therefore argue that because we are dealing with a nonrenewable and finite resource, it is imperative to incorporate sustainable development considerations in all aspects of development of a mineral project.

Further, as demand for metals grows so do proven and probable resources. This is due, in part, to the discovery of new resources through continued exploration, even as mining proceeds and, in part, due to technological advances that allow the economic processing of lower grade ores and even wastes. Examples of technological advances in primary nonferrous metals production include: froth flotation for the concentration of nonferrous metal sulfide ores; Solvent Extraction-Electrowinning (SX-EW) for leaching of copper ores and cyanide heap leaching for gold ores. Each of these advances has added considerably to the inventory of economically mineable reserves.

Additionally, continued exploration by mining companies concurrent with production and research, development, and adoption of new technologies can further the sustainability of mining.

While we mine today to satisfy current demand for metals, we provide future generations with knowledge on where the ore bodies are and the methods for extracting them.

Another attribute contributing to sustainability of metal mining is the recyclability of metals. Many metals once produced are capable of being recycled, smelted time and again to their original elemental form and refined to demanding specifications. Furthermore, recycling can conserve energy since recycling processes are much less energy intensive than the primary metals production processes. Recycling also contributes to the conservation of natural resources by providing an above ground virtual mine. As a result of recycling, the majority of all copper ever mined is still in use, as is at least 99 % of the gold ever produced and upward of 60 % of the silver mined is still in existence as bullion, coins, or fabricated products. Metals are through recycling, infinitely renewable and, therefore, useful as a store of value to future generations.

Once a metal has been mined and refined, it is available at the end of its life cycle to augment the supply of metal that is available to society in the future. It will become increasingly more important to tap these sources of supply especially as primary mineral and metal resources diminish. However, there are challenges associated with recycling, especially if the metal is in the form of all alloy or mixed with other scrap-containing impurities that may be difficult to remove. In the future, cost-effective processes have to be developed that can make it feasible to recover primary metals from alloys and mixed scrap-containing deleterious impurities. If recycling involves adding these materials as supplementary feed to the primary metal production process they may have to be added to unit operations at the front end of the primary metal production process. Although the energy used and associated cost in the recycling process can be small when compared to primary metal production, the scrap collection chain, which in most cases is complicated, results in additional costs principally labor, thereby increasing the cost of recycling.

According to the Copper Development Association Inc., "of the world's reserves of copper about one-quarter of the deposits are economically recoverable now or in the near future. Of this reserve base, about 16 % (198 billion pounds of copper) is in the United States. Every year roughly three billion pounds are withdrawn from the earth as US mine production, a barely discernible amount compared to the reserve base. The copper already mined through history amounts to 700 billion pounds, most still in recycling use. Interestingly enough, although copper is continuously mined and put into use, the estimated US reserve base has stayed relatively constant in recent years, and has increased fourfold from estimates made in 1952 as new deposits have been found and, even more important, because better extraction techniques have allowed leaner deposits to be added to the reserve base. There is every reason to believe that these dynamics will continue well into the twenty-first century. Three additional factors will also influence copper supply: US self-sufficiency, energy efficiency, and recyclability" (Copper in the United States—Bright Future Glorious Past: http://www.copper.org/education/history/us-history/g_fact_future.html).

Mining is sustainable when it is conducted in a manner that balances economic, environmental, and social considerations, i.e., by paying attention to the "triple bottom-line." It should be noted that these are not three separate bottom lines but a single integrated one. Therefore, sustainable mining practices are those that promote this balance. These practices should begin with exploration, continue with mine development, metallurgical facility design, through operations and until mine reclamation and metallurgical facility closure is completed. The emphasis throughout a mining project, which could last several decades, is one of communication with the local community and all other stakeholders, employing technologies for efficient production and minimizing environmental impacts and worker exposure throughout the production chain, including while a mineral or metallurgical process is being developed. We also need to incorporate tools such as life cycle analysis so that environmental, economic, and social implications can be comprehensively evaluated.

13.1.2 Importance of Environmental Management and Environmental Management Systems

Environmental Management Systems (EMS) provide a useful framework that helps companies in achieving environmental goals through consistent control of its operations. It does so by instituting "plan," "do," check," and "act" approach, with the objective of continual improvement. The expectation is that incorporation of EMS will translate to better environmental performance for the company. For more details, see references such as (Von Zharen 2001).

13.1.3 Regulatory Context: Applicable Environmental Laws and Regulations

In considering environmental impacts of processes, it is necessary to have an understanding of the regulatory framework and the applicable environmental regulations. This will vary with the jurisdiction where the process being developed will be implemented. A discussion of the regulatory framework in the United States is provided by Parameswaran (2005) and is summarized in this section.

In the United States, the Clean Air Act, the Clean Water Act, and the Resource Conservation and Recovery Act and their implementing regulations govern the control of gaseous and particulate emissions, the control of effluents and stormwater, and the management of solid and hazardous wastes, respectively.

13.1.3.1 Clean Air Act of 1970 and 1990 Amendments

The primary focus of the air regulations required by the Clean Air Act (CAA) is the setting, attainment, and maintenance of national standards to protect public health. Regulation focuses on two categories of air pollutants. The first category consists of six commonly found "Criteria Air Pollutants" (sulfur dioxide, nitrous oxides, particulate matter, lead, ozone, and carbon monoxide) for which air standards must be shown to be achieved in ambient air. These six pollutants are controlled by federal-based programs requiring technology-based controls on emissions from certain pieces of equipment or operations, and by additional methods as needed to keep concentrations at or below the human-health-based standards. The second set of regulated pollutants focus on a different set of 187 listed "Hazardous Air Pollutants." Unlike the first set of six, these pollutants are controlled by federal-based programs requiring technology-based controls on emissions from certain enumerated pieces of equipment or operations.

In addition, regulation of greenhouse gas (GHG) emissions is in the early stages of implementation. The U.S. Environmental Protection Agency (EPA) has made an endangerment finding for GHG emissions as it relates to climate change and finalized additional permitting requirements. However, the U.S. Supreme Court has set aside some of these requirements. There is still no consensus of how climate change would ultimately be addressed. EPA has recently proposed carbon standards for existing power plants in addition to those already proposed new power plants.

For criteria pollutants, EPA is required to set National Ambient Air Quality Standards (NAAQS), "requisite to protect public health with an adequate margin of safety." States and local air pollution control agencies have the responsibility to implement the program, usually done by establishing State Implementation Plans to implement these programs. States, along with EPA, have enforcement responsibility.

There are two main federal programs requiring technology-based limits for emissions in new mine, mineral and metallurgical processing facility construction and operation (known as New Source Review and as New Source Performance Standards) for criteria pollutants.

Additional limitations on emissions may be set for the local airshed (where the mine, mineral or metallurgical facility would be located), if needed, to maintain the NAAQS for those six criteria pollutants. Generally in a rural airshed, such additional limits are minimal but they can be more significant in urban areas where there are many more sources of pollution or in rural airsheds with a history of exceeding the health-based standards.

Under Section 112(r) of the Clean Air Act Amendments, EPA has published regulations and guidance for chemical accident prevention at facilities that use extremely hazardous substances. These regulations and guidance are contained in the Risk Management Plan (RMP) rule. The information required from facilities under RMP helps local fire, police, and emergency response personnel prepare for and respond to chemical emergencies. Making RMPs available to the public also fosters communication and awareness to improve accident prevention and emergency response practices at the local level. The RMP rule was built upon existing industry codes and standards. It requires companies that use certain flammable and toxic substances to develop a Risk Management Plans. Certain chemicals might trigger an off-site consequence assessment to model worst-case release scenario.

13.1.3.2 The Federal Water Pollution Act of 1972 (aka the Clean Water Act)

The goal of the Clean Water Act (CWA) is to restore and maintain the chemical, physical, and biological integrity of the nation's waters. The CWA set up a mechanism for federal funding of infrastructure improvements intended to improve water quality and also established two primary programs to achieve this goal: (1) a permit program—the National Pollutant Discharge Elimination System (NPDES) program—to control discharges of pollutants to navigable waters; and (2) a mandate to states to adopt (with EPA review and approval) surface water quality standards for navigable waters within their boundaries that would result in a level of water quality sufficient to protect fish and wildlife, as well as human health and recreational uses (often referred to as the "fishable, swimmable" goal). Today, administration of the NPDES program has with few exceptions been delegated to states (with EPA oversight). Applicants for NPDES permits must show that a proposed discharge will not violate the surface water quality standards set by the state for the receiving water and will comply with any technology-based effluent limitation guidelines established by EPA. Such guidelines have been established for over 50 industry sectors to date, including ore mining and dressing and nonferrous metals manufacturing—smelting and refining. The effluent limitation guidelines may establish numeric limits on the concentration of pollutants in a discharge and/or may limit when a discharge may occur.

Stormwater discharges into navigable waters from industrial activity (including mining and construction), as well as from large and medium municipalities, also are regulated under the NPDES permit program (or a state delegated program). Such discharges are commonly regulated under general permits, but in some cases an individual permit may be required.

The discharge of one class of pollutants—dredged or fill material—into navigable waters is not regulated under the NPDES program, but rather under a different permit program established by the CWA (referred to as the Section 404 permit program). The Section 404 permit program is administered primarily by the United States. Army Corps of Engineers, although EPA has a significant role in reviewing and enforcing these permits.

Over the years, the concept of "navigable waters" has been construed broadly. In addition to territorial seas, large flowing waters, lakes, and wetlands, it has been interpreted to include smaller waters that flow only on a periodic basis, including small washes that flow only briefly in response to precipitation.

13.1.3.3 The Resource Conservation and Recovery Act

The Resource Conservation and Recovery Act (RCRA) of 1976 requires EPA to regulate the management of hazardous wastes from generation to disposal, i.e., cradle to grave. It also sets forth a framework for the management of nonhazardous wastes, underground storage tanks, and medical wastes. RCRA focuses predominantly on active and future facilities and generally does not address abandoned or historical sites unless present at an active facility engaged in treatment, storage, or disposal of hazardous waste, requiring a permit. As with other environmental laws, RCRA allows EPA to delegate permitting and enforcement authority to the states.

With a few exceptions (waste in domestic sewage, irrigation return flows, discharges regulated by CWA permits, and radioactive materials regulated under the Atomic Energy Act) RCRA defines all "discarded" materials as solid waste regardless of whether a material is solid, liquid, or contained gas. The scope of discarded materials has been subject of considerable debate and litigation. EPA's regulatory definition of solid waste includes materials that are "abandoned," "recycled," or "inherently waste-like," all of which are defined in regulations. EPA's definition of "hazardous wastes," with certain exceptions includes solid wastes that:

- Are on a list of hazardous wastes: nonspecific source wastes, wastes from specific sources, and commercial chemical products.
- Exhibit characteristics of ignitability, corrosivity, reactivity, or toxicity.
- Is a mixture of a listed hazardous waste and a solid waste or is a solid waste derived from a listed hazardous waste (with some exemptions).

EPA regulations also identify the kinds of recycling activities that are classified as hazardous and those that are not. EPA published a final rule on January 2015 revising several recycling-related exclusions associated with the definition of solid waste used to determine hazardous waste regulation.

One of the exemptions to EPA's definition of hazardous wastes arising from statutory provision called the Bevill amendment, which required EPA to study extraction, beneficiation, and "mineral processing" wastes (*note that "smelting and refining" is defined in the RCRA regulations as "mineral processing"*) and make a regulatory determination on how these wastes should be regulated. EPA determined that all uniquely associated (i.e., those that derive their characteristics from the ore being processed) extraction and beneficiation wastes and 20 large volume "smelting and refining" wastes should be regulated as "nonhazardous." These 20 wastes include primary copper, lead smelting slags, copper slag tailings, and calcium sulfate wastewater treatment sludge from primary copper processing would be regulated as nonhazardous. The complete list of the 20 wastes can be found at 40 Code of Federal Regulations (CFR) §261.4 (b) (7).

EPA's hazardous waste regulations impose requirements for generators, transporters, and operators of treatment, storage, and disposal (TSD) facilities. Generator requirements include obtaining an EPA I.D. number; determining whether their wastes are hazardous; preparing their waste for transportation (according to Department of Transportation (DOT) Pipeline and Hazardous Materials Safety Administration (PHMSA) packaging and placarding requirements); preparing the Uniform Hazardous Waste Manifest (manifest); and complying with storage, training, planning, record keeping, and reporting requirements. The manifest is the key ingredient of RCRA's cradle to grave regulatory scheme. Each time a waste is transferred (from generator to transporter, from one transporter to another, or from transporter to the designated facility) the manifest must be signed to document receipt. The designated facility on receipt of the waste sends a copy of the signed manifest back to the generator. There are three categories of generators: Large Quantity Generators (LQG), Small Quantity Generators (SQG), and Conditionally

Exempt Small Quantity Generators (CESQG), depending on the quantity of waste generated per month. An SQG can generally store hazardous wastes for up to 180 days and a LQG up to 90 days, without having to obtain a TSD permit.

Transporters of hazardous wastes must obtain an EPA I.D. number, comply with manifesting requirements, meet DOT transportation requirements, must properly deal with accidental spills and accidents, and report serious accidents and spills to the National Response Center and DOT.

TSD facilities are subject to RCRA's Subtitle C permitting requirements and have to meet standards for TSD units, e.g., containers, tanks, surface impoundments, waste piles, land treatment, landfills, and incinerators. TSD facilities must comply with citing requirements. They must analyze incoming wastes, meet security requirements, inspect their facilities, and comply with employee training and record keeping requirements. Other requirements may include groundwater monitoring and reporting, control of air emissions, closure and postclosure, and financial assurance. In addition, as part of the issuance of a RCRA permit, EPA and authorized states require corrective action to remediate past releases of hazardous constituents or wastes.

The Hazardous and Solid Waste Amendments (HSWA), the 1984 amendments to RCRA, require phasing out land disposal of hazardous waste. Hazardous wastes have to be treated to meet land disposal restrictions levels, thus increasing the cost of disposal. Some of the other mandates of HSWA include increased enforcement authority for EPA and a comprehensive underground storage tank program.

13.1.4 Regulatory Context: Applicable Health and Safety Regulations

In the United States, the laws that govern workplace safety are Mine Safety and Health (MSH) Act and Occupational Safety and Health (OSH) Act and their implementing regulations. The Mine Safety and Health Administration (MSHA), an agency of the U.S. Department of Labor, has responsibility for administration and enforcement of the MSH Act of 1977, which protects the safety and health of workers employed in the nation's mines. It applies to all mines and mineral processing operations in the United States, regardless of size, number of employees, or method of extraction. The Act requires MSHA to inspect all mines each year to ensure safe and healthy work environments for miners. In addition to setting safety and health standards for preventing hazardous and unhealthy conditions, MSHA's regulations establish requirements for: (1) immediate notification by the mine operator of accidents, injuries, and illnesses at the mine and (2) training programs that meet the requirements of the MSH Act.

The Occupational Safety and Health Administration (OSHA), another agency of the U.S. Department of Labor, administers and enforces the OSH Act of 1970. Safety and health conditions in private industries, including manufacturing facilities such as smelters and refineries, are regulated by OSHA or OSHA-approved state plans. There is also a general duty to provide work and workplace free from recognized serious hazards. Federal and state regulators work to ensure worker safety and health through worksite enforcement, education and compliance assistance, and cooperative and voluntary programs.

OSHA regulations also include process safety management requirement for highly hazardous chemicals aimed at preventing or minimizing the consequences of catastrophic releases of toxic, reactive, flammable, or explosive chemicals. These releases may result in toxic, fire, or explosion hazards. These regulations also prescribe permissible exposure limits for toxic air contaminants such as inorganic arsenic, lead, cadmium, etc. These can be met by use of respirators, protective work clothing and equipment, or by implementing engineering controls. In addition, there are exposure monitoring provisions, medical surveillance, and removal provisions.

In addition, both Acts have hazard communication standards (HCS) requiring mines, smelters, and refineries to develop, implement, and maintain a written hazard communication

(HazCom) program. Operators must identify chemicals used by workers; make a hazard determination; ensure that containers have labels; make available a Material Safety Data Sheet (MSDS) for each hazardous chemical used or produced at the mine or facility; and instruct workers on the physical and health hazards of the chemicals encountered at the work place, protective measures, and the contents of the HazCom program. OSHA updated its Hazard Communication Standard in March 2012, incorporating the United Nations' Globally Harmonized System of Classification and Labeling of Chemicals (GHS). Major changes to OSHA's HCS include a new system of classifying types and degrees of hazards, changes to labeling requirements, and changes to the Safety Data Sheet (SDS, formerly called MSDS).

13.1.5 Implication for Process Development

In mineral and metallurgical process development, sustainability considerations need to be evaluated upfront. In particular, we need to evaluate opportunities for resource, energy, and water conservation. We would also need to look at measures that would be required for the control of emissions, effluents, and the management of wastes. Another important consideration is assessment of industrial hygiene and safety requirements for the protection of the health of workers, associated with the process being developed. Finally, the community needs to be informed and their views solicited regarding the new technology being developed and its benefits and how environmental impacts will be avoided or mitigated. This communication needs to occur throughout the process development cycle and community concerns need to be addressed.

In the remainder of the chapter, these considerations are discussed with illustrative examples, followed by a case study, of how such considerations are being currently addressed by industry and finally a checklist is provided to determine whether these elements have been adequately addressed by the process developer.

13.2 Opportunities for Resource Conservation

The production of many nonferrous metals involves the processing of low-grade ores and hence it is important to evaluate opportunities for resource conservation. For example, the U.S. copper mining industry works with ores that typically have around 0.5 % copper. In addition, the ores may contain minor constituents, which might include precious metals like gold, silver, platinum; valuable recoverable constituents such as lead, zinc, rhenium, molybdenum, bismuth, nickel, selenium, and tellurium, and constituents of environmental concern such as arsenic and cadmium. Rare earth elements have been detected in copper leach solutions. Ignoring opportunities for recovering these constituents means foregoing potential coproduct revenue streams and resulting in more waste generation. A few percent increase in the yield of a process can improve the process economics considerably.

It should be recognized that the composition of the ore body is determined by geological occurrence, and therefore, there is little discretion over the composition of the raw material. This restricts opportunities for source reduction with respect to impurities of environmental concern but also provides the opportunity to recover valuable constituents, providing the potential for additional revenues through coproduct sales. For example, copper ores in Arizona contain molybdenum, rhenium, rare earth metals, and zinc. Whether these are recovered will depend on the economics of coproduct recovery and will vary with metal and commodity prices. When they are not recovered, these constituents may end up in tailings or concentrates, ultimately reporting to discarded smelter slag, when the concentrates are smelted. Resource conservation in the mineral processing step, therefore, involves the evaluation of the recovery of these coproducts such as molybdenum and zinc from an economic and environmental standpoint.

Resource conservation in the copper smelting step is usually accomplished by the resmelting of copper-rich flue dusts or filter cake, as part of the smelter charge or, if they are enriched with respect to lead or bismuth, they can be shipped to off-site facilities for metal recovery. Acidic scrubber solutions can be reused for leaching copper ores. Copper-bearing secondary materials from electrorefining of copper can be resmelted at the smelter. They include: "sharp" slag, precious metals bearing furnace bricks, copper telluride leach residue, foul cathodes, and copper residues.

Some of the impurities present in copper ores such as Pb, As, Sb, and Bi accumulate in the copper electrolytic refining stage in the electrolyte and their concentrations gradually increase with time. The concentrations of these impurities have to be maintained at acceptable levels, as some of these impurities in refined copper can cause grain boundary cracks in wiredrawing. Novel technologies such as Molecular Recognition Technology (MRT) have been implemented at Asarco's Amarillo Copper Refinery to remove bismuth from copper refinery electrolyte as a bismuth sulfate product, replacing the conventional methods, i.e., bleed stream withdrawal and controlled arsenic additions to the electrolyte (Izatt et al. 2015).

In order to recover the acid and some of the other valuable metals present in the copper electrolyte such as nickel, U.S. copper refineries have installed Acid Purification Units (Sheedy et al. 2006) where, after the electrowinning process, the decopperized electrolyte bleed is sent to the Acid Purification Unit (APU), where sulfuric acid and arsenic are absorbed into the resin and then desorbed using water that is then returned to the tank house to be reused as acid makeup and to increase arsenic concentration in the electrolyte. A by-product stream is generated that is high in nickel and other valuable metals that can be further processed and the metals recovered. This process eliminates the production of black acid and the use of evaporators with their associated high amounts of energy consumption, and also 70 % of the arsenic has been reported to be returned with the sulfuric acid. Nickel carbonate product can be precipitated by treating the deacidified electrolyte with sodium carbonate. The pH selective, two-step process first recovers a residue (at pH 6–6.2) containing copper, iron, arsenic, antimony, and bismuth that is returned to the copper smelter to recover copper. The final approximately pH 10 precipitate is the nickel carbonate product that is sold.

During the leaching of anode slimes, tellurium dissolves into the leach liquor together with copper and must be removed before the solution is sent to liberator cells for copper recovery. Leaching is conducted in atmospheric pressure tanks instead of a pressurized vessel and tellurium is recovered from the leach liquor by cementation using elemental copper.

In considering these opportunities, the regulatory status of the secondary materials needs to be considered, as further discussed in Sect. 13.6.

13.3 Opportunities for Energy Conservation

Mineral and metallurgical processing is inherently energy intensive—comminution, smelting, refining, and ancillary facilities such as scrubbers, electrostatic precipitators, double contact acid plants, and oxygen plants, which themselves are large energy consumers. Therefore, it is important that opportunities for energy conservation be thoroughly explored—blast optimization, energy efficient grinding methods, waste heat recovery in smelting, utilization of energy efficient equipment to reduce energy consumption and using renewable energy sources, where appropriate. Such assessment will impact not only the process economics but also achieve emissions reduction from the reduction in fuel and electricity consumption.

According to the Copper Development Inc., there are wide variations in the energy used to recover metals from the earth's crust. "Copper ranks near the middle for energy required for extraction—higher than iron, zinc or lead, but at considerable advantage to aluminum, titanium and magnesium, which require much larger quantities of energy to break down the ore (or seawater and brines in the case of magnesium) into metallic form." (Copper in the United States—Bright Future Glorious Past: http://www.copper.org/education/history/us-history/g_fact_future.html).

In the past, the objective of the blasting operation in hard rock mining was to convert an in situ rock mass into a muck pile with proper fragmentation size and shape for the excavation equipment (Oloffson 1988). Over the subsequent decades, it has come to be recognized as one of the most important comminution considerations because of its influence on the downstream steps of crushing and grinding and blast design has emerged as an important research focus for mine-to-mill optimization (Kim 2014). Over the years considerable efforts have been expended in the development of more energy efficient crushing and grinding technologies.

13.3.1 Energy Use in Copper Smelting Processes

In a paper in the Proceedings of Copper, Diaz (Diaz 2010) discusses technological advances, increased use of oxygen, pursuit of improved productivity, changing geography of smelting capacity, energy consumption, and greenhouse gas emissions associated with various copper smelting technologies. He notes that over the last three decades, increased oxygen usage in copper smelting and converting and the implementation of computerized process control have led to increased smelter concentrate processing capacity, reduced energy consumption, increased sulfur dioxide capture and that there has been a major realignment of world smelting capacity to China, India, and other Asian countries.

Coursol, Mackey, and Diaz (Coursol et al. 2010) in the same proceedings using thermochemical modeling and industrial data compare the energy consumption in copper smelting for various smelting routes. They reference the seminal paper (Kellogg and Henderson 1976), as pioneering life cycle energy analysis applied to copper extractive metallurgy. The analysis includes not only the fuel and electricity used for the various smelting alternatives but also the energy required to produce the main supplies. These are also referred to as Level 1 and Level 2 energy analysis. Other researchers have utilized similar methodologies to compare energy consumption in various copper processes (Nadkarni and Parameswaran 1975; Parameswaran et al. 1981). The latter paper estimated the energy requirements for a number of copper pyrometallurgical and hydrometallurgical processes using industrial operating data and material balances. The energy analysis included all operations from mining through production of refined copper. INCO and Outokumpu Flash Smelting/Batch Converting and Mitsubishi Continuous Smelting/Converting are in the lowest range of energy consumption because of the utilization of the fuel value of sulfur and iron in concentrates and the utilization of tonnage oxygen. The energy consumption ranges from 19.93 to 21.25 MMBTU/ton of cathode copper. To obtain the total energy required to produce a ton of cathode copper, the energies for mining and concentrating need to be added. For a 98.7 % recovery in the smelting operation, the estimated mining and concentrator energy consumption are 20.13 MMBTU/ton and 42.57 MMBTU/ton of cathode copper, respectively.

Coursol et al. estimate the energy consumption for four copper pyrometallurgical processing routes as follows: (1) Flash Smelting/Flash Converting 10,784 MJ/ton, (2) Isasmelt/Peirce Smith Converting 11,078 MJ/ton, (3) Mitsubishi Continuous Smelting/Converting 11,006 MJ/ton, and (4) Noranda/Teniente 12,746 MJ/ton. The authors compared their estimates of energy requirements of current processes with those studied by Kellogg and Henderson. Their comparisons show that the decreases resulting from technology advances in oxygen production and increased productivity as a result of increased oxygen usage in smelting and converting have more than compensated for increased energy requirements associated with the more stringent environmental controls that are currently required. They also note that the average electrical energy (processes 1–3) is not substantially higher since the time of publication of the Kellogg and Henderson study. Changes since that time include the increased use of tonnage oxygen in copper smelting and increased electricity consumption associated with use of double contact acid plants and the greater degree of control and capture of converter fugitive gases. Electricity consumption reduction is associated with the substantially lower consumption associated with the production of tonnage oxygen (285 versus 397 KWH/ton) and higher power plant efficiencies (38 versus 32.5 %). They find noteworthy the much lower fossil fuel consumption that is due to the utilization of the fuel value of iron and sulfur in concentrates. They attribute improved furnace design and operating practices, computerized process control, and higher furnace feed throughput for these achievements.

Compliance with more stringent sulfur dioxide NAAQS in the United States is likely to increase the electrical energy consumption as certain U.S. smelters install the necessary air pollution control equipment to meet these standards. The increased electricity consumption at the smelter would result in additional emissions of criteria and greenhouse gas emissions at the power plant.

As climate change regulations mature there will be greater emphasis on reducing greenhouse gas emissions resulting from restrictions on fossil fuel usage or imposition of a carbon tax. Life cycle energy use and costing will be increasingly used to properly evaluate these sustainability considerations.

The carbon footprint of a primary copper smelting process depends on the fuel mix used for heating and that used for electricity generation (Diaz 2010). He estimates that the fuel consumption to generate electricity is four times higher than the fuel directly consumed in mining, smelting, and refining and that three quarters of the electrical energy is consumed in milling. Based on the fuel mix at Chilean smelters, he estimates the average CO_2 equivalent emitted by Chilean smelters at 0.86. Diaz also notes that by contrast China and India where copper smelting capacity is rapidly expanding, thermal power (over 70 % coal) accounted for over 80 % of the electricity generated in 2006–2007. He identifies the following opportunities for reducing the carbon footprint at metallurgical processing facilities: (1) greater heat recovery from smelting, converting and anode refining off-gases; (2) greater heat recovery from the acid plant operation with high strength sulfur dioxide gas feed and (3) the implementation continuous anode refining technology.

13.3.2 Challenges and Opportunities Associated with Waste Heat Recovery from Copper Smelting Processes

There are several challenges and opportunities associated with recovering sensible heat from primary copper smelting off-gases (Safe and Russell 2010). The challenges presented by metallurgical off-gas heat recovery are due to high dust loadings, nature of the dust, corrosive gas, and thermal cycling in batch processes. The opportunities relate to the potential recovery of sensible heat that is currently being wasted. The authors discuss common heat exchanger technologies in use such as waste heat boilers and air to gas heat exchangers and the potential for utilizing heat recovery technologies employed in other industrial applications, such as thermal oil heat recovery, Organic Rankine Cycle heat recovery, and power generation to recover sensible heat from batch processes and lower temperature operations.

Copper smelting, converting, and anode refining produce significant off-gas volumes at high temperatures, usually in the range of 1000–1200 °C. Off-gas heat recovery is therefore an important means of reducing energy consumption and reducing operating costs. Smelting furnaces smelt copper concentrates to produce matte. Typically waste heat boilers are used with Outokumpu flash smelting furnaces, continuous converting furnaces, and Top Submerged Lance (TSL) smelting units such as the Ausmelt and Isasmelt processes, usually ahead of a hot Electrostatic Precipitator (ESP). The boiler includes a radiative section that cools the gas to about 700° C and a convective section that cools the gas below 400° C, which is suitable for entry into a hot ESP. The boiler uses the recovered heat to produce superheated or saturated steam for local power generation or for process uses. The Outokumpu flash furnace off-gas at 55,000–60,000 Nm^3/h contains around 7–9 MW of sensible heat at the ESP outlet and is relatively free of dust. The INCO Flash furnace uses technically pure oxygen and the low off-gas volumes do not justify installation of a waste heat boiler. The INCO flash furnace off-gas at 25,000 Nm^3/h contains around 3–4 MW. The off-gas is generally cooled further and cleaned in a wet gas cleaning system before it is sent to an acid plant for sulfuric acid production.

Converters are used to produce blister copper from copper matte. They are either continuous converting furnaces like Mitsubishi or Kennecott-Outotec flash *or* batch like Peirce-Smith or Hoboken converters. Modern Peirce-Smith converting departments employ three converters with two blowing and a third on standby. Often a fourth converter is offline undergoing repair or rebuild. The process gas exits a blowing converter at or slightly below the converter air blowing rate and at temperatures of 1100–1200° C. This is captured by a primary hood, which allows 100–120 % infiltration air into the hood, doubling the off-gas flow rate. The gas is cooled by evaporative or radiative cooling to below 400° C before entering a hot ESP. Some copper smelters have employed waste heat boilers or gas to air heat exchangers with limited success. Like the smelting furnace gases, the converter off-gas is cooled and cleaned in a wet gas cleaning system before being sent to the acid plant. Converter off-gas at a blowing rate of 30,000 Nm^3/h contains 8–12 MW of sensible heat at the ESP outlet per blowing converter and is relatively free of dust.

Flash converting and Mitsubishi converting both utilize waste heat boilers to cool the off-gas prior to dust removal in an ESP. Flash converting utilizes quite high levels of oxygen enrichment since the feed is solidified copper matte. This in turn yields a very high strength sulfur dioxide off-gas. A typical flash converting furnace off-gas will be 20,000 Nm^3/h at 35–45 % SO_2.

Anode furnaces used at copper smelters are generally batch operations to fire refine blister copper to anode copper, before it is electrorefined in a refinery. Most modern anode departments have two or three anode furnaces with one or two operating at any time. Anode furnace operations are batch operations, with each batch requiring 3 or more hours, consisting of an oxidation stage of 1 h with a reduction stage of 2–3 h. After completing a cycle, the furnace may sit for several hours in a hot standby mode. Process off-gas from an anode furnace is generally less than 10,000 Nm^3/h at 1100–1200° C. During the reduction stage, the off-gas can contain up to 25 % combustibles comprising carbon monoxide and hydrogen. The gas is captured by a primary hood with sufficient air infiltration to ensure combustion of combustibles. Generally, there is significantly additional air infiltration into the primary hood that further dilutes the process gas. In the oxidation stage, all of the air infiltration dilutes the process gas entering the primary hood. The gas is cooled by evaporative cooling to 200° C and then routed to a baghouse for particulate removal. Despite cooling the gas to below 200° C for the baghouse, it still contains 3.5 MW of sensible heat exiting the baghouse. In some smelters, anode furnace off-gas is sent directly to concentrate dryers for direct heat recovery and a common gas cleaning system.

The oxidation stage may take a few hours longer if processing higher sulfur blister copper from the Mitsubishi continuous converting process. Flash converting produces a lower sulfur blister copper and at the Kennecott smelter a 500 tonne charge of blister copper containing 0.2 % S can be oxidized, slag skimmed if required and reduced using a mixture of natural gas and superheated steam in less than 3.5 h. The chemical efficiency of the steam-gas refining is so high that the off-gas does not contain significant uncombusted hydrocarbons or hydrogen. In this facility, only pure oxygen-fuel burners are used which reduces the off-gas volume to the point that any heat recovery is uneconomic.

Waste heat boilers have been extensively used to recover waste heat from continuous smelting and converting processes. However, heat recovery from batch processes such as Peirce-Smith and Hoboken converters and anode furnaces has been largely overlooked.

The challenges presented by metallurgical off-gas heat recovery are due to high dust loadings, nature of the dust, corrosive gas, and thermal cycling in batch processes. Most waste heat applications involve dust-laden gas, which can lead to fouling and plugging of heat exchanger surfaces and pipes. Sticky dust in smelting gases is of particular concern. Fouling and plugging can result in reduced heat transfer, increased pressure loss, and can reduce fume capture efficiency. It can also lead to localized corrosion. Heat recovery equipment therefore needs to be designed with streamlined gas flow to avoid dropout or impingement onto heat transfer surfaces and rappers or sonic horns need to be incorporated to periodically knock off dust from the heat exchanger surfaces. At times air is added to smelter off-gas to convert sulfide in the dust to sulfates reducing its stickiness.

Corrosion occurs because as the gases cool, water and acid condense onto surfaces. Thermal cycling, fouling, or poor process control can lead to corrosion, which can increase maintenance cost and reduce equipment life. In designing heat recovery systems, it is important to have a good understanding of the water and acid dew points of the gas streams over the full range of operating conditions and the surface temperature profile throughout the heat recovery equipment. In addition, effective process control is necessary to ensure that gas and surface temperatures are maintained above the dew point to prevent condensation and attendant corrosion.

Thermal cycling is associated with processes that produce variable flow rates and temperatures and is of particular concern for batch processes. The expansion and contraction associated with temperature variations can cause metal fatigue, cracking, and leakage. Lower temperatures can result in condensation and corrosion. Heat recovery equipment should be designed to allow for expansion and process control implemented to minimize the range of thermal cycling and maintenance practices instituted to minimize leakage and maintain equipment integrity.

Batch processes because of variable off-gas characteristics present additional challenges resulting in inconsistent heat recovery with adverse impact on steam or power generation or preheated air availability. Usually this can be addressed utilizing thermal storage or common heat exchangers for multiple batch units such as converters. With thermal storage, the heat from the off-gas is transferred to a thermal medium such as oil. The quantity of stored heat will rise and fall, while a constant stream of heat (less than the operating heat recovery rate) would be bled from the storage system for production of steam, power, or preheating air. The other approach is to connect multiple batch units to a common heat recovery unit. For example, in a four converter operation with two converters blowing at any time, each converter operates 45 % of the time. By installing the heat exchanger after the converter off-gas ducts combine, that heat exchanger can operate with a steady off-gas stream from two blowing converters 90 % of the time. Batch processes provide some advantages as well. The reduced operating hours of individual units and the process flexibility it affords provide opportunities to optimize performance of the unit and time to clean out the unit and make necessary repairs. Process availability due to the heat exchanger is improved since maintenance on the unit can be performed during normal downtime for the vessel, while the smelter operations continue.

Low-grade heat from off-gas temperatures ranging from 50 to 300 °C is difficult to recover economically because of the small temperature gradient between the off-gas and the heat transfer fluid. Temperatures are not high enough to boil water and raise steam and air preheat is limited by the inlet gas temperature. Conventional waste heat boilers and air-to-gas heat exchangers are not suitable for this application.

The direct benefit of heat recovery is to reduce fuel consumption and associated costs. The ability to utilize the recovered heat to generate electricity presents an opportunity to control power costs, especially since electricity costs are expected to rise in the future as power plants are modernized to meet more stringent emission control regulations. Indirect benefits include increased production, if it is limited by energy availability and reduced greenhouse gas emissions.

13.3.3 Opportunities for Cogeneration

Opportunities for cogeneration should be considered where it makes business sense. For example, at the Kennecott copper refinery in Garfield, Utah, a natural gas fired turbine and combined cycle cogeneration produces 6.2 MW of electricity, while the waste heat from the combined cycle is used to heat the electrolyte at the refinery. At the Kennecott smelter waste heat from the flash smelting and flash converting furnaces is superheated and used to power the two main sulfuric acid plant blowers (2.5 MW each). The waste steam from the backpressure turbines is combined with steam recovered from the Monsanto Heat Recovery System (HRS) boilers and used to generate about 26 MW of electrical energy. This is sufficient to provide 65 % of the smelter's total electrical demand ((Newman et al. 1998) and (Newman et al. 1999).

13.3.4 Opportunities for Utilization of Renewable Energy

Opportunities for incorporation of renewable energy such as solar and wind power should also be considered as part of process development. Such approaches besides dealing with issue of energy costs which is a significant business risk for an energy-intensive business also enhance the sustainability credentials, promotes corporate social responsibility, and improves a company's public image.

Mining companies have not considered renewable energy as a viable alternative to conventional power generation because of its costs and because it is not compatible with the demand profile of a mining or metallurgical facility. Companies in the industrialized world have access to national and regional grids and their electricity costs are low because of the large loads. Although costs of renewable energy projects are coming down, projects require tax incentives and imposition of renewable energy mandates to make them viable. However, there are remote mining areas where renewable energy is already competitive with power generated from diesel.

A study by Accenture and the U.N. Global Compact (Accenture and U.N. Global Compact 2012) notes:

> "The metals and mining industry is uniquely positioned to not only drive business value related to energy efficiency and increased use of renewable energy, but also increased access to modern energy services as companies in this industry can be a catalyst for sustainable development in areas with little or no existing energy infrastructure. This characteristic—the operation of facilities in remote areas, provides the opportunity to link access to energy with core strategic business value drivers, such as risk management by protecting the "license to operate" and brand enhancement though community outreach and collaboration. This opportunity is one of collaboration and partnership as it relates to broader development concerns—of which access to energy is a primary enabler, and is reflective of metals and mining companies."

The study recommends the following industry actions: (1) Partner with local governments and utilities to provide energy services to communities surrounding operational locations; (2) improve the energy efficiency of current operations; (3) build advanced energy considerations into the design and development of new assets and operations; (4) diversify the portfolio to develop products and generate materials that drive energy efficiency; (5) use waste and process outputs as fuel sources; and (5) use more renewable energy sources to support operational power needs. The business levers are brand enhancement, cost reduction, and risk management and the objectives are energy access, energy efficiency, and renewable energy utilization.

Mining companies have large land holdings, most of which are not utilized. Some of the disturbed lands could be suitable for locating a renewable energy project. Asarco at its Mission mine in South Tucson has partnered with solar developers Clenera LLC, Panasonic Eco Solutions, and Coronal Group and the local utility Tucson Electric Power to host a 35-MW utility scale photovoltaic (PV) solar project called the Avalon Solar Project that commenced power generation in December 2014. A second phase to be implemented by year end will increase the solar farm capacity to 56-MW. This facility is located on disturbed agricultural land that Asarco acquired for water rights purposes.

EPA (EPA 2011) has conducted a preliminary evaluation of the potential for renewable energy development on Asarco reclaimed tailings areas in the San Xavier District on Tohono O'dham Nation lands. The study found:

> "Amidst this complexity, the ASARCO Mission Mine tailings area offers unique and potentially innovative opportunity to accommodate and potentially expand a utility-scale solar energy project at a single location. The tailings area has existing transmission capacity, roads, industrial zoning and other critical infrastructure in place for a utility-scale PV project. Additionally, installing a solar generation plant and associated infrastructure on a formerly impaired area can help take development pressures off of undeveloped, open land ("greenfield") areas."

In the future, Asarco would like to evaluate installation of PV solar project on reclaimed tailings and waste rock deposition facilities.

The primary copper industry operates electrowinning and electrorefining facilities, where the copper is plated onto cathodes. These are significant electricity consumers requiring direct current. It might make sense to evaluate solar and wind power to provide part of the electricity needs. The electrolyte in copper electrowinning and electrorefining needs to be heated and solar heaters may be considered instead of fossil-fuel fired boilers.

13.4 Opportunities for Water Conservation

Water is becoming an increasingly precious commodity. In arid and semiarid areas of the world such as the Southwest United States, South Africa, Australia, and Chile, where many large mining operations are located, water supplies are scarce. The ability to secure adequate water supply can very well determine whether a mining project involving large water use can be implemented.

A paper in the International Mineral Processing Congress (IMPC) Proceedings (Dunne 2010) reports on water stewardship initiatives undertaken by the mining industry over the last two decades and provides examples of water treatment projects that have been implemented. Although the mining industry uses considerably less water than other sectors such as agriculture, mining consumes significant quantities of water and is in competition with other more valued uses such as agriculture, livestock, and human consumption. Dunne notes that "water rights and availability are extremely contentious issues in mining regions around the world," as are "the amount and quality of water that a mine is able to discharge." He advocates "developing a responsible, sustainable and transparent water management strategy that is recognized as such by all stakeholders," in order to build confidence with the public.

The majority of the water used in the copper mining industry is for beneficiation, smelting, and refining operations. Most of the beneficiation water is used in the flotation process, followed by leaching of copper ores. It is estimated that water consumption by the mining industry is in the range of 0.6–1.0 cu.m./ton of ore processed by flotation (Brown 2003; Norgate and Lovel 2006; and Wiertz 2009), compared with the consumption for the Leach, Solvent Extraction, and Electrowinning route of 0.13 cu.m./ton of ore (Wiertz 2009) and smelting in the range of 2–4 cu.m./ton metal (Norgate and Lovel 2006).

The principal sources of water at a mine are groundwater and surface water from rivers, lakes, and reservoirs and, at some locations even seawater. Groundwater can have varying salinity and pH and can contain heavy metals of natural origin. Mine dewatering can also result in high salinity and metal contamination. Acid mine drainage occurs due to the oxidation of sulfides in waste rock and tailings and has very low pH, high metal content, and sulfates. Precipitation can result in run-on water which is usually directed around the mine so that the water is not impacted by the mining operation. Run-off water is usually impacted by the operation and therefore retained at the site to the extent possible, as it can have a slightly acidic pH, low metal content, and high suspended solids due to erosion. Suspended solids are controlled by settling in sedimentation ponds.

A mine site water balance is an important component of a good water management program. Water consumption at mines can be reduced by incorporating good water management practices such as increased recycling of process water, reducing water losses by evaporation and seepage. Most conventional tailings storage facilities recycle a considerable proportion of the tailings decant water from the tailings storage facility. Generally, water treatment of a bleed stream is needed only if the water is saturated with gypsum. Advances in thickener technologies over the last two decades have made it possible to produce higher underflow densities and allowed for the treatment of "difficult to dewater ores" (Schoenbrunn et al. 2009). The development of large capacity vacuum and pressure filter equipment (Mathewson et al. 2006) has made it possible to store tailings in the unsaturated state. The filtered tailings can be transported by conveyor or truck and placed, spread, and compacted to form an unsaturated dense and stable tailings stack. Alternatives to conventional tailings management and factors that favor dry stack tailings facilities are discussed in the literature (Davies and Rice 2001) (AMEC 2008).

When reusing process water, it is important to consider any adverse impacts of organic and inorganic contaminants on the metallurgy of the unit operation where the recycled water is used. A number of researchers have studied the effects of deleterious residual reagents and their decomposition products on the metallurgy of flotation and leaching (Knapp 1973; Frossberg and Hallin 1989; Rao and Finch 1989; Klimpil 1996; Johnson 2003; Schumann et al. 2009). Mixing of waters of different pH in a flotation circuit can lead to precipitation of hydroxides or impact the pH control needed for selective flotation. Because recycle water can contain both dissolved and suspended solids it may not be ideally suited for water uses such as pump gland sealing, water cooling, and water sprays (Cooper et al. 2006). However, a positive impact is the return of residual reagents, which lowers reagent costs in addition to the lower cost of purchased water or groundwater pumping.

In some instances, desalination processes for brackish groundwater or seawater may need to be implemented to supply or supplement water needs of a mineral or metallurgical process. Examples include utilization of hyper-saline groundwater (Dunne 2010) for milling and leaching applications at most western Australian gold mines and in nickel and gold flotation plants in the region. The author also references operating experience at Batu Hijau Mine in Indonesia, where sea water is used 7–8 months of the year and a mix of runoff from stockpile, waste, and mine dewatering. Based on a number of years of operating experience, it was noted that corrosion was considerably less with seawater only as compared to the blend and that there were no deleterious effects on froth flotation process or copper recovery (McCaffrey 2010). In Chile, the Michilla copper mine has been using seawater since the 1990s for copper leaching (Wiertz 2009).

The first big mining company to use seawater in Chile was Minera Esperanza, a joint venture between Antofagasta Minerals and the Marubeni Corporation. The company's copper mine uses untreated seawater, transported through a 145-km-long pipeline, in all of its processes. Seawater currently accounts for 30 % of all of the water the mine uses. The National Copper Corporation of Chile (CODELCO) will use seawater for the first time to exploit the sulfide reserves of the Radomiro Tomic (RT) mine. The Radomiro Tomic (RT) Sulphides project will extract seawater and desalinate it through reverse osmosis, a process that uses pressure to force water through a membrane which retains the dissolved solids. The treated water will be transported to the mine's facilities, located 3000 m above sea level, through a pipeline stretching 160 km. The operation will entail an expenditure of $2.60 per cu. m. (http://www.ipsnews.net/2013/08/mining-industry-plans-massive-use-of-seawater-in-arid-northern-chile/).

In copper smelters, the use of acid plants to control sulfur dioxide emissions in sulfide smelting results in the blowdown from the acid plant scrubber of weak acid streams containing heavy metals. Another source of wastewater results from cooling water from the acid plant cooling tower and from furnace cooling. Wastewaters from a copper refinery include bleed streams from the tank house, by-products department, and rod line which may need treatment prior to recycle or disposal.

Opportunities for recycle reuse of process water are therefore an important consideration and are more attractive if they can be done without further treatment. For example, the U.S. primary copper industry utilizes acidic scrubbing solution in the leaching of ores and hydrometallurgical processing of flue dusts (Gabb et al. 1995).

13.5 Effluent Management

Most mining, mineral, and metallurgical processing operations, at least in the United States are designed to have zero or minimal discharge. However, where discharge is necessary and permitted, the effluent must be treated to ever stricter discharge limitations promulgated by regulatory agencies. In response to the increasing number of environmental regulations all over the world in the last 30 years, nonferrous smelters and refineries have implemented treatment processes for acidic bleed wastewater streams containing heavy metals. Such treatment usually results in a gypsum-saturated solution containing salts such as alkali metal sulfates, chlorides, and fluorides.

The rising cost of fresh water and the need to meet increasing stringent discharge limitation regulations has made water recycling and reuse essential. The metallurgical industry has used water as a medium for dissolving and/or rejecting metals and chemicals in addition to removing heat from processes such as smelting equipment and acid plants. Any excess process effluent is collected and sent to end-of-pipeline treatment and ultimately discharged to a surface water body for discharge under a permit (Ramachandran 1997).

Several alternatives for treatment of wastewaters from nonferrous smelters and refineries streams containing high levels of total dissolved solids with emphasis on sulfate removal have been reviewed in the literature (Ramachandran 2012). They include: (a) chemical treatment for removal of various constituents that make up the total dissolved solids, (b) processes for sulfate removal—thermal and nonthermal—to recover and reuse treated water, and (c) brine use and treatment options.

The conventional and probably the cheapest way of treating acidic metal-bearing process effluent streams is neutralization of free sulfuric acid with lime, caustic, or soda ash and to precipitate the heavy metals as metal hydroxides. Any arsenic present in the water precipitates as calcium arsenite, if lime is used as the neutralizing agent. Additional heavy metals removal can be removed by sulfide polishing using sodium sulfide to produce a treated effluent that meets the regulatory agency limits, i.e., NPDES limits in the United States. Such treatment results in a stream containing low dissolved solids. Further, the precipitation and sulfide polishing steps result in precipitation of metal hydroxides and sulfides as a residue that has to be managed in accordance with solid and hazardous waste regulations, i.e., whether they can be legitimately recycled or whether they are hazardous or nonhazardous, if discarded. See Sect. 13.6.

The dissolved solids that remain are normally comprised of inorganic salts. The possible constituents are cations, calcium, magnesium, sodium, and potassium. The anions are sulfate, chloride, fluoride, nitrate, carbonate, and bicarbonate. In addition to these constituents, the treated effluent streams contain very small amounts of heavy metals, oxyanions such as arsenites, arsenates, selenites, and selenates. Process streams from gold operations contain residual cyanide. Small amounts of ammonium ions could also be present in these streams based on the pH of the solution.

With increasing stringent effluent limitations, it may be necessary to augment a conventional treatment plant with technologies for the removal of specific metals (such as Se, Sb, Cd, Pb, Mn, and Al), oxyanions, other anions, organics, or suspended solids prior to discharge. These technologies include: Ion Exchange, Reverse Osmosis, Electrodialysis Reversal (EDR), Nanofiltration Activated Carbon Adsorption, Deep Bed Sand Filtration, Biofix Beads, and pH adjustment.

The recovery of treated water as a distillate by evaporation results in a brine solution, the quantity of which is dependent on the amount of soluble salts present in the treated water. This is an expensive option because evaporation costs are generally high. Mechanical Vapor Recompression Evaporator (MVRE) technology uses mechanical vapor recompression to recover water for reuse from brines at a fraction of the energy needed in other types of evaporators (Bostjancic and Ludlum 1996). Typical energy costs are estimated to be around 60–100 KWH per 1000 gal, the equivalent of a twelve-effect crystallizer (Bostjancic and Ludlum 1996). In a typical MVRE unit, the necessary driving force is generated by mildly compressing the evolved vapor (steam) to increase its temperature and pressure and returning it to the steam chest. The evaporator can be adjusted to work on a small temperature difference so that the energy used for steam compression can be kept low. A distillate with less than 10 mg/L total dissolved solids (TDS) and a concentrated brine—with concentration up to 250,000 mg/L TDS can be obtained. The brine can be dried in a pond; crystallized or spray dried and the resulting salts sent to a land fill. The process is economical only when the boiling point rise is less than or equal to approximately 5° F.

13.6 Solid and Hazardous Waste Management

Metallurgical processing of many nonferrous metals generates large volume, low toxicity wastes, such as tailings and slag. Under the Resource Conservation and Recovery Act, EPA has determined that these wastes are to be regulated as solid wastes. Most mining and beneficiation wastes associated with the processing of ores and a few smelting wastes, such as slags, are classified as nonhazardous. As noted in Sects. 13.2 and 13.5, with regard to secondary materials destined for recycling or reuse, their regulatory status depends on how the materials are characterized, i.e., whether they fit the definition of by-products, sludges or spent material, scrap metal, etc., and the manner in which they are recycled, i.e., whether they are being reclaimed or are being used or reused as ingredients in an industrial product or as effective substitutes for commercial products (40 CFR § 261.2).

The difference in the cost of solid and hazardous waste management can be significant. In process development, there is a need to examine the regulatory regime in areas where the process will ultimately be implemented. Therefore, means for reducing the amount and character of the waste can have an important bearing on process economics as well as reducing environmental impacts.

Another approach to waste management is waste utilization, especially large volume wastes such as copper mine tailings (MT) and copper smelter slag (SG) as substitutes for naturally occurring raw materials. Ahmari, Parameswaran, and Zhang (Ahmari et al. 2014) have studied alkali activation of copper mine tailings and low-calcium INCO flash furnace slag to produce a geopolymer that can substitute for Ordinary Portland Cement (OPC). The results show that the addition of SG significantly improves the Unconfined Compressive Strength (UCS) and microstructure of the geopolymer. The improvement is mainly attributed to the high solubility of silica in the SG and the fine particle size of the SG. The inclusion of SG also leads to a decrease of the optimum curing temperature (i.e., the temperature at the highest UCS) because of its higher reactivity than MT. In addition, the MT/SG-based geopolymer sets fast and gains a major portion of its ultimate strength within only 7 days. Based on the results, it can be concluded that the MT/SG-based geopolymer is a promising sustainable construction material for civil engineering applications.

13.7 Control of Gaseous and Particulate Emissions

Control of emissions is of particular importance in the development of pyrometallurgical processes. It mainly involves control of particulate matter, acid mist, metal fumes, sulfur dioxide, nitrogen oxides, and carbon monoxide. In process development, it is important to determine the emission rates that can be anticipated in a commercial process, so that suitable capture and control methods can be identified. If the off-the-shelf control equipment is not available, collaboration with air pollution control vendors might be necessary. A good understanding of the mining plan is also important, since sulfur and impurity concentrations can fluctuate from year to year based on the ore grade mined. If it is anticipated that there will be periods where high sulfur or high impurity concentrations will occur, then the gas handling systems have to be designed to effectively handle these periods and comply with regulations.

Although attention to potential emissions is needed at all stages of process development, it is particularly important at the pilot plant stage. Designing the pilot plant to obtain dust emissions rates and the characteristics of the dust will help in selecting the air pollution control equipment and in designing heat recovery equipment. Dry dust collection technologies, such as hot ESP, enable easier dust collection and recycling. Using a hot ESP can also allow segregation of the dust to enable recycling of the desirable constituents such as copper in the first few fields and a bleed stream for the impurities such as lead and bismuth in the last fields. The air pollution control systems should be designed to be robust for changing operating profiles such as various air blowing rates and oxygen enrichment in sulfide smelting. Process development should also consider the potential for utilization of the sensible heat in off-gas and where the collected dust can be optimally reused. Baghouses generally have the highest collection efficiency but must operate at lower temperatures than an ESP. For this reason, ESPs are commonly used for the hot primary gases from smelting and converting operations, and baghouses are commonly used for dryer and anode furnace primary gases as well as fugitive and secondary gases from the smelting furnace and the converters.

In addition to point source controls, it is often necessary to capture and control fugitive emissions. Most fugitive and secondary emissions systems in the primary copper industry now employ at least particulate control. This is typically done using a baghouse. In addition, fugitive emissions that are not captured could result in worker exposure and this can be dealt with by implementing engineering controls or through the use of respirators, as discussed in Sect. 13.8. In sulfide smelting applications, SO_2 controls are almost mandatory. This can be done using traditional dry or wet scrubbing processes or potentially with regenerative processes depending on the SO_2 strength and other factors.

Computational fluid dynamics (CFD) modeling provides a powerful tool to assist with the design of ventilation and fume control systems. It is used to predict and evaluate:

- Performance of ventilation systems
- Airflow pattern and contaminant migration paths
- Hood designs and configuration for the optimum capture of contaminants
- Worker heat and contaminant exposure limits
- Combustion efficiency for a given geometry

CFD modeling has been applied to analyze the off-gas flow pattern exiting the mouth of a converter into a water-cooled hood and a drop-out box in order to optimize hood design (Safe and Stephens 2000).

System pressure loss modeling is used to size and specify fan performance requirement and optimize ductwork configuration, routing and damper controls to provide the desired exhaust flow distribution.

Dispersion modeling is used to evaluate the impact of emissions from various sources on ground level concentrations to ensure that a facility can comply with ambient air regulations.

13.8 Industrial Hygiene and Safety

As new processes are being developed attention must be paid to industrial hygiene and safety considerations. Both physical and health hazards associated with the new process have to be assessed. Process risk assessment must be an integral part of management philosophy and should involve evaluation of consequences of a catastrophic release both from worker safety and community perspectives. As noted in Sects. 13.1.3.1 and 13.1.4, the objective is to prevent or minimize the on-site and off-site consequences of catastrophic release of toxic, reactive flammable, or explosive chemicals.

The other main concern is the release of toxic air contaminants such as inorganic arsenic, lead, or cadmium in the workplace. The pilot scale testing and commercial demonstration protocol should include testing to obtain employee exposure estimates, as well as the feasibility of implementing engineering controls to minimize and reduce the exposure. Designing the appropriate measurements of exposures is a key to doing epidemiology assessments. There must also be consideration given to conducting occupation epidemiology on the workforce, especially if these constituents are already present in the workplace prior to introduction of a new process. Employee protection requirements can be met by the use of respirators, protective work clothing and equipment, or by implementing engineering controls. In addition, there are exposure monitoring provisions, medical surveillance, and removal provisions that have to be addressed in commercial operations.

13.9 Case Study: Kennecott-Outotec Flash Converting

The preceding sections illustrate how environmental, economic, and social considerations need to be considered in metallurgical process development so that the industry continues to be sustainable. The following case study illustrates how some of the considerations being advocated here were implemented in the development of the Kennecott-Outotec Flash Converting Process (David George and Rio Tinto 2015).

Kennecott-Outotec Flash Converting is a process developed in the 1980s and first applied as the basis for the large Rio Tinto Kennecott Copper smelter near Garfield, Utah. The technology has now been adopted by three new Greenfield copper smelters in China.

The process is based on the concept that if molten copper–iron–matte (~60–75 % Cu) is solidified and then ground and introduced into a flash smelting type furnace with oxygen, the resulting converting reactions will supply the heat to sustain the process. Since the flash converting process is decoupled from the smelting step, there is greater flexibility than the Mitsubishi Process, where the Converting Furnace is closely coupled to the smelting furnace.

The resulting off-gas from Flash Converting is a small volume, around 20,000 Nm^3/h, and very high strength, over 35 % SO_2. The Kennecott smelter, constructed in 1995, incorporated a very high level of heat recovery. The two flash furnaces each have waste heat boilers producing 6000 KpA saturated steam which is superheated and used to power the two main compressors in the sulfuric acid plant. The back-pressure turbines are each rated at 2.5 MW. The exhaust steam at ~1000 KpA is sent to a steam superheater in the sulfuric acid plant along with the waste heat steam recovered from the third Pass converter outlet gas and from the two Heat Recovery System (HRS). The HRS is a Monsanto Enviro-Chem process that recovers 1000 KpA steam from the acid absorption circuit. The combined superheated steam from the acid plant is sent to a 33 MW condensing steam turbine powering an electric generator. In normal operation, the electrical power generated from waste heat and some limited natural gas fired superheaters and auxiliary boilers provide 65 % of the smelters' electrical requirement, about 26 MW.

The development of flash converting followed a traditional route with initial pilot scale testing to confirm the concept followed by contract testing at the Outotec Pori Pilot plant and development laboratory in Finland. While the concept of making blister copper directly from very high grade (or low iron) concentrate was being developed at the same time as flash converting, there were some special considerations necessary for flash converting. Since one of the key steps in flash converting is the granulation of copper matte, full-scale tests were carried out at the Outotec Harjavalta copper smelter in Finland using their nickel matte granulator. Grinding the matte to a size suitable for feeding the pilot scale flash converting furnace was done using pilot laboratory equipment.

The actual pilot testing was done at a 1 t/h scale with some limited testing at 2–4 t/h. Since this pilot furnace had been used many times to test copper and nickel concentrate smelting, the scale up from test results to commercial scale furnace design had already been proven over several decades.

The design of the Kennecott smelter also includes special provisions to manage impurities such as bismuth, arsenic, lead, and cadmium. A hydrometallurgical treatment plant was conceived and progressed through laboratory testing to final design in 18 months. This plant also treats the various bleed streams from the copper refinery. The process is based on sequential separation and precipitation of bismuth to a waste, copper to a copper sulfide, and arsenic and cadmium to a combined As/Cd cake for ultimate management.

The Kennecott smelter is also unique because it integrates the impurity management at the copper refinery into the smelter waste treatment circuit. At the Refinery the Precious Metals Refinery (PMR), which recovers gold, silver, selenium, and other minor metals from the tankhouse slimes, was designed to control the recycle of bismuth and arsenic and provide an outlet for the lead. This is accomplished through a novel hydrometallurgical PMR circuit developed especially for Kennecott (Hoffman et al. 1995). The Hoffmann process is based on a hydrochloric acid and hydrogen peroxide leach of the decopperized slimes to dissolve gold and other elements. The gold is recovered using a solvent extraction technique and di-butyl carbitol as the extractant. The gold is directly precipitated from the loaded organic using an organic acid or salt such as oxalic acid or formic acid. The plant produces selenium and lead carbonate, both of which are sold for further processing.

13.10 Checklists

Presented below is a checklist for determining if environmental and sustainable development considerations have been addressed:

13.10.1 Conceptual Stage

Preliminary identification of whether the process is likely to be a significant water, energy consumer, based on the basic chemistry, theoretical considerations, and a conceptual flow sheet. At this stage there may be not be enough information to do a cost analysis.

13.10.2 Bench-Scale Testing

Bench-scale tests should be planned to verify proof of concept. The objective is also to develop a flow sheet for pilot scale tests. Consideration should be given to whether process has the potential to generate gaseous and particulate emissions, effluents, and process residues and that they may have to be properly disposed. With regard to gaseous and particulate emissions, it will be increasingly important to consider GHG emissions and how they will be controlled as climate change regulations mature. Consideration should also be given to see if the residues can be utilized beneficially. At this stage preliminary economics of the process should be done.

13.10.3 Pilot Plant Testing

The objective of this testing is to prove the process on a scale that is larger than bench scale. Bench-scale testing will identify potential environmental impacts; pilot design and testing should attempt to obtain better estimates of emission rates, effluent volumes, and residues generated. It should also be possible to make preliminary estimates of employee exposure to toxic air contaminants. Consideration should also be given to process control with a view to process optimization. Conversations should be initiated with environmental control vendors so that the necessary information for designing control equipment is obtained during pilot plant testing. The process economic information is further refined.

13.10.4 Commercial Demonstration

The objective of this stage is to verify the scalability of the process including the environmental control equipment, engineering controls for minimizing employee exposure to toxic air contaminants, and obtaining all other information that would be used to engineer a commercial operation. At this stage there should be a better handle on utility consumption. Process control and optimization is also important. Potential for water, energy, and resource conservation should be evaluated. Enough information is available from a technical and economic standpoint to determine the technical and economic feasibility for commercial operation. Up to this stage the focus has been on environmental, industrial hygiene, and safety and economic considerations. The community interaction thus far involves keeping the community briefed on community activities including new process development. It is essential that the community be kept informed on progress with respect to the various stages of process development and to consider any concerns that are raised.

13.10.5 Commercial Operation

During the planning process, prior to commercial operations, it is important to have the community thoroughly apprised of the new process that is to be implemented; its benefits; the environmental impacts; and how they are being avoided, mitigated, and/or managed. Due considerations need to be given to adequately address community concerns.

13.10.6 Impact on Upstream Operations

A new process may involve certain unit operations in a stage of the primary metal production process. Upstream impacts of a pyrometallurgical process on beneficiation, mining, and exploration need to be evaluated to determine if any operational changes are warranted. For example, a new continuous copper smelting process producing blister with higher impurity levels may impact where concentrates for the smelter are sourced.

13.10.7 Impact on Downstream Operations

Similarly, the impacts of downstream operations need to be evaluated. Again, the objective here is to see if downstream process changes are needed. For example, a new continuous copper smelting process might produce blister with higher impurity levels that might require modification to fire refining and/or electrorefining steps.

13.10.8 Use

The impact of the product in use is generally known because it is being produced to required specification, unless a new metal or metallic compound is being produced. Usually literature sources including Occupational and Safety Administration (OSHA) Material Safety Data Sheets (soon to become Safety Data Sheets (SDS) to comply with OSHA Globally Harmonized System) are available for this purpose.

13.10.9 Ultimate Disposal

These considerations are needed only if a new metal or metallic compound is being produced. Again MSDS and SDS are a useful source for this information.

References

Accenture & The United Nations Global Compact. (2012). Sustainable energy for all: Opportunities for the metals and mining industry. Retrieved October 31, 2012, from http://www.accenture.com/SiteCollectionDocuments/PDF/Accenture-Sustainable-Energy-All-Opportunities-Metals-Mining-Industry.pdf.

Ahmari, S., Parameswaran, K., & Zhang, L. (2014). Alkali activation of copper mine tailings and low-calcium flash-furnace copper smelter slag. *Journal of Materials in Civil Engineering, 27*, 6. doi:10.1061/(ASCE) MT.1943-5533.0001159. 04014193.

AMEC Earth and Environmental Inc. (2008). *Filtered tailings dry stack: Current state of practice*, Final Report to Rosemont Copper Company.

Bostjancic, J., & Ludlum, R. (1996). *Getting to zero discharge: How to recycle that last bit of really bad waste water*. Paper presented at 57th annual International Water Conference. Bellevue, Washington, Retrieved October 21–23, 1996.

Brown, E. T. (2003). Water for a sustainable mining industry-a review. In *Proceedings of Water in Mining* (pp. 3–12). AusIMM.

Cooper, C., Bresca, R., McCullum, A., & Scealy, J. (2006). Improving raw water use efficiency at the Endeavor Mine-Cobor—A case study. In *Proceedings of Water in Mining* (pp. 41–46).

Coursol, P., Mackey, P. J., & Diaz, C. M. (2010). Energy consumption in copper sulphide smelting. In *Proceedings of Copper 2010* (pp. 2543–2562), Vol 2-Pyrometallurgy I, Hamburg, Clausthal-Zellerfeld, Germany: GDMB. Retrieved June 6–10, 2010.

David George & Rio Tinto. (2015). Private Communication.

Davies, M. P., & Rice, S. (2001). An alternative to conventional tailings management-dry-stack filtered tailings. In *Proceedings Tailings and Mine Waste '01* (pp. 411–420).

Diaz, C. M. (2010). Copper sulphide smelting: Past achievements and current challenges. In *Proceedings of Copper 2010: Vol. 7. Plenary* (pp. 649–668), Clausthal-Zellerfeld, Hamburg, Germany: GDMB. Retrieved June 6–10, 2010.

Dunne R. (2010). Water recycling and frugal water use. *XXV International Mineral Processing Congress (IMPC) 2010 Proceedings*, Brisbane, Queensland, Australia. Retrieved September 6–10, 2010.

Frossberg, K., & Hallin, M. I. (1989). Process water recirculation in lead-zinc plant and other sulphide flotation plants. In K. V. S. Sastry, & M. C. Fuerstenau (Eds.), *Proceedings Challenges in Mineral Processing* (pp. 452–456), SME.

Gabb, P. J., Howe, D. L., Purdie, D. J., & Woerner, H. J. (1995). The Kennecott smelter hydrometallurgical impurities process. In W. C. Cooper, D. B. Dreisinger, J. E. Dutrizak, H. Hein, & G. Ugarte (Eds.), *Proceedings of Coppper-95 Cobre -95 International Conference* (Vol. 3, pp. 591–603), Electrorefining and Hydrometallurgy of Copper.

Hoffman, J. E., Sutliff, K. E., Wells, B. A., & George, D. B. (1995). Hydrometallurgical processing of Kennecott Refinery Slimes. In W. C. Cooper, D. B. Dreisinger, J. E. Dutrizak, H. Hein, & G. Ugarte (Eds.), *Proceedings of Copper-95 Cobre -95 International Conference* (Vol. 3, pp. 41–57). Electrorefining and Hydrometallurgy of Copper.

Izatt, R. M., Izatt, S. R., Izatt, N. E., Krakowiak, K. E., Bruening, R. L., & Navarro, L. (2015). Industrial applications of molecular recognition technology to separations of platinum group metals and selective removal of metal impurities from process streams. *Green Chemistry, 17*, 2236. doi:10.1039/C4GC02188F. The Royal Society of Chemistry, 2015.

Johnson, N. W. (2003). Issues in maximization of recycling of water in a mineral processing plant. In *Proceedings Water in Mining* (pp. 239–246), AusIMM.

Kellogg, H. H., & Henderson, J. M. (1976). Energy use in sulfide smelting of copper. In J. C. Yannapolis et al. (Ed.), *Extractive Metallurgy of Copper* (Vol. 1, pp. 373–415). TMS-AIME.

Kim, K. M. (2014). Blast-induced rock damage and optimized blast design in a hard-rock mine. *Transaction of the Society for Mining, Metallurgy and Exploration, 336*, 435–440.

Klimpil, R. R. (1996). The effect of water chemistry, reagent type, and other environmental factors on the performance of industrial grinding and flotation processes involving sulfide minerals. SME, pp. 11–14.

Knapp, J. B. (1973). Water reclamation and reuse at Brenda Mines Ltd. In *Proceedings Ontario Industrial Wastewater Conference* (Vol. 20, pp. 195–209).

McCaffrey. (2010). Private communication from Batu Hijau Copper mine, Indonesia.

Mathewson, D., Norris, R., & Dunne, M. (2006). Cost effective screening, dewatering and water treatment. In *Proceedings Green Processing Conference, Proceedings Water in Mining* (pp. 125–132), AusIMM.

Nadkarni, R. M., & Parameswaran, K. (1975). Energy consideration in copper, lead and zinc smelting, Energy use and conservation in the metals industry. In Y. Austin, W. M. Danver, & C. M. Cigan (Eds.), *Proceedings of a Symposium sponsored by the Extractive Metallurgy Division*, (pp. 271–298). New York, NY: The Metallurgical Society of AIME, Retrieved February 16–20, 1975.

Newman, C. J., Probert, T. I., & Weddick, A. J. (1998). Kennecott Utah Copper smelter modernization. In J. A. Asteljoki, & R. L. Stephens (Eds.), *Sulfide smelting'98 current and future practices* (pp. 205–215). The Minerals, Metals & Materials Society, 1998.

Newman, C. J., Collins, D. N., & Weddick, A. J. (1999). Recent operation and environmental control in the Kennecott Utah Copper Smelter. In D. B. George, W. J. Chen, P. J. Mackey, & A. J. Weddick, Copper 99 – Cobre 99, *Smelting operations and advances*

(Vol. 5, pp. 29–45). Warrandale, PA, U.S.A: The Minerals, Metals and Materials Society.

Norgate, T. E., & Lovel, R. R. (2006). Sustainable water use in minerals and metals production. In *Proceedings Water in Mining* (pp. 331–340), AusIMM.

Oloffson, S. O. (1988). *Applied explosives technology for construction and mining*. Arla, Sweden: Applex.

Parameswaran, K., Nadkarni, R. M., Pitt, C. H., & Wadsworth, M. E. (1981). Energy requirements in copper extractive metallurgy. In D. M. George, & J. C. Taylor (Eds.), Copper smelting-an update, *Proceedings of a Symposium sponsored by the Pyrometallurgical Committee* (pp. 265–321), Dallas, TX: The Metallurgical Society of AIME.

Parameswaran, K. (2005). Current status of mining practice: Americas. In V. Rajaram, S. Dutta, & K. Parameswaran, *Sustainable mining practices-A global perspective* (pp. 14–22). London, UK: A. A. Balkema Publishers-Taylor & Francis Group plc.

Pring, G. (1998). Sustainable development: historical perspectives and challenges for the 21st century, United Nations Development Programme (UNDP) and United Nations Revolving Fund for Natural Resource Exploitation (UNRFNRE). Workshop for Sustainable Development of Natural Resources Towards the 21st Century, New York, Retrieved October 15–16, 1998.

Rajaram, R., & Parameswaran, K. (2005). What is sustainable mining? In V. Rajaram, S. Dutta, & K. Parameswaran (Eds.), *Sustainable mining practices-A global perspective* (pp. 1–11). London, UK: A.A. Balkema Publishers-Taylor & Francis Group plc.

Ramachandran, V. (1997). Wastewater treatment at nonferrous smelters and refineries: A review. *Transaction of the Indian Institute of Metals, 50*(6), 515–520. Dec. 1997 (Special issue on fifty years of Metallurgy: Retrospect and Prospect).

Ramachandran, V. (2012). Removal, control and management of total dissolved solids from process effluent streams in the non-ferrous metallurgical industry-A review. In C. Q. Jia, V. Ramachandran et al. *Proceedings of Water, Air and Land: Sustainability Issues in Mineral and Metal Extraction (WALSIM II) Symposium*, 51st Annual Conference of Metallurgists, Niagara Falls, Canada, October 2012, pp. 101–117.

Rao, S. R., & Finch, J. A. (1989). A review of water reuse in flotation. *Mining and Engineering, 2*(1), 65–85.

Safe, P., & Russell, M. (2010). Heat recovery and energy optimization in smelter gas cleaning. In *Copper 2010 Conference Proceedings* (Vol. 1, pp. 497–515).

Safe, P., & Stephens, R. L. (2000). Peirce-Smith converter hood design analysis using computational fluid dynamics modeling. In P. R. Taylor (Ed.) *Proceedings EPD Congress 2010* (pp. 51–61), Warrandale, PA: The Minerals, Metals & Materials Society (TMS).

Schlesinger, M. E., King, M. J., Sole, K. C., & Davenport, W. G. (2011). *Extractive metallurgy of copper* (5th ed.). Oxford, England: Elsevier.

Schumann, R., Levay, G., & Amitov, I. (2009). The impact of water recycling on process water quality in mineral processing. In *Proceedings Water in Mining* (pp. 79–88), AusIMM.

Schoenbrunn, F., Neiderhauser, M., & Baczek, F. (2009). Paste thickening of tailings: Process and equipment design fundamentals relative to deposition goals. In M. Deepak, P. R. Taylor, E. Spiller, & M. Le Vier (Eds.), *Recent advances in mineral processing plant design* (pp. 455–465), SME.

Sheedy, M., Pajuen, P., & Westrom, B. (2006). Control of copper electrolyte impurities-overview of the short bed ion exchange technique and Phelps Dodge El Paso case study (Tech. Rep. 183). Ontario, Canada, Retrieved October 2006.

U.S. EPA. (2011). Renewable energy development opportunities, ASARCO Mission Mine Tailings Area. San Xavier District, Tohono O'odham Nation.

Wiertz, J. V. (2009). When best water use efficiency is not enough, What can the mining industry do. In *Proceedings Water in Mining* (pp. 13–18), AusIMM.

World Commission on Environment and Development. (1987). *Our common future*. Oxford, England: Oxford University Press.

Von Zharen, W. M. (2001). *ISO 14001: Positioning your organization for environmental success*. Rockville, MD: Government Institutes.

Part VI

Steps to Commercialization

14 Process Development, Execution, Owner's Responsibility, and Examples of Innovative Developments

V.I. Lakshmanan, Raja Roy, David King, and Ram Ramachandran

As the main theme of this book is to discuss the different aspects of developing innovative processes for the metallurgical industry, the objective of this chapter is to outline the steps that are typically required to commercialize an innovative process. This covers from the conceptual testing and study of a process through the detailed design, procurement, construction, and commissioning and start-up of the facilities to postcommissioning. The total life span of the commercialization development is discussed in this chapter.

The main challenge with the implementation of an innovative process is that there is no similar flowsheet in operation to use as a benchmark, so it makes it much more difficult to convince financial organizations to invest in the project. This will typically mean that the different phases of flowsheet development will require additional work which can extend the duration of the project as well as adding to the cost. Consequently, particular attention will be paid in this chapter to some of the differences that may be necessary for the commercialization of an innovative process compared to the normal steps that are required to execute a project that utilizes a well-established technology. Consideration will also be given to using innovative processes for all the phases of the project not just to the process.

14.1 Process Development

14.1.1 Flowsheet Development

A flowsheet is like a roadmap for a process. Just like a roadmap shows how to go from point A to point B, the flowsheet shows how to make a product B starting from a raw material A. The flowsheet development involves the testing of the process at the bench scale, pilot scale, and demonstration plant levels. A flowsheet should be economically viable and meet environmental regulations. A flowsheet needs to be robust and forgiving. A flowsheet should also provide a competitive advantage in the knowledge economy of the twenty-first century. Additional factors that make the flowsheet viable are: (a) ore grade and (b) relative—possibly high—price of the final product.

V.I. Lakshmanan (✉) • R. Roy
Process Research ORTECH Inc.,
2350 Sheridan Park Drive, Mississauga,
ON, Canada, L5K 2T4
e-mail: llakshmanan@processortech.com

D. King
Process Specialist (Retired),
Unit 210, 102 Bronte Road, Oakville, ON, Canada,
L6L 6J5

R. Ramachandran
Consulting Engineer,
9650 E. Peregrine Place, Scottsdale, AZ 85262, USA

V.I. Lakshmanan et al. (eds.), *Innovative Process Development in Metallurgical Industry*,
DOI 10.1007/978-3-319-21599-0_14

The need for a new flowsheet can arise due to many reasons. A new flowsheet may be required if the type of ore changes. As nonrefractory gold ores are exhausted, the need to extract gold from refractory and double refractory ores has arisen. Gold producing companies are responding to the challenge by developing new flowsheets for processing refractory and double refractory ores. A new flowsheet may also be necessitated to comply with stricter environmental regulations. A new flowsheet for production of zinc using pressure hydrometallurgy was developed to avoid the emission of sulfur dioxide resulting from the roasting of zinc. A new flowsheet may become feasible with advances in materials of construction. The development of polymeric materials has led to the development of flowsheets with aggressive lixiviants such as chlorides.

Development of a flowsheet starts with identification of raw material(s) and potential product(s) and by-product(s). The first step in the development of flowsheet is to consider various process options and prepare a conceptual flowsheet. The conceptual flowsheet is then tested at bench scale for feasibility of the concept.

14.1.1.1 Initial Conceptualization

The development of an innovative process is typically initiated by one or more of the following reasons:

- The feed material is very complex and none of the existing processes can give good results
- Process simplification
- Cost and environmental constraints limit the use of the existing processes
- The grade of the feed material is too low to be economical to treat with the existing processes
- Markets for products change necessitating major changes to a process, for example, the tolerance for specific impurity levels
- Location of a project imposes severe logistical problems on the supply of reagents or the disposal of waste products
- The successful application of a process used in one area triggers the possibility for applying it to another area with potentially beneficial effects

When the need for the development of an innovative process arises, the first activity for flowsheet development is to carry out a preliminary desktop investigation to identify potential process options. This will typically include the consideration of:

- Mineralogical characterization of the feed material
- Physical, chemical, or thermal separation techniques
- For physical separation techniques, evaluate the whole range of physical characteristics including:
  - Appearance (for optical separation)
  - Friability (for separation by screening)
  - Particle size
  - Specific gravity
  - X-ray fluorescence
  - Magnetic
  - Flotation
- For chemical separation techniques, evaluate the use of
  - Acid or base treatment and, for each, a range of different chemicals
  - Temperature and pressure
  - Ion exchange for upgrading and refining
  - Precipitation and electrolysis for final product generation
- For thermal separation techniques, evaluate the use of:
  - Roasting
  - Smelting

The evaluation of different process options will typically include a database search for existing operations of a similar nature, but for different products.

Once a number of potential process options are identified, then testing of the processes is required to establish which of them offers the potential to be viable. This is typically carried out at progressively greater levels of detail as the number of options gets reduced. This test work performed at the bench scale, pilot, and demonstration plant levels is outlined briefly below and was addressed in greater detail in Chap. 10. A key to successful flowsheet development is to ensure that the feed materials tested are representative of what will be processed in the commercial process.

It is very important for the successful development of an innovative process that the following people or organizations be involved in the execution of the phases of the process development work:

- The initial developers of the innovative process to ensure that the developments stay true to the initial concepts
- Experienced testing organizations are retained to carry out the different phases of the test work
- Experienced engineering companies are used to execute the different study phases with team members who have worked on similar developments.

Regular contact between these organizations should be maintained throughout the project.

14.1.1.2 Bench-Scale Test Work

Bench-scale test work is the initial phase of flowsheet testing and the primary objective is to obtain a good understanding of the process chemistry involved and to investigate the operating parameters for the process. This phase applies to the development of both existing and new innovative processes. However, for existing processes, there is a historical background that provides a good starting basis for selecting the optimal operating conditions, whereas, for innovative processes, it is necessary to start at a more preliminary stage thereby increasing the amount of work required.

The size of the equipment that is used is small and, consequently, only limited quantities of sample are typically required for the test work program. Where an ore is being tested, this quantity of sample can usually be obtained by drilling. An advantage of this scale of testing is that a large number of tests can be carried out quite readily to investigate a wide range of operating parameters for a modest cost. An objective of the bench-scale test work is to reduce the number of process options that need to be taken forward for further evaluation at the pilot plant scale. The test work is typically performed by testing organizations that are well established with the required equipment, analytical facilities, and personnel. There are a number of these organizations located in strategically important locations around the world.

Scoping Study

First step in the bench-scale test work is to carry out a scoping study. During scoping study, key steps of the conceptual flowsheet are tested using selected process parameters. Let's suppose that a hydrometallurgical conceptual flowsheet involves crushing, grinding, acid leaching, solvent extraction, and precipitation. A scoping study for this case will first include carrying out a limited number of bench-scale tests at certain grind sizes and acid concentrations. Results will then be analyzed to calculate the recoveries of elements of interest. If the recoveries are poor, then a decision has to be made whether a new conceptual flowsheet is required or a few more tests are to be conducted by varying process parameters further. If the recoveries are acceptable, a few more tests may be conducted by varying process parameters to improve the recoveries of elements of interest. The results are then analyzed again and a set of process parameters will be chosen that has given the best recoveries. A fixed number of leaching tests are then conducted using the selected process parameters to generate enough pregnant leach liquor to be used for solution purification such as solvent extraction or ion exchange. Now a limited number of solvent extraction tests will be conducted at different contact times and organic/aqueous ratios. Results will then be analyzed to calculate the recoveries of elements of interest. If the recoveries are acceptable, a limited number of precipitation tests will then be conducted at different precipitation temperatures. At the end of scoping tests, it will either be decided that conceptual flowsheet is not feasible or a preliminary flowsheet will be constructed, which will be further defined during detailed bench-scale test work.

Detailed Bench-Scale Test Work

During detailed bench-scale test work, extensive testing is conducted to investigate the effect of process variables and optimize the process. The key objective of the detailed bench-scale test work is the optimization of the process. To achieve this goal, first a systematic investigation of all process variables is conducted. Recoveries of elements of interest are plotted against each variable. During the next step, a number of tests are conducted at selected combinations of process variables, which are expected to optimize the process. The results are analyzed, correlations are made between process variables, and another set of tests is conducted. This process is iterated till the whole process flowsheet has been optimized.

The bench-scale tests are typically carried out on a batch basis. The problem with batch testing is that it does not investigate the impact of recycle streams. One way that is used to overcome this problem is to use locked-cycle testing in which a series of batch tests are performed and some of the materials from the early tests are used as recycles and added to the feed materials for the later tests. The results of the final tests of the program are taken as being a realistic representation of what will be achieved with a continuous operation. This approach is typically used for mineral processing flowsheets without chemical reactions.

For hydrometallurgical operations, a different approach is required as any recycle streams from the latter part of the flowsheet will frequently consist of solid and liquid phases, which can have a profound impact on the operation of the front part of the flowsheet. So a small-scale continuous circuit can be set up for part of the circuit with the necessary streams recycled as required. It is particularly useful to start developing a computer model of the process at this stage and to progressively incorporate the results of the test work into the model. The model can also be used to investigate some different possible flowsheet options and to provide direction to the test work program.

Pyrometallurgical processes are frequently developed by organizations with the intention of licensing the technology to interested parties or that intend to use the technology for their own purposes. These organizations frequently have their own testing facilities for the development of the processes. There are also some testing organizations that are well established for performing pyrometallurgical bench-scale testing and have a range of different equipment types available. These processes are more difficult to test on a small scale as they involve the handling of hot materials, both molten and solid, and hot off-gases frequently containing hazardous compounds. The size of the equipment used for bench-scale testing of a pyrometallurgical processing will typically be larger than that used for mineral processing and hydrometallurgical processes.

At the end of detailed bench-scale test work, a detailed process flowsheet is constructed, which is tested under continuous operation at a much larger scale than bench-scale test work during pilot scale test work.

14.1.1.3 Pilot Plant Test Work

Pilot scale test work is conducted to confirm the flowsheet developed during detailed bench-scale testing. Data generated during the pilot run is used to make a process flow diagram and carry out detailed mass, energy, and water balances, which form the basis for estimating capital and operating costs and developing plant design criteria. Pilot scale test work also generates samples for product evaluation and testing for compliance to environmental regulations.

The objective of this phase is to set up a small circuit to represent the process flowsheet and to operate the plant continuously in order to validate the optimum operating parameters that were established from the bench-scale test work and to evaluate recirculating loads. There is no standard size for a pilot plant, but, obviously, the larger the plant, the larger the quantity of sample required for the test work program. If an ore is being tested, then obtaining a sample may require some means other than drilling if the sample required is large as the cost of drilling is quite high. Pilot plants are usually carried out at the facilities of established metallurgical testing organizations as they have a large inventory of small process equipment items that can be installed to form a representative circuit. Also, the organizations have the required analytical facilities and the operating staffing that are required for a continuous pilot plant operation working 24 h/day.

If a relatively simple mineral processing flowsheet is to be tested, then there is a growing trend to use a mini-pilot plant which requires much less sample. These circuits enable the impact of recycles to be demonstrated with greater confidence than with locked-cycle tests. However, when the mineral processing flowsheet is more complicated, and, particularly if there are numerous stages with successive stages treating concentrated streams from earlier stages, then a larger pilot plant is required to ensure that there is sufficient material generated for testing the later stages. In some cases, where the sample quantity produced from the later stages is very small, it may be necessary to collect the material and campaign it through a small continuous circuit or even perform the final tests on a batch basis. Not only is it important that the feed samples are representative of what the commercial plant will treat, but that the process water which will be used is also representative and used during the test program.

For a hydrometallurgical process, it is essential that a continuous pilot plant be operated so that the impact of recycle streams can be properly evaluated. The same comments that were made under the discussion on mineral processing pilot plants about the representativeness of the feed and water samples used for the test work are equally relevant for a hydrometallurgical process. Depending on the reagents that are used in the process, the materials of construction for the equipment in the circuit can be quite challenging. It is important that the process model continues to be developed in parallel with the pilot plant campaign.

Solid–liquid separation is typically a very important part of both mineral processing and hydrometallurgical processes and so it is essential that settling and filtration test work be carried out on samples from the pilot plant. The samples should be fresh to obtain the most realistic results, so it is usual that the solid–liquid separation tests are performed at the facility where the pilot plant campaign is being carried out. The solid–liquid separation test work is typically performed by vendors of this type of equipment or by organizations that specialize in this type of test work.

Running a pilot plant is a very important step toward reducing the risk of any unforeseen complication in operating the actual plant. It provides an opportunity to minimize technical, operational, and financial risks. Pilot plant can be run with a variety of feed stocks and under different operating conditions to check the robustness of the flowsheet. During pilot scale test work, flowsheet can be fine-tuned depending on test results. Pilot plant can also be used for training the employees for the full-scale production plant.

For innovative pyrometallurgical processes, it is typically necessary to perform pilot plant testing so that the process can be tested on a continuous basis where required. The comments made about the bench-scale testing of pyrometallurgical processes also apply to the pilot plant phase. The number of organizations that have equipment available for piloting of pyrometallurgical processes is more limited than for bench-scale testing.

After the pilot scale test work is completed, the next step could be either the construction of a full-scale production plant or a demonstration plant.

14.1.1.4 Demonstration Plant

The need for a demonstration plant is often required for an innovative process and this may cover not only the verification of the metallurgical characteristics but also the equipment required including materials of construction, special designs, etc. Once again, there is no standard size for a demonstration plant. Some of the factors that influence the capacity and degree of complexity of the demonstration plant are as follows:

- Concerns regarding the scaling up of the process from the pilot plant to the commercial plant capacity
- Size required to satisfy the requirements of financing organizations
- Ability to produce sufficient product to supply samples to potential users
- Concerns with the process consistently producing a product of satisfactory quality

Because of the larger size of equipment used in a demonstration plant, it is normally necessary to install a purpose-built facility so this will require the design of the plant and purchase of the equipment and related materials. The quantity of feed material required for the running of a demonstration plant is quite large and also the amount of waste materials is large too, so these two factors impose constraints on where the demonstration plant can be located. The preferred situation is to install it at an existing operating facility to take advantage of the waste disposal, assaying, and maintenance facilities. However, if the innovative process is for a green field site, then the use of an existing operation may not be possible and so another option has to be considered. Some of the organizations that provide testing facilities for bench- and pilot-plant scale can also be used for the assembly of a demonstration plant and there have been quite a few examples of this. If the proposed scale of the demonstration plant is too large, then it may be necessary to establish a complete operation including the plant as well as waste disposal, utilities production, maintenance, analytical facilities, and administration facilities. The latter option obviously would be quite expensive.

A demonstration plant is not typically needed for validation of a mineral processing flowsheet as it can be satisfactorily tested at the pilot plant scale and there are fewer concerns with scaling up the process to a commercial size. For pyrometallurgical processing, the cost of such a facility would be very high.

If the innovative process includes the use of key proprietary equipment, then it is vital that the supplier of this equipment is involved with the demonstration plant design and operation to ensure that the knowledge gained is incorporated into the commercial plant equipment design.

14.1.2 Technical Evaluation

This phase covers the technical evaluation of the process from the metallurgical and engineering perspectives and is carried out in parallel with or preferably subsequent to the execution of the test work. There are three basic levels of studies that are typically executed to evaluate a process:

- Conceptual or scoping study
- Prefeasibility study
- Definitive feasibility study

The level of detail of the evaluation increases as the test work progresses from the bench scale through to the pilot and demonstration levels. At the conceptual/scoping level, it is common that a wide range of technical options be considered for the process and then the range is narrowed at the end of the study. The prefeasibility study will typically evaluate the narrowed range of options and select one option to take forward to a more detailed evaluation at the definitive feasibility study level. An important aspect of the technical evaluation of innovative processes can be the practice of benchmarking the use of similar processes or parts of processes that have been applied for other duties. The continuity of key personnel, particularly process engineers and metallurgists, through all the different study phases of the development of an innovative process is crucial to ensure the successful development of the process and this includes the participation of the initial developers of the process.

As part of the evaluation, capital and operating cost estimates for the process facility are developed and the level of accuracy of these estimates increases with the increased level of detail of the technical evaluation. The capital and operating cost estimates for the Definitive Feasibility Study are typically used as a basis for making a go/no go decision on whether to proceed with the design, procurement, and construction of the industrial plant.

The following sections outline the typical steps that are involved in performing the different level of studies.

14.1.2.1 Conceptual/Scoping Study

This is the initial level of study for the development of an innovative process. This is a very vital stage as it is the first time that the innovative process is actually realized as a preliminary design and, also, the first idea is obtained of the capital and operating costs, so that an initial understanding of the financial viability of the process can be achieved. The level of detail of the preliminary design and cost estimates is low, so the cost for this is very modest, which allows a number of variations to the process to be evaluated quite economically. In developing an innovative process, it is quite possible that the findings of this study might result in some significant changes being made to the process in order to improve the project economics or to overcome some technical problems that have been identified, thus requiring further evaluation work at this level. The duration of this level of study is typically up to 3 months depending on the complexity of the process.

The work will typically be carried out by process engineers or metallurgists with some assistance in developing cost estimates from estimating specialists and with only limited input from other engineering disciplines.

The objective is to develop documentation for each of the options considered for the process which typically includes:

- Process block diagram
- Preliminary process design criteria
- Preliminary process model/mass balance
- Preliminary equipment list of major items
- Conceptual plant layouts
- Capital cost estimate
- Operating cost estimate
- Simplified process description
- Relative ranking of options including preliminary assessment of health, safety, and environmental aspects

The process block diagram and preliminary process design criteria are typically based on the results of the bench-scale testing. The preliminary process model is established to develop mass and energy balances so that these can be used to generate a preliminary equipment list showing major items. The conceptual plant layout is produced to give a general understanding of the arrangement and size of the facility.

Preliminary capital cost estimates are developed for each of the process options under consideration. A capital cost estimate for this level of study is considered a factored estimate and is developed primarily from a mechanical equipment cost estimate that is based on the preliminary equipment list and historical information for similar facilities. The estimate is typically produced by developing estimated costs for each of the mechanical items in the equipment list from a combination of ballpark vendor prices and in-house information. Then factors are applied to the mechanical equipment cost to estimate the cost for other items such as civil works and structures, piping, electrical, and instrumentation to produce an estimate for the direct costs for the project. Additional factors are applied to cover indirect costs including indirect construction costs, engineering/procurement/construction management costs. Then an overall contingency is applied to the estimate to allow for the preliminary level of detail used in the study and is quite high for this level of estimate. The factors used to develop the estimate are based on historical information for similar facilities.

Preliminary operating cost estimates are also developed for each of the process options under consideration. An operating cost estimate covers the costs for labor, reagents and consumables, electrical power, maintenance materials, contractor services, and administration. At this level of study, the estimate is developed using information from the mass balance, capital cost estimate, and other historical information. The capital and operating cost estimates for this level of study typically have an accuracy of ±50 %.

The above information is then evaluated to develop a relative ranking of the most viable options. This evaluation would also include a preliminary assessment of the technical, health and safety, environmental, and other risks of each option. The selected options will then be tested more thoroughly at the bench scale in preparation for the prefeasibility study.

14.1.2.2 Prefeasibility Study

This is the second level of study for the development of an innovative process. The level of detail is higher for this study level than the conceptual/scoping study; so the cost for this work is higher but is still quite modest and again allows some variations to the process to be investigated quite economically. The details of the process are firmer during this phase and the confidence in the capital and operating estimates increases. The duration of this level of study is typically up to 6 months, depending on the complexity of the process. This is a very crucial phase for the development of an innovative process as, at the end of it, an option will be selected which will be the basis for the definitive feasibility study after which the decision will be taken on whether to proceed with the process or not.

The work will typically be carried out by process engineers or metallurgists with some input from other engineering disciplines and with the capital cost estimates developed by estimating specialists.

The objective is to develop documentation for each of the options considered for the process which typically includes:

- Preliminary process flowsheets
- Expanded process design criteria
- Elaborated process model/mass balance
- Elaborated equipment list of mechanical items
- Overall plot plan and preliminary plant general arrangements for each plant area
- Simplified electrical single-line diagrams and preliminary electrical load list
- Capital cost estimate
- Operating cost estimate
- Preliminary project schedule
- Simplified process description
- Financial analysis
- Relative ranking of options including preliminary assessment of health, safety, and environmental aspects

All of these documents are developed to a greater level of detail than for the conceptual/scoping study.

Process flowsheets and expanded process design criteria are developed for each of the process options with the former showing the mechanical equipment and the latter including sizing criteria for all of the plant areas and the individual items of equipment based on the results of the more thorough bench-scale testing. The preliminary process model is expanded to include all the unit operations of the process and to include a more detailed understanding of the reaction chemistry involved. The model generates mass and energy balances so that these can be used to develop a more complete equipment list showing all the mechanical items and the relevant equipment sizes, materials of construction, and electrical motor sizes, where required.

An overall plot plant is developed to demonstrate the relative location of the different areas of the plant, the size of the facility, and also the related facilities such as the tailings storage facility, and then preliminary plant general arrangements, typically just plans, are produced for each area of the plant.

Preliminary electrical single-line diagrams are developed for the plant based on the mechanical equipment list and the motor sizes shown for the different items. A preliminary electrical load study is then developed based on this information.

Preliminary capital cost estimates are developed for each of the process options under consideration. A capital cost estimate for this level of study is also considered a factored estimate developed primarily from a mechanical equipment cost estimate that is based on the preliminary equipment list. The estimate is typically produced by developing estimated costs for each of the mechanical items in the equipment list by obtaining order-of-magnitude vendor prices for all major equipment with in-house information used for smaller items. The estimate for the major electrical items is produced from the electrical load study and then factors are applied for the other electrical costs. The costs for other items such as civil works and structures, piping, and instrumentation are developed by applying factors to the mechanical equipment cost to estimate the complete direct costs for the project. As an alternate, structural costs are estimated using cost per square foot based on takeoffs from the plan drawing. Civil costs can also be estimated based on takeoffs from the drawings; in this case, a cost per cubic yard is assumed. Then further factors are applied to cover indirect costs including indirect construction costs, engineering/procurement/construction management costs. The overall contingency that is applied to the estimate for this level of study is lower than for a conceptual/scoping study due to the higher level of detail used in the study. The factors used to develop the estimate are also based on historical information for similar facilities.

Preliminary operating cost estimates are also developed for each of the process options under consideration. An operating cost estimate covers the costs for labor, reagents and consumables, electrical power, maintenance materials, contractor services, and administration. At this level of study, the cost for labor is developed by producing a preliminary manpower list for the management, supervision, operation, and maintenance of the facility and applying costs for these different positions based on typical levels of reimbursement for the region in which the plant is installed. The costs for reagents and consumables are developed by producing a list of all the different items, estimating the individual consumptions using information from the mass balance and in-house data, and then applying costs for each item from supplier pricing wherever possible and in-house information. The cost for electrical power is estimated from the electrical load study for each area and applying the unit cost that is applicable to the region in which the plant is located. The costs for maintenance supplies are estimated by applying factors to the direct capital cost estimates for each area. The costs for contractor services, administration, etc., are typically taken from historical information. The capital and operating cost estimates typically have an accuracy of ±25 %.

The above information is then evaluated to develop a relative ranking of the options in order to determine the most viable option. This evaluation would also include a higher level assessment of the technical, health and safety, environmental, and other risks of each option. The selected option will then be tested more thoroughly at the pilot plant scale in preparation for the definitive feasibility study.

14.1.2.3 Definitive Feasibility Study

This is the third, and typically highest, level of study for the development of an innovative process. The detail of the work for this is much higher than for a prefeasibility study requiring the involvement of a broader range of expertise. So the cost for the execution of this level of study is much higher and the schedule may be as long as a year, depending on the complexity of the process. So it is vital to select the optimum variant of the process as the basis for this study, although it is quite typical to carry out some trade-off studies early during the study to evaluate any smaller opportunities to improve the process.

The work will be carried out by a complete range of process engineers or metallurgists and other engineering disciplines and with the capital cost estimates developed by estimating specialists. There will also be a major input from scheduling and planning, procurement, project management, and construction management personnel. It is vital for this level of study that the process flowsheet is frozen and that pilot plant-scale test work has been carried out to provide greater confidence in the process design criteria used for the plant.

The objective is to develop documentation which typically includes:

- Process flow diagrams (PFDs)
- Process design criteria
- Process model/mass balance
- Detailed equipment list of mechanical items
- Equipment specifications for major mechanical items
- Equipment datasheets for commodity equipment items such as pumps, cranes, HVAC, etc.
- Piping and instrumentation diagrams (P&IDs)
- Process control philosophy
- Overall plot plan and plant general arrangements for each plant area
- Electrical single-line diagrams, electrical load list, electrical equipment list and specifications, cable and conduit list, and material takeoffs
- Preliminary designs and material takeoffs for civil/structural, piping and instrumentation and control
- Vendor enquiries for the supply of equipment and materials and for construction subcontracts
- Project execution schedule

- Capital cost estimate
- Operating cost estimate
- Process description
- Technical, health and safety, environmental, and other risk reviews

All of these documents are developed to a much greater level of detail than for the prefeasibility study.

The list of the process deliverables for this level of study is similar to that for a prefeasibility study but with the addition of equipment datasheets, P&IDs, and control philosophy. The other items in the list are developed to a higher level of detail. Preliminary P&IDs are developed based on the process flowsheets and these are then passed to piping engineering for further elaboration.

An overall plot plant is developed to a higher level of detail to demonstrate the relative location of the different areas of the plant, the size of the facility, and also the related facilities such as the tailings storage facility, and then plant general arrangements, both plans and elevations, are produced for each area of the plant. There is a growing trend to develop the layouts using 3D CADD.

For mechanical engineering, the major activity is to develop equipment specifications for all the major equipment items which include information from the process equipment datasheets, where available and material takeoffs for plate work, e.g., tanks, bin, chutes, etc. Mechanical engineers work closely with procurement to obtain multiple quotes for each equipment item. Quotes are evaluated and equipment is selected. Often long lead equipment is purchased based on these quotes.

For civil and structural engineering, preliminary designs are developed using sketches which are used together with the plant general arrangements as the basis for developing preliminary designs for the different areas, and then material takeoffs are developed of the amount of site work, concrete, structural steel, etc.

For electrical engineering, single-line diagrams are developed for the plant based on the mechanical equipment list and the motor sizes shown for the different items. A more detailed electrical load study is then developed from this information. Preliminary electrical designs are produced for the different plant areas and then electrical equipment specifications are developed for major items, with material takeoffs being developed for the other electrical requirements based on the plant general arrangements and the preliminary designs. Electrical engineers work with purchasing to obtain equipment quotes and evaluate them.

For piping engineering, the preliminary P&IDs developed by process are elaborated and, also, preliminary piping specifications are produced. These documents together with plant general arrangements are then used as the basis for producing material takeoffs for piping, valves, and fittings.

For instrumentation engineering, preliminary instrumentation datasheets are developed and preliminary designs specifications will be produced for the overall control system. Then material takeoffs for the other instrumentation are developed based on the P&IDs, plant general arrangements, and process control philosophy.

For the required level of accuracy for a definitive feasibility study, it is necessary to obtain vendor and supplier pricing for most of the equipment and bulk commodities and also for construction subcontracts. Consequently, procurement specialists are used to obtain this pricing based on the equipment specifications and bulk material takeoffs developed by the different engineering disciplines together with standard procurement terms and conditions.

A more developed project schedule is produced covering all aspects of engineering, procurement, and construction.

The required level of accuracy for a capital cost estimate for this level of study requires that each item be estimated from the relevant equipment list plus bulk materials from the material takeoffs for mechanical, civil/structural, piping, electrical, and instrumentation. The majority of the costs are developed by obtaining firm vendor and supplier prices for all major equipment and bulk materials. For this level of study, the indirect costs including construction indirect costs, engineering/procurement/construction management costs are developed from detailed analyses of the individual components of each estimate category. The overall contingency that is applied to the estimate for this level of study is lower than for a prefeasibility study due to the higher level of detail used in the study and typically is developed from an analysis of the level of confidence in the different areas of the estimate.

An operating cost estimate is also developed. At this level of study, the cost for labor is developed by producing a manpower list for the management, supervision, operation, and maintenance of the facility and applying costs for these different positions based on typical levels of reimbursement for the region in which the plant is installed. The costs for reagents and consumables are developed by producing a list of all the different items, estimating the individual consumptions using information from the mass balance and typical data from existing information, and then applying costs for each item from supplier pricing wherever possible and in-house information. The cost for electrical power is estimated from the electrical load study for each area and applying the unit cost that is applicable to the region in which the plant is located. The costs for maintenance supplies are estimated from a detailed analysis of the equipment and materials that are used for each area. The costs for contractor services, administration, etc., are typically developed by analyzing the different elements of these categories. The capital and operating cost estimates typically have an accuracy of ±15 %.

This evaluation would also include a more detailed assessment of the technical, health and safety, environmental, and other risks of the process. As part of this assessment, a hazard and operability (HAZOP) review will be carried out which is an activity involving the owner's management team including members of the operation and maintenance groups and the engineering company's lead engineers for each discipline. HAZOP reviews are always done in chemical plants, oil refineries, and gas plants. Current belief in the mining industry is that the benefits of carrying out a HAZOP review have little value. A constructability review will also be performed using construction management specialists together with key members of the engineering and project management groups. A formal risk review is also carried out including most of the people involved in the other reviews. All of these review meetings will preferably be facilitated by people who are experienced in these areas and have not been directly involved in the execution of the study.

If the results of the definitive feasibility study are positive, the decision may be taken to test the process at the demonstration plant level as the basis for basic engineering.

14.1.3 Capital and Operating Cost Estimation

Projects in the metallurgical industry are capital intensive. Any metallurgical process starts as a conceptual idea; this is followed by (a) scoping experiments to ascertain feasibility, (b) laboratory scale test work, (c) pilot scale test work, (d) demonstration plant—if needed, and (e) data collection for the design of a full-scale plant. Based on this data, a complete mass and energy balance and detailed description of all the process steps should be established prior to the design and sizing of all process equipment. All these items were described in Chap. 10.

Based on the above information, the next step is to develop equipment design for the full-scale plant. This was described in Chap. 12.

In 1958, the American Association of Cost Engineers published five definitions of capital cost estimate types as follows:

(a) Order of Magnitude Estimate (variable accuracy over plus or minus 30 %)
(b) Study Estimate (factored estimate, accuracy up to plus or minus 30 %)
(c) Preliminary Estimate (generally for authorization, accuracy plus or minus 20 %)
(d) Definitive Estimate (more detailed information, accuracy plus or minus 10 %)

(e) Detailed Estimate (contractor's estimate, accuracy plus or minus 5 %)

Over the years, the accuracy of some of these estimates have changed based on actual experience and variations in available data. Specifically, the following changes are noted here:

(a) Order of Magnitude Estimate (variable accuracy; plus or minus 50 %)
(b) Preliminary Estimate (accuracy plus or minus 25 %)
(c) Definitive Estimate (accuracy plus or minus 15 %)

The full capital cost of the project consists of three components shown below:

(a) Plant and equipment cost, including costs of direct engineering, procurement, and delivery to site. This cost is part of (b).
(b) Total erected and installed cost (TEC), i.e., the cost of the complete plant erected and ready to operate—including constructional engineering, labor and materials, all services, buildings, and so on. This cost is part of (c).
(c) Total capital employed (TCE), which also includes the working capital and financing charges.

Capital costs are only one aspect of a full cost of a given project. In addition, it is normal practice to calculate operating costs and internal rate of return (IRR). Sometimes, sensitivity analyses on the most critical process design parameter in each unit operation are run to establish its effect on IRR. In the current section, only details of calculating operating costs will be discussed.

14.1.3.1 Estimating methods:

There are four methods used for estimating costs: They are:

(a) Step count estimating.
(b) Power or exponential law.
(c) Factorial method.
(d) Detailed methods.

The reader is referred to "Guide to Capital Cost Estimating" by Gerrard, A.M. for details of the above four estimation methods (Gerrad 2000).

An estimating checklist consists of the following items:

(a) Site.
(b) Process Plant.
(c) Service Plant and Equipment.
(d) Civil works.
(e) Overhead Costs
    - Engineering costs
    - Temporary facilities required for construction.
    - Direct construction costs.
    - Permitting cost.
    - Miscellaneous overhead items.
    - Owners cost.

14.1.3.2 Updating Estimates

In estimating the costs of equipment or process plants, it is always necessary to have a means of adjusting the cost of an item at one time to the estimated cost of the same item at another time. This is normally done by using cost indices. A cost index is given a value of 100 at a particular date chosen by the index compiler.

The US cost indices are listed below:

(a) ENR Construction Index applies mainly to civil engineering works and consists of construction material and unskilled labor factors. This does not really apply to process industries. This is the oldest of the inflation indices with a base data of 1904 at 100.
(b) CE Plant Cost Index: The Chemical Engineering (CE) Plant Cost Index is published by McGraw Hill . It is a complex multicomponent index. The base index for year 1957–1959 is 100. In addition, they publish indices for seven categories of equipment, construction labor, buildings, plus engineering and supervision.
(c) M and S Equipment Cost Index: The Marshall and Swift Equipment Cost Index is an "all industries" measure consisting of a composite

value of indices from various industries. This index is published in *Chemical Engineering* together with component indices. The base index for year 1926 is 100. *Chemical Engineering* also publishes indices for eleven categories of air pollution control equipment. These indices are based on the assumption that it is 100 in the year 1994.

14.1.3.3 Sources of Cost Information

Sources of cost information are listed below (Gerrad 2000):

(a) Survey of sources: Estimators should have access to comprehensive and accurate cost information for preparing good cost estimates.
(b) Supplier's detailed quotations.
(c) Supplier's budget quotations.
(d) Company Records.
(e) Trade Literature.
(f) Unit prices for measured work; published annually.
(g) Technical literature and textbooks.
(h) Personal cost books.
(i) Government Departments.
(j) Computer data.
(k) Data Sources in Trade Journals.

The starting point for detailed capital cost estimation is to compile a complete list of equipment with specifications as listed below:

(a) Item Number based on the location in any specific operating unit.
(b) Description of equipment.
(c) Operating parameters for that equipment.
(d) Approximate size and shape.
(e) Materials of construction.
(f) Quantity required.
(g) Item cost.

For a hypothetical waste water treatment plant, it is assumed that the total equipment cost is $500,000.00. An order of magnitude cost estimate is shown below:

Order of Magnitude Cost Estimate

	Equipment	Material	Labor
Equipment cost	500,000		
Labor to install @ 20 %			100,000
Foundation: @ 5 %		25,000	
Labor: @ 133 %			33,250
Structures: @ 4 %		20,000	
Labor: @ 50 %			10,000
Buildings: @ 8 %		40,000	
Labor: @ 100 %			40,000
Insulation: @ 2 %		10,000	
Labor: @ 150 %			15,000
Piping: @ 15 %		75,000	
Labor: @ 50 %			37,500
Electrical: @ 8 %		40,000	
Labor: @ 75 %			30,000
Instrumentation: @ 6 %		30,000	
Labor: @ 40 %			12,000
Painting: @ 2 %		10,000	
Labor: @ 300 %			30,000

	Equipment	Material	Labor
Miscellaneous: @ 5 %		25,000	
Labor: @ 80 %			20,000
Sub totals	500,000	275,000	327,750
Note: Material costs are percentages of equipment costs.			
Labor costs for material are percentages of material costs.			
Total direct cost:			1,102,750
Engineering/purchasing @ 10 %			110,275
Field supervision: 6 months @ 10,000/ month			60,000
Field indirects; 50 % of direct Labor			163,875
Subtotal:			1,436,900
Escalation			0
Subtotal:			1,436,900
Contingency @ 15 %			215,535
Subtotal:			1,652,435
Fee: 8 % of material and labor			48,220
Taxes: 5 % of equipment and material:			38,750
Project estimated cost:			1,739,405

Note: The accuracy of an order of magnitude cost estimate such as this one is considered to be +/− 50 %.

As a rule of thumb, the total project estimated (installed cost) is around 3.3–3.7 times the equipment cost. In the above case, the calculated factor is 3.48.

The following items are excluded from the above sample calculation. However, they must be factored in before a final decision is made about the viability of the project.

(a) Treatment facilities for or impoundment of wastes.
(b) Any costs related to *connection* to utilities; power, water, natural gas, or sewer.
(c) Permits of any kind.
(d) Land costs.
(e) Escalation costs.
(f) Owner's administrative costs.

Estimate of Operating Cost

Operating Schedule

- General operations: 3 shifts/day, 7 days/week
- Operating time of 8000 h per year (91 % operating time)
- Volume of water to be treated per year.

Summary of estimated operating costs are shown below:

(a) Labor, excluding benefits.
(b) Employee benefits; 30 % of labor (consists of F.I.C.A., vacation, sick leave, insurance, holidays, etc.).
(c) Chemical reagents (such as chemicals, flocculants, gaseous reagents, etc.)
(d) Operating supplies (fuel for mobile equipment, shipping bags, filter cloth, and miscellaneous items).
(e) Maintenance supplies.
(f) Power (based on demand, consumption rates, and unit cost during peak and nonpeak hours).
(g) Fuel (essentially natural gas; based on usage and unit cost).
(h) Protective clothing and safety supplies.
(i) Process water (based on usage and unit cost).
(j) Steam (if needed for heating; based on usage and unit cost).
(k) Cost of feedstock.
(l) Value of product.

The operating costs do not include:

(a) Amortization and depreciation
(b) Sales expense
(c) Taxes and Insurance
(d) Owner's overhead expense
(e) Cost of Sales
(f) Escalation
(g) Cost of processing and/or impounding of waste streams

The capital cost equipment sizing was based on treating waste water (volume~25 gpm using a specific process operation). However, no exact dollar figures have been assigned for the estimation of operating cost as quite a few other variables, such as location, variation in prices of chemicals, utilities etc., are involved based on location.

14.2 Project Execution

The execution of a project for the design and construction of a metallurgical plant is typically executed by an engineering, procurement, project management, and construction management organization that specializes in this type of project. When there is an innovative process involved, then there is an even greater importance to the experience of that organization on projects using similar technology. The organization will need to involve a full range of engineering, procurement, project management, and construction management capabilities.

If the technology is a proprietary process provided by a specialist development organization, then it is essential that they are involved in the key phases of the project to ensure that the project as completed is consistent with the requirements of their technology.

The steps involved for the execution of project using an innovative process are very similar to a project using a well-established technology as the bulk of the challenges with a new process are addressed during the process development stage. So the discussion of the different phases of project execution primarily addresses some of the areas that need particular attention with an innovative process.

The use of 3D CADD for the design of the plant has numerous advantages, particularly for facilities that include a large amount of piping, as the software can be used to run interference checks between civil works, structural steel, piping, electrical cabling, and instrumentation. This can result in the elimination of reworking on site as the interference clashes were identified during the installation of the facilities. Also, the model is invaluable in the execution of HAZID and HAZOP reviews and also the training of operating personnel. Modern 3D CADD systems can also have smart P&IDs integrated with them. These documents can then be used to develop the material takeoffs for piping, electrical cabling, etc.

14.2.1 Basic Engineering (Feasibility Engineering)

Basic Engineering is the initial phase of the execution of the design, procurement, and construction of the industrial plant. It is sometimes referred to as front-end engineering. It is not always executed as a separate phase but is often integrated into the detailed engineering. This phase typically takes 10–15 % of the total engineering man-hours for the project. After this phase, the level of activity on the project increases substantially and the cost magnifies significantly with the procurement of the equipment and materials and the awarding of the construction subcontracts, so it is crucial that the plant design is frozen before proceeding to detailed engineering as the cost of making changes to the process after this phase would be extremely high.

The engineering of a project involves the input of a number of different disciplines as follows:

- Process
- Mechanical
- Civil/Structural
- Piping
- Electrical
- Instrumentation and Automation

In addition, there is major involvement from the following nonengineering groups:

- Project Management
- Procurement
- Project Planning and Scheduling
- Project Cost Control
- Project Accounting
- Document Control
- Cost Estimation
- Construction Management

The main objective of this phase is to substantially complete the process design for the project and to develop the basic criteria, standards, specifications, and drawings for the different engineering disciplines. It is very common to produce a detailed capital cost estimate for the project at the end of this phase with a typical accuracy of ±15 %.

This phase of the project is always very important as the process design is fully developed during this time which basically determines how the plant will perform ultimately. The importance of this phase is heightened with an innovative process and the execution of HAZID and HAZOP analyses is crucial to identify potential risks and how to mitigate them.

There is a growing trend for the Basic Engineering to be carried out in a location where the relevant detailed experience exists with the technology used in the project, frequently in the more developed countries, and then for the Detailed Engineering to be executed in another location where the costs of engineering services are much lower. This can result in substantial savings in the cost for engineering. This is a strategy that has been applied to many projects, but the key to its success is to ensure that there are excellent communications between the basic and detailed engineering teams. The development of online communications has greatly facilitated the easy transfer of documentation between widely separated locations in different countries if required. However, a key factor is to ensure that the different execution teams fully understand one another, so it is preferable to transfer key people between the different locations during the various phases of the project.

14.2.2 Detailed Engineering and Procurement

This phase covers the completion of the detailed engineering for the project by each of the disciplines and the procurement of all of the equipment and materials, and services for the construction of the project. The different engineering disciplines and the nonengineering groups involved in this phase are similar to basic engineering, but the level of effort is much higher as it involves the completion of the other 85–90 % of the engineering together with the buying, expediting, and inspection of the different equipment and materials for the project and the awarding of construction subcontracts. This phase typically takes 12–18 months.

As part of the procurement activity, there is a growing trend to modularize part or all of a project and to transport the preassembled modules to the construction site. This can reduce the cost of constructing facilities on site and also it is possible to test the different systems on the modules at the modularization facility before shipping the units. This is extremely useful for a remote site providing that there are no major problems with transportation.

Also, there is a growing trend to procure equipment from locations other than the developed nations where the costs are lower and the quality has improved substantially.

14.2.3 Construction

This phase addresses the preparation of the project site and the installation of all of the different equipment and facilities. This is a very labor intensive phase, although the amount of labor can be reduced substantially by the supply of preassembled process modules. The construction risk in terms of the safety of construction labor can be significantly reduced by moving a significant portion of the construction to a more controlled environment and by reducing the number of construction workers needed on site. The successful execution of this phase requires very careful planning and close control of the quality of the site preparation and installation work.

Special requirements may be necessary with innovative processes such as hazard identification and mitigation, special construction techniques, and constructability and operability issues. This will probably require the involvement of specialist construction supervision and vendor representatives for any specialist equipment. This is often a contractual obligation on the part of the vendors, and the cost of providing commissioning expertise may be built in to the capital cost of the project. The duration of this phase is typically 12–18 months or more.

14.2.4 Commissioning and Start-up

This phase covers the "dry testing" (aka cold commissioning) of the different systems in the plant to ensure that all the mechanical equipment, electrics, and instruments function correctly, and then the "wet testing" (aka hot commissioning) of the plant with the addition of water or other inert materials, i.e., without the addition of feed materials and reagents. It is essential that the activities are well planned and that comprehensive procedures are developed for these operations. When this phase is complete, the project is handed over to the owner for operation. This will probably also require the involvement of specialist commissioning/start-up personnel and vendor representatives for any specialist equipment. This is often a contractual obligation on the part of the vendors. Sometimes, the cost of providing commissioning expertise may be built in the capital cost of the project.

14.3 Owner's Responsibilities

This phase covers some of the aspects of the role of the owner that will use the innovative process and is required to ensure that the process is fully brought to successful implementation.

14.3.1 Owner's Team

The establishment of an Owner's Team early in the development of the innovative process is a key to a successful outcome. The team will expand as the project moves through the different phases of the process development program. In addition to the typical management positions to look after project management, project controls, accounting, HR, administration, HSE and legal, the team should also include representatives who will eventually have a major role in the operation of the plant using the innovative process, so that there is early buy-in to the technology and there is continuity throughout the project program.

Ideally, key members of the Owner's process team should be involved from the beginning of the bench-scale test work right through to commissioning and start-up. They would have a monitoring role during the bench-scale and pilot-plant work, but would typically be involved very deeply in the design and operation of a demonstration plant.

14.3.2 Staffing of Operating and Maintenance Personnel

It is important to hire the key members of the operating and maintenance teams as early as possible in the project schedule. As far as possible, they should have relevant experience on operations of a similar nature. They should be involved in the operation of the demonstration plant so that they obtain an in-depth understanding of the innovative process. This will allow them to take leadership roles in the training of the rest of the operating and maintenance teams at the appropriate time.

14.3.3 Training of Operating and Maintenance Personnel

Training of operating and maintenance personnel is always very important for the successful ramp-up and operation of a metallurgical plant but is much more so when an innovative process is involved.

The training should be a very structured program and it is recommended that the training effort be organized and led by a team of experienced trainers. The training program will need to be established well in advance of the actual training effort as there will be considerable work involved in the preparation of schedules, systems, and documentation. Detailed operating instructions for the whole plant and individual operating procedures for each plant area are required to provide the basis for the training. Also, the use of the 3D-CADD model of the facility is an invaluable aid so that the personnel can understand the locations of the equipment, piping, and instrumentation.

14.3.4 Ramp-up

The time required after start-up and commissioning to ramp-up the plant production to the design capacity can vary enormously and this can have a major impact on the financial return for the project. Historical experience has shown that, the more detailed the level of test work and development work carried out, the fewer the number of problems experienced with the ramp-up. Also, maximizing the level of experience of the operating and maintenance teams and the quality of the training will have a marked impact on facilitating this process.

Wherever possible, it is advisable to use a higher feed grade for the first few years in order to improve the plant revenues and thereby reduce the payback time of the project. However, it is better to have a stockpile of lower grade feed material available for the first few months of operation until the plant operating conditions have been stabilized and acceptable extractions achieved for the materials targeted, as this minimizes the losses due to poor extractions that will typically be experienced until the plant settles down.

14.4 Examples of Innovative Developments

Two examples of innovative developments are presented below.

14.4.1 Copper Solvent Extraction

For many years, copper ores that were unsuitable for treatment by flotation and smelting were treated by acid leaching either in vats for high-grade oxide ores with the copper being electrowon from the leach solution producing an impure copper product, or in heaps and dumps of low-grade oxide and sulfide ores with the solution being treated by cementation on iron scrap to produce an impure copper precipitate. Solvent extraction had been used for many years in uranium processing flowsheets, but the lack of a copper-specific reagent prevented its application to copper processing. However, in 1968, a major breakthrough was initiated by General Mills Chemicals, who developed a copper-specific organic extractant.

This was first applied commercially in 1968 at the Ranchers' Bluebird mine to treat leach solution from a heap leach operation treating mined oxide ore. This resulted in a high quality copper stream being produced which could then be sent to a tank house where the copper could be deposited on to copper starter sheets, thereby producing an LME-grade product on site.

This development has had a major impact on the production of copper and there are now numerous copper solvent extraction circuits in use around the world.

14.4.2 Alternatives to Solid–Liquid Separation

For the leaching of some ores and concentrates, the use of solid–liquid separation can be rather expensive and also very difficult because of the poor setting and filtration characteristics of the leach material. So there have been numerous innovative processes developed to overcome this difficulty. These processes can be loosely divided into two categories:

- Treatment of unclarified solutions
- Treatment of pulps

Fixed bed ion-exchange and solvent extraction were used in the 1950s and 1960s for uranium extraction, but both of these technologies required a feed solution with a very low suspended solids content. For some ores, this was a significant problem due to poor solids settling characteristics and this was before the development of modern flocculants. One method that was developed was to use coarser ion-exchange resin beads and create a bed in a perforated basket that was suspended in a soluble uranium bearing slurry stream. The uranium was adsorbed on to the resin and then periodically the resin basket was removed from the slurry and the resin was washed to remove any remaining slurry and then it was eluted in a fixed-bed column.

In the 1970s, another approach was developed involving feeding leached uranium slurry to a solid–liquid separation circuit to carry out a primary separation of the solids and liquids and then the unclarified leach solution was sent to a fluidized-bed ion-exchange column. The column had a number of stages separated by perforated plates and a resin bed was collected on each plate. The solution passed up through the column and the beds of resin in the column were fluidized and the uranium in solution was adsorbed on to the resin. Periodically, the flow of solution was stopped and the resin beds allowed to settle and then a valve was opened at the bottom of the column and a bed of resin was allowed to discharge by gravity and was collected in a vessel where it was ultimately eluted. At the same time, a bed of eluted resin was fed to the top of the ion-exchange column. Then the resin discharge valve was closed and the leach solution flow was reestablished. This technology is still offered today.

Also, in the 1970s, there was a development program to extract gold from leached slurry by countercurrent contact with activated carbon in a series of agitated tanks. Initially, this was achieved by pumping a slurry stream from each agitated tank over an external screen with the slurry being sent to next tank in the series and the carbon being sent to the previous tank in the series. Eluted carbon was periodically added to the last tank in the series and loaded carbon was removed from the first tank in the series and sent for elution. The use of external screens resulted in significant carbon losses due to abrasion and so there was an extensive program to develop in-tank screens, and that is what is typically used today. This process is now referred to as carbon-in-pulp or carbon-in-leach and is very widely used around the world. This technology has also been adapted for the extraction of uranium using a resin and that is referred to as resin-in-pulp.

14.5 Summary

From the earlier sections of this book, it should be evident that the development of an innovative process from concept to operation introduces numerous additional challenges over and above those that are typically experienced with a well-established process. So these challenges will require a greater level of effort in the different phases of the project which will probably extend the time required for execution.

Process development: Process development is the phase that is most significantly affected by the development of an innovative process. During this phase, the process moves from an initial concept through to the development of a plant design and capital and operating cost estimates that are suitable for making a decision about whether to proceed with the execution of the project or not.

Thorough testing of the innovative process through a program of test work that gets progressively more detailed is the basis for the successful development of the process. As part of the test work program, bench-scale and pilot plant test work are essential requirements and, frequently, it is also necessary to assemble and operate a demonstration plant to prove the process at a larger scale with equipment that is truly representative of what a commercial plant will use and also to satisfy the requirements of financing organizations and purchasers of the product.

In parallel with the test work program, the technical evaluation and cost estimating efforts progressively grow in detail from the conceptual/scoping phase through the prefeasibility phase to the definitive feasibility study phase. At the end of the process development phase, the innovative process has been developed into an engineered facility which provides a good basis for proceeding with the project execution phase if the studies show that it is technically and economically viable.

Project execution: The project execution phase of an innovative process is more routine, but there are still areas that need careful attention compared to more widely used processes.

The execution of a project for the design and construction of a metallurgical plant is typically executed by an engineering, procurement, project management, and construction management organization that specializes in this type of project. When there is an innovative process involved, then there is an even greater importance to the experience of that organization on projects using similar technology. The organization will need to involve a full range of engineering, procurement, project management, and construction management capabilities.

In addition to the use of an innovative process, there is a growing trend to use some innovative work practices including the use of integrated 3D CADD, detailed engineering in third-world high value centers, plant modularization and supply of equipment from China, etc.

If the technology is a proprietary process provided by a specialist development organization, then it is essential that they are involved in the key phases of the project to ensure that the project as completed is consistent with the requirements of their technology.

Owner's Responsibilities: The owner of the project that utilizes the innovative process has several areas of responsibility in ensuring the successful implementation of the project. These areas include the assembly of an experienced owner's team for the execution of the different project phases, staffing and training of operating and maintenance personnel, and ramp-up of the plant.

Reference

Gerrad, A. M. (2000). *Guide to capital cost estimating* (4th ed., p. 115). Warwickshire, UK: Institute of Chemical Engineers.

Part VII

Financing

15 Investing, Financing and Harvesting Innovation and Technology

Michael Dehn

When considering the development of new innovation and technology, financial resources required to advance the process from idea to commercialization are not typically easy to come by. Unlike the development of new software which requires less capital, and computer hardware which seems to have a near unending source of research and development financing, marketing, and commercialization, the availability of research and development financing for new technology in the physical sciences is often as hard to come by as the scientific breakthrough itself. While small entrepreneurs have the most significant breakthroughs in technology, for example, the Canadian Titanium Limited (CTL) process versus the Chloride Process and Sulfate Process for Titanium Dioxide production, the initial innovation came from previous projects and personal funds that developed a front-end leaching process. Followed by new capital in the form of public equity for a solvent extraction process, and an "off-the-shelf" solution for pyrohydrolysis, successful bench and pilot plant scale testing have brought the project to the point where commercialization capital is being sought. The CTL process seems to be an exception versus the norm for successful financing; however, financing for the final CTL plant is ongoing and is not yet completed.

This is by no means a complete coverage of all funding sources/mechanisms but should give a good overview.

15.1 Self-Financing/Friends and Family

Typically the easiest form of financing for a new technology is self-financing or friends and family financing. If the innovation or technology is your own idea, and you have the personal means to finance that idea, naturally you are the simplest and fastest financier. You're the boss, it's your money, and win or lose you are in total control.

The next strongest support level comes from friends and family where quite often little confidence in the innovation or technology is required, but only confidence in the individual. Often there is a clear path in the developers mind but it is difficult to convey these ideas to others. Hence seeking support of the individual versus the idea is easier thus going to those who know you best and/or have known you the longest is the easiest form of financing.

The most difficult portion part of approaching friends and family for financing is the risk. You will typically only want to seek funding from friends and family who can afford it if they were

M. Dehn (✉)
Avanti Management & Consulting Limited, 600 Orwell Street, Unit 14, Mississauga, ON, Canada L5A 3R9
e-mail: michael@avantimac.com

V.I. Lakshmanan et al. (eds.), *Innovative Process Development in Metallurgical Industry*, DOI 10.1007/978-3-319-21599-0_15 to lose the investment. As most innovation or technology fails, the risk for investors is high. It is often mentally easier to accept capital from a professional investor who can usually manage risk through diversified investing versus friends and family, as professional investors fully understand the risks as their business is investing.

A reasonable friends and family financing would be to form the investment as a convertible debenture provided at a discount rate and an upper cap on valuation. The valuation is dependent on, and will vary according to, the amount of capital raised, and the ultimate value of the opportunity if commercialized.

15.2 University/State Research Facilities/State Research Grants and Funding

In Canada, a mechanism administered by the Natural Science and Engineers Research Council of Canada, or NSERC, offers a number of programs, grants, and fellowships that "depend upon partnerships among the academic, private and public sectors."

15.3 Corporate-Funded Private Research: Internal/External

A lot of technology is developed in house by large corporations to support existing and expected consumer and commercial markets, sometimes in collaboration with research labs. A good, and recent, example is the development of hydraulic fracturing (or fracking) which has come from entrepreneurs, government agency, academics, and corporations. Prior to the 2014 drop in oil prices, innovation and technology development has moved at a very rapid pace, and despite the drop in oil prices, the innovation and technology that was under development is even now coming to market.

The corporate funding of innovation and technology research ends up with the investors owning the intellectual property. The technology developer may be well compensated, but there is no ongoing compensation or royalty from the innovation and technology. Innovation and technology often builds upon previous research. Corporate-funded research often gets the fastest commercial results. Funds can be directed internally to a corporate research facility or externally to private or university research centers. Pittsburgh Plate Glass (PPG), for example, started its first research center in 1910, and now has three major research and development centers in Pittsburgh, Pennsylvania, with more internationally.

There are also some hybrid combinations of Corporate and University/State funding opportunities available. Examples of these include the Ontario Centres of Excellence (OCE) in Ontario, Canada, where government and industry partners work with academic institutions and/or individuals. The OCE's mandates are to "create significant economic impact for Ontario," to work with "high-potential early-stage entrepreneurs and innovative enterprises to help these emerging businesses grow to the point where they can attract private investment and ultimately become sustainable, global competitors," and "foster student entrepreneurship and support entrepreneurial activities by students and youth across Ontario."

Another example of these hybrid combinations is the organization MaRS, named after Medical and Related Sciences. MaRS works with private and public sector partners to help entrepreneurs launch and grow innovative companies. It is focused on working with Canadians, primarily in Ontario.

15.4 Public and Private Company Financings

15.4.1 Equity Financing

In an equity financing, capital is obtained by selling a partial interest or shares in the company to investors. The ownership position is the "equity" that investors receive for their investment. There is no direct obligation to repay the funds, but investors become partial owners of the business and thus are able to have a degree of control or say in the operations and direction of the innovation or technology, depending on the amount or percentage of ownership.

Equity investors accept more risk than debt financiers and expect to earn a higher return on their investment for the higher risk. Investors make money on equity investments by selling their shares at a higher price than they were originally purchased for. In some cases investors will make a return on their investment from a profitable innovation or technology, in the form of a dividend.

15.4.1.1 Sources of Equity Financing

For equity financing, possible sources may include friends and family, private investors, investment clubs, customers and suppliers, employees, investment banking firms, insurance companies, hedge funds, venture capital firms, closed-end investment companies, large corporations, government-backed investment vehicles (such as Investment Quebec) and angel investors (wealthy entrepreneurs/investors) who may work alone or as a group.

15.4.2 Debt Financing

Debt financings are difficult for nonrevenue generating companies for innovation and technology development at the start-up level, but once commitments to production or technology purchases have been signed, debt financing may be a viable alternative. Debt financings come in the form of loans. These loans are repaid over time with interest, although some government loans are interest free. Short-term loans are less than 12 months, long-term loans are considered for more than 12 months. The main sources of debt financings are banks and government agencies. But often friend and family, wealthy investors who prefer a guaranteed return on their investment will take debt versus equity.

Debt financing interest payments can be deducted from income earned, but in the early days of innovation and technology development there is generally no revenue and only expenses, so there is no tax advantage. However, debt financing leaves control, except in the case of a default on the debt, in the innovation and technology developers control once the debt has been paid as the lender does not receive any ownership in the operation.

Debt financings potentially are not a viable option for innovation and technology development as innovation and technologies often fail to reach commercialization in a "reasonable" timeline. This often leads to no income availability to make regular payments on loans. Economic downturns and interest rates, that are also typically higher due to higher risk to lenders, also affect most innovation and technology.

Having too much debt can also lead to future difficulty raising other forms of equity, as the debt on top of the commercialization risk of the innovation or technology, making the opportunity less attractive to potential investors.

15.4.3 Debentures

Debentures are debt instruments issued by a company to raise capital to pay for expenses. Debentures are typically transferable, just like equity, but come with no rights to influence the innovation or technology that a typical shareholder would have. These types of debt are unsecured loans, and the company has more flexibility than with traditional debt.

There are two types of debentures: convertible and nonconvertible.

15.4.3.1 Convertible Debenture

Convertible debentures pay a set or fixed income stream in the form of interest payments as well as, at or in some cases before maturity, the principal amount or conversion of the principal amount into equity at a specific price. The convertible debenture is often more attractive to the investor, having the option of taking back the principal amount in cash or in equity, but usually yields a lower interest rate than a nonconvertible debenture.

15.4.3.2 Nonconvertible Debentures

Nonconvertible debentures pay a set or fixed income stream in the form of interest payments at maturity, the principal amount.

Nonconvertible debentures do not convert into equity. Typically nonconvertible debentures have a higher interest rate payable when compared to a convertible debenture.

15.4.4 Bonds

Bonds are debt instruments that are issued to cover expenses and expansions. A bond is considered a loan taken by a borrower from an investor with no equity component. Unlike debentures, bonds are secured loans and can yield very low interest rates. Bonds are redeemed at a specified time determined at the issuance on the bond. Bonds may be used interchangeably with debentures as some debentures may be secured.

15.5 Strategic Partnerships (Joint Venture)

A strategic partnership is a collaborative agreement between two or more parties in which all parties work together for a common goal. The parties share the risks, responsibilities, resources, skills, and benefits. In the case of technology and innovation, a funding partner could collaborate with the innovation/technology developer, and jointly advance the project while sharing the risks and rewards.

15.6 Crowdfunding

Crowdfunding is the use of internet and web-based social networks as a sales platform to reach a pool of nontraditional investors and offers access to investments that traditionally may not have fit the risk profile of investment banks or hedge funds. Leveraging a crowdfunding network is quite often now the starting place for many entrepreneurs to raise capital for innovation and technology and is leading to more completed financing. There are three models to crowdfunding which will be described: Donation (Reward), Lending (Debt), and Investment (Equity).

15.6.1 Donation

Currently, the largest crowdfunding model is the donation model. In a donation-based model, capital is raised with no expectation of a return on investment beyond the goodwill to support a charitable cause or a reward (a mug, t-shirt, or other token of appreciation). The donation model currently is the fastest way to complete a crowdfunding.

15.6.2 Lending

The second largest crowdfunding model is the lending model. The lending model can be broken down into three main categories: Presales, traditional loan, and forgivable loan.

15.6.2.1 Presales

In presales, which is related to product-related crowdfunding, a finished product, assuming successful innovation and technology is commercialized, is offered in return for financing. In the case of a larger contribution, multiple copies of the product are typically delivered to the funder.

15.6.2.2 Traditional Loan

Crowdfunding using a traditional loan model is unique way to borrow small amount of capital from many sources. Standard loan terms are used and there is an expectation of regular interest payments and well as repayment of the principal. There is often a catch with crowdfunding via a traditional loan as loans are not always guaranteed.

15.6.2.3 Forgivable Loan

In crowdfunding a forgivable loan model, the capital is repaid to the lender in two scenarios: when the innovation and technology generate sales; or when the innovation and technology generates a profit. This is the most appealing model for entrepreneurs and the risk is minimized, but also a more difficult debt financing model as investors have no security if the innovation and technology fails.

15.6.3 Investment

The crowdfunding investment model is a new way to access investors on a small scale that traditionally would have been an investment bankers mandate to raise equity. The major differences between the traditional investment model and the crowdfunding investment model are the number of investors and securities regulations.

In an investment model for crowdfunding, capital is raised in exchange for equity in the innovation and technology. It is straightforward and allows for ownership of fractional shares of the innovation and technology. There are some hurdles with securities regulations including number of shareholders, and it appears that the securities regulators are dealing with these issues in a manner that will be positive for both the investors and the entrepreneur.

15.7 Streaming Financing

In a streaming financing, a "streaming company" pays an upfront fee used to advance development of a project in exchange for the right to purchase a commodity at a fixed price. While streaming royalties are very attractive to a streaming company, the innovation or technology company can get caught in a low price environment where their operation becomes uneconomic and the company is forced to close. This scenario also eliminates upside in a rising price environment to the innovation or technology company as their selling price is fixed to the streaming company. The streaming company provides a one-time upfront payment when the royalty acquisition is made, and then is not responsible for any required capital expenditures, operating costs, or sustaining costs.

15.8 Royalty Financing

In a royalty financing, a "royalty company" pays an upfront fee used to advance development of a project in exchange for the royalty of part or total production from a given innovation or technology. Royalties vary, but may include a SSR (Sliding Scale Royalty), NSR (Net Smelter Return), GSR (Gross Smelter Revenue), NPI (Net Profit Interest), CMR (Contained Metal Returned, which specifies individual metal or product or a group of products). These types of royalties can often mitigate risk, and with certain types of royalties can provide upside to both the innovation or technology company and the royalty company. The royalty company provides a one-time upfront payment when the royalty acquisition is made, and then is not responsible for any required capital expenditures, operating costs, or sustaining costs.

15.9 Licensing Fee Financing

In a licensing fee model, an end user pays a fee, typically up front and annual, to the innovation and technology company, for a defined period of time with renewal clauses for the right to use typically proven and commercial technology.

15.9.1 Types of Licensing Fee Agreements

A long-term license is usually more beneficial if the licensee is a funded small business with limited cash. The initial payment is usually relatively small, with subsequent royalty payments forming the greater part of the financial compensation, usually after the company earns the bulk of money from the technology.

In the case of a short-term license, the greater portion of the compensation is made via an initial payment. A company that prefers to pay for a license in cash instead of equity may find this short-term agreement useful.

15.9.2 Exclusive and Nonexclusive Licensing Fee Agreements

An exclusive licensing fee agreement allows for only one company to license the innovation.

In a nonexclusive licensing fee agreement, more than one company may purchase the right to use the licensed technology. Nonexclusive licensing fee agreements are often more palatable for a widely used innovation or technology (e.g., velcro or paint balls). Lower licensing fees and reduced royalty fees can reduce the cost of the product which may increase market appetite for the product. Improvements in the technology made by one licensee can also benefit the innovation and technology developer as well as the other licensees.

Part VIII

Case Study Examples

Innovative Case Study Processes in Extractive Metallurgy

16

V.I. Lakshmanan and Ram Ramachandran

The objective of this chapter is to give a brief review of various innovative processes that have been developed in the area of process metallurgy in the last 100 years.

At a symposium held by Society of Mining Engineers (SME) in 2013 and the Proceedings published in 2014 titled “Mineral Processing and Extractive Metallurgy—100 years of Innovation” (Anderson et al. 2014), Dr. Fathi Habashi reviewed the History of Innovations in Extractive Metallurgy (Habashi 2014). The proceedings also had chapters on innovation in various branches of Mineral Processing and Extractive Metallurgy. Innovations in Mineral Processing have been dealt in Chap. 2.

The current chapter summarizes the innovative processes developed in three main areas of extractive metallurgy, viz., pyrometallurgy, hydrometallurgy, and electrometallurgy. This concept is illustrated using the metallurgy of copper as an example. Some reference is made to the metallurgy of lead, zinc, and a few other metals. These three metals have been chosen for illustrating some innovations in extractive metallurgy as the authors have considerable experience in these areas. A summary of innovative processes for other metals has been listed toward the end of the chapter. Details of these processes are described in some detail in the Proceedings of the Symposium on “Mineral Processing and Extractive Metallurgy—100 years of Innovation” published by SME in 2014 (Anderson et al. 2014).

Additionally, innovative ideas have played a major role in: (a) process “intensification” and (b) solving environmental issues in process metallurgy. A few examples from the area of pyrometallurgy are included to illustrate these solutions at the end of this chapter.

16.1 Copper: (Partelpog 2014)

16.1.1 Pyrometallurgy

About 75–80 % of the copper produced in the world is by pyrometallurgical processes. The balance is produced by hydrometallurgical processes, viz., leaching of oxide and low-grade sulfide ores, solvent extraction, and electrowinning (EW).

About 90 %+ of the copper produced by pyro processes are done either by flash smelting and

V.I. Lakshmanan (✉)
Process Research Ortech Inc.,
2350 Sheridan Park Drive, Mississauga,
ON, Canada L5K 2T4
e-mail: llakshmanan@processortech.com

R. Ramachandran
Consulting Engineer, 9650 E. Peregrine Place,
Scottsdale, AZ 85262, USA
e-mail: ramvasanti@aol.com

V.I. Lakshmanan et al. (eds.), *Innovative Process Development in Metallurgical Industry*,
DOI 10.1007/978-3-319-21599-0_16

bath smelting. Out of this 90 %, about 49 % is produced by flash smelting followed by 40 % using bath smelting. Peirce-Smith converters continue to be the workhorse for converting copper matte with two plants using flash converting and two other plants using Mitsubishi converting.

In this chapter, only flash and bath smelting are described in some detail as most of the copper is produced essentially by these two smelting units.

The success of these two innovative ideas of copper smelting technology, viz., flash smelting and bath smelting can be attributed to: (Mackey 2013)

- Extensive testing of the idea at every level., viz., bench, pilot, demonstration scale; all these levels of testing followed by design, construction, and commissioning of the full-scale plant have been described in detail in Part II.
- Excellent process understanding, optimization, and control.
- Use of tonnage oxygen in the early 1970s.
- Advances in Furnace Integrity.

16.1.2 Flash Smelting

- Flash smelting for copper concentrates.
- Outotec Flash Smelting. Outotec process uses oxygen enriched air; involves a slag cleaning step.
- Outotec Direct to Blister.
- INCO Flash Smelting. INCO process used 100 % O_2; discard smelter slag.

The first Outotec flash smelting unit was started in 1949. As of 2013, there are about 59 Outotec flash smelting units in service. Of these, 45 units are copper flash smelters, five are copper converters, six are Ni smelting units, three are direct to blister copper, and one unit is for pyrite smelting. During its history, the process has undergone a considerable number of improvements such as: (a) use of tonnage oxygen (since 1971), (b) high matte grade, (c) improvements in furnace construction (described later), etc. Of these three improvements, use of oxygen made a great impact as shown below:

- Decreased Energy Consumption
- Decreased investment costs
- Additional operational advantages
- Decreased pollution and better working atmosphere
- Higher concentration of SO_2 in the off-gas (also lower volume) and its better capture

The Outotec flash smelting technology has been successfully used in three plants to produce blister copper from concentrates having a high Cu/Fe ratio.

The following three figures illustrate some aspects of the Outotec Flash Smelting units (Figs. 16.1, 16.2 and 16.3).

Normal oxygen enrichment used in flash smelting varies from 45 to 85 % depending on the desired matte grade. The INCO flash smelter—four of them in operation—uses 100 % oxygen; two of the smelters process Cu concentrates while the other two smelters at Sudbury, Canada treats Ni–Cu–Co–Fe–S concentrates to produce a Ni–Cu–Co matte. INCO flash smelter produces a discard slag with low levels of Cu; the Outotec flash smelter slag contains high levels of copper; it is normally treated in an electric furnace or slow cooled and milled for additional copper recovery.

For reasons unknown, the INCO flash smelting technology was never marketed aggressively and hence explains their small presence in the smelting world in comparison to Outotec flash smelting technology.

16.1.3 Bath Smelting for Copper Concentrates

- Top submerged lance (TSL) Xstrata's ISASMELT process for smelting for Cu and Pb concentrates.
- Outotec/AUSMELT process for smelting copper and lead concentrates.
- Noranda's Reactor for smelting copper concentrates.
- Mitsubishi smelter for smelting copper concentrates; involved an intermediate slag cleaning step (Slag/Matte separation) that made the process somewhat continuous.
- El Teniente Converter technology for smelting copper concentrates.

Fig. 16.1 A view of the Outotec flash smelting unit showing feed system, burner, reaction shaft, settler, and uptake. Reproduced with permission from Outotec Inc

Fig. 16.2 Simplified rendering of the flash smelting unit. Reproduced with permission from Outotec Inc

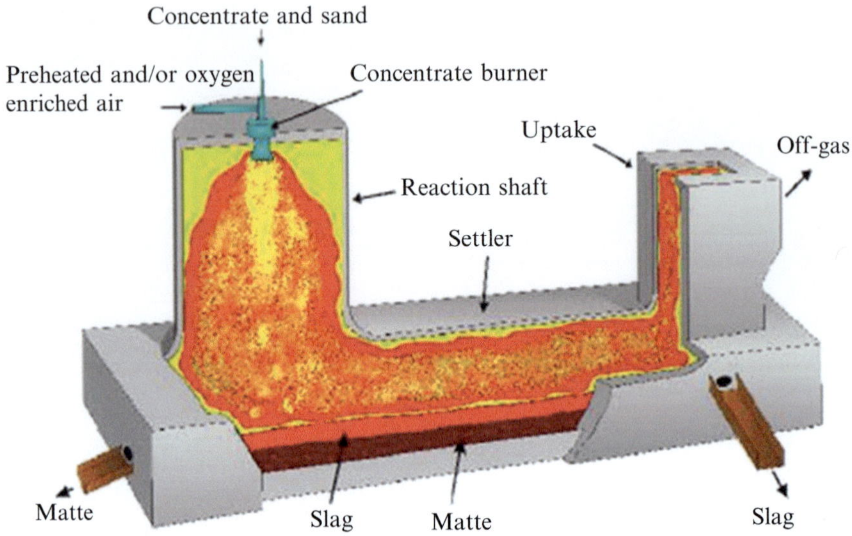

Fig. 16.3 A cross section of the Outotec flash smelting unit. Reproduced with permission from Outotec Inc

- SKS (Shuikoushan) smelting copper and lead concentrates—developed by SKS Mining Bureau, China.

Since the advent of Xstrata's ISASMELT at Mt. Isa in the 1980s, about 2.5–3.0 million tons of copper are being produced today by this technology. This innovative technology seems to handle complex copper concentrates with impurity elements somewhat better than flash smelting. As an example, a cross section of the ISASMELT is shown in Fig. 16.4.

However, one new emerging bath smelting technology, viz., Shuikoushan (SKS)—for treating both copper and lead concentrates deserves a mention here. Currently, there are a number of SKS lead smelters in China and one in India. Application of the SKS process technology for copper smelting is slowly on the rise in China. The SKS copper smelter in Dongying, China employs highly oxygenated tuyeres, similar to Savard-Lee type tuyeres—similar to Lurgi's QSL direct lead smelting process. With about 70 % oxygen enrichment, a matte grade with 70 % Cu is produced for conversion in a PS converter. As of today, the maximum capacity of a SKS copper smelter is about 150,000 t/year. Technical data on one of the important features of the technology, viz., shrouded oxygen tuyere are not available in sufficient detail to make a complete evaluation of the process. Also, environmental data, viz., (a) level of gaseous and particulate emissions, (b) effluent discharge levels, (c) sulfur fixation, etc., are not available for an informed technical evaluation. Its success is dependent on: (a) producing very high-grade mattes and (b) high tonnage copper at least 250,000–300,000 tpy in one single smelting unit.

The history of the development of the flash smelting technology at Outotec is well documented in a book titled "A Flash of Knowledge" by Tuomo Sarkikoski and published by Outotec company (Sarkikoski 1999).

16.2 Converting of Copper Matte

The modern version of the Peirce-Smith converter has been the dominant technology and workhorse for copper matte converting. The great flexibility and simplicity of the Peirce-Smith converter, its ability to handle scrap copper (both old and new), modern design for converter hoods and

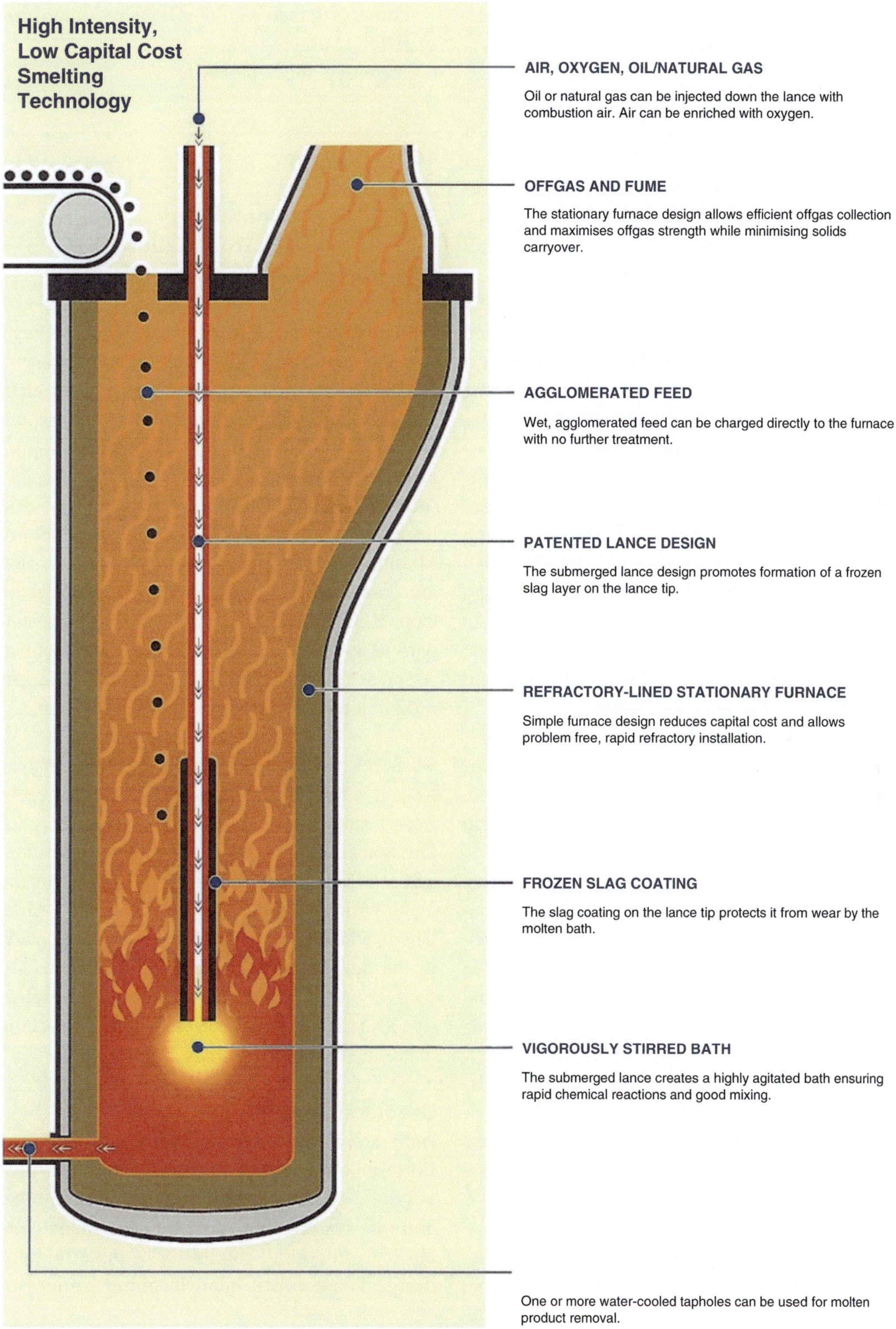

Fig. 16.4 Cross section of ISASMELT bath smelting furnace for copper and lead production. Reproduced from Xstrata Technology Brochure, with permission from Glencore, Australia

cost still make a compelling case for its use in copper matte converting.

The modern trend toward mega-plants, viz., plants with potential capacity of 400,000 t/year and above may edge future plants toward flash converting which would need fewer vessels and lower off-gas volume. The newest "double flash smelter" in China—designed to produce 400,000 t/year of copper—has one flash smelter and one flash converter (Mackey 2014).

Oxygen enrichment is used in converters (maximum 29 % O_2) to (a) convert high-grade mattes, (b) increase converting rate, and (c) increase melting of solids such as flux, reverts, and scrap. Above 29 % oxygen enrichment, refractory erosion becomes excessive. Pierce-Smith converter technology will continue to dominate the matte converting scene; however, there may be inroads by the new continuous converting technologies—particularly for large-scale plants.

The development of continuous flash converting technology—developed and adopted at the Kennecott copper smelter in Utah, USA—has made a breakthrough in copper smelting—specifically in reducing emissions from the smelter.

Typically, the flash converter treats dry, crushed granulated matte using oxygen enrichment up to 80 % oxygen. Molten matte of size 0.5 mm granules are crushed to 50 μ and dried prior to charging to the flash converter.

In the Proceedings referred to at the beginning of this subchapter, Partelpog (2014) has summarized all other innovations related to copper pyrometallurgy that has resulted in considerable improvements in smelting operations and increased metal production. They are listed below:

1. Flash Smelter burner system to improve distribution of process air
2. Concentrate feed system to the flash furnace
3. Furnace cooling with external water cooling jackets
4. Improved gas cooling technology
5. Anode Refining Innovations: (a) porous plug technology, (b) increased use of oxy-fuel burners (Praxair's Dilute Oxygen Combustion (DOC) burner)
6. Increased anode mold life

16.3 Copper

16.3.1 Hydrometallurgy and Electrometallurgy: (Hiskey 2014)

A major innovation in copper hydrometallurgy was the use of solvent extraction technology to produce concentrated and purified copper electrolyte that could be electrowon to make pure copper cathodes equivalent to refined copper cathodes. The process was piloted at Bagdad (now Freeport McMoran, Bagdad) in 1965 and later commercialized at Rancher's Bluebird, Miami. Some details of the copper SX process are described in Chaps. 2 and 10. Since then, Copper SX process has become the industry standard in hydrometallurgical production of copper. At present, about 20–25 % of primary copper production is produced by the SX-EW process.

16.3.1.1 Heap and Stockpile Leaching

Recent advances in heap and stockpile leaching—a step prior to the SX step—have been: (a) Ore Stacking, (b) Acid Cure and Agglomeration, and (c) Drip Irrigation. Since the advent of the alpha-hydroxy-oxime by General Mills Chemicals, considerable progress has been made in the development of various SX reagents for recovery of copper. In recent times, more selective SX reagents have been developed for specific operating regimes.

Ore Stacking

Innovative idea of Modern Automated Super Portable technology featuring all-track propulsion, onboard generators, articulated head and tail systems has increased the maneuverability and overall flexibility of material handling. The Safford heap leaching project of

Freeport-McMoran has employed this system in their operations. In addition, this innovation has improved the permeability of the heap pad.

Acid Cure and Agglomeration

Bouffard has done an excellent review of the fundamental and practical considerations regarding agglomeration for both precious metals and copper ores (Bouffard 2005). He summarized the process of agglomeration as "Agglomeration" was a technical and economic breakthrough (innovation) technology for heap leaching of clayey ore and ore containing high fines. Researchers headed by Heinen H.J. and others at U.S. Bureau of Mines also played a major role in this development. McClelland at USBM was responsible for transferring this innovative technology to plant operations. In summary, this innovation has greatly helped heap leaching of copper ores in the following ways:

- Improved heap physical structure
- Improved leach chemistry
- Improved environmental impact

Drip Irrigation

Among the various methods of solution application to any heap and stockpile of ore, the innovative use of drip irrigation—an idea borrowed by agricultural irrigation—plays a major role in copper ore leaching. Drip systems have helped optimize metal recovery. In addition, the following advantages have made it an industry standard for leaching ores. They are:

- Operates well in extremely dry climates where evaporation rates are very high
- No drifting of spray mist
- Less ice-build up and fewer freeze-up during cold weather
- Better control at very low application rates
- Better working conditions for the employees

In summary, a number of innovations—developed over a long period by the metallurgical industry—have played a major role in the hydrometallurgical treatment of low-grade copper ores, viz., oxidic ores and low-grade sulfidic ores. This accounts for about 20–25 % of the world copper production today.

16.3.2 Copper Electrowinning and Electrorefining

Productivity improvements in copper EW—mainly due to use of higher current density—have occurred due to the use of:

1. Stainless Steel Permanent Cathode Technology—developed by both Kidd and Mt. Isa (both of Xstrata—now Glencore).
2. Air Sparging under the cathode in an EW cell.
3. Use of fiber reinforced plastic electrode frames for improved alignment of cathodes.
4. Use of PGM-coated Ti mesh anodes for energy savings by reducing the anode overvoltage.

Of the four innovations listed above, the stainless steel permanent cathode technology has become the "industry standard" for both EW and ER copper tank houses. In addition, the joint efforts of Xstrata and Falconbridge resulted in the development of a cathode stripping machine (CSM) that has a high degree of automation and advanced robotics.

16.3.3 Hydrometallurgy of Copper Concentrates

A considerable number of hydrometallurgical processes have been developed and tested for treating copper concentrates. They have been reviewed extensively in a paper by Ramachandran et al. (2007). However, none of the processes have evolved as a competitor to pyrometallurgical option of smelting copper concentrates. The reasons for this limitation are:

1. Scale of operation.
2. Relative costs of production of cathode copper through the smelting/refining option vs. leaching/SX-EW option.

3. Various operational issues that have not been resolved to date.

Among the various lixiviants that were used for leaching copper concentrates, chloride assisted, sulfuric acid, and oxygen/sulfate systems have been reasonably successful. In spite of its limited success, hydrometallurgy of copper concentrates will contribute only a small fraction of copper to the overall copper production. None of the processes to date have tested dirty copper concentrates or concentrates containing Se, Te, and PGM metals. Further, treatment of process effluents from such hydrometallurgical processes has not been fully addressed. All these issues are a ripe area to come up with innovative ideas to establish the hydrometallurgical treatment of copper concentrates—for small-scale production—as a viable option in locations far away from urban areas.

16.4 Process Metallurgy of Lead: Pyro, Hydro, and Electrometallurgy

16.4.1 Pyrometallurgy

Details of new smelting processes for production of lead have been described in detail in a very good review in Proceedings of Lead and Zinc'05 (Stephens 2005a). They are:

(a) QSL (Queneau–Schumann–Lurgi) process
(b) Kivcet process
(c) ISASMELT process
(d) AUSMELT process
(e) Shuikoushan (SKS) process

Four processes from the above list are bath smelting units; KIVCET is a "flash smelting" unit. Unlike copper, lead pyrometallurgy is not easily amenable to "flash smelting" as the vapor pressure of PbS and PbO is quite high at smelting temperatures thereby resulting in considerable loss of lead in the gas phase. Also, none of the above processes—except KIVCET process—produce slags with lead levels low enough—in comparison to the lead blast furnace slag—for discard. Invariably, the high PbO slags are reduced to recover some additional Pb prior to slag fuming for zinc recovery.

A typical cross section of the KIVCET lead smelting unit is shown below (Fig. 16.5):

The latest innovation in the production of lead is the Shuikoushan (SKS) process, developed by Shuikoushan Mining Bureau in China. The process uses a horizontal vessel resembling a Noranda Reactor with oxygen injection tuyeres (air shrouded) like QSL. Lead bullion is produced in the SKS vessel and the PbO slag is reduced separately in a blast furnace. A considerable number of SKS plants are in operation in China and one in India. Little published information is available on this process—especially on the details of shrouded oxygen tuyere.

16.4.2 Hydrometallurgy

Lead production by hydrometallurgical processes has never been a success to achieve commercial scale operations. Examples are the Dextec chloride process, USBM chloride process, and the recently developed Engitec™ process using ferric fluoborate as a leachant. No new hydrometallurgical process for lead production is on the radar screen—in spite of the environmental pressure on the established pyrometallurgical processes.

Any innovative process to produce lead by a hydrometallurgical process will go a long way in possibly eliminating the environmental issues related to the pyrometallurgical process of producing and refining lead.

16.4.3 Electrometallurgy

Electrorefining (ER) of impure lead—specifically containing large amounts of bismuth—was developed and patented by Anson Betts in 1901. The Betts Electrolytic Process (BEP) stands as a testament for electrometallurgy of Pb. Teck Resources, Trail, Canada pioneered the commer-

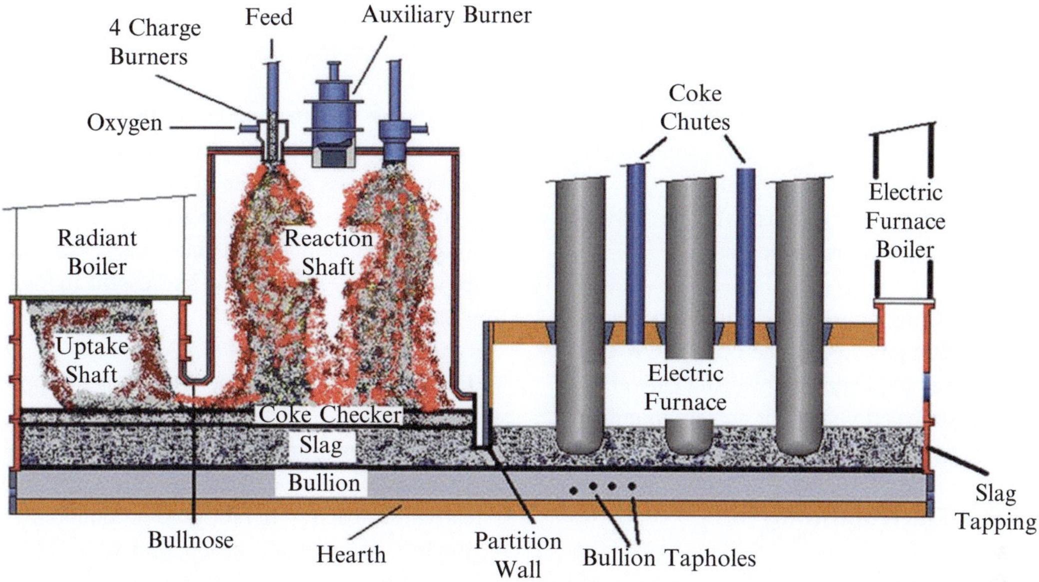

Fig. 16.5 KIVCET lead smelting unit. Reproduced from Pb/Zn Short Course, Vancouver, B.C., October 3, 2010, Modern Lead Smelting, Courtesy: Dr. Rob Stephens, Teck Metals Ltd, British Columbia

cial application of the BEP process with continuous innovation focused on productivity improvement and quality of refined lead. The details of the BEP process are described in some detail in a review article on Innovations in Zinc and Lead Hydrometallurgy (Robinson and Anderson 2014).

16.5 Process Metallurgy of Zinc: Pyro, Hydro, and Electrometallurgy: (Robinson and Anderson 2014)

16.5.1 Pyrometallurgy

The production of zinc by pyrometallurgical processes is limited to the use of Imperial Smelting Furnace (ISF) process wherein zinc–lead concentrates are treated to produce both lead and zinc. However, zinc so produced is not as pure as Special High Grade (SHG) zinc and has limited industrial use. The ISF furnace is fraught with considerable amount of environmental issues related to acceptable work place atmosphere. There are only about 6–8 ISF plants in operation today in the world.

16.5.2 Hydrometallurgy

Two major innovative processes in the hydrometallurgy of zinc have been:

1. The jarosite process for iron removal and increased zinc recovery.
2. Sherritt's pressure leaching of zinc concentrates as an alternate and/or add-on to roast-leach process.

The jarosite process is well documented in literature and is widely used by many zinc refineries. Similar processes for iron control and improved zinc recovery were the Goethite process (used by Vieille Montagne) and Hematite process (used by Dowa Mining and Ruhr Zink).

Tecnicas Reunidas of Madrid, Spain pioneered the solvent extraction of zinc with the ZINCEX process. The process is well suited for raw materials such as zinc silicate ores and secondary zinc bearing materials such as electric arc furnace dust containing halides.

Any future innovations will be driven by energy savings: Some examples are:

1. New amorphous-coated titanium anodes—recently pioneered in the copper EW area by Freeport and Outotec/Doshisha University.
2. Real-time cell voltage or current monitoring.

16.5.3 Electrorefining

In the electrowinning of zinc from its sulfate electrolyte, the overall cell voltage is around 3.2–3.5 V. Out of this about ~1.0 V is for the anode reaction for the evolution of oxygen. Any innovation wherein the anode voltage can be reduced considerably—either by using a new anode material or a change in the anode reaction by the addition of some reagent will go a long way in reducing the cell voltage and the energy used for the electrowinning of zinc. Of course, the change also has to be economically feasible.

16.6 Innovations in the Process Metallurgy of Other Metals

Innovations in the extractive metallurgy of various other metals are described in some detail in the Proceedings of the Symposium on "Mineral Processing and Extractive Metallurgy—100 years of Innovation" published by SME in 2014 (Anderson et al. 2014).

The other metals covered in this Symposium Proceedings are:

1. Gold and Silver Extraction and Recovery.
2. Nickel and Cobalt—hydrometallurgy.
3. Bayer's process for Alumina production—still the workhouse with various improvements—since its discovery in 1888.
4. Rare Earth—Hydro and pyrometallurgy.
5. Biohydrometallurgy—described in Chap. 6.
6. Uranium process metallurgy.

A review of the metallurgical processes developed in the last 100 years shows that considerable progress has been in the area of extractive metallurgy. Some processes have been "evolutionary" and others have been "revolutionary." Some examples of successful innovations are:

1. Flash and bath smelting for copper concentrates.
2. Solvent extraction for low-grade copper ores.
3. Cyanidation and carbon-in-pulp (CIP) and carbon-in-leach (CIL) for gold production.
4. Bayer's process for production of smelter-grade alumina (the workhorse of aluminum industry since 1888).
5. Hall–Heroult electrolytic process for production of aluminum (the work horse of aluminum industry since 1889).
6. Pressure hydrometallurgy for zinc concentrates, copper concentrates, refractory gold ores, and nickel laterite.

Hydrometallurgical processes have met the challenge from pyrometallurgy in the production of certain metals. The reasons for this were (a) to reduce energy costs and (b) meet regulations on waste emissions from the gas streams.

16.7 Chloride-Based Chemistry

Chloride-based chemistry has been successful in production of titanium and magnesium and has been well documented in literature.

An innovative disruptive technology has been developed by Process Research Ortech Inc. (PRO) for the production of pigment-grade titanium dioxide from low-grade ilmenite ores. The flow sheet includes mixed chloride leaching using $HCl–MgCl_2$ based lixiviant, selective removal of Fe by solvent extraction, thermal precipitation of TiO_2, and reagent recycle. The process along with flow sheet is described in detail in Chap. 18.

Historically, Ni–Co matte from Falconbridge (now Glencore) was leached with chlorine in their Norway plant; Ni and Co were separated using solvent extraction and the respective metals recovered by electrowinning. Sumitomo Metal Mining Co., Ltd (SMMC) treats mixed Ni–Co sulfide precipitate from laterite ores using the Matte Chlorine Leach-Electrowinning Process at their Niihama, Japan plant. Also, the Ni–Co sulfide precipitate produced at Coral Bay Nickel Corporation, Philippines is being processed to nickel and cobalt at the Niihama by the chlorine leach process.

16.8 Ammonia Leaching

Ammonia leaching for treating Ni–Co concentrates was developed by Sherritt. However, these lixiviants have not been competitive for production of "tonnage" copper or lead.

Demonstration plants—built to evaluate chloride-based chemistry.

CYMET's chloride leaching and Aribiter's ammonia-based leaching for the production of copper were not successful on a commercial basis due to operational and economic issues.

16.9 Process "Intensification" in Process Metallurgy

"Intensification" means (a) increased production, (b) lower costs, and (c) environmental improvements using the "existing" equipment with some changes in the process steps. It could also imply the replacement of multiple units with one properly sized big unit and/or conversion of a batch or multistage process into a continuous process.

Some examples of common threads among process intensification are:

- Use of tonnage oxygen
- Improved process understanding and control
- Advances in furnace integrity

The use of oxygen enrichment to increase production is well known and not described here.

Improved process understanding and control consists of:

1. Proper management of knowledge
2. Use of process visualization tools such as: (a) mass and energy balance, (b) FACTSage-METSIM Interface (both trade mark names), (c) molten phase properties, and (d) CFD and FEA modeling. Process control involves online feed analysis and use of sensors.

Advances in furnace integrity have played a key role in process intensification. They can be summarized as:

- Refractory improvements such as materials, installation, and operation.
- Cooling blocks—material selection and casting quality
- Improvements in tap holes thereby increasing furnace campaign lives
- Process telemetry and understanding and its use in predicting performance of cooling jackets, tap holes, etc.

Details of process intensification have been discussed in some depth in his paper titled: "Process Intensification in Pyrometallurgy" (Stephens 2005b).

16.10 Innovative Ideas for Solution of Environmental Issues

Environmental awareness has been one of the most dramatic drivers in developing innovative ideas to minimize pollution of atmosphere and water. Some major advances are listed below:

1. Sulfur dioxide
    - Emissions of SO_2 have decreased considerably since the 1980s
    - Sulfur capture has increased from 90 to 98 %
    - Highest sulfur capture—99.9 % (equivalent to 3.5 kg/t of copper)
    - Advances in acid plant technology now produce tail gas with <100 ppm SO_2 (standard is 650 ppm)

- Acid plant tail gas scrubbing reduces SO_2 emissions to 40–60 ppm
2. Heavy Metals
   - Mercury and selenium removed from gas streams
   - Other impurities bled through dusts; dusts treated by hydrometallurgical processes to fix impurities in a stable form
3. Particulates
   - Wet Electrostatic Precipitator

The cleaning up of air always results in water pollution—not to mention bleed streams from hydrometallurgical process plants. This has resulted in development of innovative water treatment schemes to remove undesirable impurities. The treated water is either recycled back into the plant or discharged under an environmental permit. Considerable work has been done to develop "zero water discharge" treatment plants to conserve water—a valuable commodity in the mining and metallurgical industry today.

16.11 Summary

In summary, innovations in the extractive metallurgy industry have had a pretty good run for the last 100 years and are reaching a plateau in the development of new and innovative processes. It is believed that future innovations are expected to be more "evolutionary" rather than "revolutionary." For example, the development of Ni pig iron in China using old iron blast furnaces and laterite as feed was mainly due to the rising cost of nickel for the production of stainless steel.

In the last 100 years, the corporate Research and Development (R&D) has been the prime mover for generating new ideas. The active cooperation and alliance between engineers and corporate management was an important factor in developing an innovation portfolio. However, execution of innovative ideas for a process is a very different problem. Successful implementation needs considerable financial support and cheer leading—in the form of a champion—from the corporate management.

The demise of dynamic research organizations has been a major setback in the development of innovative processes and a cause for concern. It is a wake-up call to the leaders of all the metallurgical industries to channelize efforts and resources to build and/or rebuild an ongoing, vibrant R&D department to generate new innovative processes for improving the company's bottom line and for the benefit of society. The project should be titled: "Technology for Survival."

References

Anderson, C. G., Dunne, R. C., & Uhrie, J. L. (Eds.). (2014). *Mineral processing and extractive metallurgy—100 years of innovation*. Englewood, CO: Society of Mining, Metallurgy and Exploration (SME).

Bouffard, S. C. (2005). Review of agglomeration practice and fundamentals in heap leaching. *Mineral Processing and Extractive Metallurgy Review, 26*, 233–294.

Habashi, F. (2014). History of innovations in extractive metallurgy. In C. G. Anderson, R. C. Dunne, & J. L. Uhrie (Eds.), *Mineral processing and extractive metallurgy—100 years of innovation* (pp. 29–47). Englewood, CO: SME.

Hiskey, J. B. (2014). Innovative strategies for copper hydrometallurgy. In C. G. Anderson, R. C. Dunne, & J. L. Uhrie (Eds.), *Mineral processing and extractive metallurgy—100 years of innovation* (pp. 347–360). Englewood, CO: SME.

Mackey, P. J. (2013). *Copper smelting technologies in 2013 and beyond.* Plenary paper presented at Copper 2013, Santiago, Chile, December 2013. Santiago, Chile: with permission from Instituto de Ingenieros de Minas de Chile.

Mackey, P. J. (2014, February). Evolution of the large copper smelter-1800s to 2013. In P. J. Mackey, E. J. Grimsey, R. T. Jones, G. A. Brooks (Eds.), *Celebrating the Mega-Scale: Proceedings of the Extraction and Processing Division Symposium on Pyrometallurgy in Honor of David. G. C. Robertson* (pp. 17–37). Warrendale, PA: The Metals and Materials Society. Copyright 2014, quoted with permission from The Minerals, Metals and Materials Society.

Partelpog, E. H. (2014). Innovations in copper smelting. In C. G. Anderson, R. C. Dunne, & J. L. Uhrie (Eds.), *Mineral processing and extractive metallurgy—100 years of innovation* (pp. 515–534). Englewood, CO: SME.

Ramachandran, V., Lakshmanan, V. I., Kondos, P. D. (2007). Hydrometallurgy of copper sulfide concentrates: An update from Copper 2007. In: P.A. Riveros, D. G. Dixon, D. B. Dreisinger, & M. J. Collins (Eds.),

Proceedings of Sixth International Copper Conference, Toronto, Canada: Vol. 4. The John. E. Dutrizac International Symposium on Copper Hydrometallurgy, Book 1 (pp. 101–128).

Robinson, T., & Anderson, C. G. (2014). Innovations in zinc and lead hydrometallurgy. In C. G. Anderson, R. C. Dunne, & J. L. Uhrie (Eds.), *Mineral processing and extractive metallurgy—100 years of innovation* (pp. 389–401). Englewood, CO: SME.

Sarkikoski, T. (1999). *A Flash of knowledge: How an Outokumpu innovation became a culture* (p. 304). Espoo, Finland: Outokumpu Oyj.

Stephens, R. (2005). Advances in primary lead smelting. In T. Fujisawa, J. E. Dutrizac, A. Fuwa, N. L. Piret, & A. H. J. Siegmund (Eds.), *Lead and Zinc'05: Proceedings of the International Symposium on Lead and Zinc Processing*. Tokyo, Japan: The Mining and Materials Processing Institute of Japan.

Stephens, R. (2005, July 3–6). Process intensification in pyrometallurgy. In M. Nilmani, & W. J. Rankin (Eds.), *John Floyd International Symposium on Sustainable Developments in Metal Processing* (pp. 115–124). Melbourne, VIC, Australia: AusIMM.

Development of a New Technology for Converting Iron-Bearing Materials to Nodular Reduced Iron for Use in Various Steelmaking Operations

17

Donald R. Fosnacht, Iwao Iwasaki, Richard F. Kiesel, David J. Englund, Rodney L. Bleifuss, M.E. Mlinar, and David W. Hendrickson

17.1 Introduction

The current trend in the steel industry is an increase in iron and steel produced in electric arc furnaces (EAF) and a gradual decline in conventional ironmaking from blast furnaces in the USA which employ iron ore pellets made from Minnesota taconite iron ore. Currently, iron ores from Canada, Minnesota, and Michigan are pelletized and shipped to the lower Great Lakes ports as blast furnace feed. The existing transportation system and infrastructure is configured to handling these bulk materials. In order to expand the opportunities for the existing iron ore mines beyond their blast furnace customer base, a new material is needed to satisfy the market demands of the emerging steel industry while utilizing the existing infrastructure and materials handling capabilities.

Several processes have been proposed as alternatives to blast furnace ironmaking and significant activity on a worldwide basis continues in developing these alternative ironmaking processes. Mesabi Nugget LLC recently commercialized Kobe Steel's ITmk3 (Kobayashi et al. 2001; Hoffman and Tsuge 2004; US Department of Energy et al. 2004) process by producing "iron nuggets" in a 500,000 mt/year plant in Hoyt Lakes, Minnesota. A pilot plant campaign has also been demonstrated outside of Tokyo, Japan using JFE Steel's Hi-QIP process (Sawa et al. 2000; International Report Japan 2001; Ishiwata et al. 2004). As with any new process or technology, much opportunity exists for further cost reduction and continued quality improvement to create added incentive for commercial development.

The University of Minnesota Duluth—Natural Resources Research Institute has been developing an advanced ironmaking technology in conjunction with one of the leading US EAF steel producers. The development, first funded by the University and the US Economic Development Administration and subsequently by the US DOE, has been summarized in a technical report entitled "Next Generation Metallic Iron Nodule Technology in Electric Arc Steelmaking—Phase II," DE-FG36-05G015185, US Department of Energy (Fosnacht et al. 2010). The process, developed by NRRI, considered for commercial development uses a specialized linear furnace

D.R. Fosnacht (✉) • I. Iwasaki • R.F. Kiesel
D.J. Englund • R.L. Bleifuss • M.E. Mlinar
D.W. Hendrickson
Natural Resources Research Institute,
University of Minnesota Duluth, Duluth, MN, USA
e-mail: dfosnach@nrri.umn.edu

V.I. Lakshmanan et al. (eds.), *Innovative Process Development in Metallurgical Industry*,
DOI 10.1007/978-3-319-21599-0_17

design (LHF) with oxy-fuel combustion that affords several process and environmental advantages. Further expansion of the capabilities of the oxygen-fuel combustion to utilize pulverized solid fuels for ironmaking technologies, including carbon char produced from woody biomass or agricultural products, affords unique opportunities with respect to carbon neutrality and reduced greenhouse gas (GHG) emissions. The LHF to NRI process uses fine iron oxide concentrates rather than fired pellets as required in the most prevalent gas-based, shaft DRI (direct reduced iron) systems in use today.

The work reported in this chapter focuses on demonstrating the application of various processing parameters including: oxygen-fuel combustion technology, various atmosphere control methods and mix designs, and the use of alternative reductants produced from biomass while employing the best technology and processing conditions for converting iron oxide resources to high-quality **N**odular **R**educed **I**ron (NRI). The resulting product will: (1) contain less gangue, (2) contain less sulfur, (3) be more resistant to reoxidation, and (4) use the existing transportation infrastructure and material handling systems. High-quality NRI potentially can be universally acceptable feedstock across the steel industry, electric arc furnace (EAF), submerged arc furnace (SAF), basic oxygen furnace (BOF), iron foundries, or as supplementary iron units to the blast furnace (BF).

17.2 Background

17.2.1 The Scientific and Technical Merit of the Technology

Currently, most ironmaking processes require agglomeration with induration and sintering of iron-bearing materials prior to processing into an alternative iron product, especially if the iron-bearing material is a very fine material. The iron ore materials from the United States and Canada fall into this category of iron-bearing material. The process of converting iron ore or iron-bearing waste oxides to metallic iron containing between 2 and 4 % carbon has been developed by others over the course of the last decade. The technology differs from that of Midrex/Kobe or JFE in that a linear hearth furnace concept is considered instead of the rotary hearth technology of other developers.

For this application, oxygen-fuel burners offer many advantages over air-fuel burners. They are inherently more stable throughout a wide range of operating conditions and excess oxygen. They provide good turndown performance. They can be designed and operated to produce either compact, high velocity, low luminosity flames, or, long, highly luminous, low velocity flames. Oxygen-fuel burners can produce a wide range of oxidizing or reducing products of combustion streams.

Oxygen-enriched combustion can:

- *Increase efficiency*. Flue gas heat losses are reduced because the flue gas mass decreases as it exits the furnace. There is less nitrogen to carry heat from the furnace.
- *Decreased emissions*. Certain burners and oxy-fuel fired systems can achieve lower levels of nitrogen oxide, carbon monoxide, and hydrocarbons.
- *Improve temperature stability and heat transfer*. Increasing the oxygen content allows more stable combustion and higher combustion temperatures that can lead to better heat transfer.
- *Increase productivity*. When a furnace has oxygen-enriched burners installed, throughput can be increased for the same fuel input because of higher flame temperature, increased radiative heat transfer to the load, and reduced flue gas.
- *Localized atmosphere control*. The velocity of the products of combustion are reduced by an order of magnitude by eliminating the nitrogen content and subsequent gas velocity associated with typical air-fuel combustion systems.

One distinct advantage of this technology is that it can utilize solid fuel rather than natural gas, where the cost and/or the effect of the combustion products on the furnace gas atmosphere are problematic. The technology development has focused on providing the necessary heating requirements to the endothermic process through employment of oxygen-fuel and oxygen-coal combustion. In furnaces governed by convective heat transfer, the furnace gas velocity might drop because the convective heat transfer coefficient might decrease in a larger proportion than the increase in gas temperature. If this happens, the conversion would do little to increase the overall heat transfer, so reducing flue gas temperature to pre-conversion level may not be possible.

The test program was initiated using a laboratory tube furnace with subsequent testing in a slightly larger box furnace. Over 6000 laboratory, pilot and demonstration scale tests have been conducted over the course of the development, which was initiated in 2001. Major parameters investigated included such raw materials as:

1. Taconite iron ore concentrates with different levels of silica content,
2. Various carbonaceous reductants including Eastern anthracite, low-, medium-, and high-volatile bituminous and Western subbituminous coals as well as their carbonized char and coke,
3. Various additives, such as balling binders and some specific additives for slag fusion temperature reduction and iron nodule sulfur control.

Tests were conducted in three furnaces of increasing size including: a laboratory tube furnace of conventional design, a specially designed, electrically heated, batch, two-zone box furnace, and a larger gas-fired, semi-continuous, pusher-type, linear hearth furnace with three heated zones and a cooler. Furnace operating conditions included temperature, residence time, furnace atmosphere, hearth layer materials, iron nodule, and slag chemistries as well as iron nodule size. A major difference between the test conditions of the three-zone LHF and a two-zone, electrically heated box furnace is the high CO_2, low CO concentrations, and the high turbulence of the burner combustion products. From computational fluid dynamics (CFD) modeling of the LHF, the furnace gas was noted to be circulating vigorously within each zone, and while the temperature at the surfaces of trays in Zone 3 was relatively uniform 1427 °C (2600 °F), furnace gas velocities approached 1–3 m/s (3–10 ft/s) in localized regions at tray level. In the box furnace, on the other hand, the furnace gas velocities were estimated at as much as two to three orders of magnitude less, 0.03–0.003 m/s (0.1–0.01 ft/s). Furnace atmosphere profoundly influenced the temperature and residence time needed to form fully fused iron nodules. Increasing concentrations of CO_2 resulted in reoxidation of reduced iron and FeO-rich slags, creating high sulfur content. Higher temperatures improved the fusion behaviors of iron nodules, which became less sensitive to the presence of CO_2 above 1400 °C. In laboratory tests, fully fused iron nuggets could be formed at as low as 1325 °C (2417 °F) under a N_2-CO atmosphere, and sulfur in iron nuggets could be decreased to less than 0.01 %. Thus, from both a product quality standpoint and from an operating standpoint, furnace atmosphere control becomes a key control variable and must be considered in design of the overall furnace operating conditions.

The choice and amount of addition of carbonaceous reductants was found to be an important factor in iron nodule formation. While anthracite, low- and medium-volatile bituminous coal as well as coke worked well both in dry balled feed and a feed without prior agglomeration, subbituminous coal was totally unsatisfactory in balled mixtures, and its char generated inordinately large amounts of micro-nodules under similar conditions. The optimum level of carbonaceous reductants was determined to be 75–85 % of the stoichiometric requirement for metallization for the reaction mixture, based on fixed carbon analyses, when the furnace atmosphere consisted of N_2-CO mixtures.

Certain additives were found to be effective for lowering the fusion temperature of NRI, while some other additives lowered sulfur in iron nodules to as low as less than 0.01 %. The generation of micro-nodules, which need to be recycled, had been a drawback to the newly developed approach, but proprietary methods to control the generation of micro-nodules have been developed (see Appendix).

17.2.2 Innovation, Originality, and Feasibility of the Technology

17.2.2.1 Linear Hearth Furnace Design Considerations

The alternative processes previously described utilize a Rotary Hearth Furnace (RHF) design, where the NRI process makes use of a Linear Hearth Furnace (LHF) that affords some unique opportunities:

1. The LHF creates an exit end which provides opportunity to recuperate sensible heat in a heat recuperation zone that cannot be done with an RHF. Combustion gas exhaust temperature in the LHF will be lower (inherent "self" recuperation) than in the RHF, since the exhaust gas duct will be located away from the hot fusion zone.
2. The LHF has the advantage of the charging and discharging being accomplished outside of the kiln, in a more accessible location. External charging and discharging has the added advantage of simplified dust control, outside of the kiln. Additionally, under ambient conditions for unloading, the product, slag, and hearth layer can be removed magnetically, dumped, scraped, vacuumed, or with a non-cooled auger—options not available in the RHF, where the only practical discharge mechanism is an expensive water-cooled screw taking all material off at one time.
3. It is a simple matter to incorporate an in-line drying zone with the LHF, utilizing recoup-heat from the process; this cannot be done directly with an RHF.
4. The LHF affords segregation of combustion gas atmosphere from reducing atmosphere in reduction/fusion zones. This would be very difficult to engineer, build, and maintain in an RHF.
5. For all but very large RHFs, the product is not symmetrically loaded and side to side uniformity is affected. This causes the production tonnage to be greater at the outer diameter than the inner diameter, requiring asymmetrical application of heat. Exhaust flow around the RHF tends to follow the inner diameter.
6. The RHF is less efficient in building space use and therefore more capital intensive. An LHF can be constructed on grade, whereas the RHF requires an elevated hearth to allow access to the inner circle of the furnace.
7. Uniform feeding, product removal, and temperature distribution have to be considered in the RHF. Symmetrical loading affects side to side uniformity. Also, exhaust flow around the RHF tends to follow the inner diameter. An LHF is simply more uniform. In addition to the side to side uniformity discussed above, the co-current flow from hottest to coolest product is more easily controlled in an LHF. Since the charge and discharge are adjacent in an RHF, any seal that is less than 100 % at this point allows exhaust flow to short circuit between these two points.
8. A change in length requirements to obtain increased production is an expansion option for an LHF.
9. The LHF design facilitates the creation of side by side reduction-smelting furnaces that allow pallet cars to be shifted to the adjacent furnace in a close and compact manner for processing material in the reverse direction in order to effectively increase facility productivity.
10. The LHF has the capability to add and remove hearth cars, in situ to perform refractory maintenance, without shutting down and therefore increasing overall furnace availability.

When a fuel is burned, oxygen in the combustion air chemically combines with the hydrogen and carbon in the fuel to form water and carbon dioxide, releasing heat in the process. Air is made up of 21 % oxygen, 78 % nitrogen, and 1 % other gases. During air-fuel combustion, the chemically inert nitrogen in the air dilutes the reactive oxygen and carries away significant amounts of the energy in the hot combustion exhaust gas. An increase in oxygen in the combustion air can reduce this energy loss in the exhaust gases and increase heating system efficiency.

17.2.2.2 Oxygen-Fuel Combustion Systems

Oxygen-fuel burners can produce a wide range of oxidizing or reducing products of combustion. The burners can be designed and operated to produce either compact, high velocity, low luminosity flames, or long, highly luminous, low velocity flames. For the LHF, an oxygen-fuel burner capable of producing an optimum atmosphere in the furnace, along with low emissions and a low momentum, highly radiant flame, is the desired goal.

The principle of CO_2 capture by oxy-fuel combustion is to burn fuel with oxygen rather than air so that the flue gas consists mainly of CO_2 and water vapor with little nitrogen. When CO_2 capture is not required, oxygen-firing is inherently more expensive than combustion with air using current state-of-the-art technologies. Potential advantages of oxygen-firing deriving from smaller equipment size are offset by costs related to cryogenic air separation and flue gas recycle necessary to maintain acceptable temperature levels in the equipment (boiler/heater/gas turbine). New and lower cost oxygen production methods are under active development which means that the overall cost of oxygen-fuel concepts, i.e., those using flue gas recycle, should fall significantly. Combustion in pure oxygen or in oxygen-enriched air in special high temperature furnaces is widespread in the metallurgical, glass, and other industries, and therefore the operational and safety issues of oxygen combustion are well understood (US Department of Energy 1993; Riley et al. 2001).

17.2.2.3 Oxygen-Biomass Combustion

Biomass as an energy source has several important advantages. Renewability and an infinite list of plant, by-products, and waste materials that can be used as feedstock are very attractive features. Also, biomass consumes atmospheric CO_2 during growth and therefore may have no net generation of CO_2 and therefore no negative impact on greenhouse gases. Wood is a clean and renewable fuel which, compared to coal, contains little ash, sulfur, and nitrogen, it is not an ideal fuel for gasifiers. Its optimum gasification temperature is rather low (below 700 °C) due to the high O/C ratio. As a result, wood is generally over-oxidized in gasifiers leading to thermodynamic losses. It is possible to reduce these thermodynamic losses by prior thermal pre-treatment in the range of 250–300 °C, i.e., wood torrefaction. If the heat produced in the gasifier is used to drive the wood torrefaction reactions, the chemical energy preserved in the product gas has been shown to increase provided that both torrefied wood and volatiles are introduced into the gasification process (Veranth and Krishnamoorthy 2002; Prins 2005). In addition, the char produced from torrefaction potentially can be a good carbon and hydrogen source for iron ore reduction either through gasification or direct use of the char. The opportunity exists to use torrefied woody biomass or agricultural by-products as low cost carbon-neutral fuels for iron production using existing technology for oxy-fuel combustion and pulverized coal-firing techniques. In addition to the benefits afforded by the concentration of CO_2 gas, the woody biomass products are also relatively free of sulfur oxides, creating more opportunities for CO_2 capture technologies.

17.2.3 Potential Energy, Carbon Emissions Reduction, and Environmental Benefits

The total energy requirement for a competitive technology was estimated at 16,282 MJ/t (14.0 MM BTU/st). This compares favorably with recently published data of the ITmk3 demonstration project of 15,689 MJ/t (13.49 MM BTU/st) (Hansen 2005). An analysis for the proposed LHF process led to overall energy savings of 3,489–5,815 MJ/t (3–5 MM BTU/st-MIN), or 23–37 % reduction over competitive processes. This benefit is in addition to reported energy savings of 30 % for similar technologies over the current integrated steelmaking process(Kobayashi et al. 2001; Hoffman and Tsuge 2004). Furthermore, the proposed technology reduces emissions over ITmk3's reported reduction in emissions of more than 40 %(US Department of Energy et al. 2004).

The principle is to burn fuel with oxygen rather than air so that the flue gas consists mainly of CO_2 and water with little nitrogen. Oxy-firing has the unique advantage to generate an effluent stream of the necessary purity for separation from this stream by simple water condensation. In addition, replacing air-fuel burners with oxygen-fuel burners is reported to save natural gas by 50–60 % in a steel reheating furnace (Riley et al. 2001). This fuel savings can be directly related to reduced CO_2 emissions. Therefore, the principle behind CO_2 reduction is twofold: (1) reduced products of combustion by reducing fuel requirements by 40–50 % and (2) the absence of nitrogen in the combustion system makes the CO_2 more available for capture by containment and reduced separation requirements. Typical oxygen-fuel combustion results in very high combustion temperatures without the nitrogen diluent from air. Recent designs of oxygen-fuel burners allow for staging of the combustion oxygen deeper into the flame zone and further out into the fired chamber, thereby reducing flame temperatures. This oxygen staging capability produces NO_x emissions that are *80 % lower* than conventional, non-staged designs. In fact, NO_X emissions decreased from 2.5 kg to 0.4 kg/t (5.0–0.8 lb/st) (84 % decrease) in a glass furnace (US Department of Energy 1993).

17.2.4 Previous Laboratory Development

Tests were conducted in three stages: laboratory-scale tests in a typical tube furnace, laboratory-scale batch tests in a specially designed two-stage box furnace, and pilot-scale, semi-continuous tests in a three-zone, pusher-type, linear hearth (LHF) furnace. During the course of the laboratory investigation, various combinations of variables were studied in order to determine the best combination of mix chemistry and processing conditions that would result in the production of high-quality metallic NRI that contain little residual gangue, possess low sulfur levels (<0.05 %), and contain high amounts of carbon (>2.5 %). The reduction of iron ore with coal and fluxes is simple in concept, but various factors must be controlled in order to effectively hit target chemistries and at the same time minimize energy requirements and avoid potentially catastrophic processing problems (e.g., aggressive slag attack on refractories and equipment). Various phenomena occur in the process of carbothermic reduction and these phenomena are summarized in Table 17.1.

17.2.4.1 Laboratory Tube Furnace Tests

The test program was initiated using a tube furnace with a 50.8 mm diameter × 1168 mm long mullite tube, which takes 25.4 mm wide × 101.6 mm long and 25.4 mm high graphite boats. This furnace is capable of achieving

Table 17.1 Sequence of events

		Mixture temperature range	
Event	Description	°C	°F
1	Dehydration of free water	100	212
2	Emission of volatile matter from coal	350–500	662–932
3	Dehydration of bound water from $Ca(OH)_2$	427	800
4	Reduction of magnetite and hematite to iron	>827	>1521
5	Calcination of calcium and magnesium carbonates	~848–900	~1558–1652
6	Melting of fayalite ($2FeO{\cdot}SiO_2$)	1177–1204	2150–2200
7	Melting of carbon saturated iron	1150	2102
8	Melting of slag (CaO, MgO, Al_2O_3, SiO_2)	>1311	>2392
9	Melting of wustite (depending on dissolved oxygen content)	1373 through 1426	2503 through 2599
10	Melting of fluorspar	~1418	~2584
11	Melting of pure iron	1538	2800

temperatures up to 1450 °C. These tests were used to screen multiple test conditions for use in subsequent laboratory box and pilot plant linear hearth furnaces. To control the furnace atmosphere, N_2 and CO were supplied to the combustion tube via respective rotameters. Tests were carried out with a mixture consisting of N_2 and CO mixture at 2 and 1 L/min, respectively.

Various raw materials have been investigated:

1. Multiple forms of iron oxide such as different magnetite concentrates with different levels of silica content, natural hematite ore, and multiple waste oxides from industrial plant operations;
2. Carbonaceous reductants including Eastern anthracite, low-, medium-, and high-volatile bituminous coals, Western subbituminous coals including their carbonized char, coke and coke breeze, and biomass-derived carbon sources;
3. Different additives, such as balling and briquetting binders and some specific additives for slag fusion temperature reduction and NRI sulfur control.

In addition, furnace operating conditions, such as temperature and time at temperature, furnace atmosphere, hearth layer materials, NRI and slag chemistries as well as NRI size, were varied. Taconite iron ore concentrates with different levels of silica indicated that magnetite concentrates with 6 % SiO_2 produced NRI more readily than a more expensively produced flotation concentrate of 4 % SiO_2, or super-concentrate of 2 % SiO_2.

The choice and amount of addition of carbonaceous reductant was found to be an important factor in quality NRI formation. While anthracite, low- and medium-volatile bituminous coal as well as coke worked well, Powder River Basin (high-volatile subbituminous) subbituminous coal was totally unsatisfactory with agglomerated (briquette or pellet) mixtures due to low strength development. Subbituminous coal char resulted in generating inordinately large amounts of micro-NRI (−2.54+0.85 mm) when used in stoichiometric amounts under similar conditions. As a result, the optimum level of carbonaceous reductants was determined to be 75–85 % of the stoichiometric requirement for NRI formation, based on fixed carbon analyses and minimum generation of micro-NRI, when the furnace atmosphere consisted of N_2-CO mixtures and a carbon hearth layer was used.

Certain additives were found to be effective for lowering the fusion temperature of NRI, while some other additives lowered sulfur in NRI to as low as less than 0.01 %. Furnace atmosphere profoundly influenced the temperature needed to form fully fused NRI. Increasing concentrations of CO_2 required higher temperatures, but the fusion behaviors of NRI became less sensitive to the presence of CO_2 over 1400 °C. The laboratory apparatus used for the studies are shown in Figs. 17.1 and 17.2. The tube furnace was used to screen variables and the box furnace was used to confirm test conditions on a larger scale.

17.2.4.2 Laboratory Box Furnace Tests

Laboratory investigations were followed up from the tube furnace trials using an electrically heated box furnace. This furnace shown in Fig. 17.2 consisted of two 304.8 mm × 304.8 mm × 304.8 mm heating chambers with the two chambers capable of controlling temperatures up to 1450 °C independently, preceded by a water jacketed cooler (which is used as the feeding and cooling chamber for products), separated by a manually controlled flip door. This furnace accepts a 127 mm wide × 152.4 mm long × 38.1 mm high graphite or ceramic fiber board trays. To control furnace atmosphere, N_2, CO, and CO_2 were supplied to the furnace in different combinations via respective rotameters. Total gas flow could be adjusted in the range of 10–50 L/min. In most tests, graphite trays were used, but in some tests, trays made of fiber board with a thickness of 12.7 mm were used. After introducing a tray into the cooling chamber, the furnace was purged with a gas, typically a mixture of N_2 and CO at 18 and 2 L/min, respectively, for 15 min to expel the air when a tray was introduced into the cooling chamber. Initially, the tray was pushed just inside of the flip-up door, held there for 3 min for preheating, then pushed into the first chamber, held at 1149 °C for 5 min, and

Fig. 17.1 Laboratory tube furnace

Fig. 17.2 Two-zone, electrically heated laboratory box furnace

then pushed into the second chamber, held at 1400 °C for selected periods of time. After the test, the tray was pushed to the back of the flip-up door and held there for 3 min, and then into the cooling chamber. After cooling for 10 min, the tray was removed from the cooling chamber for visual inspection to see if NRI was formed.

It was found that various sizes of NRI could be routinely produced, ranging from 8.38 to 63.5 mm. Box furnace tests provided an opportunity to further develop methods which showed promise in controlling the generation of micro-NRI and for achieving desired sulfur levels in the metallized product. A major emphasis was placed in developing methods to produce larger-sized

NRI by feeding mounded raw material mixtures in an attempt to circumvent costly agglomeration and drying steps. Modifications of hearth materials as well as proper selection of additives to feed mixtures were studied. It was found that high-quality NRI could be produced using non-agglomerated mounds and various agglomerated mixtures (pellets or briquettes) at smelting temperatures <1450 °C. During the course of the laboratory investigation, it was also found that control of slag chemistry and the atmosphere above the reduction mixtures was essential to allow complete smelting of the iron oxides to metallic iron and for separation of the gangue components to an easily removable slag phase. The atmospheres within the tube furnace and box furnace were much less turbulent than that encountered during pilot-scale testing in the larger linear hearth furnace that will be discussed later. In addition, the degree of mixing of the off-gas from the reduction process was very low relative to a gas-fired furnace. These differences proved to be very significant in evaluating the actual performance. The laboratory tests were extremely useful for developing mix compositions and hearth carbon (carbon on refractory holding reaction mixture) and cover compositions (carbon placed on top of mixture to protect reduced metal) that could be employed at the next scale, but these furnaces could not duplicate the conditions that actually occurred in the larger gas-fired linear hearth furnace.

17.2.4.3 Findings from Laboratory Tests

The optimum size of coal added as a reductant was determined to be −0.210 mm to +0.149 mm. Finer ground coal formed NRI just as effectively, but the amount of micro-NRI increased somewhat. The use of coarser coal required increased amounts of coal for forming fully fused NRI, suggesting that a certain amount of fine coal was necessary. Optimal size of coke or anthracite char used as hearth materials was found to be less than 2 mm.

Effective fluxing of the feed mixtures could be achieved using manganese oxide and fluorspar with a lime to silica ratio (1.5–1.7) that produced a fusion temperature of the slag under 1400 °C. Other fluxes such as MnO and borax were investigated, but fluorspar was found to be most effective. It was also found that the height of a reaction mound or the size of the agglomerate was a very significant factor in controlling total process time. This indicates that the time needed for radiant heat transfer was significantly impacted by the surface area exposed to the radiant heat source. The mechanism for NRI formation was also investigated in some detail in the laboratory studies. Formation of fully fused NRI depends not only on the effectiveness of the radiant and conductive heat transfer, but also on the rate of carburizing of the sponge iron with carbon coming from the hearth layer and also from the cover layer. Briquettes heated at 1400 °C for different periods of time showed that slag was observed to form initially at the bottom of the agglomerates.

17.2.5 Linear Hearth Furnace

17.2.5.1 Description

The Linear Hearth Furnace (LHF) can be best described as a moving hearth reduction furnace. The furnace is 12.2 m long, consisting of three individual heating zones and a final cooling section (Fig. 17.3). The LHF has undergone several stages of development, transitioning from a walking beam, natural gas-air fired furnace to one with a continuous moving car system and three distinct combustion systems that can be used individually or in combination. It has routinely been used to test a variety of the variables shown to be important from tube and box furnace tests. The primary goal of the program was to develop sufficient understanding of the controlling variables associated with iron oxide reduction and smelting using coal-based reductant. The research has allowed sufficient knowledge to be developed so that nodular reduced iron (NRI) can be routinely produced with low levels of tramp impurities using various carbonaceous reduction materials. The laboratory furnaces allow very precise manipulation of key variables under very controlled experimental conditions. The LHF

Fig. 17.3 Pilot-scale linear hearth furnace

facility allows these basic studies to be expanded to a significantly larger scale and to create bulk samples of product for further testing. The conditions studied in the course of this project have shown that nodules of iron can be produced with various additives and operating conditions by manipulating the correct variables.

Atmosphere and Combustion: A major difference between the laboratory tube and box electrically powered furnaces and the LHF is the high turbulence associated with the natural gas–air burner combustion products. In the natural gas-fired LHF, operating under sub-stoichiometric gas and air mixtures simulates the required reducing conditions for reduction and smelting. The resulting furnace gas atmosphere contains a relatively low ratio of $CO:CO_2$ (approximately 1:5). Partially metallized iron ore from the reduction zone directly contacts the high CO_2, low CO mixture, and this action is further enhanced by the highly turbulent furnace gas flow at high temperature as this metallic material enters the melting zone. Exposure of the partially metallized feed mixtures to this atmosphere causes rapid loss of added reductant carbon and formation of high FeO slag. The FeO content in the slag controls the oxidation state, and consequently, makes sulfur removal to the slag less favorable. The furnace atmosphere and the high FeO content of the slag coupled with the operating temperature, typically 1450–1550 °C as claimed in previous patents, appears to lead to some difficulty in lowering sulfur in iron nodules to below 0.1%S. Processing of high sulfur nodules in the EAF would lead to higher steelmaking costs and extra energy use, as more slag forming compounds would be needed to purify the steel. The results indicate that atmosphere control in the smelting section is extremely important for the production of high-quality NRI.

A simple Computational Fluid Dynamic (CFD) model of the furnace was built and used to demonstrate the impact of nitrogen introduced using air-fired burners, on turbulence in the furnace. Output from the CFD model using air-fired burners shows the velocity scale ranges from 0 to 2.5 m/s. With the same level of energy input, and using oxygen at 90 % purity, the velocity range is decreased to 0–0.8 m/s. This reduction of turbulence reduces the interaction of high-CO_2-containing furnace gases and aids the metallization process efficiency. As a result, the LHF furnace has been equipped with three distinct combustion systems that can be operated separately or in combination:

- Natural Gas-Air Fuel Combustion System
- Oxygen-Natural Gas Combustion System
- Dilute-Phase Pulverized Coal-Oxygen Combustion System

Thus, from both a product quality standpoint and from an operating standpoint, furnace atmosphere control is a key control variable and must be a key parameter in design of the overall furnace operating conditions. Various concepts for atmosphere control were tested. One method that was found to be very useful was to use a carbonaceous cover on top of the reaction mixture. This effectively shielded the reaction mixtures from the highly turbulent and oxidizing gas atmospheres. Other control methods included use of various reducing gas mixtures during the reduction stage of the process (>1093 to ~1250 °C). Finally, the use of oxygen-fuel burners reduces the volume of flue gas, thereby alleviating the turbulence within the furnace and conserving the energy associated with heating chemically inert nitrogen(Fosnacht et al. 2004). Turbulence may also be further reduced through flame shape characteristics.

Oxygen-Fuel Burners: Natural gas-air fired linear hearth furnace (LHF) tests generated high CO_2 (10 %CO_2, 2–4 %CO) and highly turbulent furnace gas as compared to the electrically heated box furnace. This difference made it difficult in the LHF to produce satisfactory iron nodules consistently and the nodules produced often had sulfur contents that were undesirably high (0.1–0.3 %S). The LHF remodeled with an oxygen-fuel combustion system was tested initially by comparing the effect of oxygen-fuel and air-fuel burners on fusion time using briquettes containing bituminous coal. The results show fusion time was shorter by 10–30 % when oxygen-fuel burners were used as compared to air-fuel burners. This difference was directly related to the high turbulence of the furnace gas with air-fuel burners and their effect on the endothermic carbon solution reaction. NRI quality at fusion time analyzed 3.0–3.6 %C and 0.04–0.05 %S under the conditions tested using the oxy-fuel burner systems.

Oxygen-Coal Combustion: A 590,343 kJ/h oxygen-coal burner was positioned to fire horizontally from the end of the furnace, down the length of the LHF. The coal is fed by a variable feed Acrison screw conveyor/hopper through an eductor system for dilute-phase coal injection into the burner. Oxygen is then monitored through a Programmable Logic Control (PLC) system to match coal addition and adjust stoichiometry. Atmosphere control was investigated while simultaneously controlling temperature by minimizing airflow, operating the burner in a substoichiometric manner and controlling furnace zone pressures to prevent heat transfer into adjacent zones. The installation resulted in the LHF operating successfully on the oxygen-coal burner system, controlling both atmosphere and temperature. This was accomplished by reducing the ratio of conveying air to coal with increased fuel flow rates and using sub-stoichiometric oxygen-gas burners to control oxygen content. The oxygen-coal burner system was capable of controlling the temperature of 1413 °C in Zone 3 while maintaining good atmosphere control for the production of NRI. The coal type used was a low volatile bituminous coal. A maximum loading of 0.27 kg/min of coal or an equivalent of 507,063 kJ/h was used as a base energy load while the oxygen-gas system was used to trim and control the temperature. While the coal system was in operation, the natural gas system was operating at less than 10 % of full fire, and frequently less than 1 % on a single burner.

For this project, the furnace was relined with new firebrick refractories (see Fig. 17.4). In addition, the exhaust ducts from two zones (Zones 2 and 3) were consolidated to isolate the exhaust gases to a single duct. This allowed process gases

Fig. 17.4 Refractory reline in zones 2–3 of the LHF

to be directed co-current to the flow of the LHF and the gas analysis to be concentrated to a single stream for evaluation of the process emissions under each of the combustion systems. Figure 17.4 shows the finished refractory lining just prior to staged curing.

17.3 Use of Alternative Fuels to Produce Nodular Reduced Iron Products

The energy intensity and greenhouse gas emissions were evaluated in the linear hearth furnace (LHF) using a solid fuel-oxygen fired combustion system. The solid pulverized fuels are identified as coals and torrefied woody biomass char materials. The successful production of high-quality NRI provides opportunities for recycling steelmaking waste products, diversification of iron ore processing, and a new pure iron feedstock for the electric arc furnace steelmaking industry, while affording unique opportunities with respect to reducing greenhouse gas emissions (GHG) while providing process advantages. The objectives of the test work are described below:

1. Establish the baseline energy and GHG intensity of the nodular reduced ironmaking process using standard air-fired natural gas burners
2. Establish the energy and GHG intensity of the nodular reduced ironmaking process using oxygen-fired natural gas burners
3. Measure the energy and the GHG intensity of the nodular reduced ironmaking process using oxygen-fired solid fuel burners and a variety of solid fuel types
4. Determine the feasibility of using thermally processed biomass as a reductant carbon source for advanced ironmaking
5. Evaluate the feasibility of using thermally processed biomass as a fuel for oxygen-fired solid fuel burners in the nodular reduced ironmaking process

All combustion systems, fuel types (natural gas, coal, and bio-coal), and conditions were tested. GHG were recorded during the production of quality NRI from the baseline briquette chemistry with a full furnace of anthracite char to simulate a true representation for the performance.

17.3.1 Development of an Alternative Biobased Reductant Through Torrefaction

An investigation was undertaken to determine if treated biomass could be used as a whole to partial substitute for coal in the reduction process. The reason was to have a material with less mercury, sulfur, and gangue input that would reduce the environmental consequences of a carbothermic reduction process.

Torrefaction is a relatively new technology that partially carbonizes biomass, making it moisture-free and friable. Torrefaction is a mild pre-treatment of biomass at a temperature between 200 and 300 °C under low oxygen conditions to remove moisture, light volatiles, and accumulate the carbon component. Torrefying biomass has the potential to process diverse biomass materials with different physical and chemical properties and convert the materials into a more uniform product with better handling and processing characteristics. During torrefaction, the biomass properties are changed to obtain an enhanced fuel quality for combustion applications. Torrefaction of biomass is an effective method to improve the grindability of biomass to enable more efficient co-firing with fossil based fuels. It can then be used as a complete, or more likely partial, replacement for coal in pulverized coal facilities.

The feedstock utilized for the creation of a torrefied solid fuel consisted of mixed hardwood biomass donated by a company in Minnesota. Woody biomass was chosen as the research feedstock due to its relatively low ash content and ready availability in Northern Minnesota. The feedstock was chipped and screened to −9.525 mm (3/8″) before arriving at the Natural Resources Research Institute (NRRI). The raw woodchip moisture content was typically 7–10 %

Table 17.2 –3/8″ Wood chip screen structure

Mesh (Tyler)	Weight %	Cumulative % passing size (%)
8 (2.38 mm)	22.5	77.5
10 (2.0 mm)	6.7	70.8
14 (1.41 mm)	17.1	53.8
20 (0.841 mm)	12.1	41.7
28 (0.595 mm)	12.5	29.2
100 (0.149 mm)	26.3	2.9
200 (0.074 mm)	2.1	0.8
–200 (0.074 mm)	0.8	0.0
Total	100.0	–

Table 17.3 Wood chips –9.53 mm (–3/8 in.) ash chemistry

Compound	Weight %
SiO_2	5.54
Al_2O_3	0.88
CaO	51.67
MgO	5.59
Total C	0.89
Fe^{++}	1.65

with a bulk density of approximately 200 kg/m^3 (12.lb/ft^3). Hardgrove Grindability Index for the raw wood chips was approximately 10–15, and fuel value of the dried wood chips (as determined by calorimetry) was 19.05 MJ/kg (8192 BTUs/lb). Table 17.2 displays a screen analysis for the wood chips.

Ash chemistry for the woodchip feedstock can be viewed in Table 17.3.

The raw woodchip feedstock was fed into the torrefier apparatus to create the solid fuel used in the LHF.

17.3.2 Calciner/Torrefier Apparatus and Testing Conditions

The continuous process consisted of a retrofitted calciner unit. Dimensions of the reactor tube were 178 mm (7″) in diameter by 9.53 mm (3/8″) wall thickness by 2134 mm (7″) in length. The unit was fed through a variable frequency screw auger and was electrically heated by three heating elements, each 2.5 kW. The reactor tube could be rotated between 1.5 and 10 rpm. Torrefaction gases were taken away via an air blower and either condensed or flared. Figures 17.5, 17.6, and 17.7 display the torrefier diagram, a photo of the device, and the feeder/temperature controller, respectively.

The torrefaction operating parameters included feed rate, feedstock preprocessing, internal reactor temperature, and reactor rotational speed and angle. Feedstock sizing and moisture content, along with the other operating parameters, were held constant throughout each run with the exception of the reactor temperature. The baseline operating conditions were as follows:

- *Feeder setting*: Variable frequency drive setting of 50 %
- *Raw material throughput*: ~10.9 kg (~24 lb.)/h or [(0.18 kg (0.4 lb)/min)]
- *Feedstock moisture content*: ~7–9 %
- *Drum rotational speed*: 9 rpm
- *Device alignment*: 0.5° decline from feeder to discharge end

Almost the entirety of the sample was processed between 250 and 300 °C, with approximately 50 % or more being thermally treated at 275 °C in steady state. Figures 17.8 and 17.9 below show the biomass before and after roasting to produce the bio-char.

17.3.3 LHF Application and Testing Conditions

After thermally processing approximately 385.9 kg (850 lb.) of raw feedstock, the entire sample was ground via an Allis Chalmers vibrating ball mill. Fineness was judged upon a plus/minus sieve analysis. After thermally processing and grinding, the sample was greater than 60 % passing a 100 mesh sieve (0.149 mm). A sample was cut and sent to Standard Labs, Inc. for a chemical analysis. Table 17.4 displays the proximate and ultimate analyses of the ground torrefied biomass which is now bio-char.

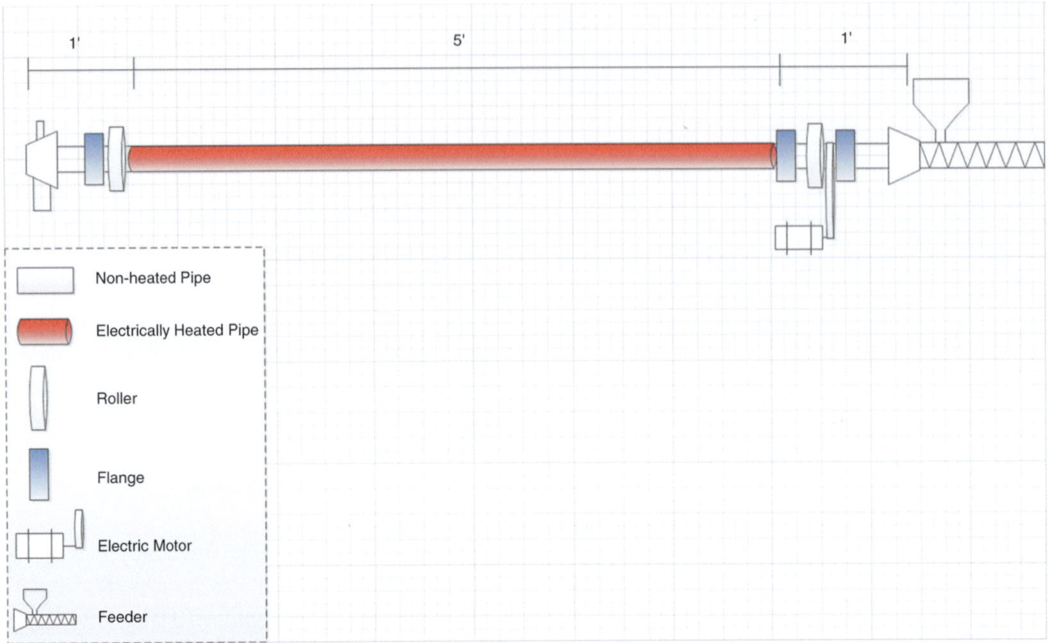

Fig. 17.5 Calciner torrefier device diagram

Fig. 17.6 Completed rotary calciner/torrefier at NRRI

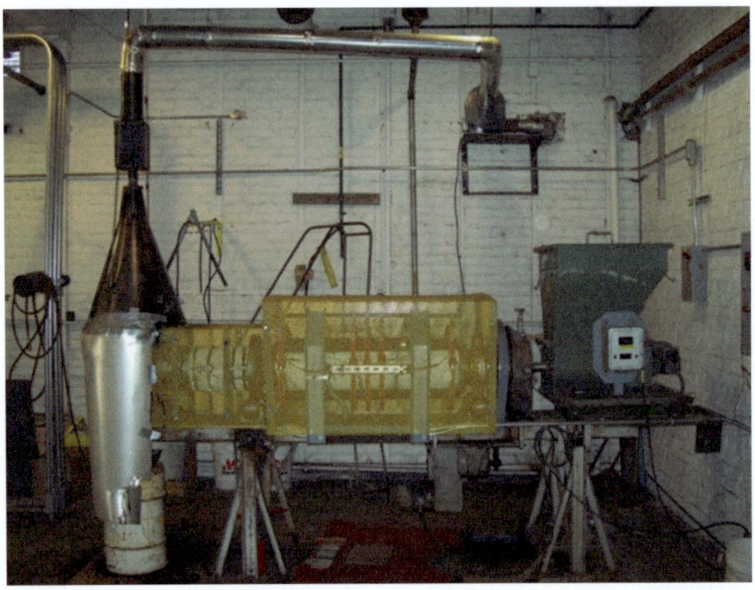

Fig. 17.7 VFD feeder and temperature controller

Fig. 17.8 Raw feedstock –3/8″ wood chips

Fig. 17.9 Thermally processed bio-coal (275 °C)

Table 17.4 Bio-char proximate and ultimate analyses

	As received	Dry basis
Moisture content	4.52 %	–
Volatile content	68.54 %	71.79 %
Fixed carbon	23.73 %	24.85 %
Ash	3.21 %	3.36 %
Sulfur	0.10 %	0.10 %
Carbon	53.55 %	56.09 %
Hydrogen	5.26 %	5.51 %
Nitrogen	0.17 %	0.18 %
Oxygen	33.19 %	34.76 %
BTU/LB	9150	9583
MJ/kg	21.283	22.290

The ground torrefied bio-char was fed into the linear hearth furnace (LHF) via a coal feeder, a feed hopper, and an air conveyance system.

17.4 Comparison of Operating Results with Different Fuels and Reductants

17.4.1 LHF Operation with Alternative Solid Fuel Combustion

Coal types tested were an eastern medium volatile bituminous coal, a southern high-volatile bituminous coal, and western high-volatile subbituminous coal. The bio-char solid fuel was prepared using the torrefaction system previously described to a dry solid weight loss (DSL) of 30 %. All baseline tests were conducted on briquettes consisting of 73.5 % magnetite concentrate, 18.0 % medium volatile bituminous coal, and fluxing components consisting of 6.5 % hydrated lime and 2.0 % fluorspar. Results showed that residence time increased slightly by approximately 15 % when using solid fuel combustion. This result was attributed to the loss of radiant energy from the idled oxygen-gas burners and difficulty in controlling atmosphere due to the generation of CO_2 and H_2O from the conveying airflow.

The use of thermally processed bio-char was successful for the production of NRI, however, the steady-state operation of the solid fuel-oxygen burner system in the LHF required the bio-char to be processed prior to use. The successful use of the fuel was found to be dependent on reducing the moisture content to less than a few percent, and sizing the bio-char to a size distribution of −20 Mesh (−0.841 mm). The present solid fuel combustion system uses an eductor to convey the solid fuel, and with the current configuration this system was prone to plugging. The relatively small size of the pilot LHF prevents the use of a larger diameter pipe to control the solid fuel addition and minimize air contamination within the furnace. It is believed that a larger system on a more industrial scale would not exhibit this phenomenon; however, a dense phase conveyance system is desired to eliminate air contamination. The trends for comparison of CO_2 emissions by individual zone in the LHF are shown in Figs. 17.10, 17.11, 17.12, and 17.13. In each case, the trend is representative of the full

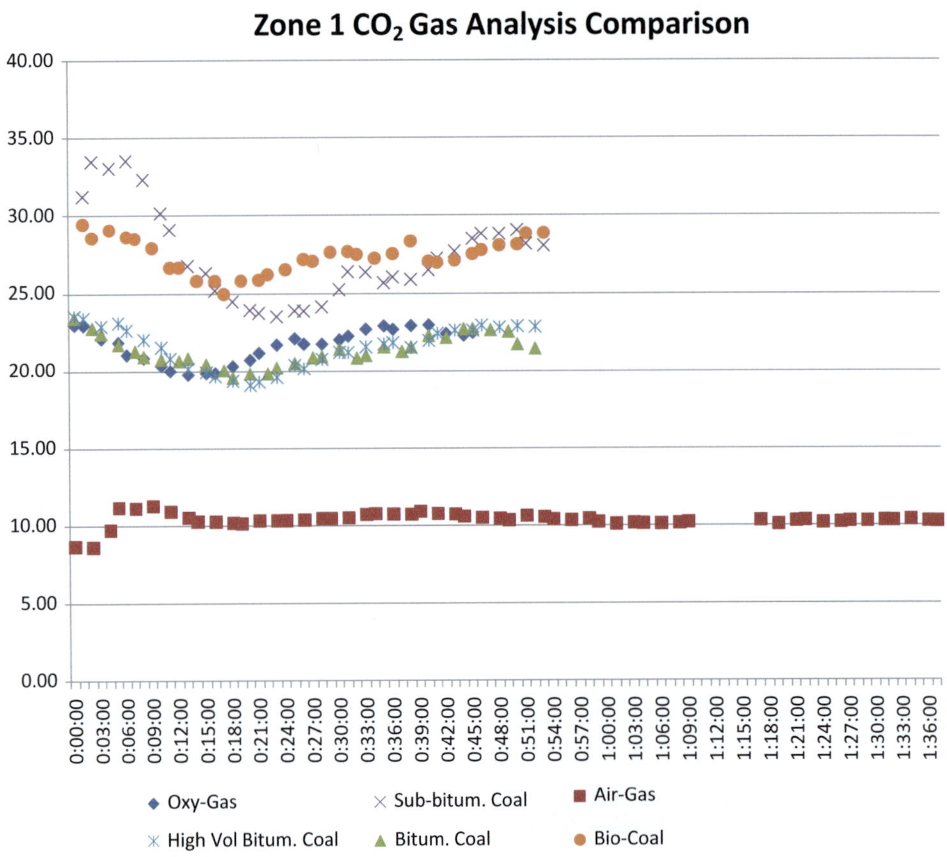

Fig. 17.10 LHF Zone 1 CO_2 measurements

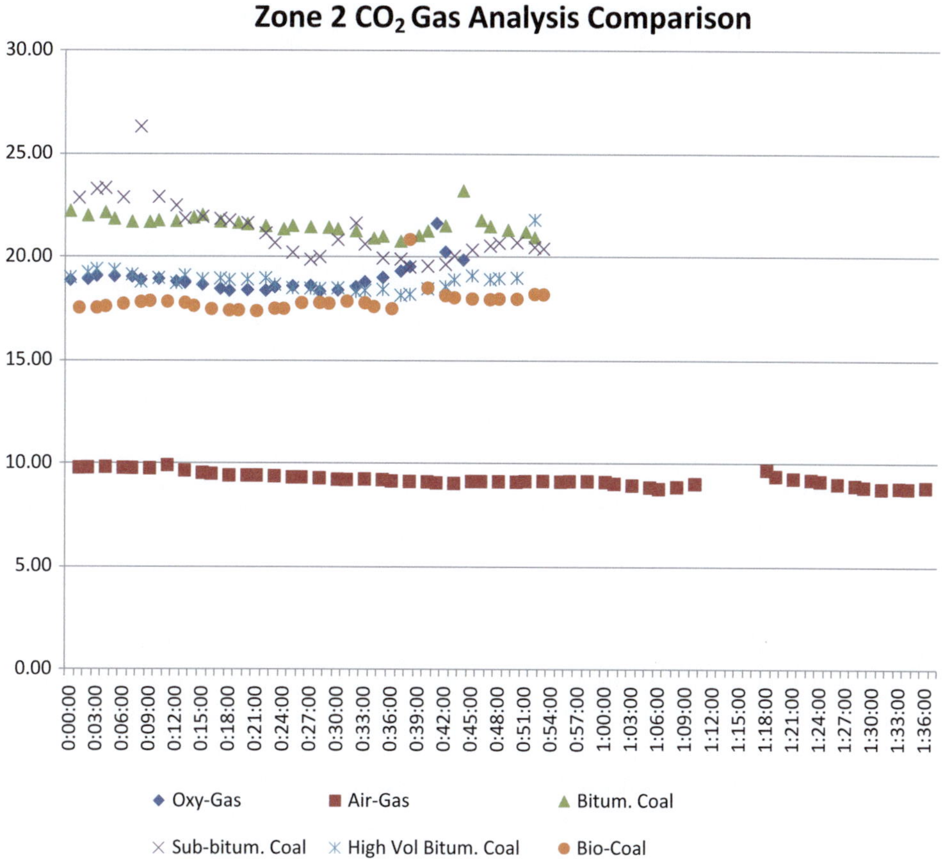

Fig. 17.11 LHF Zone 2 CO_2 measurements

furnace cycle, including the feed vestibule and the cooling section of the LHF.

These trends show the increased concentration of CO_2 in the off-gases when firing the solid fuels over the conventional air-fired system. In addition, the emissions analysis shows an overall 57 % reduced CO_2 reduction from the oxygen-fired fuel options versus the air-based system, on the basis of kg CO_2/t of iron. The balances of the gases are primarily CO, N_2, and H_2O. It should be noted that the N_2 and H_2O content are estimated on the basis of a dew point calculation with the balance assumed to be N_2. For the intent of this study, N_2 and H_2O were assumed constant and comparisons were made on a relative basis. Although the use of solid fuels, including the thermally processed bio-char, was successful for the production of NRI, the configuration of the burner system resulted in air contamination, therefore minimizing the impact on reduced GHG (kg CO_2/mt) to 12–15 %. The oxy-solid fuel combustion system uses an air swept eductor to convey the fuel. Although the emission analysis for the coal-based testing and the bio-based tests were essentially equivalent within the error capabilities of the equipment, the carbon footprint for the bio-based fuel is significantly lower. Figure 17.14 shows a comparison of the average CO_2 production based on per metric ton of NRI produced in the LHF.

17.4.2 Atmosphere Control Using Late Stage Bio-char Addition

Tests were conducted to evaluate the feasibility of late stage addition of a coarse (+16 mm) bio-char cover layer at a relatively high temperature

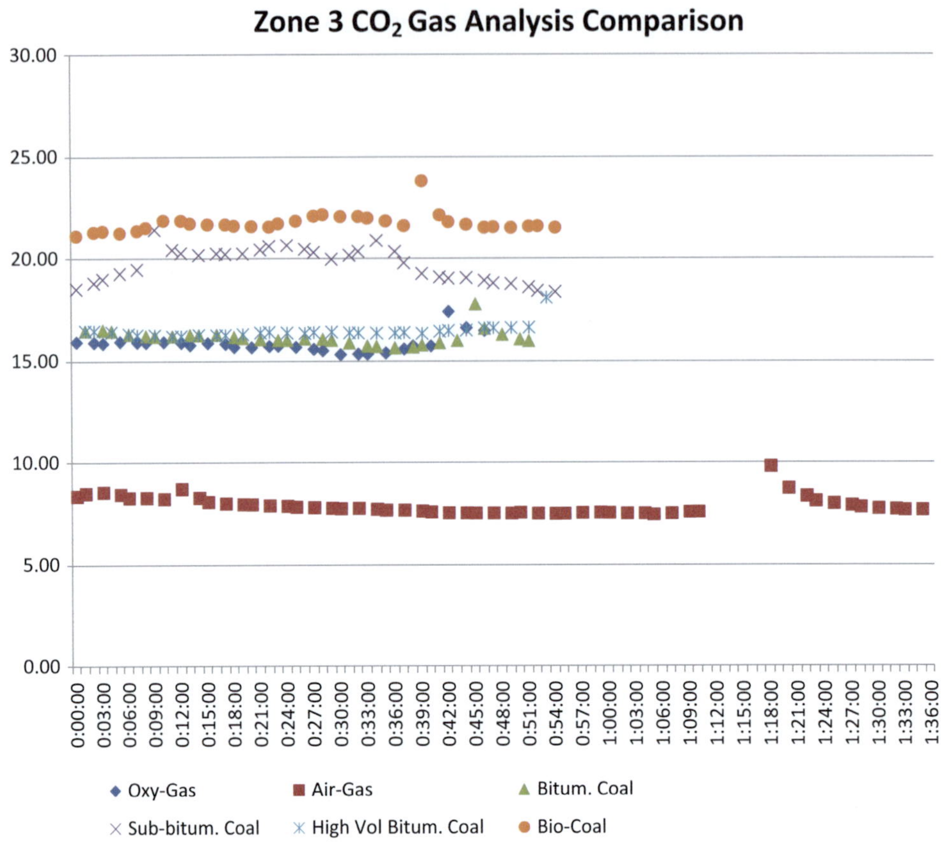

Fig. 17.12 LHF Zone 3 CO_2 measurements

~1260 °C (2300 °F) to modify the atmosphere in the LHF during critical stages of fusion to increase productivity. Previous work had shown that the use of partial cover layer of carbon helped protect the partially reduced iron products from reoxidation by furnace gases, but the carbon addition also reduced heat transfer rates to the reaction mixture. The objective was to determine if waiting to place the carbon layer until the briquettes had reached a relatively high temperature, and metallization of the iron was essentially complete, would improve productivity on the LHF. Previous testing determined that direct exposure of the briquettes to radiant and convective heating without the cover layer will accelerate both heat up and reduction. The addition of the cover layer at this point has several advantages.

The briquettes are less than 150 °C (300 °F) below the temperature needed for fusion.

1. The reduction of the briquettes is largely completed, and therefore the endothermic reduction reaction is not absorbing energy.
2. Hot briquettes will accelerate the heating of the cover layer and promote the Boudouard reaction necessary to provide the high CO levels needed to prevent back oxidation of the iron.
3. The gases from the reduced mass have greatly subsided, and a protective atmosphere is now needed to prevent reoxidation of the reduced iron.

The object of the tests was to determine if biochar could be substituted for coke or anthracite as a cover layer. The feeder system delivered a uniform layer. It was observed that the char contained a high amount of volatiles, about the same, or perhaps a bit more, than we observed when western US subbituminous coal was used as a cover layer

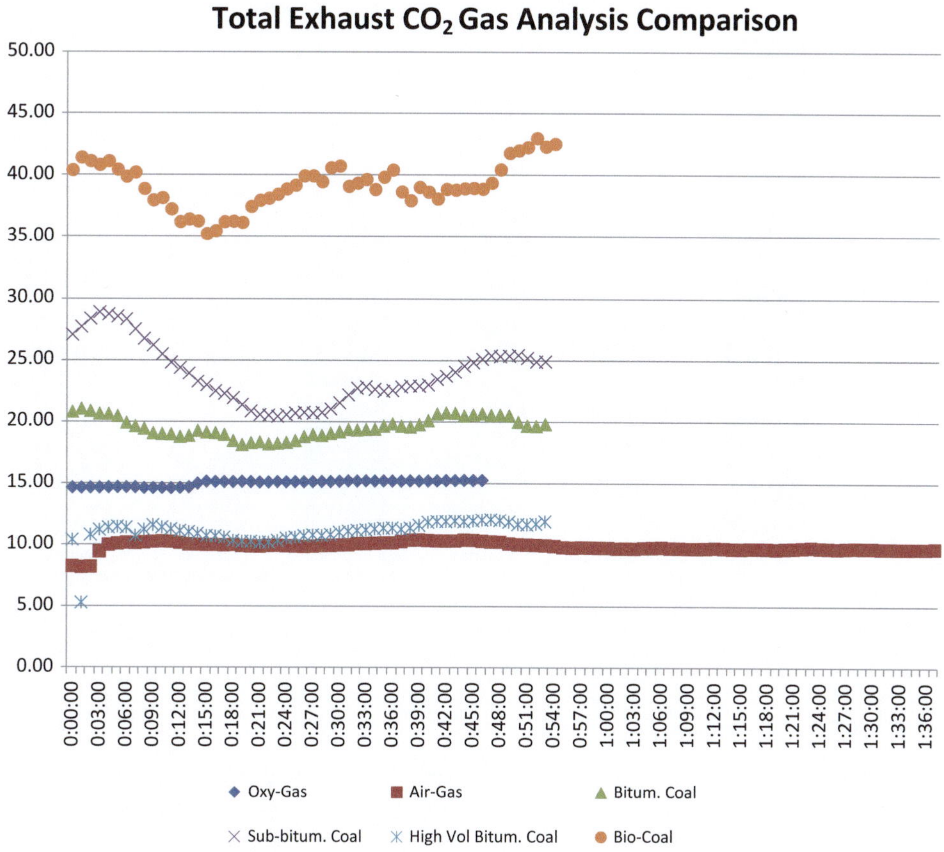

Fig. 17.13 LHF combined total flue gas CO_2 measurements

in prior studies. The tests demonstrated that wood char is an effective atmosphere cover layer and has the potential to increase productivity by 15 % based on processing times measured.

17.4.3 Bio-char as a Reductant

The tube and box furnace tests were used to identify the benefits of using bio-chars in the agglomeration mixture as a reductant. Results are compared to results obtained at a residence time of 5 min at 1400 °C for a baseline briquette composition. Briquettes were prepared using a common molasses binder (4 %). Each of the tests conducted for comparison of the bio-char addition as reductant were done using the designated $L_{1.5}$ FS_2 fluxing system described at the end of this chapter.

17.4.3.1 100 % Bio-char Reductant

Box furnace tests indicated that the fusion behavior using biomass-derived carbon reductant resulted in significantly reducing fusion time by over 30 % while reducing sulfur content of NRI to 0.028–0.031 %. Prior results showed that a stoichiometric ratio of 0.85–0.95 fixed C:Fe was optimal for the production of quality NRI at low sulfur concentration. Using biomass reductant at comparable stoichiometric ratios increased the volume of reductant carbon in the mix as a result of the decreased fixed carbon and lower bulk density. Coupled with the relatively high-volatile matter in the bio-char, an increase in the generation of micro-NRI (−6 Mesh (3.36 mm) +20 Mesh (0.841 mm) fraction) was obtained. Reducing the ratio to 0.75 resulted in decreasing the amount of micro-NRI. The results of these tests are shown in Table 17.5. These results are

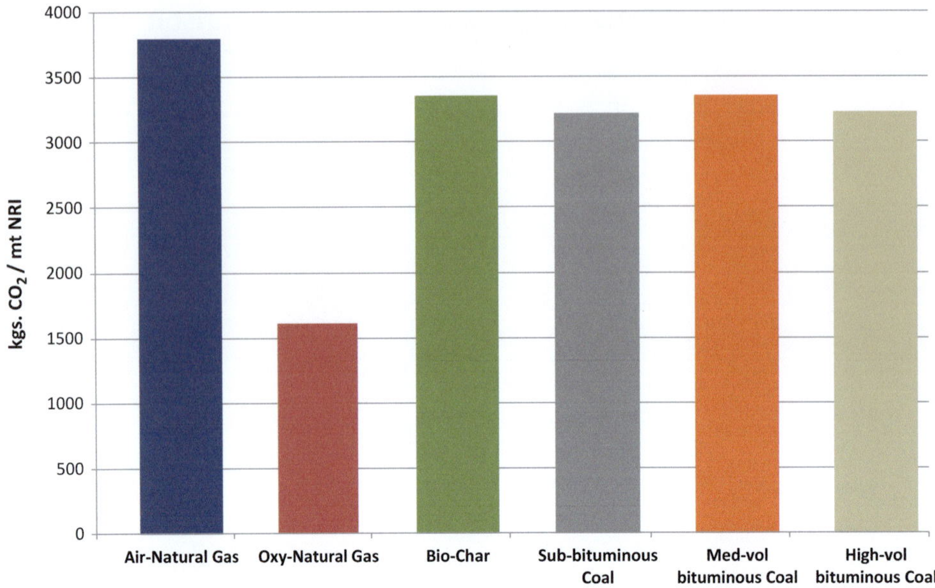

Fig. 17.14 Greenhouse gas emissions comparison by fuel type

Table 17.5 100% Bio-char test results

Bio-coal % stoichiometry	B_2	Time at 1400 °C min	Micro-NRI (%)	% C	% S
65	1.85	5.5	0	2.93	0.031
75	1.87	5	7.3	2.83	0.028
85	1.89	3	43.5	NA	NA
100	1.92	3	96.7	NA	NA

consistent with a study conducted for DOE (Department of Energy), for biomass application to the rotary hearth furnace, where replacing coal with wood charcoal as reductant can provide significant productivity gains, from 33 to 46 %, a carbon source virtually free of net CO_2 emissions (Fruehan and Fortini 2004).

This shows that 100 % fused metallics can be formed in about 60 % of the time when compared to the baseline mix of 5 min to fusion at 1400 °C. However, these were primarily micro-NRI. Carbon and sulfur analyses were not conducted on sample mix at 85 % and at 100 % stoichiometry due to an insufficient quantity of actual NRI produced because of the high amounts of micro-NRI. In addition, due to the relatively low fixed carbon level (25–30 %) in bio-char, significant quantities were required to achieve the stoichiometric levels required. This reduced the density of briquettes by 30 %, and therefore reduced productivity of NRI.

17.4.3.2 Bio-char and Coal Reductant Blends

In an effort to reduce micro-NRI while attaining the benefit observed with the bio-char reductant, blends of bio-char with both medium and high-volatile bituminous coal at 10, 30, and 50 % were conducted. Two series of box furnace tests were carried out at 75 and 85 % stoichiometric ratios of fixed carbon to iron. Table 17.6 shows the results of these tests.

This shows that 100 % fused metallics can be formed with the same 85 % stoichiometric ratio determined in previous testing, reducing residence time requirements by 30 %, reducing

Table 17.6 Bio-char with medium-volatile coal test results

Char %	Coal %	Stoich %	B_2	Time at 1400 °C min	Micro-NRI (%)	% C	% S
10	90	75	1.54	10	0	3.34	0.030
30	70	75	1.58	7	0.2	3.20	0.030
50	50	75	1.63	6	0.7	2.95	0.029
10	90	85	1.53	5.5	0	3.21	0.025
30	70	85	1.57	4.5	0	2.81	0.025
50	50	85	1.62	3.5	0	2.66	0.024

Table 17.7 Bio-char with high-volatile coal test results

Char %	Coal %	Stoich %	B_2	Time at 1400 °C min	Micro-NRI (%)	% C	% S
10	90	75	1.57	6.5	0	2.82	0.047
30	70	75	1.61	6	0	2.93	0.042
50	50	75	1.65	5	0	2.94	0.034
10	90	85	1.56	4	2.6	2.78	0.029
30	70	85	1.60	3.5	7.8	2.48	0.027
50	50	85	1.65	3.5	13.2	2.61	0.029

sulfur content while increasing briquette density proportionately with the amount of bio-char and coal addition.

For comparison, blends were also made from the bio-char with high-volatile bituminous coals. Briquettes were prepared with the same proportions of bio-coal to coal at 10, 30, and 50 % bio-coal and were tested in the box furnace with both 75 and 85 % stoichiometric ratios. Table 17.7 shows the test results of these blends.

Again, this shows that 100 % fused metallics can be formed with 85 % stoichiometric ratio, while also reducing residence time requirements by 30 %. The amount of micro-NRI generated at this ratio increased slightly over that attained with the medium-volatile coal, indicating that the generation of micro-NRI is proportional to the amount of volatile matter contained within the reductant. This is also consistent with a previous study using high-volatile subbituminous coals as reductant.

17.4.3.3 Micro-Nodule Generation

To further investigate the generation of micro-NRI, a sequence of box furnace tests was conducted at timed intervals with blends of bio-coal and medium-volatile bituminous coal. Briquettes prepared for previous tests ranging from 50 to 100 % bio-coal were heated at 1400 °C in a 10 % CO atmosphere (balance N_2) at 1-min intervals up to 6 min. The products from the processing of various mix chemistries noted in Table 17.8 were sampled at various time intervals. The samples were then magnetically separated and screened on the standard 6.3 mm, and 20 mesh (0.833 mm) sieves. The non-magnetic portion is reported as slag. Table 17.8 shows the results of the interval testing.

The data in Table 17.8 show that the generation of micro-NRI, defined as the % −6.3 mm + 20 Mesh (0.833 mm) size fraction, occurs very early in the process. At the 2-min interval, with 100 % bio-char micro-NRI accounts for 35 % of the metallic portion of the sample. It's reasonable to assume that a portion of these micro-NRI reform into metallic NRI since, at the 3-min interval, all samples showed some signs of micro-NRI; however, this portion decreased in subsequent intervals. It's also important to note that the 100 % bio-char sample was near full fusion at the 2-min time interval. In addition to the work discussed above, other coal types including high-volatile bituminous coal, subbituminous coal, and char made from

Table 17.8 Sequence of micro-nodule generation for various blends

% Bit. coal	% Bio-char	Time (min)	Magnetics (%) +6.3 mm″	Magnetics (%) −6.3 mm +20 M	Magnetics (%) −20 M	Non-mag (%) Slag	Observation
0	100	1	29.3	16.0	54.7	0.0	Not fused
0	100	2	16.3	35.1	48.6	0.0	Primarily fused, partial reduced iron
0	100	3	13.2	38.5	36.9	11.4	All fused
0	100	4	27.2	32.6	30.9	9.4	All fused
0	100	5	16.5	50.2	22.1	11.3	All fused
0	100	6	11.6	49.3	27.4	11.6	All fused
90	10	1	100.0	0.0	0.0	0.0	Not fused
90	10	2	100.0	0.0	0.0	0.0	Not fused
90	10	3	89.4	10.6	0.0	0.0	Not fused
90	10	4	82.4	10.6	0.0	17.6	All fused
90	10	5	82.7	0.0	0.0	17.3	All fused
90	10	6	85.3	0.0	0.0	14.7	All fused
70	30	1	100.0	0.0	0.0	0.0	Not fused
70	30	2	100.0	0.0	0.0	0.0	Starting to show fusion, primarily DRI
70	30	3	90.2	3.4	0.0	6.4	Some fusion, primarily DRI
70	30	4	81.5	0.0	0.0	18.5	All fused
70	30	5	81.8	0.0	0.0	18.2	All fused
70	30	6	82.7	0.0	0.0	17.3	All fused
50	50	1	100.0	0.0	0.0	0.0	Not fused
50	50	2	100.0	0.0	0.0	0.0	Approx. ½ fused
50	50	3	75.7	9.6	1.4	13.2	Partially fused, partial reduced iron
50	50	4	82.4	0.0	0.0	17.6	All fused
50	50	5	80.2	1.9	0.6	17.3	All fused
50	50	6	81.3	0.0	0.0	18.7	All fused

subbituminous coal were investigated. The results from these tests indicated that high-volatile bituminous and char produced from subbituminous coals can be substituted for medium volatile bituminous coals.

17.5 Mass Balances on LHF Furnace Tests

Six mass balances, shown in Table 17.12, were developed for furnace fuel tests. The mass balances were based on inlet feed streams and as such, serve mainly for relative comparison. Solid feed (briquettes) loading and briquette composition were fixed, and furnace speed (residence time) was changed as required to yield complete NRI conversion.

Table 17.9 provides a summary of loading conditions and components in the briquettes. Feed rates to furnace were determined from mass of briquettes per unit area and tray speed through the furnace. For these balances, the feed was assumed constant during the tests.

Table 17.10 provides composition information for the molasses binder, both on a dry basis and

Table 17.9 Baseline feeding conditions—briquette and raw materials loading

Briquette loading conditions	
Kg/tray	1.85
Briquette bed width (mm)	460
Briquette bed area (mm^2)	209,012
Kg by component	
Fe	0.89
SiO_2	0.61
Al_2O_3	0.01
CaO	0.1
MgO	0.005
H_2O Hydrate	0.028
CaF_2	0.034
Molasses	0.07
Fixed C	0.22
Coal volatile and free water	0.07
Oxygen in solid iron oxide	0.34
Totals	1.84

Table 17.10 Binder components and addition

Molasses binder composition			
	Dry %wt.	By component in briquette	
Dextrose	9.45	Sugar % C	33.94
Glucose	5.36	Sugar % H_2O	50.49
Sucrose	65.18	% Ash (slag)	1057
Fructose	8.88	Free water (final after drying)	5.00
Ash	11.13		100.00
Totals	100.00		

Mass weighted-free water in briquette, %wt 0.41

on a partially dried basis. Molasses was diluted with nominal 25 % water. The briquettes were dried, but in some cases might have absorbed moisture during storage. The mass balance assumptions included 5 % moisture remaining in the molasses as fed to the furnace +1.2 % moisture in the briquette reductant coal. On a total mass basis, these assumptions result in 0.4 % moisture in the briquettes as fed to the furnace.

Anthracite coal was used for hearth and cover layers with a combined loading of 10.84 kg/m^2 (0.0154 lbs/in^2).

Oxygen and natural gas flow rates were measured directly for all tests except the air-natural gas test. In the air-natural gas baseline burner, percent-of-full fire was recorded, and airflow was determined using the same ratio of oxygen to natural gas used in the oxygen-natural gas burner. Natural gas and air flow were back calculated from the percent-of-full-fire values for each burner. Coal mass flow rates were measured, and carrier air flow was determined pressure drop over an orifice. Mass flows were determined from these measurements to complete the inlet flow into the furnace. Inlet feeds were broken down into elemental components, C, H, O, etc. from which a total elemental composition was determined and used in an equilibrium calculation to determine flue gas composition. All reductant carbon was assumed to gasify, except for 3 % remaining in the briquettes. No hearth carbon was assumed to gasify as there were no measurements taken to indicate that hearth carbon losses to furnace atmosphere occurred. Table 17.11 gives an example for the bio-char test.

Equilibrium inputs shown in Table 17.11 were used in FactSage software (FactSage Thermochemical Software) to determine an equilibrium gas composition at 1093 °C (2000 °F). Table 17.12 contains a summary of these compositions. The carbon dioxide emissions were determined from these compositions. Estimated energy contents in the flue gas based on hydrogen and carbon monoxide are found at the bottom of the table. The Testo gas analyzer measurements for CO_2 were also be compared, and plotted in Fig. 17.15.

On a relative basis using the information from Figs. 17.14 and 17.15 and Table 17.12, with the Air-Natural Gas as the baseline, the following observations are made:

(a) Oxygen-natural gas burners resulted in a 57 % reduction in carbon dioxide emission in the flue gas.
(b) Oxygen-natural gas in combination with solid fuels yielded a carbon dioxide reduction of 10–15 % relative to air-natural gas.
(c) The greatest reduction in fuel consumed occurred for oxygen-natural gas–high-volatile subbituminous coal at 62 % of the air-natural gas baseline.
(d) Oxygen-natural gas condition demonstrated a 52 % reduction in fuel relative to the air

Table 17.11 Bio-char mass balance on inlet feeds

	Furnace feed, kg/min	Volatiles									
		Fe	Slag (Ash)	Fix C	Water	C	H_2	N_2	O_2	S	Row totals
Zone 1	N. gas	–	–	–	–	0.04669	0.01466	0.00446	0.00035	–	0.06616
	Oxygen	–	–	–	–	–	–	0.00076	0.17384	–	0.17460
Zone 2	N. gas	–	–	–	–	0.07427	0.02332	0.00709	0.00055	–	0.10523
	Oxygen	–	–	–	–	–	–	0.00121	0.27516	–	0.27637
Zone 3	N. gas	–	–	–	–	0.06606	0.02074	0.00630	0.00049	–	0.09360
	Oxygen	–	–	–	–	–	–	0.00107	0.24254	–	0.24361
	Bio-char Burner fuel		0.00380	see vol C		0.06414	0.00583	0.00019	0.03680	0.00011	0.11088
	Air							0.18683	0.05676		0.24358
	Oxygen							0.00227	0.51689		0.51916
	Briquettes	0.30041	0.07188	0.08263	0.02276	0.01284	0.00487	0.00179	0.11703	0.00060	0.61481
	Cover & hearth	–	0.13982	0.57882	0.00000	0.00743	0.01308	0.00781	0.00171	0.00588	0.75455
											3.20256
	Column totals	0.30041	0.21550	0.66145	0.02276	0.27144	0.08251	0.21977	1.42212	0.00659	3.20256
		Solids fixed C to gases via reduction 0.07361731 (assumed no carbon solution loss)									
					Equilibrium inputs kg						
					Water	C	H_2	N_2	O_2	S	kg/min
					0.0228	0.3451	0.0825	0.2198	1.4221	0.0066	2.0988

Table 17.12 Mass balance and gas composition summary

	Air-natural gas	Oxy-Nat gas	Bio-char	Sub-bit. coal	Med-vol bit. coal	High-vol bit. coal
Natural gas MJ/mt NRI	137,899	60,011	115,268	42,095	52,313	52,276
Reductant coal MJ/Mt NRI	11,244	11,244	24,794	11,244	11,244	11,244
Solid carbon fuel MJ/Mt NRI			18,308	2822	20,333	14,713
Total fuel consumed, MJ/Mt NRI	149,144	71,255	158,370	56,162	83,891	78,233
% Deviation from air N. gas baseline	0.00	−52.22	6.19	−62.34	−43.75	−47.54
kg CO_2/Mt NRI	3788	1605	3347	3215	3348	
% Deviation from N. Gas Fuel Rate	0.00	−57.63	−11.64	−15.12	−11.61	−14.97
Inlet mass, kg/min						
Briquettes	0.3074	0.7173	0.6148	0.6148	0.6148	0.6148
Hearth & cover C	0.3773	0.8803	0.7546	0.7546	0.7546	0.7546
Natural gas	0.4361	0.4428	0.2663	0.2663	0.3309	0.2663
Oxygen	4.8819	1.1535	0.6946	0.6946	0.8604	0.6946
Solid carbon burner fuel			0.1109	0.0365	0.1928	0.1438
Oxygen			0.5192	0.5192	0.5192	0.5192
Air			0.2436	0.3294	0.4601	0.4026
Totals	6.0026	3.1940	3.2038	3.2152	3.7328	3.3958
Outlet mass, kg/min						
NRI	0.1547	0.3610	0.3094	0.3094	0.3094	0.3094
Slag	0.0359	0.0839	0.0719	0.0719	0.0719	0.0719
Hearth C	0.2894	0.6753	0.5788	0.5788	0.5788	0.5788
Burner coal and hearth ash	0.0699	0.1631	0.1436	0.1422	0.1588	0.1501
Flue gas	5.4506	1.9086	2.0988	2.1116	2.6123	2.2843
Totals	6.0005	3.1918	3.2026	3.2139	3.7312	3.3946
Equilibrium gas composition at 1093 °C vol %						
H_2O	15.14	33.66	48.02	48.18	35.40	39.52
H_2	9.14	27.97	5.2965	3.28	12.54	8.85
CO	7.42	22.43	6.5991	3.93	15.86	11.51
CO_2	6.13	13.48	29.867	28.87	22.34	25.65
N_2	62.13	2.23	9.9573	15.48	13.64	14.20
H_2S	0.05	0.24	0.22552	0.16	0.22	0.26
HS	1.82E-04	5.52E-04	1.19E-03	1.07E-03	7.59E-04	1.05E-03
NH_3	2.15E-04	2.18E-04	3.79E-05	2.30E-05	1.62E-04	9.78E-05

(continued)

Table 17.12 (continued)

	Air-natural gas	Oxy-Nat gas	Bio-char	Sub-bit. coal	Med-vol bit. coal	High-vol bit. coal
SO_2	1.30E-04	1.19E-04	3.36E-02	1.01E-01	1.35E-03	5.53E-03
Totals	100.00	100.00	100.00	100.00	100.00	100.00
Calc. m³ 15.5 C	5.14	2.31	0.85	1.85	2.49	2.09
Measured m³ at 15.5 CF	1.94	3.74	4.45	5.05	4.48	4.79
Measured (testo data) % CO_2	9.92	15.06	39.21	23.92	19.61	11.13
Avg. LGA N_2 Ports 1–4, % N_2	71.61	21.48	37.13	42.49	28.43	25.37
Flue gas energy content, MJ/min						
H_2	4.81	6.62	0.46	0.62	3.21	1.90
CO	4.56	6.20	0.67	0.87	4.73	2.88
MJ/Mt NRI						
H_2	31,113	18,351	1,484	2,010	10,358	6,128
CO	29,467	17,181	2,159	2,817	15,284	9,298
Totals	60,579	35,532	3,643	4,827	25,642	15,426

Note: 1 BTU = 1.055 kJ

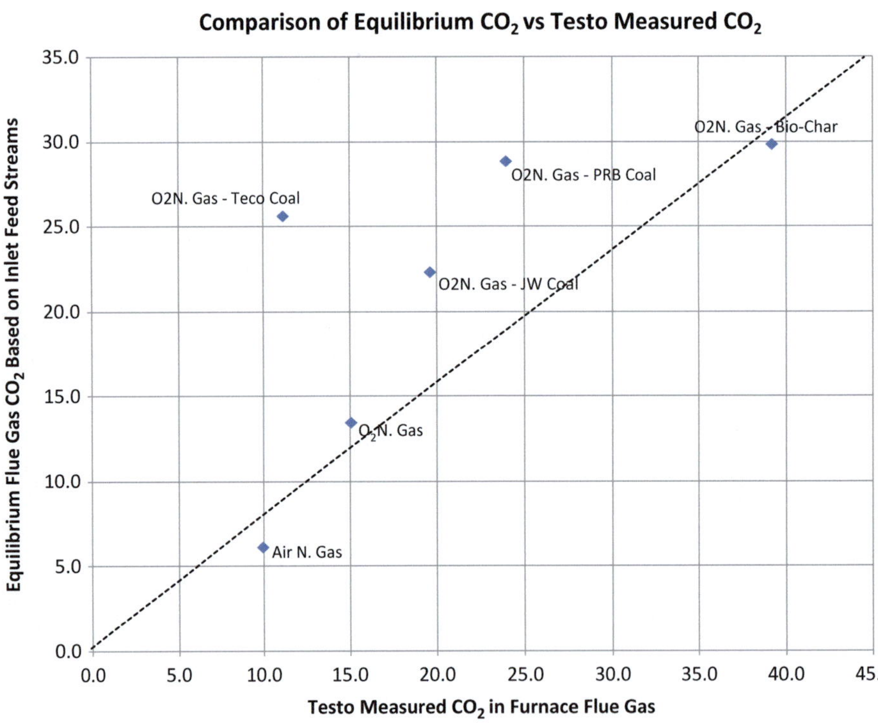

Fig. 17.15 Comparison of equilibrium of CO_2 vs. measured

baseline; similarly using bio-char reduces fuel by 51 %, High-volatile bituminous coal reduced fuel by 47 % and J W Coal reduced fuel consumed by 43 %, all relative to air.

17.6 Commercialization and Market Acceptance of Nodular Reduced Iron (NRI)

17.6.1 Value in Use

The products from this process development are targeted to provide high quality, low impurity iron units to electric arc furnace (EAF) steel manufacturers, but can also be used to enhance blast furnace productivity and meet basic oxygen furnace coolant and scrap requirements, and can be used in various iron foundry applications. NRI consists of approximately 96.5–97 % metallic iron, 2.5–3 % carbon, and minimal tramp impurities. NRI can be handled using conventional material handling techniques and is very dense and can easily penetrate steel slag. It is anticipated that NRI will be used at rates up to 30 % of the metallic charge in a high-powered electric furnace and can be added to the furnace on either an intermittent basis or using continuous charging practices. The contained carbon provides valuable chemical energy to displace electrical power requirements during steel processing when oxygen blowing practices are employed in the EAF operation. The product quality from pilot plant operations at the NRRI's Minerals' Laboratory was evaluated by a leading electric furnace-based steel company in the United States who found that NRI would be equivalent to or better than purchased pig iron that is routinely used in their various plants today. NRI technology must be compared to other viable technologies such as natural gas-based DRI production to determine the best overall fit for the future in producing iron from iron ore or other iron-bearing materials. The availability of merchant natural gas and coals will play a key role in determining the best path forward.

17.6.2 Economic Analysis

Depending on the cost of the incoming iron oxide materials, a preliminary economic analysis of the cost of iron nodule production by the development team indicates that iron nodule production costs can range from $190 to $250/t using the data generated from the pilot-scale testing. The biggest cost items are the cost of iron ore and coal required for the process. A key need for the process demonstration is to refine the economic analysis of the process using a facility design that is much closer to commercial size compared to the pilot furnace at NRRI.

17.6.3 Market Share

The amount of steel produced by electric arc furnaces (EAF) on a worldwide basis is enormous. Over 393,000,000 t of crude steel was produced in 2007 from steel manufacturers using this type of steel melting facility. In the United States, over 63 % of all crude steel was made using this steel processing method. Europe, the Middle East, North America, India, and Africa also utilize EAF steel production extensively. The volume of electric arc steel manufactured in Asia is also very high even though the blast furnace/basic oxygen converter process is the predominant steel manufacturing technology employed. The key iron raw materials used in electric furnace steelmaking are scrap, direct reduced iron, and purchased pig iron. Based on discussions with steel producers, a reasonable target for NRI use in the metallic charge to an electric arc furnace is estimated to be approximately 30 % of the total metallics. If this technology were widely adopted on a worldwide basis, approximately 118 million metric tonnes of iron nodule or equivalent product could be utilized based on 2007 production levels and a 30 % market penetration using the proposed technology. This would amount to 236 NRI production modules with each module sized at 500,000 metric tonnes of production.

17.6.4 Barriers to Commercialization

The chief barriers to commercialization are: (1) Confirmation of the technical feasibility of the pilot-scale test results on a prototype level. This includes establishment of a cost-effective operating regime that will simultaneously achieve the desired yield of NRI meeting metallurgical quality requirements and the product size characteristics desired for electric arc furnace consumers. (2) The desired level of engineering detail must be developed as well so that commercialization issues can be minimized when full-scale modules are constructed. (3) The reliability of the various subprocesses including material preparation, exhaust gas handling, and product removal also need to be established so that working ratios for system availability are well understood. (4) The costs of the raw materials for the process are within control levels of the original assumptions so that the attractiveness of the new process remains favorable compared to alternative technology options for pig iron including conventional blast furnace iron production, charcoal mini-blast furnace iron production, or direct reduced iron processes based on natural gas systems or iron-smelting processes.

17.7 Conclusions

The objective of development effort was to develop a robust process for producing NRI that would also allow environmental benefits to be attained through displacement of traditional carbonaceous fuels. During the development process, various factors ranging from furnace design to mix development, to processing parameter control, and fuel type were investigated. The results from this development show that a high purity iron product can be routinely produced and used routinely as a scrap substitute in a variety of steelmaking processes, especially the electric arc furnace. The research examined the use of solid fuel-oxygen fired combustion system and compared the results for this system with both oxygen-fuel and air-fuel combustion systems. The solid pulverized fuels tested included various coals and a newly created bio-char produced through torrefaction of woody biomass in a specially constructed pilot-scale torrefaction reactor at the NRRI's Laboratory. The various fuels were tested to determine if high-quality NRI could be produced under varying operating conditions. In addition, the properties of bio-char were evaluated both as combustion fuel and as a metallurgical agent used for atmosphere control or as a reducing agent for iron ore conversion to metallic iron. The results establish that a variety of reductant materials and iron ore mixtures with appropriate flux additions can allow routine production of high-quality, low sulfur metallized nodules. The results also illustrate that atmosphere control is vitally important to produce high-quality NRI and that both torrefied biomass, subbituminous chars and high-volatile bituminous coals can be used as potential reductant materials. The actual blend of materials used is dependent on the type of reductant and some limitations in terms of reducing the production of micro-NRI. The results from this study indicate that the approaches taken can reduce both energy intensity and greenhouse gas emissions associated with the Linear Hearth Furnace process for converting iron ore to NRI. The process results coupled with earlier reported developments indicate that this process technique should be evaluated at the next level in order to develop parameter information for full-scale process design. Implementation of the process to full commercialization will require a full cost production analysis and comparison to other reduction technologies and iron production alternatives. The technical results verify that high-quality NRI can be produced under various operating conditions at the pilot level.

17.8 Recommendations

The results show that the reduction and smelting process can be used to produce high-quality NRI under a variety of conditions and that oxygen-fuel systems can result in reduced energy requirements for the production process. In addition, torrefaction can be used to produce an interesting bio-char product that has very useful metallurgical properties. The results of the program were conducted using both laboratory and pilot-scale equipment. The development needs to be brought to the next level of scale so that detailed engineering can be based on a process demonstration closer to that required for commercialization. This would be the next logical step for continued development of this technology. The results from laboratory and process modeling can facilitate the next level of development for this process.

17.9 Intellectual Property and Patents Relating to NRI Development

In the course of the development program supported by the US Department of Energy and prior to that by the US Economic Development Agency, the following patent portfolio has been developed:

Title	Grant reference	Status
High luminosity burner	DOE S-111,957 & DE-FG36-05GO15185	Abandoned
Linear hearth furnace	DOC/EDA 06-69-04501	US 7,413,592; US 7,666,249; US 7,875,236
Magnetic removal system used with linear hearth system	DOE DE-FG36-05GO15185	Disclosure
Method and system for producing metallic iron nuggets (coarse coal cover)	DOD/EDA 06-69-04501 & DE FG36-05GO15185	PCT/US07/74471; CA2,658,897; US 12/359,729
Method and system for producing metallic iron nuggets (late stage addition of coarse cover)	DOE DE FG36-05GO15185	US 8,021,460
Method and system for producing metallic iron nuggets	DOC/EDA 06-69-04501	US60/633,886
Method and system for producing metallic iron nuggets (domes, cones, and mounds)	DOC/EDA 06-69-04501	US 7,695,544; AU 2005313001; Multiple applications PCT/IB05/054110
Method and system for producing metallic iron nuggets (70–90 stoichiometric carbon)	DOC/EDA 06-69-04501	US 7,628,839; AU 2005312999; multiple applications PCT/IB05/054108
Method and system for producing metallic iron nuggets (stoichiometric carbon with additives)	DOC/EDA 06-69-04501	US 7,641,712
Method and system for producing metallic iron nuggets (micro-agglomerates)	DOC/EDA 06-69-04501	US 7,632,335
Method and system for producing metallic iron nuggets (multiple layer and subbituminous material)	DOC/EDA 06-69-04501 & DE FG36-05GO15185	PCT/US09/032519; US 12/847,591
Production of iron from metallurgical waste (recovery of metallurgical waste)	DOC/EDA 06-69-04501 & DE-FG36-05GO15185	PCT/US10/21790; US 13/187,937; AU 2010 206718
Refractory coating for hearth	DOE DE FG36-05GO15185	Disclosure
System and method for cooling and removing iron from a hearth	DOC/EDA 06-69-04501 & DE-FG36-05GO15185	PCT/US08/50855; US 12/522,867; CA 2,675,311;
System and method for making iron with reduced CO_2 emissions (shrouded LHF with CO_2 control)	DOE DE FG36-05GO15185	US 61/246,817; PCT/US10/50547
System and method for producing metallic iron (O_2 plus coal consumption including flue gas recirculation)	DOE DE-FG36-05GO15185	PCT/US10/50743; US 61/246,739

(continued)

(continued)

Title	Grant reference	Status
System and method for producing metallic iron (injection Under the hood in conversion zone)	DOC/EDA 06-69-04501	US 60/828,171; PCT/US07/80364
System and method for producing iron (wedge linear hearth furnace)	DOE DE FG36-05GO15185	PCT/US10/507 30; US 61/246,787
Use of bimodal carbon distribution in compacts for producing metallic iron nodules	DOE DE FG36-05GO15185	US 12/977,035
Use of manganese oxide with fluorspar	DOE 117,356 & DE FG36-05GO15185	Disclosure

17.10 Glossary

17.10.1 Abbreviated Notation of Lime and Fluorspar in Slag

In order to simplify the notation of the fluxing additive of lime and fluorspar, the following notation was used in this report. Composition (L) is located in the low fusion temperature trough near $(CaO)/(SiO_2)$ of 1.2 in the CaO-SiO_2-Al_2O_3 phase diagram. The slag compositions were abbreviated by indicating the amounts of additional lime used in percent as a suffix, for example, $L_{0.5}$ and L_1 indicated lime additions of 0.5 % and 1 %, respectively, over that of Composition (L). The amount of fluorspar (abbreviated to FS) added in percent was also indicated as a suffix, for example, $L_{0.5}FS_{0.25}$, which represented that 0.25 % by weight of fluorspar was added to a feed mixture with Slag Composition of $L_{0.5}$.

17.10.2 Atmosphere Control

Atmosphere control in the LHF refers to the control of the atmosphere directly above the feed compositions, specifically CO and H_2, with the absence of oxygen, and the volumetric turbulence of the combustion products that have an influence on these compositions.

17.10.3 Basicity

$$B_2 = \frac{CaO}{SiO_2} \qquad B_4 = \frac{CaO + MgO}{SiO_2 + Al_2O_3}$$

17.10.4 Bimodal Super Stoichiometry

Bimodal Super Stoichiometry or BMSS is a term given to the practice of using a coarse and fine size fraction of carbon reductant in the reaction mixture for the production of NRI. The fine fraction provides the reductant source for conversion of iron oxides to reduced iron while the coarser fraction survives the furnace atmosphere to allow the carbon solution reaction to sufficiently reduce the melting temperature to enhance the kinetics for the production of metallic iron.

17.10.5 Bio-char

Bio-char is defined as a coal or carbon substitute produced from a biomass resource.

17.10.6 Fusion Time

Fusion time for NRI, also referred to as residence time, is defined as the time required in the box furnace or LHF to fully reduce a composition consisting of iron oxide, carbon reductant, and fluxes to metallic iron. Reducing fusion time correlates to an increase in productivity for a given furnace dimension.

17.10.7 Micro-NRI

Both in box furnace and LHF tests, NRI with a range of sizes formed depending on the test conditions used. The magnetic products after each test were collected with a hand magnet and screened into +6.35 mm (1/4″), −6.35 mm (1/4″), +20 mesh (0.833 mm), and −20 mesh (0.833 mm) fractions. Plus 6.35 mm (1/4″) fractions were fully metallic when the products were judged to be fused. Minus 6.35 mm (1/4″) +20 mesh (0.833 mm) fractions were essentially all metallic and referred to as "micro-NRI." Minus 20 mesh fractions had large amounts of fine carbon particles to which small metallic iron particles were attached. These are generally considered as unacceptable product as they typically contain higher levels of sulfur and their relatively small size can negatively influence EAF productivity.

17.10.8 Stoichiometric Amount

In an attempt to quantify the amount of coal, coke, or char needed as a reductant in feed mixtures, the amount of carbon required to reduce iron oxides to metallic iron with the formation of CO was calculated and termed "stoichiometric amount" according to

$$FeO + C = Fe + CO$$

$$Fe_3O_4 + 4C = 3Fe + 4CO$$

$$Fe_2O_3 + 3C = 2Fe + 3CO$$

Fixed carbon from proximate analysis was used in the calculation.

17.10.9 Torrefaction

Torrefaction is a mild pre-treatment of biomass at a temperature between 200 and 300 °C under low oxygen conditions to remove moisture and light volatiles and accumulate the carbon component. The resulting product is a partially carbonized biomass, essentially moisture-free and friable which allows it to be used similar to coal. Torrefaction of biomass has the potential to produce chars from various biomass sources (e.g., wood by-products, agricultural by-products, grasses, energy crops) giving them similar material handling and processing capabilities.

References

Bale, C. W., Pelton, A. D., Thompson, W. T., reiksson, G., Hack, K., Charthand, P., Decterov, S., Jung, I-H., Melancon, J., & Petersen, S. *FactSage thermochemical software, center for research in computational thermochemistry*, www.crct.polymtl.ca, www.gtt-technologies.de, and www.factsage.com.

Fosnacht, D. R., Iwasaki, I., & Bleifuss, R. L. (2004). *Metallic iron nodule research* (Final Report to the Economic Development Administration, Project #06-69-04501, March 25, 2004, 15 p).

Fosnacht, D. R., Iwasaki, I., Kiesel R. F., Englund D. J., Hendrickson D. W., & Bleifuss, R. L. (2010). *Next generation, metallic iron nodule technology in electric arc steelmaking—Phase II* (Natural Resources Research Institute, University of Minnesota Duluth, Technical Report NRRI/TR-2010/32, p. 365).

Fruehan, R. J., & Fortini, O. (2004). *Sustainable steelmaking using biomass and waste oxides—Evaluation of a new process for ironmaking: A productivity model for the Rotary Hearth Furnace* (Carnegie Mellon University, US DOE-DE-FC36-97ID13554, C-1 to C-25).

Hansen, J. A. (2005). Mesabi nugget—The new age of iron. *Iron and Steel Technology, 2*(3), 149–153.

Hoffman, G., & Tsuge, O. (2004). ITmk3—Application of a new ironmaking technology for the iron ore mining industry. *Mining Engineering, 56*(10), 35–39.

International Report Japan. (2001). KSC develops new ironmaking process. *Iron and Steelmaker, 28*(10), 8.

Ishiwata, N., et al. (2004). Pilot plant test for production of iron pebble (Development of new reduction and smelting process on coal bed—1). *CAMP-ISIJ, 17*, 150.

Kobayashi, I., et al. (2001). A new process to produce iron directly from fine ore and coal. *Iron and Steelmaker, 27*(9), 19–22.

Prins, M. J. (2005). *Thermodynamic analysis of biomass gasification and torrefaction*. Eindhoven, The Netherlands: Technische Universiteit Eindhoven, Proefschrift. ISBN 90-386-2886-2.

Riley, M. F., et al. (2001). Praxair's dilute oxygen combustion technology for pyrometallurgical applications. *Journal of Metals, 53*(5), 21.

Sawa, Y., Yamamoto, T., Takeda, K., & Itoya, H. (2000). New coal-based process, Hi-QIP, to produce high quality

DRI for the EAF. *2000 Electric Furnace Conference Proceedings, vol 58, Orlando, FL* (pp. 507–517).

US Department of Energy. (1993). *Oxygen-enriched combustion can reduce emissions and fuel use in energy-intensive industries.* Retrieved DOE/CH10093-198, DE93000063, September, 1993, from http://es.epa.gov/techinfo/facts/o2-nrich.html

US Department of Energy, Energy Efficiency and Renewable Energy, Industrial Technologies Program. (2004). *Steel success story. Ironmaking: Quality and supply critical to steel industry.* Retrieved September 21, 2004, from http://www.eere.energy.gov/industry/steel/

Veranth, J. M., Krishnamoorthy, G. (2002). Oxygen-enriched coal combustion with carbon dioxide recycle and recovery: Simulation and experimental study—final report (July 01, 2001–May 30, 2002, DOE Award DE-FG26-00NT40828, Department of Chemical and Fuels Engineering, University of Utah).

18 Innovative Process for the Production of Titanium Dioxide

V.I. Lakshmanan, Raja Roy, and M.A. Halim

18.1 Introduction

Titanium (Ti) is the ninth most abundant element in the earth's crust and forms a very stable oxide, titanium dioxide. Titanium dioxide occurs in three different forms naturally: anatase, rutile, and brookite. The brookite phase is relatively rare compared to anatase and rutile. Rutile phase is the most stable phase and is widely used in the paint industry. With suitable heat treatment, anatase and brookite can be converted to rutile phase. In addition to stable rutile, anatase, and brookite phases, there are three metastable phases that can be produced synthetically (monoclinic, tetragonal, and orthorhombic) and five high-pressure forms (α-PbO_2-like, baddeleyite-like, cotunnite-like, orthorhombic, and cubic phases) (http://en.wikipedia.org/wiki/Titanium_dioxide).

Since its first commercial production in 1916, titanium dioxide has found several uses and is currently used in paints, plastics, rubber, paper, inks, textiles, and a number of other miscellaneous applications. More than 50 % of the six million tpa of TiO_2 produced is used as pigment in paints. Rutile phase is preferably used for application as pigment; however, the anatase phase has recently gained interest as it can be used for making nanomaterials (Lakshmanan et al. 2014). Currently, nanosize TiO_2 is being investigated for application in the cosmetics industry. Inorganic sunscreens use TiO_2 or ZnO particles, which reflect and scatter ultraviolet light. TiO_2 nanomaterials are also finding application as smart construction materials and as coatings in photocatalytic converters. Nano TiO_2 can catalyze the dissociation of NO_x gases and other automotive emissions, thereby reducing the smog in urban centers. After the construction material is exposed to sunlight or ultraviolet light, TiO_2 is photoactivated and degrades the pollutants.

Due to growing market demand, the TiO_2 price has increased substantially over the past few years as shown in Fig. 18.1 (Cianfichi 2012). There are two commercial processes for the production of TiO_2, the chloride process and the sulfate process. Both of these processes have several environmental challenges. While the sulfate process suffers from large-scale generation of acidic and solid waste as well as toxic emissions, the chloride process suffers from CO_2 and other toxic emissions. These processes have not been designed to recycle the reagents and discharge the effluents into the environment causing an environmental concern. With this in view, an innovative process has been developed and patented by Process Research Ortech Inc. (PRO) and ownership transferred to Canadian Titanium Limited (CTL). This process uses chloride metallurgy for the production of pigment-grade TiO_2.

V.I. Lakshmanan (✉) • R. Roy • M.A. Halim
Process Research Ortech Inc.,
2350 Sheridan Park Drive, Mississauga,
ON, Canada, L5K 2T4
e-mail: llakshmanan@processortech.com

V.I. Lakshmanan et al. (eds.), *Innovative Process Development in Metallurgical Industry*,
DOI 10.1007/978-3-319-21599-0_18

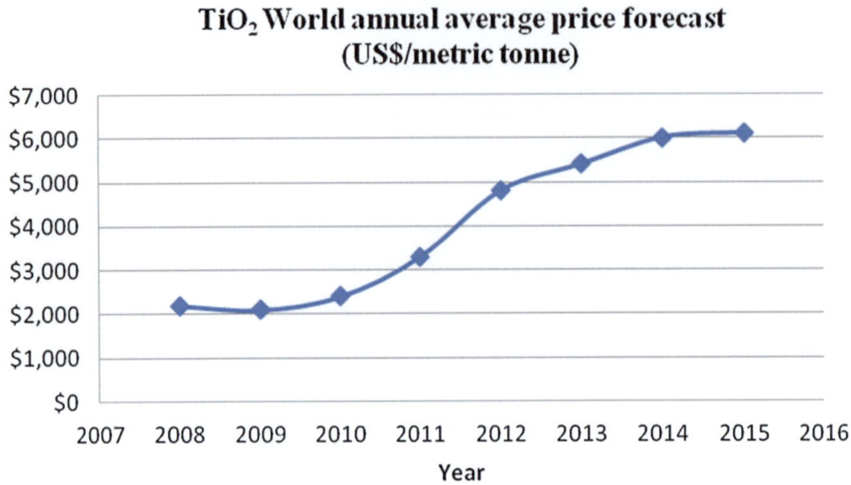

Fig. 18.1 TiO_2 price history and forecast (based on Cianfichi 2012)

Table 18.1 Titanium minerals and their chemical compositions with TiO_2 contents (based on Lakshmanan et al. 2014)

Mineral	Composition	TiO_2 content (%)
Rutile	TiO_2	94–99
Anatase	TiO_2	90–95
Brookite	TiO_2	90–100
Brookite	TiO_2	90–100
Ilmenite	$FeTiO_3$	40–65
Leucoxene	$Fe_2O_3TiO_2$	60–90
Perovskite	$CaTiO_3$	40–60
Titanite	$CaTiSiO_3$	30–42
Titaniferous magnetite	$(Fe \cdot Ti)_2O_3$	2–20

18.2 Raw Materials for Titanium Dioxide Production

Table 18.1 shows the most common titanium minerals along with their chemical compositions (Lakshmanan et al. 2014). Titanium dioxide is produced from the following ores: Ilmenite ($FeTiO_3$), rutile, anatase, and leucoxene ($TiO_2 \cdot xFeO \cdot yH_2O$). Other sources of raw material for the production of titanium dioxide are slag produced during iron smelting and synthetic rutile produced from ilmenite.

Ilmenite is the most abundant titanium dioxide-bearing ore containing 40–65 % TiO_2, depending on its geological history. Leucoxene ($Fe_2O_3 \cdot nTiO_2$) contains typically more than 65 % TiO_2 and is a natural alteration product of ilmenite. Ilmenite and Leucoxene may contain magnesium, manganese, vanadium, aluminum, calcium, chromium, and silicon as impurities. Table 18.2 shows the chemical composition of ilmenite ores found in different locations around the world (Kothari 1974). Major igneous deposits of ilmenite are found in Brazil, Canada, Norway, the Russian Federation, and the Ukraine, while the heavy mineral sands are found along the eastern and western coasts of Australia, the eastern coast of South Africa, the southeastern coast of the United States, the west coast of South Island, New Zealand, the eastern coast of China, the northeastern coast of Sri Lanka, at various locations along the southern coast of India, in coastal Malaysia and in alluvial deposits in Sierra Leone (Chang 2002). World reserves of ilmenite in terms of TiO_2 content are shown in Table 18.3 with major reserves being in China (31 %), Australia (15 %), India (13 %), South Africa (10 %), Brazil (7 %), Madagascar and Norway (6 % each), as well as in Mozambique (2 %) (http://minerals.usgs.gov/minerals/pubs/mcs/2013/mcs2013.pdf).

Table 18.2 Chemical composition of ilmenite ores (in %) (Kothari 1974)

	North and South America			Europe			Asia		
(%)	U.S.A.	Canada	Brazil	Norway	Russia	Portugal	India	Malaysia	Sri Lanka
TiO_2	43–50	35–73	48–61	37–44	44	52.2	52.2	51–54	53.61
FeO	35–39	31–33	26–27	32–36	32.4	42.1	9–26	35–38	20.67
Fe_2O_3	1.6–13.8	20	14–15	11–13	16.9		12–27	3–6	20.96
SiO_2	1.4–3.0	0.8–4.0	1.4	0.6–3.0	1.84	0.27	0.9–1.4	0.5–1.11	0.38
Al_2O_3	0.2–1.21	1.05–1.7	0.25	0.85–1.8	0.2	0.29	1.0–1.11	1.09	0.54
P_2O_5	0.07–1.01			0.01–0.2	0.15	0.03	0.17–0.26	0.09	
ZrO_2	0.05–0.55		0.25–0.3	0.3–1.0			0.60–2.1		
MgO	0.6–2.35	1.0–2.0	0.3–0.35	1.6–3.0	2.76	0.03	0.65–1.01	0.21	0.92
MnO	0.1–0.52	0.03–0.04	0.1–0.2	0.2–0.3	0.72	5	0.40–0.48	3.5	0.95
V_2O_5	0.05–0.27	0.2–0.35	0.06–0.2	0.2–0.3	0.09	0.09	0.03–0.26	0.04	
Cr_2O_3	0.02–0.28	0.15	0.1–0.5	0.03–0.07	0.02	0.02	0.03–0.26	0.02	0.05
REE Oxide	0.08	0.07	0.08	0.068	0.07		0.08	0.079	

	Africa		Australia	
(%)	Senegal	Sierra Leone	West Coast	East Coast
TiO_2	54–56	42.3	52–55	45–55
FeO	14–Jul	28	20–23	28–33
Fe_2O_3	28–30	25	15–19	1.5–5
SiO_2	0.9–1.3		1.41	1.2
Al_2O_3	0.5		0.25	1
P_2O_5	0.14–0.15		0.15	
ZrO_2	2.37		0.08	
MgO	1.9		0.01	
MnO	1.32		1.44	1.25–1.5
V_2O_5	0.27	0.38	0.13	0.23–1.0
Cr_2O_3	0.23		0.03	1.84–4.9
REE Oxide	0.06		0.089	0.12

Rutile containing about 95 % titanium dioxide with small amounts of iron and other impurities is the second most commonly available ore for the production of titanium dioxide. Rutile is often found as beach sands and is the simplest to mine and concentrate in a form suitable for titanium dioxide production. Rutile sand deposits are found in Australia, Sierra Leone, and South Africa. Production of titanium dioxide from rutile now accounts for less than 10 % due to diminishing reserves as compared to 20 % in the 1970s. Major rutile reserves are located in Australia (43 %), South Africa (20 %), India (18 %), Sierra Leone (9 %), and Ukraine (6 %) as shown in Table 18.3 (http://minerals.usgs.gov/minerals/pubs/mcs/2013/mcs2013.pdf). The largest reserves of anatase are found in carboniferous intrusions in Brazil.

18.3 Beneficiation and Upgrading of Ore

18.3.1 Beneficiation and Upgrading of Ilmenite Ore

Beneficiation of ilmenite ores typically includes the following process steps: crushing, grinding, sorting, sizing, gravity concentration, and magnetic separation. The flowsheet for beneficiation of ilmenite ore is shown in Fig. 18.2 (Gueguin and Cardarelli 2007). The majority phases in an ore are ilmenite and hematite, while the minority phases include silicates, oxides, and sulfides. The major part of the silica, alumina, and calcium oxide comes from gangue minerals such as plagioclase, feldspar, and andesine. The aim of

Table 18.3 World reserves and production of ilmenite and rutile in 2012 (in kilotons) (http://minerals.usgs.gov/minerals/pubs/mcs/2013/mcs2013.pdf)

	Ilmenite		Rutile	
Country	Reserves	Production	Reserves	Production
United States[a]	2000	300	–	–
Australia	100,000	940	18,000	480
Brazil	43,000	45	1200	5
Canada	31,000	700	–	–
China	200,000	700	–	–
India	85,000	550	7400	25
Madagascar	40,000	280	–	–
Mozambique	16,000	380	480	8
Norway	37,000	350	–	–
Sierra Leone	–	–	3800	100
South Africa	63,000	1.030	8300	131
Sri Lanka	–	60	–	–
Ukraine	5900	300	2500	60
Vietnam	1600	500	–	–
Other countries	26,000	40	400	17
World total (rounded)	650,000	6.200	42,000	830

[a]United States rutile production and reserve data are included with ilmenite

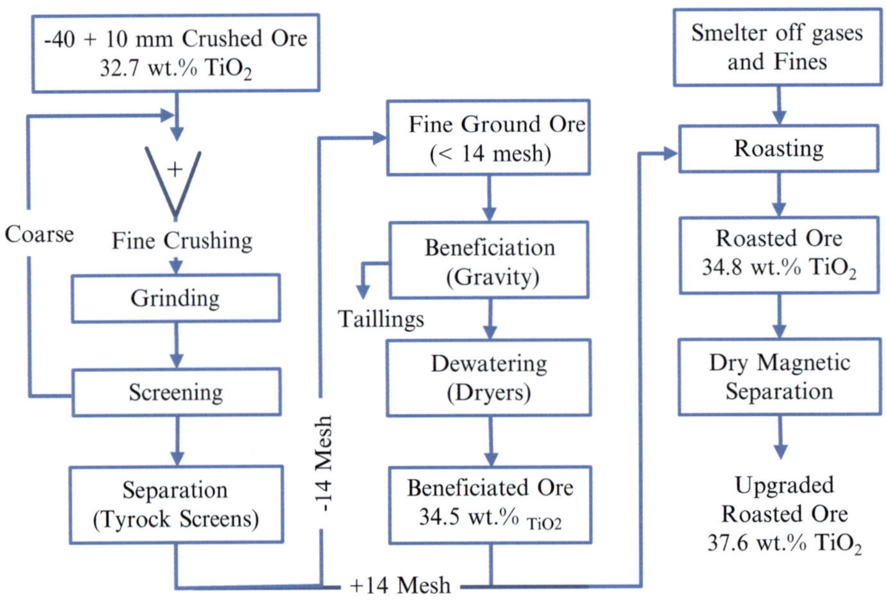

Fig. 18.2 A simplified flowsheet for the beneficiation process of an ilmenite ore (based on Gueguin and Cardarelli 2007)

beneficiation and upgrading processes is to generate a concentrate with high Ti content suitable for further processing.

After primary crushing and screening of the ore at the mine site using gyratory and cone crushers, the ore is further crushed and sized at a secondary milling facility. The ore is ground to a fine powder and subjected to gravity flotation, which separates lighter gangue (nonmetallic or nonvaluable rock) material from the heavier valuable mineral based primarily on differences in the specific gravities of the materials. The next step

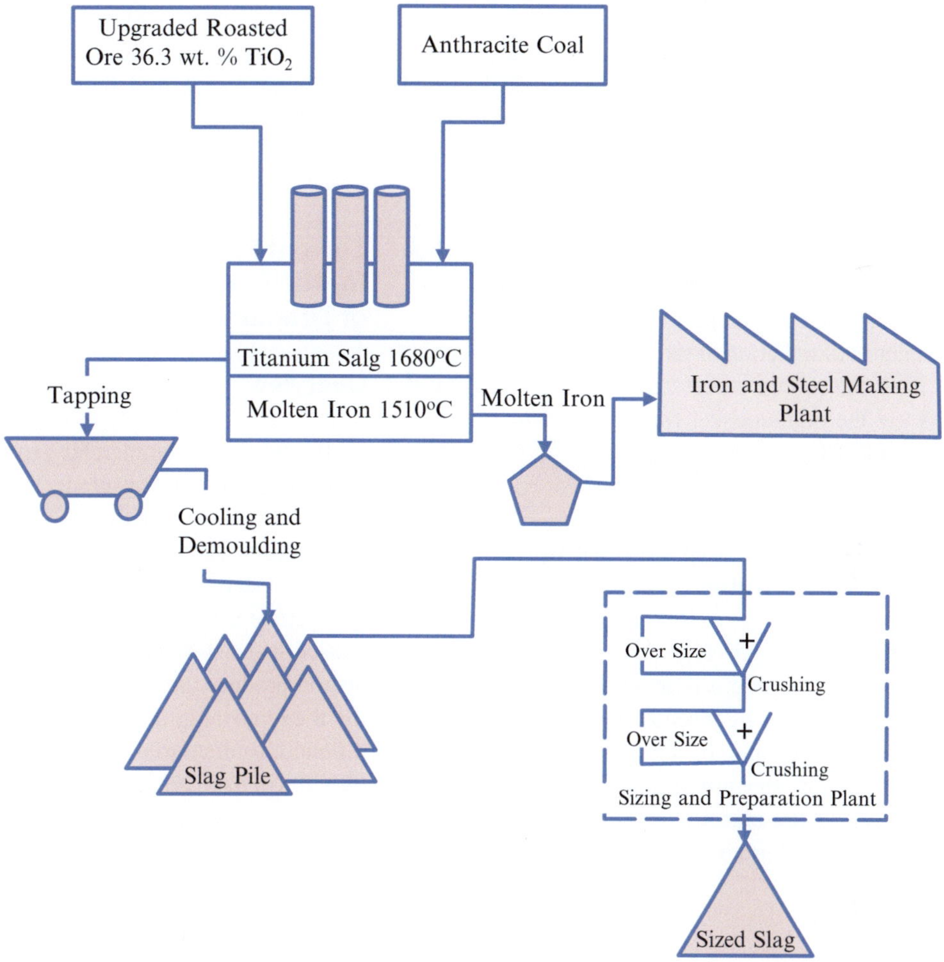

Fig. 18.3 A process flowsheet for production of titanium slag from upgraded roasted ilmenite ore (Gueguin and Cardarelli 2007)

is oxidative roasting to prepare a so-called gravity concentrate for magnetic separation. The following reaction takes place to produce magnetite (Fe_3O_4) and pseudobrookite (Fe_2TiO_5) during oxidative roasting:

$$FeTiO_3 + 2Fe_2O_3 = Fe_3O_4 + Fe_2TiO_5 \quad (18.1)$$

Further upgrading is carried out by magnetic separation to generate an upgraded roasted ilmenite ore, which is then used for the production of titanium slag.

Titanium slag of TiO_2 content between 75 and 86 % is produced by smelting ilmenite at high temperature (approximately 1650–1700 °C) with a suitable carbonaceous reductant in electric furnaces. The smelting process can be described by the following reactions (Chase Jr 1998):

$$Fe_2O_3 + C = 2FeO + CO \quad (18.2)$$

$$FeO + C = Fe + CO \quad (18.3)$$

$$2TiO_2 + C = Ti_2O_3 + CO \quad (18.4)$$

During the smelting process, Ti(IV) is partially reduced to Ti(III), as indicated by the high Ti_2O_3 content of the bulk titanium slag. Figure 18.3 shows the process flowsheet for the production of titanium-rich slag from upgraded roasted ore (Gueguin and Cardarelli 2007). During the high-temperature smelting process, liquid iron settles to the bottom of the furnace forming a pool of a molten iron–carbon alloy, while a thick layer of titanium-rich slag forms above the molten iron bath. The titanium-rich slag is poured directly into steel wagons lined with crushed slag bedding and subsequently cooled under water sprays for several hours to build a thick protective skull. It is then air cooled till complete solidification of the inner core takes place. Both the sulfate process and the chloride process use the titanium slag for the production of high-grade TiO_2.

18.3.2 Beneficiation and Upgrading of Natural Rutile

Beneficiation of rutile ores typically includes the following process steps: crushing, grinding, magnetic separation, gravity concentration, flotation, and electrostatic separation. The flowsheet for the beneficiation of rutile ore is shown in Fig. 18.4 (Kothari 1974). After crushing and grinding, the iron-rich magnetic phase is removed by magnetic separation. Gravity separation methods and flotation are used to separate silica, silicate, and aluminates. Zircon is removed by electrostatic separation to obtain high-grade titanium dioxide from beach sands.

18.4 Current Processes for the Production of Titanium Dioxide

18.4.1 Overview

There are currently two commercial processes for the production of pigment-grade titanium dioxide: the sulfate process and the thermochemical chloride process. These two processes differ in both their chemistry and raw material requirements. In the sulfate process, concentrated sulfuric acid is used to leach high-grade ilmenite or titanium slag to produce titanium sulfate. This is followed by the removal of iron through crystallization of titanium sulfate solution to precipitate titanium dioxide. In the thermochemical chloride process, titanium slag, natural rutile, or synthetic rutile is reacted with petroleum coke and chlorine

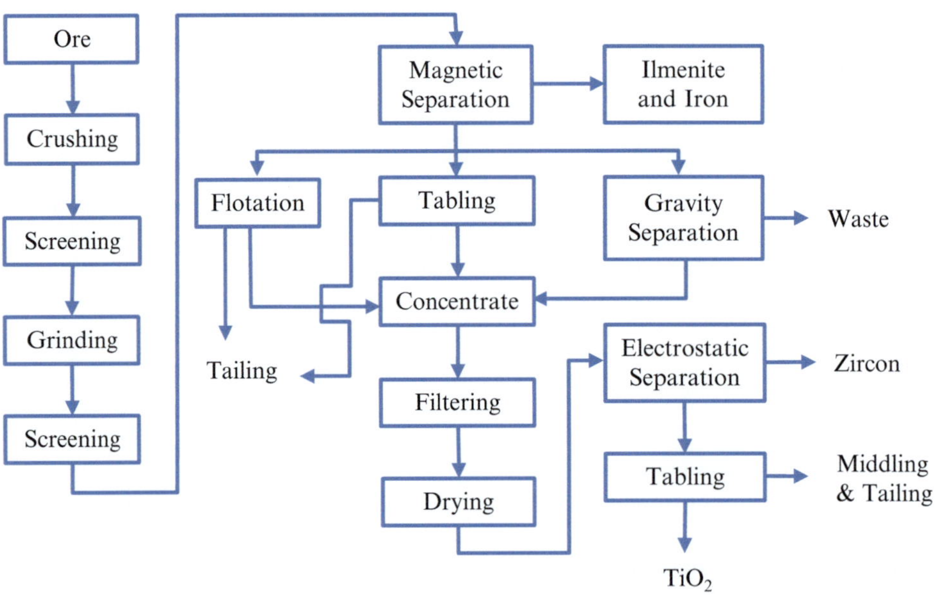

Fig. 18.4 A flowsheet for processing natural rutile from beach sands (Kothari 1974)

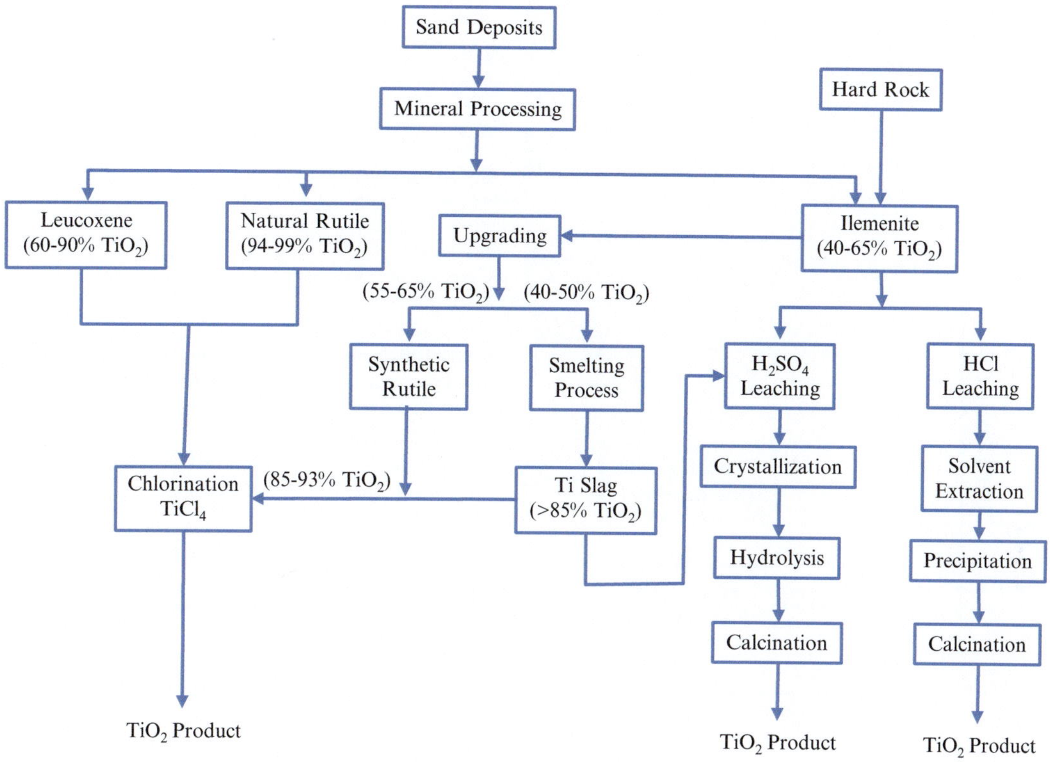

Fig. 18.5 Overview of the titanium mineral processing industry from the raw material to the final product (Lakshmanan et al. 2014)

gas at high temperatures to form titanium tetrachloride ($TiCl_4$) gas, which is reacted with oxygen to produce pigment-grade titanium dioxide.

A new process, CTL process, is currently being commercialized to produce pigment-grade titanium dioxide (Lakshmanan et al. 2002a, b, 2004, 2010a, b). The process involves leaching of ilmenite ore in a mixed chloride lixiviant ($HCl+MgCl_2$), followed by solid–liquid separation and successive solvent extraction stages for the production of high purity iron and titanium pregnant strip liquors. Using concepts of solution chemistry, hydrated titanium oxide is thermally precipitated from the Ti-bearing pregnant strip liquor at controlled pH and Eh conditions. It is subsequently calcined to produce titanium dioxide. Iron oxide is produced by pyrohydrolysis of the Fe-bearing pregnant strip liquor as a by-product. HCl is recovered in this step and recycled to the leaching stage. Figure 18.5 shows the various processing routes for the production of high-grade TiO_2 from different titanium-bearing minerals. A brief description of the sulfate and the thermochemical chloride processes is provided below, which is followed by the description of the innovative mixed chloride process (CTL process) developed by Process Research Ortech Inc.

18.4.2 The Sulfate Process

The sulfate process was the first process developed for producing TiO_2 pigment. It was first used in Norway in 1915. It has been undergoing continuous improvements over the years. The sulfate process uses H_2SO_4 and is sensitive to the price of acid. Figure 18.6 shows the flowsheet of the sulfate process for production of TiO_2 pigment (McNulty 2007). The sulfate process consists of the following steps:

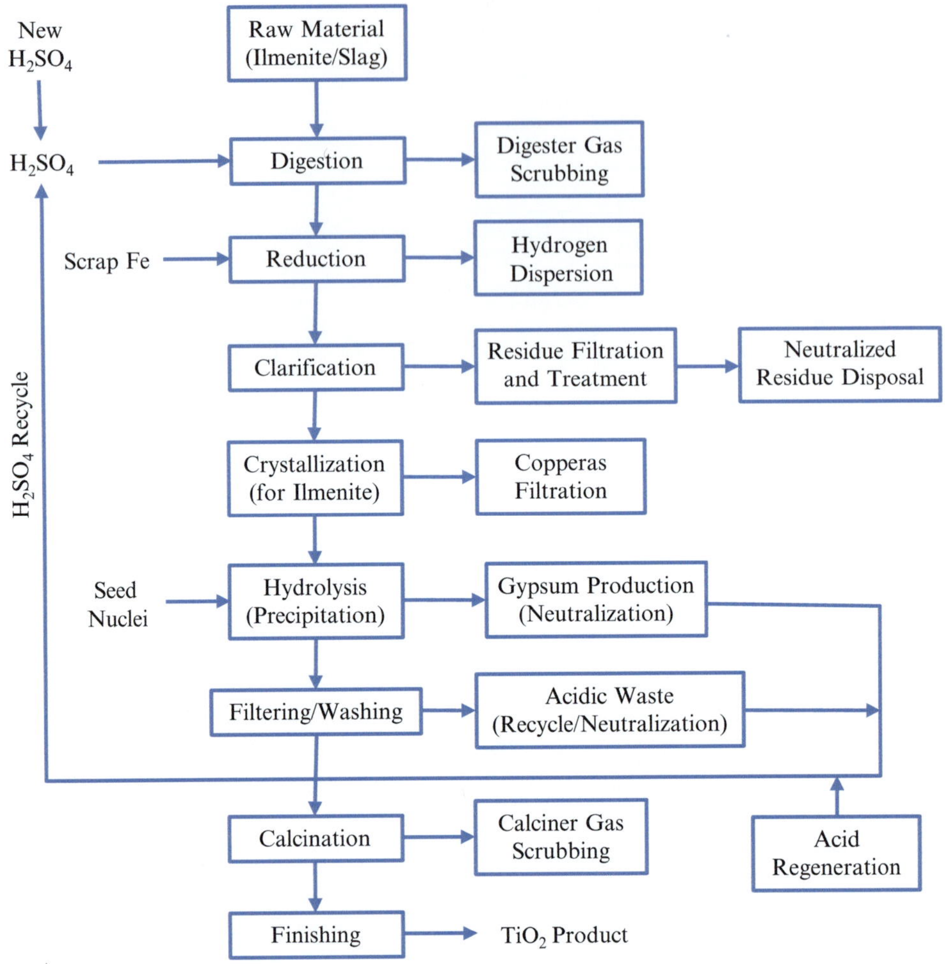

Fig. 18.6 The flowsheet for the production of TiO_2 pigment from ilmenite ore or slag by the sulfate process (based on McNulty 2007)

18.4.2.1 Acid Digestion and Clarification

The raw material (ilmenite ore or synthetic titanium slag) is dried and finely ground (<40 μm) in ball mills before digestion in sulfuric acid. Excess free acid is used to increase digestion efficiency. The temperature initially increases to 50–70 °C due to the heat of hydration of the sulfuric acid and subsequently the exothermic reaction increases the temperature to 170–220 °C. After completion of the leaching reaction, the filter cake is further leached with cold water or diluted sulfuric acid recycled from the process and scrap iron is added to reduce any remaining Fe(III) to Fe(II). Due to the addition of scrap iron, some of the Ti(IV) is reduced to Ti(III). This prevents the reoxidation of iron during subsequent process steps. The leaching and reduction reactions are given by the following reactions:

$$FeTiO_3 + 2H_2SO_4 = TiOSO_4 + FeSO_4 + 2H_2O \quad (18.5)$$

$$Fe_2(SO_4)_3 + Fe = 3FeSO_4 \quad (18.6)$$

The solid residue is separated from the pregnant leach liquor that contains titanyl sulfate ($TiOSO_4$) and ferrous sulfate by flocculation, sedimentation, and filtration.

18.4.2.2 Crystallization

Iron sulfate in the pregnant leach liquor is then separated from titanyl sulfate by crystallization. Pregnant leach liquor is cooled to less than 15 °C under vacuum to crystallize ferrous sulfate heptahydrate ($FeSO_4 \cdot 7H_2O$), also known as copperas, which has several commercial applications including water purification, Fe_2O_3 pigment production, agriculture, and chromate reduction for cement production. Copperas is separated from the concentrated titanyl sulfate solution by filtration or centrifugation.

18.4.2.3 Precipitation and Purification

Hydrated titanium oxide is precipitated from the titanyl sulfate solution by thermal hydrolysis. This is done by heating the solution to 95–110 °C. The feed solution is seeded with $Ti(OH)_2$ to accelerate the precipitation process. The thermal hydrolysis reaction is shown below:

$$TiOSO_4 + 2H_2O \rightarrow TiO(OH)_2 + H_2SO_4 \quad (18.7)$$

After hydrolysis, the precipitate is filtered and repulped with Ti(III) solution ($Ti_2(SO_4)_3$) and water to remove impurities including Fe, Cr, Cu, Mn, and V. Repulping can also be performed with dilute sulfuric acid at 50–90 °C. The filtrate from the thermal hydrolysis contains 20–28 % sulfuric acid, which is not recycled as explained below.

18.4.2.4 Doping, Calcination, and Grinding

In order to produce specific pigment grades, purified hydrated titanium dioxide is treated with alkali metal compounds (for rutile) and phosphoric acid (for anatase). Hydrated titanium dioxide is calcined to produce crystalline titanium dioxide. Anatase is formed at a final temperature of 800–850 °C, while a final temperature of 900–930 °C is required for rutile formation. The calcination reaction is as follows:

$$TiO(OH)_2 \rightarrow TiO_2 + H_2O \quad (18.8)$$

After cooling, the rutile/anatase TiO_2 product is milled and surface treated to form the TiO_2 base pigment.

During crystallization of ferrous sulfate, a temperature that is as low as possible is used, usually less than 15 °C. This however leaves about 20 g/L of Fe in solution. In addition, it contains other impurities such as Cr and V, which are dissolved during leaching. The presence of iron and other impurities in the solution makes the production of very pure TiO_2 difficult. This results in a slightly yellow-colored pigment. In addition, for every ton of TiO_2 produced, 1.9 t of ferrous sulfate has to be sold or disposed off. Also, the acid regenerated in reaction (18.7) is dilute and cannot be reused in the leaching reaction (18.5), which needs concentrated acid. In general, this acid is neutralized with lime producing 2.2 t of gypsum ($CaSO_4 \cdot 2H_2O$) per ton of TiO_2 and disposed off.

18.4.3 The Chloride Process

This process was developed to overcome the environmental challenges of the sulfate process such as disposal of solid waste material produced in the sulfate process. It was commercialized in the United States in 1958 (McNulty 2007) and currently accounts for more than 60 % of the world's production of titanium dioxide (TiO_2) pigment. The commercial chloride process is a continuous process. The chloride process generates significantly less waste (up to three times less) than the sulfate process (DuPont 2014). Naturally occurring rutile ore is the preferred raw material for the chloride process. As the reserves for rutile ore are diminishing, synthetic rutile, upgraded slag (UGS), and high-grade QIT (Quebec Iron and Titanium) slag are now being used as feed material for the chloride process. Figure 18.7 shows the flowsheet of a typical chloride process for production of TiO_2 pigment (McNulty 2007).

The chloride process consists of the following steps:

18.4.3.1 Chlorination

The feed is dried, mixed with coke at 250–300 kg/t TiO_2, and fed into the fluidized-bed chlorinator, a long, brick lined steel reactor, with a gas chamber and gas distributor at the bottom

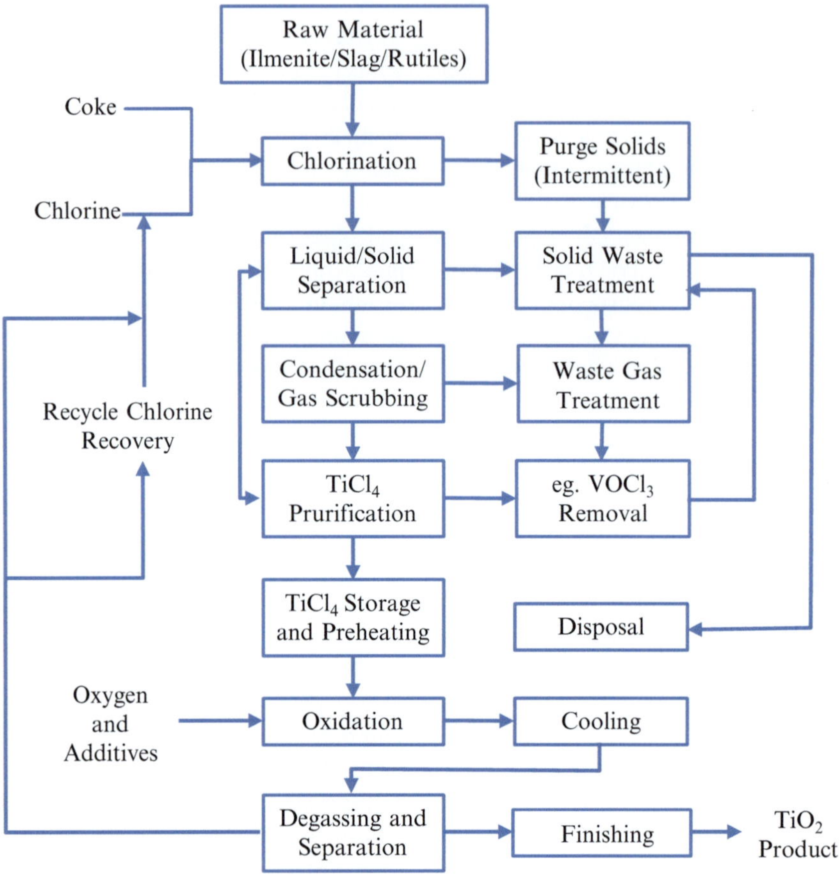

Fig. 18.7 A flowsheet for the production of TiO_2 from natural, or synthetic rutile, or slag, or high-grade ilmenite by the chloride process (based on McNulty 2007)

and a feed port and product discharge line at the top. The chlorine gas is introduced through the bottom of the chlorinator and the feed with the coke is charged from the top of the reactor. The most important chlorination reactions are shown below:

$$TiO_2 + 2Cl_2 + C = TiCl_4 + CO_2 \quad (18.9)$$

$$TiO_2 + 2Cl_2 + 2C = TiCl_4 + 2CO \quad (18.10)$$

$$C + CO_2 = 2CO \quad (18.11)$$

The gas exiting the chlorinator contains titanium tetrachloride ($TiCl_4$), carbon monoxide, carbon dioxide, nitrogen, and volatile chlorides of impurity metals contained in the feed, such as iron and aluminum. The chlorination takes place in fluid beds at 800–1000 °C. Any MgO or CaO results in liquid $MgCl_2$ and $CaCl_2$ and defluidizes the bed. Hence MgO and CaO are not acceptable constituents of the rutile or ilmenite feeds. In addition, Cr and V produce colored chlorides, which introduce color to the pigment and are not acceptable.

18.4.3.2 Condensation of Gas and Purification of $TiCl_4$

The gas stream from the chlorinator is cooled to less than 300 °C to separate impurity chlorides from $TiCl_4$ by condensation and vapor deposition. The exhaust gases (CO, CO_2 and N_2) contain small amounts of $TiCl_4$ and Cl_2, which are scrubbed using water and caustic solutions prior to venting to the atmosphere.

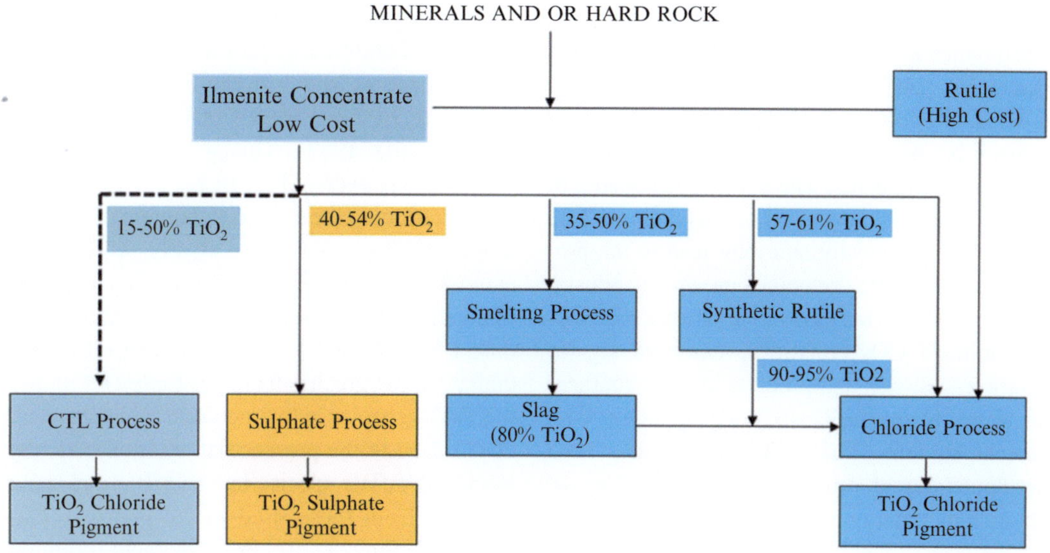

Fig. 18.8 A comparison of CTL, chloride, and sulfate processes

18.4.3.3 Oxidation of $TiCl_4$ and Recovery of Titanium Dioxide

In the next stage, $TiCl_4$ is reacted with oxygen at 900–1400 °C to form TiO_2 and chlorine gas according to (18.12).

$$TiCl_4 + O_2 \rightarrow TiO_2 + 2Cl_2 \qquad (18.12)$$

Aluminum chloride ($AlCl_3$) is added in this step to promote/catalyze the formation of titanium dioxide in the rutile form. Alternatively, phosphorus trichloride (PCl_3) or silicon tetrachloride ($SiCl_4$) is added to suppress rutile formation and promote the formation of pure anatase. Chlorine gas forms during this oxidation reaction, which is recycled to react with fresh feed material.

Impurities like Fe, which are in the rutile ore, chlorinate have to be disposed off. With decreasing availability of rutile ore and increasing demand for TiO_2 by the pigment industry, ilmenite is blended with rutile as the feed material. This chlorinates in the first-stage reactor as follows:

$$C + FeO \cdot TiO_2 + 1.5Cl_2 \rightarrow FeCl_3 + TiO_2 + CO \qquad (18.13)$$

The TiO_2 is then chlorinated according to reactions (18.9) and (18.10).

The disposal of iron chloride is an environmental problem. In the United States, it is injected in deep wells, which may not be an acceptable solution in the long term. In order to minimize the iron input for the chlorination, a process for making synthetic rutile or slag has been developed which adds cost to the raw material fed to the chloride process.

18.4.4 The CTL Process

The current processes have limitations due to impurities like Cr, V, Mg, and Ca in the ilmenite ore. This is restricting the availability of ilmenite ores suitable for the TiO_2 industry. One of the objectives of the new CTL process development was to enable the use of a wider range of available ilmenite ores for the production of TiO_2 pigment. A comparison of feed materials for different processing routes to make pigment grade TiO_2 is shown in Fig. 18.8.

The CTL process, which was originally developed by Process Research Ortech Inc. is currently owned by Canadian Titanium Limited (CTL) with Argex Titanium Inc. having a controlling stake in CTL (50.1 % ownership). The CTL technology is protected by several patents (US patent

7,736,606 B2, Canadian patent 2,513,309, Australian Patent No. 2004291568, and Indian Patent No. 249701). We foresee that the CTL process will become a third processing route and a major breakthrough in the production of TiO_2 for making pigments. The process is capable of treating low-grade ores and ores containing Mg, V, or Cr economically and may lead to the processing of ilmenite ore bodies previously considered not viable. This innovative process promises to minimize CO_2 emissions, consume significantly less energy, and produce less environmental waste due to the closed loop operation with recycle streams. The CTL process has been developed based on a deep understanding of chloride metallurgy.

18.4.4.1 Chloride Metallurgy

Chloride metallurgy is emerging as an alternative process for the production of base metals (Moyes and Houllis 2002; Moyes et al. 2004; Leimala et al. 2003; Hamalainen et al. 2003; Hyvarinen et al. 2002; Yllo and Hietala 2004; Harris et al. 2004a, 2007). The solubility of metal chlorides is generally higher than metal sulfates. The solubility of $CuSO_4 \cdot 5H_2O$ is about one-fifth that of $CuCl_2 \cdot 2H_2O$, while nickel and ferrous chlorides are also more soluble than their sulfate salts. The enhanced leachability of minerals in chloride solutions can be attributed, in part, to the high oxidizing potential of the Fe(III)/Fe(II) and Cu(II)/Cu(I) couples and the high stability of the chloro-complexes of Cu(I), Ag(I), Pb(II), Zn(II), Bi(III), and Fe(III).

Process Research Ortech has been at the forefront of technological development of chloride metallurgy (Harris et al. 2004a, b, 2006, 2007, 2010; Christies et al. 1974; Lakshmanan et al. 1974, 2008, 2009, 2011a, b, c, 2012a, b, c; Bell et al. 1977) and has been developing process flowsheets based on mixed chloride lixiviants consisting of HCl and $MgCl_2$. Having developed a proprietary atmospheric chloride leach process technology for the treatment of sulfide ores, PRO further developed the technology for the production of high purity titanium dioxide from ilmenite ore (Lakshmanan et al. 2002a, b, 2004, 2010a, b). The HCl leaching system was chosen as it is possible to regenerate the acid by pyrohydrolysis. The mixed chloride system was chosen because the presence of $MgCl_2$ in the lixiviant enhances the activity of the hydrogen ion by orders of magnitude (Jansz 1983). This makes the lixiviant very active as seen in Fig. 18.9.

According to the Pourbaix diagram of the titanium–water system shown in Fig. 18.10 (Zhu et al. 2011), TiO^{2+} could be the dominant species in an aqueous solution with acid activity greater than 1 M and Ti(IV) could form $TiO_2 \cdot H_2O$ precipitate at pH greater than five. Table 18.4 shows the Ti(IV) species in hydrochloric acid (Zhu et al. 2011). Due to the presence of chloride ions, the proposed Ti(IV) species in Table 18.4 are quite different from those used in the Pourbaix diagram of Fig. 18.10 in high acid solution ($[H^+] > 4$ M). According to both Fig. 18.10 and Table 18.4, the TiO^{2+} species is stable and predominant in the acid range of $pH < 0.3$ and under a normal solution potential of about 0.3–1.3 V.

Figure 18.11 shows calculated distribution of Ti(IV) oxy-chloride species at an activity of Ti(IV) of 0.1 M in 1 M HCl solution at 298 K (Zhu et al. 2011). A number of complexes are formed in chloride solutions by the combination of TiO^{2+} with Cl^-. The neutral $TiOCl_2$ species reaches a maximum at about 4.5 M Cl^- concentration and forms adduct complexes for extraction.

The CTL process was developed based on a deep understanding of the chloride solution chemistry. The flowsheet for the CTL process is shown in Fig. 18.12. A detailed description of the CTL process is provided below:

18.4.4.2 Leaching

The ore is first crushed, ground, and dissolved in mixed chloride ($HCl + MgCl_2$) lixiviant to bring Ti, Fe, and other metals in solution according to (18.14) and (18.15).

$$FeTiO_3 + 4\,HCl \rightarrow FeCl_2 + TiOCl_2 + 2\,H_2O \quad (18.14)$$

$$Fe_2O_3 + 6\,HCl \rightarrow 2\,FeCl_3 + 3\,H_2O \quad (18.15)$$

For dissolution of ilmenite, earlier studies at Process Research Ortech Inc. (Lakshmanan et al.

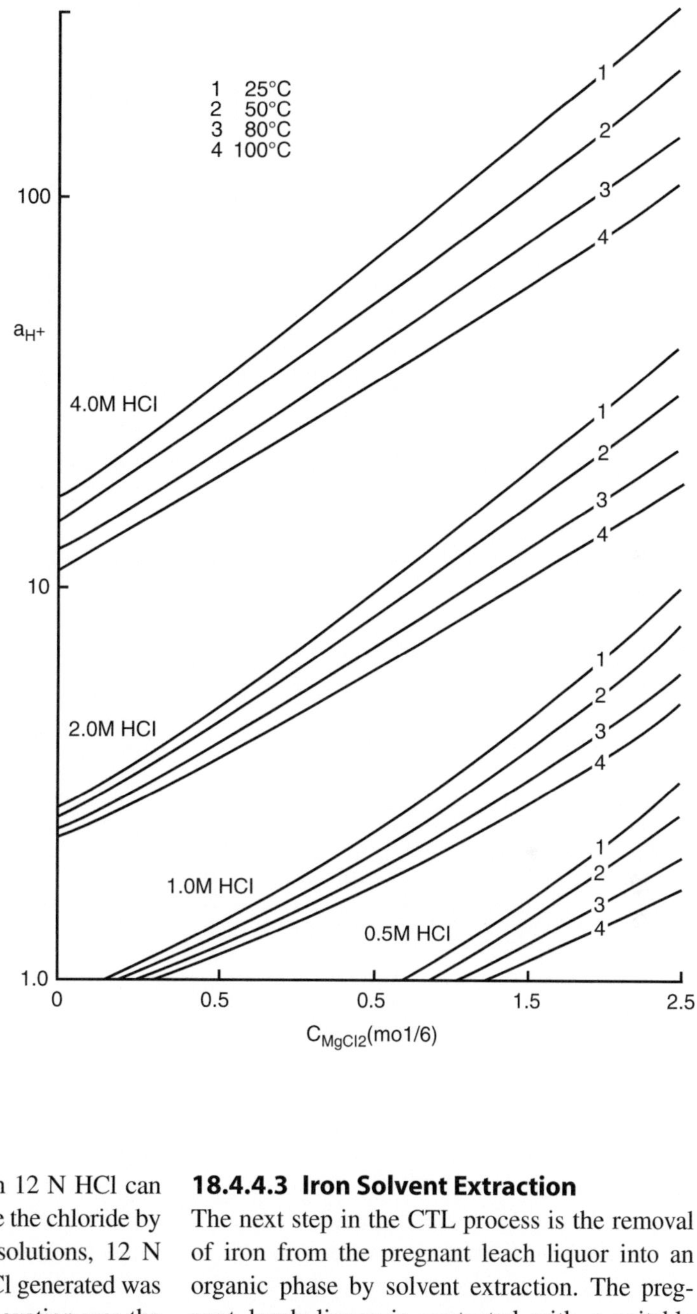

Fig. 18.9 The effect of $MgCl_2$ concentration on the activity of H^+ in HCl solutions (Jansz 1983)

2002b) showed that leaching with 12 N HCl can be carried out. However, to recycle the chloride by pyrohydrolysis of iron chloride solutions, 12 N HCl was not appropriate as the HCl generated was only 6 N in strength. A major innovation was the finding that the addition of $MgCl_2$ enhances the H^+ ion activity as shown in Fig. 18.9. This enabled the leaching of ilmenite ore with 6 N HCl at about 70 °C, which was not possible previously.

After leaching, the solution is then filtered to separate the residue from pregnant leach liquor. Leaching in mixed chloride is one of the most important aspects of the CTL process.

18.4.4.3 Iron Solvent Extraction

The next step in the CTL process is the removal of iron from the pregnant leach liquor into an organic phase by solvent extraction. The pregnant leach liquor is contacted with a suitable extractant (e.g., a mixture of ketones for salvation as anion exchanger and amines with high molecular weights) in a modifier and diluent. Iron is stripped from the loaded organic and regenerated organic is recycled to the extraction stage. The iron in the organic phase is transferred to an iron-rich pregnant strip liquor during stripping of the organic phase.

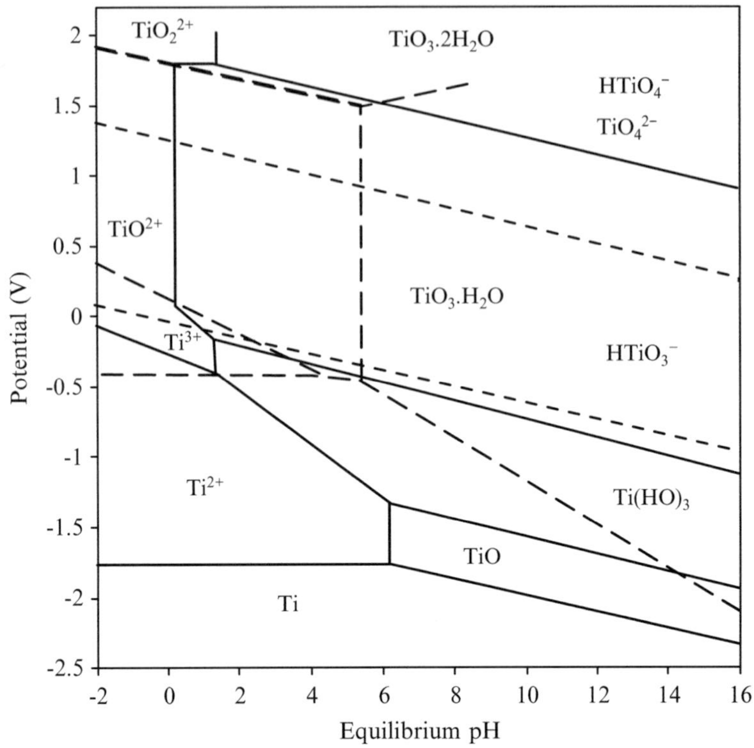

Fig. 18.10 The Eh–pH diagram of the Ti-H_2O system at 25 °C (Zhu et al. 2011)

Table 18.4 Proposed species of Ti(IV) at various pH and HCl concentration levels (Zhu et al. 2011)

Acidity	pH>3.3	pH=3.3–2.0	pH=2.0–0.3	pH=0.3–[H^+]=4 M
Species	$(Ti(OH)_4\cdot 4H_2O)^0$; $(Ti(OH)_5\cdot 3H_2O)^-$ $(Ti(OH)_6\cdot 2H_2O)^{2-}$	$(TiOOH\cdot 2H_2O)+$; $(Ti(OH)_3\cdot H_2O)^+$; $(Ti(OH)_4\cdot 2H_2O)^0$	$(TiOH)^{3+}$; $(Ti(OH)_2)^{2+}$; $(TiOOH)^+$	TiO^{2+}

Acidity	[H^+]=4 M–8 M	[H^+]=8 M–11 M	[H^+]>11 M
Species	$(TiCl)^{3+}$; $(TiCl_2)^{2+}$; $(TiCl_3)^+$; $(TiOCl)^+$	$(TiCl_3\cdot 2HCl)^+$; $(TiOCl\cdot HCl)^+$; $(TiCl_4\cdot 2HCl)^0$	$(TiCl_4\cdot 4HCl)^0$; $(TiCl_5\cdot 3HCl)^-$; $(TiCl_6\cdot 2HCl)^{2-}$

A concentration of 120 g/L of Fe is achieved by a countercurrent solvent extraction process with controlled O/A ratios, flow rate, temperature, and organics of desired composition.

The iron-containing aqueous phase can be converted into iron oxide and 6 N HCl solution by a process called pyrohydrolysis. This is practiced widely in the steel industry. Pyrohydrolysis regenerates HCl from the ferrous chloride by converting ferrous chloride to Fe_2O_3 and HCl. The hematite produced is very pure and can be used for pigments, making ferrites, and also as sinter plant feed. In CTL process, pyrohydrolysis is used to convert ferric chloride to Fe_2O_3 and HCl.

18.4.4.4 Titanium Solvent Extraction

The next step in the CTL process is the solvent extraction of titanium from the raffinate of iron solvent extraction. This raffinate is contacted with another organic phase to transfer titanium from the raffinate to the organic phase. Titanium-rich pregnant strip liquor is then produced by stripping the organic phase. Table 18.5 shows the main organophosphorus extractants, neutral extractant, and amine extractants used for the Ti(IV) solvent extraction (Zhu et al. 2011). The reaction times to reach extraction equilibrium of Ti(IV) are shown in Table 18.6 (Zhu et al. 2011). The two most important extractants for Ti(IV) are D2EHPA ($(HA)_2$) and Cyanex 923

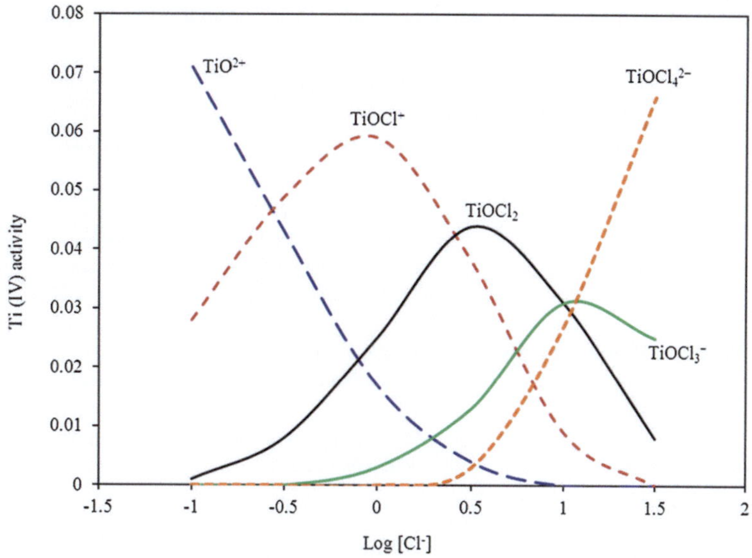

Fig. 18.11 Distribution diagram of Ti(IV) oxychloride species at 0.1 M Ti(IV) activity and 298 K (Zhu et al. 2011)

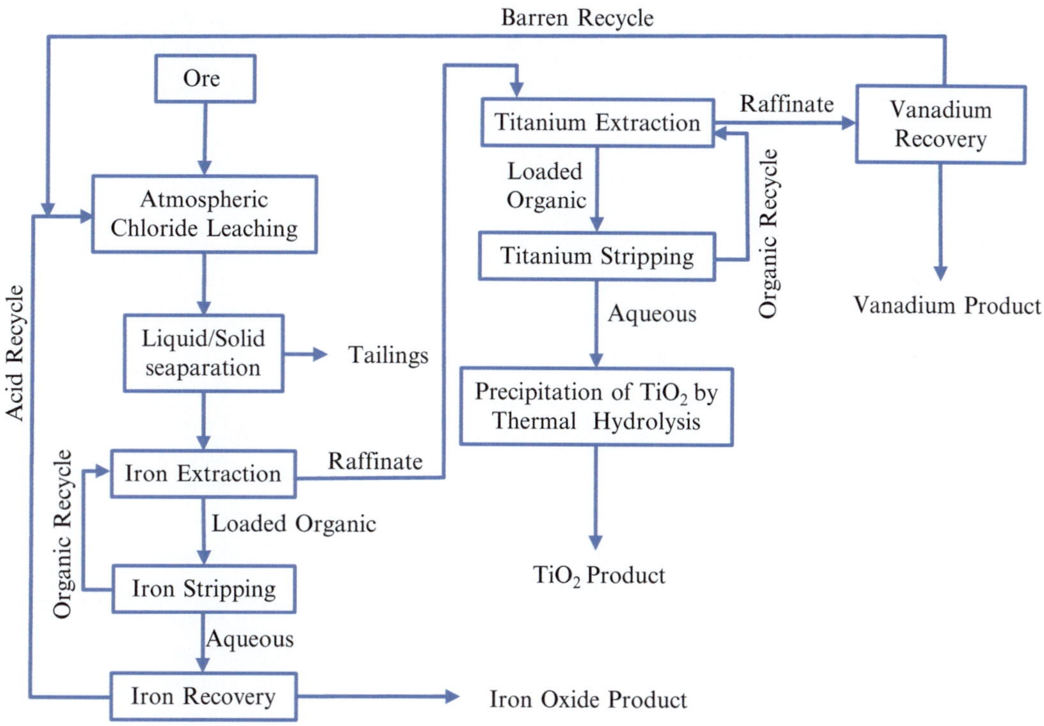

Fig. 18.12 CTL flowsheet for the production of TiO_2 from titanium-containing raw materials (Lakshmanan et al. 2012b)

(TRPO—Tri-alkyl phosphine oxide). The following equations show the reaction that takes place between the aqueous and organic phases, when these two extractants are used (Biswas and Begum 1998; Remya and Reddy 2004):

$$TiOCl_2 + 2(HA)_2 = TiO(HA_2)_2 + 2H^+ + 2Cl \quad (18.16)$$

$$TiOCl_2 + 2TRPO = TiOCl_2 \cdot 2TRPO \quad (18.17)$$

Table 18.5 Commonly used acidic organophosphorus, neutral organophosphorus and amine extractants in solvent extraction study (Zhu et al. 2011)

Type/extractant	Chemical name	Formula
Acidic organophosphorus extractants		
Alkyl phosphoric acids	Di(2-ethylhexyl)phosphoric acid (D2EHPA)	$(RO)_2(OH)P=O$; $R=CH_3(CH_2)_3CH(C_2H_5)CH_2$
Alkyl phosphonic acids	2-Ethylhexyl phosphonic acid mono-2-ethylhexyl phosphoric ester (EHEHPA)	$R(RO)(OH)P=O$; $R=CH_3(CH_2)_3CH(C_2H_5)CH_2$
Alkyl phosphinic acids	Di-2,4,4-trimethylpentyl phosphinic acid (Cyanex 272)	$R_2(OH)P=O$; $R=CH_3CH(CH_3)_2CH_2CH(CH_3)CH_2$
Thio-alkyl phosphinic acids	Di-2,4,4-trimethylpentyl monothiophosphinic acid (Cyanex 302)	$R_2(OH)P=S$; $R=CH_3CH(CH_3)_2CH_2CH(CH_3)CH_2$
	Di-2,4,4-trimethylpentyl dithiophosphinic acid (Cyanex 301)	$R_2(SH)P=S$; $R=CH_3CH(CH_3)_2CH_2CH(CH_3)CH_2$
Neutral organophosphorus extractants		
TBP	Tri-n-butyl phosphate	$(RO)_3P=O$, R=butyl
TOPO	Tri-octyl phosphine oxide	$R_3P=O$, R=octyl
Cyanex 923 (TRPO)	Tri-alkyl phosphine oxide	A mixture of $R_3P=O$, $R_2R'P=O$ and $RR'_2P=O$, R=octyl, R′=hexyl
Amine extractants		
Alamine 336	Tri-octylamine	$(C_8H_{17})_3N$
Aliquat 336	Tri-octyl methylammonium chloride	$(C_8H_{17})_3CH_3N^+Cl^-$

Table 18.6 Contact time for Ti(IV) extraction to achieve equilibrium (Zhu et al. 2011)

	Contact time (min)	Primary conditions		
Extractant		T (°C)	HCl (M)	Diluent
D2EHPA	40	20	–	Kerosene
	120	30	1.0 and 0.5	Kerosene
EHEHPA	<5	30	0.2–1.5	Xylene
	300	20	–	Kerosene
	90	30	1–2	Kerosene
	60	Room	0.5	Kerosene
Cyanex 272	120	30	0.14	Xylene
Cyanex 302	<5	30	0.1	Xylene
	<2	25±2	0.01–5	Toluene
Cyanex 301	<2	25±2	0.01–5	Tolulene
TBP	3	Room	10	16 % Decanol+Dodecane
TOPO	5	Room	8	7 % Decanol+Dodecane
Cyanex 923	<3	30	1–7	Xylene
	<2	30	1–4	Kerosene

The solvent extraction process developed by PRO for the CTL process is extremely selective and results in very pure Ti-rich strip liquor with ppm levels of impurities.

18.4.4.5 Precipitation

Hydrated titanium dioxide is produced from the Ti-rich pregnant strip liquor by thermal precipitation at 85–95 °C according to (18.18).

$$TiOCl_2 + 2H_2O \rightarrow TiO(OH)_2 + 2HCl \quad (18.18)$$

The particle size has to be ideally in the range of 200–300 nm with a narrow distribution for the best opacity or hiding power as required in pigments (http://www.titanium.dupont.com/). The particle size of the TiO_2 product is a critical component of the product quality. The particle size distribution of CTL TiO_2 product is shown in Fig. 18.13 (Lakshmanan et al. 2012c).

The particle size distribution of a reputed commercial sample is also shown in Fig. 18.13 for comparison. It can be seen that the CTL TiO_2 product better meets the narrow particle size requirement sought by pigment manufacturers. The CTL product has 80 % particles in the 200–300 nm range while the best commercial pigment has about 60 % in the range. This is a major breakthrough in the industry. The acid produced by reaction (18.18) is recycled to the process making it environmentally very attractive.

18.4.4.6 Calcination

After washing the pulp to remove impurities, hydrated titanium dioxide is dried and calcined to produce titanium dioxide according to reaction 18.19.

$$TiO(OH)_2 \rightarrow TiO_2 + H_2O \quad (18.19)$$

During calcination, the precipitate is transformed to rutile TiO_2. The particle size and other physical properties are controlled to meet specifications during the finishing stage.

18.4.4.7 Finishing

During the finishing stage, the calcined TiO_2 is ground to break up any sintering that may have occurred during calcination. Then the calcined TiO_2 is surface coated with silica and alumina to improve the UV protection and enhance the optical properties.

18.4.4.8 Product Quality

The main use of TiO_2 is for the pigment in paint industry. The technical specifications for making pigment grade TiO_2 are very strict and require extreme control of other transition metal impurity elements, which can make the color of pigment off-white and unacceptable.

The TiO_2 produced before surface finishing has to contain ideally <10 ppm for each of Fe, V, and Cr. The CTL product meets all the above specifications. The sulfate pigments on average contain about 50–100 ppm Fe (http://www.kilburnchemicals.com/) and hence attract a lower price. Also, other reputed commercial samples from the superior chloride process have 20–70 ppm of Fe.

The crystallinity of the product was evaluated using X-ray diffraction. Figure 18.14 shows the XRD pattern of the rutile TiO_2 product (Lakshmanan et al. 2012a). The distinct sharp and narrow peaks indicate a crystal structure of rutile. The XRD pattern of a commercial sample is also shown in Fig. 18.14 for comparison.

Figure 18.15 shows the morphology of the CTL TiO_2 product (Lakshmanan et al. 2012a).

The color rating used was the CIE (Commission Internationale de l'Eclairage) L*a*b* System. This system is an international standard, which was designed to approximate human vision. There are three components to this system, L*, a*, and b*. "L*" represents lightness ranging from 0 to 100 where 0 is dark and 100 is light. "a*" represents red-green balance in which positive values are more red while negative values are more green. "b*" is similar in that it represents yellow-blue where positive values are more yellow while negative values are more blue. Due to the white nature of pure TiO_2 product, it is difficult to visibly see the differences in the "a*" and "b*" values. The BYK made spectro-guide sphere was used for measuring L*, a*, and b*. Daylight D65 was used as a CIE standardized light source with 10°

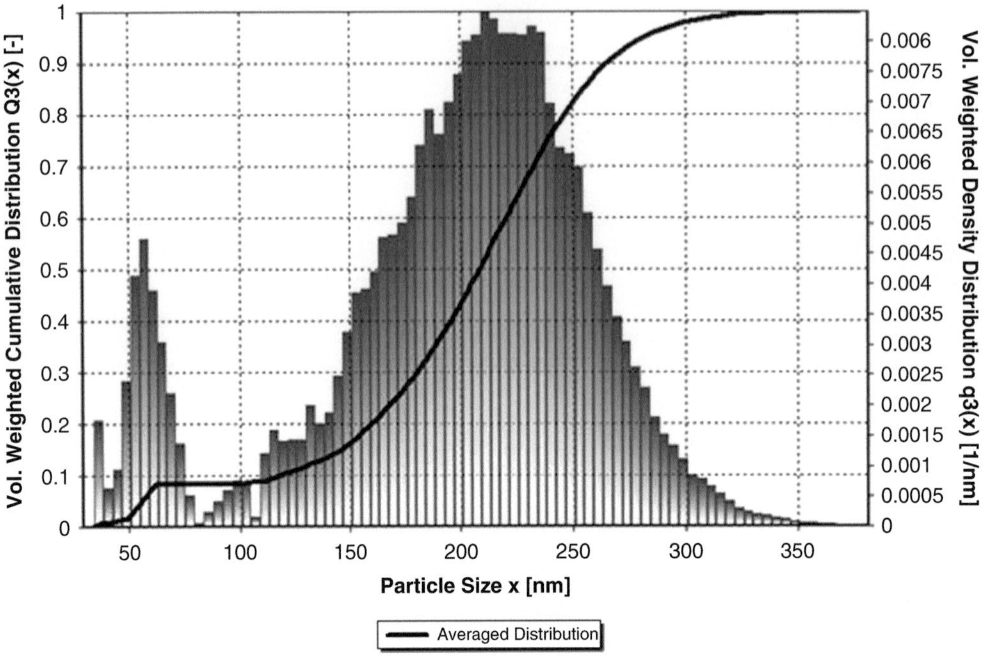

TR 92-Commerical Sample

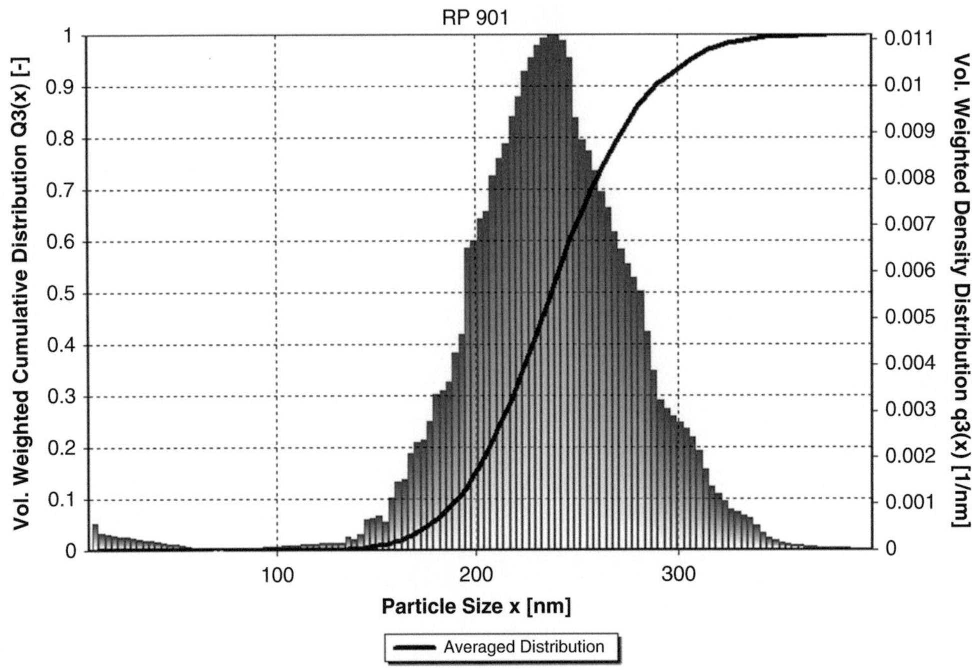

CTL Sample

Fig. 18.13 Particle size distribution of a CTL TiO_2 product and a commercial sample (Lakshmanan et al. 2012c)

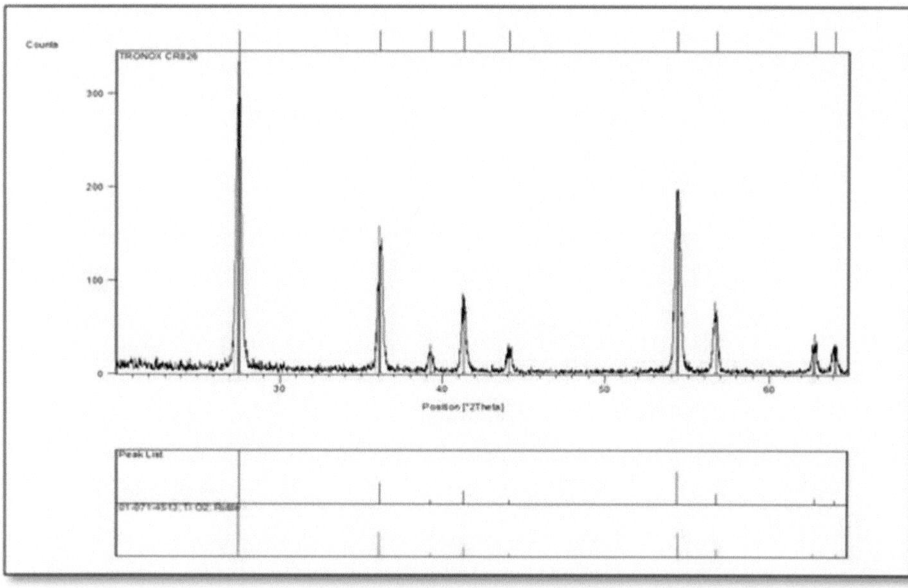

Commercial Sample (CR 826)

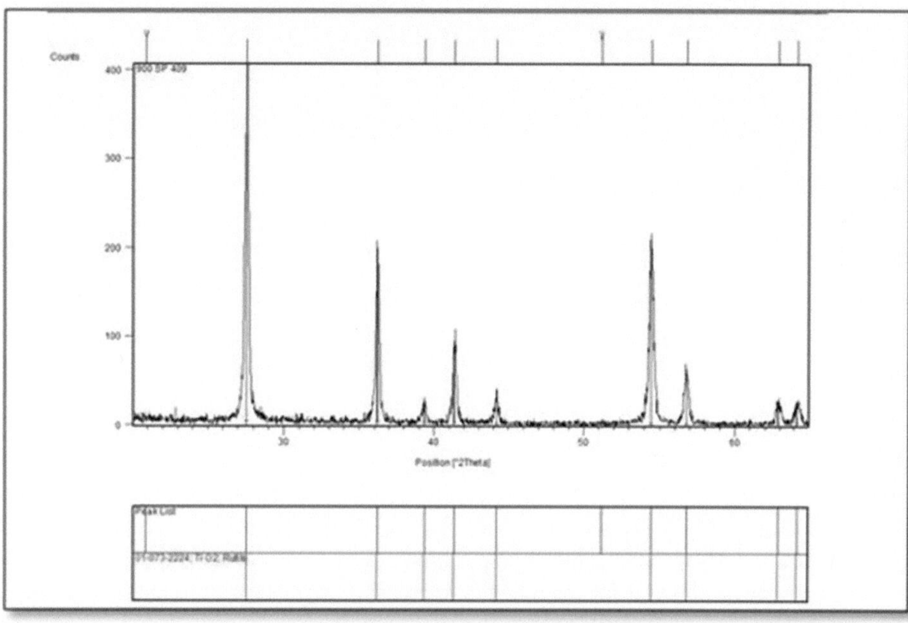

Sample produced using CTL process

Fig. 18.14 X-ray diffraction patterns of a CTL TiO_2 product and a commercial sample (Lakshmanan et al. 2012a)

observer angle. Table 18.7 shows the lightness of a TiO_2 product from the CTL process (Lakshmanan et al. 2012a). The Lightness (L*) values of some other commercial samples are also shown in Table 18.7 for comparison. It can be seen that the lightness value of the CTL product surpasses the lightness of other commercial samples.

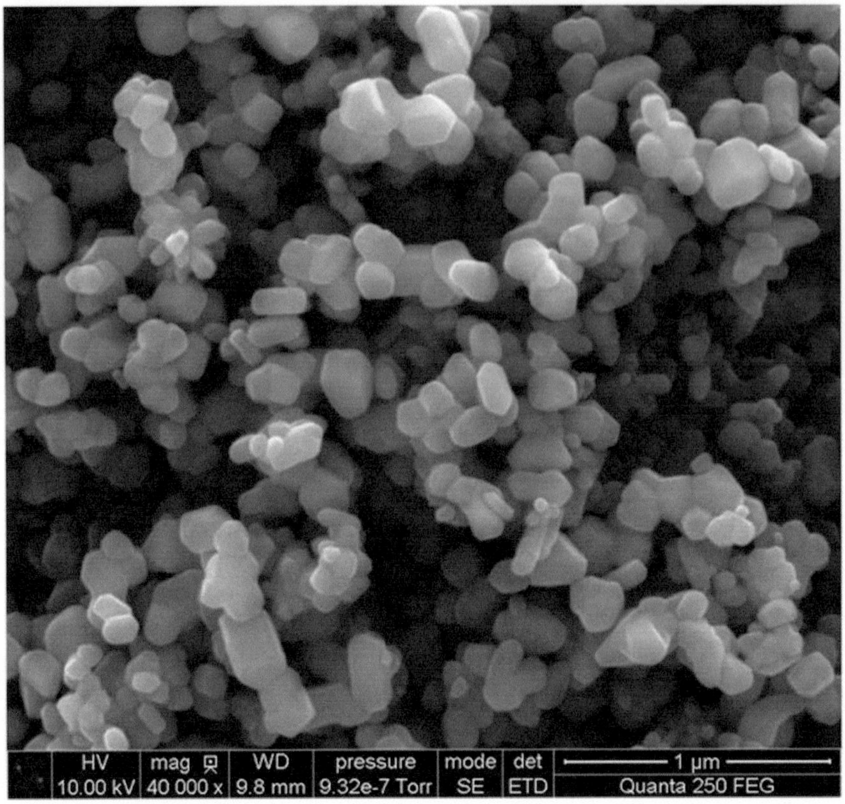

Fig. 18.15 Morphology of a TiO_2 sample (Lakshmanan et al. 2012a)

Table 18.7 Comparison of lightness of a CTL TiO_2 product with commercial samples (Lakshmanan et al. 2012a)

CIEL*	CTL PRODUCT	Commercial sample 1	Commercial sample 2	Commercial sample 3
Lightness, L*	100.48	99.56	99.63	99.39

18.4.4.9 Bench-Scale Test Work

An ilmenite ore containing 22.8 wt% of Ti, 38 wt% of Fe, 0.13 wt% of Cr_2O_3, 4.69 wt% of SiO_2, and 2.82 wt% of MgO with a mesh size of—100 was used as feed material for the recovery of titanium. Laboratory-scale leaching experiments were carried out using a solution of hydrochloric acid and magnesium chloride as a lixiviant under atmospheric pressure at 70–73 °C for 4 h. The leaching conditions and results are given in Table 18.8.

The results show that extraction of titanium as high as 96.9 % was achieved for a solid loading of 8.8 % and HCl level 1.4 times of stoichiometric amount and $MgCl_2$ concentration at 300 g/L. At stoichiometric concentrations of hydrochloric acid, magnesium chloride appeared to have significant effect on the extraction of titanium. The effects of magnesium chloride in increasing extraction of titanium were most pronounced at concentrations of hydrochloric acid that were greater (1.2× and 1.4×) than the stoichiometric amount required for the extraction of titanium and iron as $TiCl_4$ and $FeCl_3$. With an increase in magnesium chloride concentration, titanium extraction reached maxima of 79 % and 96.9 % at HCl levels of 1.2 times and 1.4 times the stoichiometric amount, respectively. At the highest concentrations of $MgCl_2$ that were tested, the $MgCl_2$ concentrations were near the solubility

Table 18.8 Conditions and results using a leaching solution of HCl-$MgCl_2$ (wt %)

Test code	1	2	3	4	5	6
Solids (initial) (g)	175	175	145	145	145	145
Solids loading (%)	10.25	9.92	10.55	10.25	10.08	9.92
HCl amount[a]	1.0	1.0	1.2	1.2	1.2	1.2
$MgCl_2$ added (g/L)	200	250	200	250	300	320
Total chloride (g/L)	362	398	362	398	436	450
Solids (final) (g)	56.5	42.0	42.0	25.3	34.6	44.9
Weight loss (%)	67.7	76.0	71.0	92.6	76.1	69.9
Ti extraction (%)	64.7	65.8	46.0	76.5	79.0	57.2
Fe extraction (%)	71.7	72.4	74.4	69.3	70.2	64.2
Test code	7	8	9	10	11	12
Solids (initial) (g)	141	125	125	125	125	124
Solids loading (%)	9.00	9.23	9.00	8.81	8.67	8.00
HCl amount[a]	1.2	1.4	1.4	1.4	1.4	1.4
$MgCl_2$ added (g/L)	386	200	250	300	320	386
Total chloride (g/L)	498	362	398	436	450	500
Solids (final) (g)	30.6	23.7	30.6	17.3	17.4	26.2
Weight loss (%)	78.3	81.0	75.5	86.2	86.1	78.9
Ti extraction (%)	42.9	53.3	78.6	96.9	93.6	77.0
Fe extraction (%)	48.9	77.9	98.9	84.8	87.2	83.6

[a]HCl amount = stoichiometric amount of hydrochloric acid according to (18.14) and (18.15), based on the amount of titanium and iron in the sample being leached

limits in the solutions used. Thus, the results indicate that there is an optimal range of concentrations of both hydrochloric acid and magnesium chloride for the extraction of titanium and iron from the ore into solution. At hydrochloric acid level of 1.4 times the stoichiometric amount, the optimal magnesium chloride concentration was 300 g/L.

In a series of comparative experiments, to identify the effect of concentration of hydrochloric acid in the absence of $MgCl_2$, a series of leaching experiments were conducted on another ilmenite ore with mesh size of—100 that contained 26.8 wt% of Ti, 35.8 wt% of Fe, 0.03 wt% of Cr, and 0.12 wt% of V. Further experimental details and results are given in Table 18.9.

Table 18.9 Conditions and results using a leaching solution of HCl (wt%)

Test code	13	14	15	16
Solids (initial) (g)	27.7	50	30	50
Solids loading (%)	5	13	5	21
Solution volume (mL)	500	308	500	158
HCl concentration (N)	4	6	8	11.6
HCl amount[a]	2.07	1.06	3.83	1.05
Total chloride (g/L)	142	215	284	416
Solids (final) (g)	7.4	13.4	8.0	13.4
Weight loss (%)	31.4	37.6	50.0	59.2
Ti extraction (%)	7.0	6.3	17.5	18.5
Fe extraction (%)	37.4	13.2	36.7	30.9

[a]HCl amount = stoichiometric amount on hydrochloric acid according to (18.14) and (18.15), based on the amount of titanium and iron in the sample being leached

The results show low titanium extraction of less than 20 % in the absence of magnesium chloride, even when the concentration of hydrochloric acid was as high as 8 N and amounting to 3.83 times the stoichiometric requirement for a solids loading of 5 wt%. It is evident that the use of a mixed chloride lixiviant of hydrochloric acid and magnesium chloride offers advantages in extracting titanium from ilmenite ore in terms of achieving high recovery. It also permits recycling of hydrochloric acid, magnesium chloride, and magnesium oxide in the process.

The use of magnesium chloride allows lowering of the hydrochloric acid concentration to no more than 20 % (mass ratio). This permits the use

Table 18.10 Comparison of TiO_2 production processes

	Chloride	Sulfate	CTL
Raw material (\$/t of TiO_2 feed)	Higher cost (\$2000+)	Lower cost Ilmenite (\$300)	Lowest cost Ilmenite (\$250)
TiO_2 product (\$/t of TiO_2)	High value (\$4500)	Low value (\$3500+)	High value (\$4500)
Capex	Highest (including front end)	Medium	Lowest
Opex	Highest (including front end)	Medium	Lowest
Environmental	Medium challenges	Major challenges	Environmentally friendly
Flexibility in processing raw material	Limitation (Ca, Cr, Mn, Mg, V, size)	Limitation (Cr, V)	Flexible
Process condition	High Temp. Chlorine (800–1000 °C)	High Temp. (140–180 °C)	Low temp., atmospheric Leaching (70 °C), precipitation (90 °C)
Technology	Old	Old	Patented, new
End to end in one location	Not practiced	Possible	Possible
Pigment production	Rutile	Rutile/anatase	Rutile/anatase
Commercially proven process	In practice	In practice	Innovatively applied, will soon be in practice
Environmental challenges	Disposal of iron and other by-product chlorides	Disposal of large iron sulfate product and dilute acid	Minimum environmental impact, iron oxide as by-product
Safety requirements	High (chlorine at high temperature)	High (high temperature acid digestion)	Low (no pressurized vessel and lower temperature)
Chlorine and carbon–carbon containing chemicals at high temperature	Challenges to handle	N/A	N/A
Energy consumption	High	High	Efficient
Sulfur price	No effect	Substantial effect	No effect

of the azeotropic HCl solution produced by pyrohydrolysis without the need to add substantial amounts of concentrated hydrochloric acid, thereby avoiding disposal of excess hydrochloric acid, which is an environmental problem. In addition, the use of lower concentration of hydrochloric acid will result in lower extraction of impurities or gangue from the ore, and thus decrease the downstream burden for the removal of impurities.

The solution purification was carried out using solvent extraction and consisted of two different stages of extraction in a countercurrent circuit. In the first stage, iron removal was carried out and in the second stage titanium was extracted. The strip solution obtained from the second stage solvent extraction was used to precipitate titanium dioxide. The barren solution was used for reagent recovery and the regenerated leach liquor was recycled. Nano-sized rutile TiO_2 powder was produced by controlling thermal precipitation conditions.

18.4.4.10 Pilot Scale Testwork

A titanium extraction pilot plant run was carried out to obtain the design criteria for capital and operating costs, samples for market evaluation, and data for environmental needs. The pilot plant run was performed on an ilmenite feed material assaying ~14 % Ti and 42 % Fe.

The pilot plant included unit operations for leaching, solid–liquid separation, solution purification, and titanium extraction by SX operation and thermal precipitation. The three leach tanks in the leach circuit were set up in a cascading fashion to allow the slurry to flow cocurrently from one tank to the next. The slurry from the third tank was discharged into a barrel and then filtered using a plate and frame filter press. The pregnant liquor obtained after filtration was used for the SX operation.

The SX and thermal precipitation circuits ran continuously for 100 cycles under steady-state conditions. The titanium concentration in the final strip liquor was analyzed to be ~17 g/L, which was used for thermal precipitation of TiO_2.

18.4.4.11 Scale-up of CTL Technology

After completing the bench-scale test work on an ilmenite ore, pilot scale test work was started at 0.3 kg/day (~100 kg/year) in the TiO_2 mini plant located at Process Research Ortech Inc. in Mississauga, Ontario, Canada. This was later scaled up to 3 kg/day (~1 tpa) TiO_2 pilot plant and then to 10 kg/day (~3.3 tpa) TiO_2 pilot plant. First commercial plant of 50,000 tpa capacity is currently under construction. This scale-up is within the accepted scale-up factor for such plants.

18.5 Comparison of Titanium Dioxide Production Processes

CTL process is compared with Sulfate and Chloride processes in Table 18.10.

18.6 Conclusion

An innovative hydrometallurgical technology (CTL process) has been developed for the production of pigment-grade titanium dioxide from ilmenite ores using a mixed HCl–$MgCl_2$ lixiviant. Innovative aspects of the process include treatment of low-grade feed materials, mixed chloride leaching, selective removal of Fe by solvent extraction, thermal precipitation of TiO_2, and reagent recycle. The process is environment friendly with recycling of reagents. Pigment-grade TiO_2 product meeting industry specifications can be produced using the patented CTL process. The process is currently producing TiO_2 for market evaluation and a manufacturing plant at 50,000 t/year TiO_2 production capacity is under construction in Quebec, Canada currently.

References

Bell, M. C. E., Sridhar, R., Bakker, H. F. (1977). *Gaseous solubilization of reduced metal values* (Canadian Patent No. 1013576).

Biswas, R. K., & Begum, D. A. (1998). Solvent extraction of tetravalent titanium from chloride solution by di-2-ethylhexyl phosphoric acid in kerosene. *Hydrometallurgy, 49*, 263–274.

Chang, L. L. Y. (2002). *Industrial mineralogy: Materials, processes, and uses; upper saddle river*. Englewood Cliffs, NJ: Prentice Hall.

Chase, M. W. Jr. (1998). NIST-JANAF Thermochemical tables (Part I and II). In *Physical and Chemical Reference Data* (Monograph No. 9, 4th Ed.) New York: Springer

Christies, P. G., Lakshmanan, V. I., & Lawson, G. J. (1974). The liquid-liquid extraction of copper (II) and iron (III) from chloride solutions using LIX 64N in kerosene. *ISEC Proceedings, 1*, 685–696.

Cianfichi, G. (2012, August). *Global TiO_2 Profile Report*. Ti Insight, LLC.

DuPont. (2014). DuPont Ti-Pure titanium dioxide—Titanium dioxide for coatings. Retrieved August 23, 2014, from http://www2.dupont.com/Titanium_Technologies/en_US/tech_info/literature/Coatings/CO_B_H_65969_Coatings_Brochure.pdf

Gueguin, M., & Cardarelli, F. (2007). Chemistry and mineralogy of titania-rich slags. Part 1—hemo-ilmenite, sulphate and upgraded titania slags. *Mineral Processing and Extractive Metallurgy Review, 28*, 1–58.

Hamalainen, M., Hyvarinen, O., & Jyrala, M. (2003). Solution purification in the Outokumpu HydroCopper™ Process. In C. A. Young, A. M. Alfantazi, C. G. Anderson, D. B. Dreisinger, B. Harris, & A. James (Eds.), *Hydrometallurgy: Vol. 1. Leaching and Solution Purification* (pp. 545–553). Warrendale, PA: TMS.

Harris, G. B., Lakshmanan, V. I., Magee, T. J., & Sridhar, R. (2004). Atmospheric chloride leaching of base metal sulfides. In *Hydro-Sulfides 2004, International Colloquium on Hydrometallurgical Processing of Copper Sulfides, Santiago, Chile* (pp. 384–398).

Harris, G. B., Lakshmanan, V. I., Sridhar, R., & Puvvada, G. (2007). *A process for the recovery of value metals from base metal sulfide ores* (Canadian Patent No. 2478516).

Harris, G. B., Lakshmanan, V. I., Sridhar, R., & Puvvada, G. (2010). *Process for the recovery of value metals from base metal sulfide ores* (US Patent No. 7736606B2).

Harris, G. B., Lakshmanan, V. I., Sridhar, R. (2006). *A process for the recovery of value metals from material containing base metal oxides* (Canadian Patent No. 2467288).

Harris, G. B., Magee, T. J., Lakshmanan, V. I., & Sridhar, R. (2004, March 14–18). The Jaguar Nickel Inc. Sechol Laterite Project Atmospheric Chloride Leach

Process. In *The Minerals, Metals & Materials Society (TMS) 2004 Annual Meeting, Charlotte, NC* (p. 22).

Hyvarinen, O., Hamalainen, M., & Leimala, R. (2002). Outokumpu HydroCopper™ Process-A Novel Concept in Copper Production. In E. Peek & G. Van Weert (Eds.), *Chloride Metallurgy 2002* (Vol. 2, pp. 609–613). Montreal, Canada.

Jansz, J. J. C. (1983). Estimation of ionic activities in chloride systems at ambient and elevated temperatures. *Hydrometallurgy, 11*, 13–31.

Kothari, N. C. (1974). Recent developments in processing ilmenite for titanium. *International Journal of Mineral Processing, 1*, 287–305.

Lakshmanan, V. I., Bhowmick, A., & Halim, M. A. (2014). Titanium dioxide: Production, properties and application. In J. Brown (Ed.), *Titanium dioxide: Chemical properties, applications and environmental effects* (pp. 75–130). New York, NY: Nova Science Publishers Inc.

Lakshmanan, V. I., Lawson, G. J., & Nyholm, P. S. (1974). The extraction of copper(II) and iron(III) with Kelex 100 from aqueous media containing chloride ions. *ISEC Proceedings, 2*, 1169–1183.

Lakshmanan, V. I., Sridhar, R., Harris, G. B., & Puvvada, G. (2010a). *Process for the recovery of titanium in mixed chloride media* (US Patent No. 7803336 B2).

Lakshmanan, V. I., Sridhar, R., Harris, G. B., & Puvvada, G. (2010b). *Process for the recovery of titanium in mixed chloride media* (Canadian Patent No. 2513309).

Lakshmanan, V. I., Sridhar, R., Rishea, M., & DeLaat, R. (2002a). *Separation of titanium halides from aqueous solutions* (US Patent No. 6500396).

Lakshmanan, V. I., Sridhar, R., Rishea, M., & DeLaat, R. (2002b). *Methods of separation of titanium from ore* (Canadian Patent No. 2289967).

Lakshmanan, V. I., Sridhar, R., Rishea, M., & DeLaat, R. (2004). *Methods of separation of titanium from ore* (US Patent No. 6699446).

Lakshmanan, V. I., Sridhar, R., Roy, R., & Di Cesare, E. (2012a, November 16–19). Innovative sustainable process for producing pigment grade TiO_2 from ilmenite ores. In *50th National Metallurgists' Day (NMD) and 66th Annual Technical Meeting (ATM) of The Indian Institute of Metals (IIM), Jamshedpur, India* (p. 21).

Lakshmanan, V. I., Sridhar, R., Roy, R., & Ramachandran, V. (2008, August 17–20). Recovery of value metals from ores by mixed chloride extraction. In *Proceedings of Sixth International Symposium, Hydrometallurgy 2008, Phoenix, AZ* (pp. 895–902).

Lakshmanan, V. I., Sridhar, R., Sheikhzeinoddin, T., Halim, M. A., & Roy, R. (2012b, March 11–15). Extraction of titanium and vanadium by chloride leach processes. In S. Wang, J. E. Dutrizac, M. L. Free, J. Y. Hwang, & D. Kim (Eds.), *T. T. Chen Honorary Symposium on Hydrometallurgy, Electrometallurgy and Materials Characterization, The Minerals, Metals & Materials Society (TMS) 2012 Annual Meeting, Orlando, FL* (pp. 295–301).

Lakshmanan, V. I., Sridhar, R., & Roy, R. (2009, August 23–26). Selective Reduction Leach Process for Laterites. In *Conference of Metallurgists, Sudbury, Ontario, Canada* (p. 11).

Lakshmanan, V. I., Sridhar, R., & Roy, R. (2011a, November 27–December 1). Application of mixed chloride technology for the recovery of TiO_2, base metals and PGM. In *Fray International Symposium, Cancun, Mexico* (p. 11).

Lakshmanan, V. I., Sridhar, R., & Roy, R. (2011b, February 28–March 3). Synthesis of TiO_2 by an innovative atmospheric mixed-chloride leach process. In *Chloride 2011 Proceedings, The Minerals, Metals and Materials Society, San Diego, CA* (p. 8).

Lakshmanan, V. I., Sridhar, R., & Roy, R. (2011c, February 28–March 3). Recovery of precious metals from base metal sulfide ores by a hydrometallurgical process. In *Chloride 2011 Proceedings, The Minerals, Metals and Materials Society, San Diego, CA* (p. 8).

Lakshmanan, V. I., Sridhar, R., & Roy, R. (2012c, May 26–June 2). Application of mixed chloride technology for the recovery of TiO_2 and base metals. In *ALTA 2012, Perth, WA, Australia* (p. 12).

Leimala, R., Hyvarinen, O., Hamalainen, M., & Jyrala, M. (2003). The Outokumpu HydroCopper™ Process-Design, Implementation and Operation of a Demonstration Plant, COPPER 2003. In P. A. Riveros, D. G. Dixon, D. B. Dreisinger, & J. M. Menacho (Eds.), *Proceedings of fifth International Conference, Metallurgical Society of CIM, Montreal, Canada: Vol. 6. Hydrometallurgy of Copper (Book 1)* (pp. 281–288).

McNulty, G. S. (2007, March 19–22). Production of titanium dioxide. *Proceedings of the 5th International Symposium on Naturally Occurring Radioactive Material (NORM V), Seville, Spain* (pp. 169–188).

Moyes, J., & Houllis, F. (2002). Intec base-metal processes realizing the potential of chloride hydrometallurgy. In E. Peek & G. Van Weert (Eds.), *Chloride metallurgy 2002* (Vol. II, pp. 577–592). Montreal, Canada: Metallurgical Society of CIM.

Moyes, J., Lam, J., Huens, J., Gabb, P., Bluck, D., & Henderson, G. (2004). The Intec Copper Process: Development of a New Electrowinning Cell and Demonstration of Oxygen-Free Wire Rod Production. Hydro-Sulfides 2004. In *International Colloquium on Hydrometallurgical Processing of Copper Sulfides, Santiago, Chile* (pp. 413–443).

Remya, P. N., & Reddy, M. L. (2004). Solvent extraction separation of titanium(IV), vanadium(V) and iron(III) from simulated waste chloride liquors of titanium minerals processing industry by the trialkylphosphine oxide Cyanex 923. *Journal of Chemical Technology and Biotechnology, 79*, 734–741.

Retrieved August 23, 2014, from http://en.wikipedia.org/wiki/Titanium_dioxide

Retrieved August 23, 2014, from http://minerals.usgs.gov/minerals/pubs/mcs/2013/mcs2013.pdf

Retrieved March 18, 2015, from http://www.titanium.dupont.com/

Retrieved March 18, 2015, from http://www.kilburnchemicals.com/

Yllo, E., & Hietala, K. (2004). HydroCopper™—A New Copper Method From Outokumpu. In *Hydrosulfides 2004, International Colloquium on Hydrometallurgical Processing of Copper Sulfides, Santiago, Chile* (pp. 363–369).

Zhu, Z., Zhang, W., & Cheng, C. Y. (2011). A literature review of titanium solvent extraction in chloride media. *Hydrometallurgy, 105*, 304–313.

19 Innovative Processes in Electrometallurgy

Tim Robinson

19.1 Introduction

Operating and capital cost reduction is the driving forces behind technology development in the electrolytic process field. Reducing energy and labor costs are central to overall cost reduction.

Large-scale electrolytic refining of copper, nickel, and lead has focused on productivity improvements including electrode handling automation to reduce costs, whereas the higher energy consuming copper electrowinning has developed energy reduction technology. Energy reduction has centered on anode technology. The zinc industry like the copper refining with its large cathode product handling requirements has been driven to automation technology and larger equipment. Recently, copper process technology developments in acid mist abatement have been transferred to nickel electrowinning. Technology transfer between similar electrometallurgical industries is an important process in reducing cell house operating costs and designing more capital efficient cell houses.

This chapter focuses on the technical developments in the electrometallurgical processing of certain metals that use aqueous sulfate and chloride electrolytes.

T. Robinson (✉)
Outotec (USA) Inc., 11288 Alameda Drive,
Strongsville, OH 44149, USA
e-mail: tim.robinson@outotec.com

19.2 Copper Electrorefining and Electrowinning

Unlike lead, zinc, and nickel recovery processes, copper recovery using electrolysis, either by refining or electrowinning accounts for almost all of world copper cathode production. This is because the main industrial use for copper is for electrical conductivity purposes and this requires very pure copper product that can only be produced on a large scale through electrolysis.

Before 1865, recovery of copper metal was by fire refining. This process produced an impure copper product in comparison to electrolytic processes. Elkington developed and patented an electrolytic copper process in 1865 using a dynamo electric machine that produced direct current for industrial use. His focus was not on copper but silver recovery but the copper produced was of the highest chemical quality. The first commercial refinery started up in 1869 in Pembrey, South Wales. So it was by accident that the first electrolytic process for producing base metals was discovered.

The first copper refinery built in the USA was built at Balbach, PA in 1875 and treated southwest USA copper ores that were high in gold and silver. In 1891 the electrical generator design was improved at a time when there was a large demand for new electrical lighting and within 10 years eleven more copper refineries were built in the USA.

V.I. Lakshmanan et al. (eds.), *Innovative Process Development in Metallurgical Industry*,
DOI 10.1007/978-3-319-21599-0_19

Copper refining technology with its large cell house footprint has focused most of the technical development on productivity and automation rather than energy reduction as energy requirements (~340 kWh/t of copper) (Moats et al. 2013) are low compared to copper, zinc, and nickel electrowinning. The biggest productivity step change in refinery design has been the development of a permanent cathode system, like zinc electrowinning. This occurred in the late 1970s at Townsville, Australia. This technology, which plated copper onto permanent stainless steel cathode plates, was more automated and removed the requirement for starter sheet section. Derivations of this more automated technology occurred in Canada in late 1980s and with Outotec of Finland in the late 1990s.

New developments include higher current density (and smaller refinery footprint) with the Mettop BRX technology. This was first demonstrated at over 400 amperes per square meter (ASM) at Brixlegg, Austria in 2005 and uses a relatively high electrolyte parallel flow by manifold installed in each cell. This gives homogenous distribution of copper concentration and temperature to the permanent cathode plating position while not disturbing the slimes fall of the dissolving copper anode.

Other productivity developments include large robotic cathode stripping machines, automated cranes that can handle cathodes and anodes simultaneously, fast anode preparation machines, and polymer concrete cells that have replaced lead or plastic lined cells and have lower maintenance costs.

Permanent cathode technology and more advanced anode preparation machines have also improved copper cathode quality with more vertical electrodes. Cathode smoothing reagent measurement such as Reatrol (Thiourea) and Collomat (Glue) has also helped improve cathode quality.

Energy efficiency development has centered on cell voltage and temperature monitoring and the early detection of short circuiting between the electrodes that reduces current efficiency. This also improves cathode quality.

Copper electrowinning (EW) as we know it today began as by-product or side processes of larger base metal plants. Boston and Montana's Great Falls copper refinery in Montana began to use copper EW cells for liberators and copper balance regulation in the late 1880s. These cells used cast lead anodes. Nikkelwerke in Kristiansand developed a copper EW process from 1905 to 1910 to treat the by-product copper from the Hybinette process to treat nickel–copper matte. This by-product recovery plant used cast lead anodes and the EW cathode was melted into Cu ingots.

In the years leading up to the First World War, two ventures in Arizona (Ajo) and northern Chile (Chuquicamata) began to plan, pilot, and test for direct electrowinning of copper from copper leach solutions. Large and valuable low-grade oxide ore reserves had proved difficult to float and so new direct EW processes were developed and piloted.

Leaching of oxidized ore from Calumet and Arizona (C&A) Copper Co's new Cornelia mine at Ajo in Arizona began in July 1912 at their smelter laboratories in Douglas, AZ. In August 1914, a pilot plant was commissioned at the mine site treating 1 t/day to confirm the leach and electrowinning process. A 40 t/day pilot plant was built in early 1915 to confirm initial findings and ultimately a 5000 t/day commercial plant was built in 1917 and operated until 1930.

EA Cappelen Smith (or Pierce-Smith converter fame) of the Chilean Exploration Co built a 15 t/day pilot plant at the Perth Amboy copper refinery in NJ to demonstrate a new EW process. In 1916 a commercial plant was built on site at Chuquicamata in northern Chile.

These first large EW tankhouse designs were based on refinery starter sheet technology. Based on these first commercial designs in Chile and Arizona, more direct EW plants would be built in Mexico, central Africa, and the USA and more automated starter sheet handling machines would be developed.

EW cathode quality has always been not up to copper ER purity due to the direct EW processing from impure leach solution. Solvent extraction technology had been used in uranium and vanadium extraction and in the 1960s General Mills Chemicals began developing organic extractants designed for copper recovery. Pilot plants were set up first at Bagdad, AZ in 1965 (now Freeport-McMoRan Bagdad), Duval (at Esperanza), and then at Ranchers Bluebird near Miami, AZ. Ranchers became the first commercial plant and copper SX technology has now become the standard in the hydrometallurgical copper industry.

Productivity gains in EW design over the past 40 years have historically tracked copper refining with the application of permanent cathode technology in South American and southwest USA copper SX-EW projects. Some copper EW tank houses are designed with relatively large electrodes (1.3 m long and similar to Zn EW) such as at Safford, Arizona to minimize capital cost.

The trend toward higher productivity with higher current densities, in larger EW plants, has led to advances in acid mist abatement technologies. Most of this development has occurred in South America with cross flow ventilation systems and Outotec and SAME cell hood technology.

Copper EW has focused more on energy reduction technology than ER because EW consumes considerably more electrical energy (over 2000 kWh/t of copper) (Robinson et al. 2013). These developments have focused on the EW anode.

A significant energy saving innovation, a PGM-coated titanium anode, was commercialized in 2007 and similar anodes have been installed at other EW facilities in North and South America. The coated titanium anode has produced a 15 % reduction in cell voltage and eliminated lead completely from the EW operations (Sandoval et al. 2010a).

Also in 2010, researchers from Freeport-McMoRan described work involving alternative anode reaction technology that would oxidize ferrous to ferric ion at the anode and reduce cell voltage by approximately 50 %. Whereas typical copper EW would produce acid and oxygen at the anode, this ferrous–ferric reaction would not produce acid mist (Sandoval et al. 2010b). Although promising, the technology has yet to be commercialized.

In short, the electrorefining copper process still dominates world copper production—with about 70–80 % of production (Moats and Free 2007). Refining technology development has focused on productivity, automation, and quality. Due to its higher electrical energy consumption, the copper EW industry has also focused on electrical energy reduction as well as improvements in productivity, automation, and quality.

19.3 Electrolytic Lead Refining

Up to the late 1800s, lead was pyrometallurgically refined. However, this process produced an impure final lead product and there were operation occupational health issues in early lead refining furnaces. First development of an electrolytic lead refining process used a lead bullion anode and an acetate electrolyte by Tommasi (1871/1891) and NS Keith (1878) (Betts 1908). The lead cathode produced in this electrolyte was spongy and difficult to clean.

The lead plated in an acetate electrolyte was unable to be used as starter sheets and this proved problematic for potential commercial scale-up. Electrolytic processes were at their early stage of development but still allowed for fast and easy construction of pilot plants made of small cells and electrodes that could test the many different electrolytes specifically for lead refining.

By the late 1800s, New York City area (including New Jersey) had become the world center of the electrolytic copper refining industry. Successful commercial application of copper refining used glue addition as a smoothing agent to produce smooth and rigid starter sheets for further plating.

Anson Betts in a New York laboratory began a series of experiments with different electrolytes in a small pilot cell arrangement consisting of four small bench scale cells that contained eight cathodes each. Electrolytes tested included acetate, fluoboric acid, and hydrofluosilicic acid and eventually hydrofluosilicic acid was chosen due to low cost, relatively higher conductivity, and high lead solubility. Betts also tested glue (gelatin) as an addition reagent and this made a dense deposit that could make a refined lead starter sheet from a lead bullion anode.

The main advantages of the electrolytic lead refining process over other lead refining processes were that it could produce a purer lead product in a single step from lead bullion that included impurities such as bismuth, antimony, arsenic, silver, and gold. Both glue addition and hydrofluosilicic acid electrolyte were key innovations that led to commercial application and led to the Betts Electrolytic Process or BEP being patented in 1901. Hygiene in the BEP was improved over the pyrometallurgical refining processes and dross recycle streams reduced.

Electrolytic processes can be scaled up faster than in pyrometallurgical applications due to the modular cell and electrode design so a small commercial BEP refining plant was built at Cominco's existing lead and zinc plant at Trail, BC in Canada in 1902. The Trail location also had a ready source of continuous and stable hydropower supply.

Trail pioneered a number of innovations and developments in the BEP process including the application of aloe as a smoothing agent that replaced glue (Mather and Lantz 1953) Aloe was a stronger and more improved smoothing agent that allowed lead refinery current density to be increased with the same plant footprint. Further innovations included low maintenance and unlined polymer concrete cells, improvements in electrode composition and dimensional quality, and new slimes treatment processes.

A significant amount of development over the last several decades has been in BEP refineries in Japan and China. These developments are mainly focused on refinery automation (Kubota et al. 1998; Hiraki and Ogata 1998). Lead refinery automation leads to higher current efficiency (cell voltage monitoring and electrode straightening), improved occupational health, increased productivity, and reduction in labor cost.

Future innovation opportunities in lead refining could implement coated titanium anode technology in lead ER liberator cells. These anodes use an amorphous PGM coatings pioneered by Freeport and Doshisha/Outotec in copper EW that do not precipitate lead oxide (Robinson et al. 2011).

Trail also piloted a bipolar lead refining technology in the 80s that enabled electrode size to be much larger. The advantages of this technology were even more automated permanent cathode system, lower energy, and smaller refinery footprint.

19.4 Nickel Electrorefining and Electrowinning

Electrolytic processing of nickel is unique within base metal processing being recovered by both electrorefining and electrowinning from a number of different electrolytes including chloride, sulfate, and mixtures of both. Nickel also typically employs separated cell technology using diaphragms to separate catholyte from anolyte which is unlike other base metal electrolytic processes. Besides electrolytic processing, final nickel product can also be produced in a metal briquette form that is produced by hydrogen reduction or metal produced by the carbonyl process.

Refining of ground nickel copper matte had been first tried in the late 1800s at Evje smelter site in southern Norway using a process with arsenic to separate the nickel from copper but test work did not lead to continuous operation. In 1905 it was decided to build a refinery at Kristiansand on the coast south of the smelter using the Hybinette Nickel refining process. Norway had cheap and reliable hydropower to supply an electrolytic process. This plant would be one of the first electrolytic nickel processes. The cells were built in pairs and used cloth cathode bags to separate the anolyte and catholyte and a sulfate electrolyte. Kristiansand would eventually test and convert to a mixed sulfate/chloride electrolyte that increased anode current efficiency and lower cell voltage. It was also one of the first copper electrowinning plants using lead anodes in the world in 1905. It ran until the early 1970s until Kristiansand converted to a chloride EW process using coated titanium anode technology.

Inco started nickel refining with impure nickel anodes and cathode bags at Port Colborne, Ontario in 1926 (later shutdown). Today the largest nickel producer, Norilsk Nickel produces nickel cathode from refineries at Norilsk and Monchegorsk in Russia. Other operations include Vale Thompson (1961) in Manitoba (60 ktpa) and Jinchang, China (100 ktpa). Electrorefining of nickel of cast impure nickel anodes to produce pure nickel cathodes is the largest producer of electrolytic nickel today. A description of the electrorefining processes can be found in literature (Boldt and Queneau 1967).

The Kristiansand refinery also operated a small electrowinning circuit in chloride electrolyte using carbon anodes that produced chlorine gas as a by-product. This was an ideal test circuit to develop a new nickel EW process in chloride electrolyte using anode bags that collect chlorine gas for leaching at the beginning of the process. Initially graphite anodes were tested but eventually new coated titanium anode technology (used in commercial chlorine production) was used in the new process in the early 70s. Plants that perform electrowinning from chloride electrolyte are also located in France and Japan. Typically, these plants operate with a cell voltage of 3 V and at current densities of approximately 230–280 ASM (Crundwell et al. 2011).

Nickel electrowinning from sulfate electrolyte was first used in 1960 in Harjavalta, Finland. Plants that use this technology are now also located in South Africa and Canada. Nickel electrowinning from sulfate electrolytes use separated cells with either the anode or cathode bag, cathode starter sheets, lead anodes, and generate oxygen gas at the anode.

In sulfate electrolytes, the catholytic pH needs to be approximately 3.5 to avoid hydrogen gas evolution on the nickel cathode and bags are used to maintain optimum current efficiency. Most operations add boric acid to help buffer the pH and/or sodium lauryl sulfate to help hydrogen gas bubble release. Typically, cells with sulfuric acid electrolyte operate at a cell voltage of 3.6–3.8 V and a current density of 200–260 A/m^2.

Developments in nickel electrowinning have focused on energy reduction and acid mist abatement. Current passing through nickel electrolytes can consume relatively high energy so reduction of electrode spacing has led to energy reduction. Converting from sulfate to chloride electrolyte and also certain electrolyte additives can reduce energy consumption.

Recent acid mist abatement developments in nickel electrometallurgy have been transferred from other copper and zinc EW technologies. Rustenburg's new Ni EW plant with sulfate electrolyte uses cell hoods to capture acid mist, has converted to a permanent cathode system and automation to further remove operators away from the cell house acid mist.

While it is necessary to employ coated titanium anodes (rather than lead) for chloride electrolyte systems, nickel EW operations in sulfate electrolytes are also interested in testing this technology to eliminate cell and bag cleaning of lead oxide sludge, producing lead-free nickel cathode, and lower cell voltage.

19.5 Zinc Electrowinning

First large-scale zinc recovery was observed in India and China. Before the late 1880s zinc retort technology produced most of the world's zinc. Zinc metal innovation and production from retorts was dominated by Germany and Belgium. Horizontal retort technology developed in Belgium became the standard commercial zinc recovery technology around the world.

In late 1800s, Frenchman Léon Letrange was the first to develop on an experimental scale, the reduction of zinc ores in a sulfate system (Letrange 1883). This process used roasting, weak sulfuric acid leaching, and electrowinning (EW). This electrowinning bench-scale process used graphite anodes and zinc starter sheets (as per copper refining process and later lead refining). This is now known as the roast-leach-electrowinning or RLE process.

Compared to the retort process, this zinc electrolytic process improved operator environmental conditions, higher recovery, and would produce a higher purity EW zinc cathode product that could be used for improved munitions fabrication.

In the 1890s, attempts were made to develop a commercial zinc recovery process using chloride leaching. The first was Carl Hopfer and he patented a process of roasting pyrite cinders with salt, purifying and then electrowinning a sodium–zinc chloride electrolyte to make zinc and chlorine gas. Two plants were built in Germany and Winnington, England (operated until 1924) using rotating cathodes. In 1894, Edgar Ashcroft built a plant in Newcastle, NSW to treat difficult Broken Hill zinc ores using chloride leaching but was dismantled within a year.

It was soon evident that due to plant materials of construction and design simplicity, that the sulfate system had the best opportunity for large-scale commercial zinc production. Numerous attempts had been made, without success, to scale-up and it was soon found that purification requirements were specific to the type of ore treated.

RLE Test plants were built in:

- Kellogg, Idaho
- Widnes, England
- Niagara Fall, NY
- Bully Hill, Colorado
- Palo Alto, California
- Nelson, BC

Shortage of brass artillery shells in 1915 at the western front gave added impetus to commercialization of the RLE Zinc process, particularly in British Commonwealth countries of Canada and Australia. This was known as The Shell Crisis (Ralston 1921). Pilot plants and eventually larger production plants were built at:

- Anaconda, Montana: 25TPD (1915) increasing to 150 TPD (1916)
- Trail, BC, Canada: 50–75 TPD (1914)
- Risdon, Tasmania: 15–100 TPD (1917–1919)

In zinc electrowinning in sulfate electrolytes, it was found that pure electrowon zinc metal could be easily stripped from an aluminum cathode plate, thereby replacing a zinc starter sheet and cast lead anodes were used to replace the graphite anodes. So zinc EW became the first permanent cathode harvesting system.

Trail's first Zn EW plant was the largest of the three at 50 TPD compared to Anaconda and Risdon. Trail had already had a strong electrometallurgical technical team and had been operating a commercial lead refinery since 1902 on the site. All demonstration plant sites had a continuous and cheap source of hydro power that was essential for the new energy-intensive Zn EW technology.

The first zinc cell houses that used manual handling of electrodes and a cascade flow system (like lead ER) in lead-lined wooden cells were small compared to today with only 24 cathodes. As current density was gradually increased the electrolyte had to be cooled by water circulation in lead coils in the cell. Trail used a Zn EW electrode size of approximately 1.2 m^2 (0.6 m^2 each side) which was the same size as the commercial lead ER size of 1.2 m^2.

After WWI, RLE plants were built at:

- Viviez, France (1922)
- Odda, Norway (1926)
- Hudbay, Canada
- Kellogg, USA
- Balen, Belgium
- Sauget, USA
- Portovesme, Italy
- Crotone, Italy
- Magdeburg, Germany

From WWII to present day, technology development has driven productivity efficiencies in the zinc cell house. Once again during wartime, during WWII, ASARCO's Corpus Christi plant was a revolutionary design that used parallel electrolyte flow to cells, larger electrodes (1.8 m^2) and external air cooling of the electrolyte to replace the lead coils in the cells of earlier designs. This was known as the Jephson design so named after its design engineer and became the standard Zn EW design for a generation. With increasing labor costs after WWII, Lurgi (now Outotec) would also use this design with automated electrode handling machines. The Japanese zinc producers at Mitsui Hikoshima and Dowa Akita also would have their own automated Zn EW plant designs.

Jephson design Zn EW plants included:

- Corpus Christi, TX, USA (1941)
- Boleslaw, Poland (1955/1969)
- Plovdiv, Bulgaria (1959)
- Kokkola, Finland (1969)
- Nordenham, Germany (1972)
- Ruhr Zinc, Germany (1968/1972)
- San Luis Potosi, Mexico (1982)

The drive to automate the stripping of zinc cathodes was higher priority than for electrolytic copper production due to the lower yield per ton than copper, relatively higher current densities, shorter cathode cycles, and shorter cell cleaning cycles to clean MnO_2 from the cells.

The late 60s saw a step change in Zn EW design with the Jumbo electrode design by Vielle Montagne (VM). This technology was first installed in Balen, Belgium in 1969 and used electrodes of 2.6 m^2 size, increased EW automation and improvements to plant layout (Caufriez et al. 1998). The VM design would still use Jephson parallel cell electrolyte flow and electrolyte cooling with air cooling towers.

VM Jumbo Zn EW plants include:

- Balen, Belgium (1969)
- Budel, Netherlands (1973)
- Penoles, Mexico (1973)
- Overpelt, Belgium (1974)
- National Zinc, USA (1976)
- Auby I, France (1975)
- VdM, Brasil (1977)
- Clarksville, USA (1978)
- Cajamarquilla, Peru (1981)

VM Superjumbo Zn EW designs increased cathodes per cell from 40 to 120, reduced electrode spacing to 80 mm, and increased electrode size to 3.2–3.8 m^2. VM Super jumbo technology was installed at Balen II, Auby II, Korea Zinc, and Skorpion SXEW Zinc plant in Namibia.

Cominco modernized its own Zn EW cell house at Trail, BC with larger electrodes (3.2 m^2) and more automated electrode handling. Asturiana de Zinc further developed the super jumbo size of electrodes (3.6 m^2) with even more automation and even larger cells at San Juan de Nieves (Spain), Flin Flon (Canada), and Mooresboro, NC.

Now, the latest Zn EW cell house super-jumbo electrode designs use automation that removed operators away from the cell house acid mist, totally automated driverless cranes, electrode sizes up to 3.8 m^2, and cells with over 120 cathodes.

Over half the operating cost of a RLE zinc plant is the power cost. Future technology innovation in the RLE Zinc plant will be driven by energy savings and include new amorphous-coated titanium anode technology from Outotec/Doshisha University which lowers the anode oxygen overpotential and produce a lead-free zinc product (Robinson et al. 2011). Water cooled double intercell contact designs that optimize current distribution and real-time cell voltage or current monitoring will continue to be installed.

Recently, there also has been technology innovation in the acid mist reduction in the zinc EW cell house. These developments include mist suppressant foams and designed forced ventilation systems.

19.6 Zinc SXEW Process

Due to large-scale commercial success of the solvent extraction (SX) in copper, zinc operators began to investigate zinc SX to produce zinc electrolyte for electrowinning. The feed for these plants is typically zinc-containing steel furnace dust residues.

Tecnicas Reunidas of Madrid, Spain initially developed this zinc SX with their ZINCEX process installed at a number of small plants on the Iberian Peninsula that focused on leaching of secondary zinc feeds from the steel industry. The experience gained from these smaller plants led to design and start-up of larger Zn SX plant installations at Akita, Japan (20 ktpa), Skorpion, Namibia (150 ktpa leaching silicate ore), and Horsehead Zinc's Mooresboro plant (150 ktpa) in NC, USA. The Zn SX process was used at Metmex, Penoles, Mexico lead smelter zinc flue dust leach (5 ktpa) to make Zn EW electrolyte and also at the Teck pilot plant at Sullivan, BC that bioleached sulfide ore.

This Zinc SX process is suited to secondary zinc feeds such as from electric arc furnace (EAF) dust that contain halides and where low-grade zinc leach solutions need to be concentrated and purified to EW electrolyte zinc concentrations.

19.7 Conclusions

Process technology development will continue to optimize the electrolytic process to recover high purity metals. Commercial application in higher current density technology and developments in large capacity robotic electrode handling machines will continue to improve productivity and operator occupational health conditions for large metal producers. Lower cost and more energy efficient catalytic coatings on electrowinning anodes will improve energy costs. Acid mist abatement technology will continue to be improved and implemented.

In short, the future looks bright for continuous improvement in the electrolytic process industry.

References

Betts, A. G. (1908). *Lead electrorefining by electrolysis*. New York: Wiley.

Boldt, J. R., & Queneau, P. (1967). *The winning of nickel*. Toronto, ON, Canada: Longmans Canada.

Caufriez, A., Dubois, M., & Lejay, E. (1998). Zinc cellhouses, concept and practice at Union Meniere. In J. E. Dutrizac, J. A. Gonzales, G. L. Bolton, & H. Hancock (Eds.), *Zinc and lead processing* (pp. 337–352). Calgary, AB, Canada: The Metallurgical Society of CIM.

Crundwell, F. K., Moats, M. S., Ramachandran, V., Robinson, T. G., & Davenport, W. G. (2011). *Extractive metallurgy of nickel, cobalt and platinum-group metals* (pp. 327–345). Oxford, England: Elsevier, 575–578.

Hiraki, T., & Ogata, M. M. (1998, August) An outline of the lead electrolysis operation at the Chigirishima refinery. In J. E. Dutrizac, J. A. Gonzales, G. L. Bolton, & H. Hancock (Eds.), *Zinc and Lead Processing, Calgary, Canada, 37th Annual Conference of Metallurgists of CIM* (pp. 379–388).

Kubota, H., Kusakabe, T., Takei, K., Takewaki, M. (1998, August) Current operation of Sumitomo Metal Mining's (SMM) Betts Lead Electrorefinery. In J. E. Dutrizac, J. A. Gonzales, G. L. Bolton, & H. Hancock (Eds.), *Zinc and Lead from Steel Mill Dusts, Zinc and Lead Processing, Calgary, Canada, 37th Annual Conference of Metallurgists of CIM* (pp. 353–366).

Letrange, L. (1883). US Patent No. 286208.

Mather, F. C., & Lantz G. B. (1953, December 29). US Patent No. 2664393.

Moats, M., & Free, M. (2007). A bright future for copper electrowinning. *Journal of Metals, 59*, 34–36.

Moats, M., Robinson, T., Wang, S., Filzwieser, A., Siegmund, A., & Davenport, W. (2013). Global survey of copper electrorefining operations and practices. In R. Abel, & C. Delgardo (Eds.), *Copper-Cobre 2013 International Conference: Vol. 5 (Book 2), Santiago, Chile* (pp. 67–108).

Ralston, O. C. (1921). *Electrolytic deposition and hydrometallurgy of zinc*. New York: McGraw Hill.

Robinson T. G., Sole, K. C., Moats, M. C., Crundwell F. K., Morimitsu, M., & Palmu, L. (2011). Developments in base metal electrowinning cellhouse design. In *Electrometallurgy, 2011, TMS, Orlando, FL* (pp. 147–156).

Robinson, T. G., Sole, K. C., Sandoval S., Moats M. S., Siegmund, A., & Davenport, W. (2013). Copper electrowinning-2013 world tankhouse operating data. In R. Abel, & C. Delgardo (Eds.), *Copper-Cobre 2013 International Conference: Vol. 5 (Book 1), Santiago, Chile* (pp. 17–50).

Sandoval, S., Clayton, C., Dominguez, S., Unger, C., & Robinson, T. (2010a) Development and Commercialization of an alternative anode for copper electrowinning. In *Copper 2010 Proceedings, Vol. 4, GDMB* (pp. 1635–1648).

Sandoval, S., Cook, P., Morales, C., & Robinson, T. (2010b). Demonstration of the ferrous/ferric anode reaction for copper electrowinning. In *Copper 2010 Proceedings, Vol. 4 GDMB* (pp. 1617–1634).

Innovations in Gold and Silver Processing

20

B.K. Gorain, Peter D. Kondos, and V.I. Lakshmanan

20.1 Introduction

As the gold mining industry is evolving, so are the challenges associated with mining. Some of the key challenges are low grade complex ore bodies at greater depths, high capital and operating costs, high energy costs, water scarcity and quality issues, complex environmental issues including tailings management, worldwide constraints on resources availability, stringent regulations impacting the permitting processes, increasing stakeholder expectations, changing demographics of mining operations and the urgent need for a better assessment of project viability, all these in a very dynamic metals market.

These challenges are compelling mining companies to look at their mining business holistically and also from different points of views. An integrated approach to mining is gaining prominence for improving project economics and also allows companies to better engage with various stakeholders for developing robust solutions to issues.

The average global gold ore grades were around 20 g/t between 1830 and 1900; thereafter the decline in gold grade has been exponential as shown in Fig. 20.1. It is estimated that by 2050, the average gold grade would be in the range from 0.9 to 2 g/t (Müller and Frimmel 2010).

Though this chapter is about innovation in gold and silver processing, the authors strongly feel that this has to be seen in the light of other disciplines and the mining value chain as a whole. Any innovation focusing on processing alone will only provide part of the solution and will miss big opportunities. The question then is "What are these big opportunities that will allow a step change in profitability of gold mining operations in a safe and sustainable manner?"

Despite significant innovation in the gold industry, mining and processing still involves the old paradigm with significant amount of waste transportation, processing and handling of waste and tailings. An ore after mining typically contains mostly gangue minerals. The amount of non-valuable gangue minerals could be more than 99.0 % for a refractory gold deposit. The lower the head grade, the larger the volume of ore and waste generated. Kennecott mine in Utah alone handles almost 100 million tonnes of ore and additional 200 million tonnes of waste every year (Ericsson 2012).

B.K. Gorain (✉) • P.D. Kondos
Barrick Gold Corporation, 161 Bay Street, Suite 3700, Toronto, ON, Canada, M5J 2S1
e-mail: bgorain@barrick.com

V.I. Lakshmanan
Process Research Ortech Inc., 2350 Sheridan Park Drive, Mississauga, ON, Canada, L5K 2T4

V.I. Lakshmanan et al. (eds.), *Innovative Process Development in Metallurgical Industry*, DOI 10.1007/978-3-319-21599-0_20

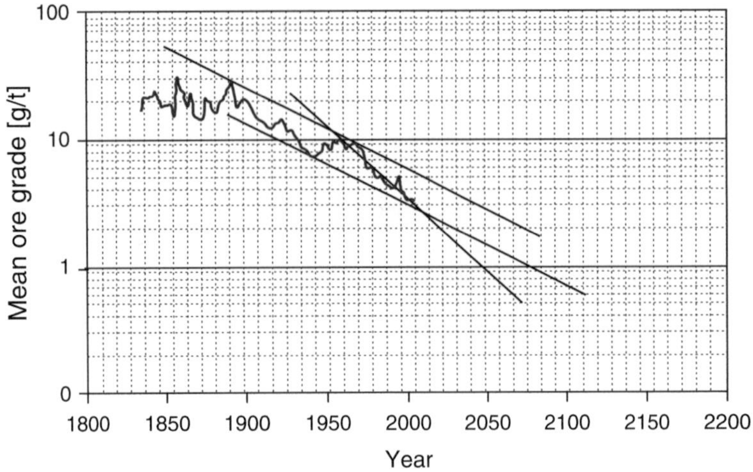

Fig. 20.1 Trend of gold ore grade since 1834 (Müller and Frimmel 2010)

To meet the needs to handle large volumes of ore and waste, bigger trucks and shovels are utilized, and the payload trucks have increased from around 200 t in 1990 to almost 350 t in 2012. At the same time, the installed power has increased by a similar factor. During the past 5–10 years, the bucket volume of wheel excavators has increased from 25 to 40 m^3. The processing plants have followed suit with large 42 ft. diameter semi-autogenous grinding mills, 500–700 m^3 flotation cells along with larger oxidative pretreatment, leaching, cyanide destruction, dewatering, and tailings handling facilities.

The trend of increasing throughout and larger equipment is continuing despite all associated challenges along with an on-going trend of extremely high waste generation. The basic extraction paradigm in the mining industry is "drill, blast, load, haul, dump, crush, grind, separate, leach, dewater, and tailings disposal." There are many variations, but fundamentally, the paradigm has not changed since ancient times (Dunbar 2014). Almost all the innovations so far have made operations in this paradigm safer, more efficient, automated, and even autonomous. What's needed is a major shift in this paradigm for a major breakthrough in the gold industry.

Since both mining and processing operations handle a significant amount of waste, the first consideration for any innovation should be to cut down on waste as early as possible in the mining value chain. The gold mining industry is now facing a dilemma and there is a strong push by companies, some governments, and stakeholders to focus on innovation to address the challenges.

The following are the main areas that have attracted serious attention for innovation in the area of gold and silver processing:

- Energy consumption and GHG reduction
- Economy of scale (capital cost)
- Efficient use of consumables (media, liner, reagents)
- Use of alternative water sources
- Environmental management of Tailings and Waste
- Seamless integration with Value Chain
- Breakthrough technologies (focus on step change such as pretreatment, cheaper and environmental friendly alternative to oxidative treatment and cyanide leaching and detoxification)

This chapter will review the trend in gold and silver processing innovations that have already resulted or have potential to make a significant contribution to the mining industry. Principles of operation and detailed design criteria are not the main focus of this chapter as they are readily available for reference in various publications (Mular et al. 2002; Deschenes et al. 2005; Adams 2005; Anderson et al. 2014).

The following major areas in gold and silver processing will be subject of discussion in this chapter:

1. Ore body knowledge
2. Comminution and classification
3. Pre-concentration and ore beneficiation
   (a) Pre-concentration
   (b) Gravity separation
   (c) Flotation
   (d) Magnetic separation
   (e) Electrostatic separation
4. Cyanidation
   (a) Cyanide leaching
   (b) Cyanide detoxification
   (c) Cyanide recovery
   (d) Cyanide treatment of copper-gold ores
5. Oxidative pretreatment
   (a) Roasting
   (b) Pressure oxidation
   (c) Bio-oxidation
   (d) Ultrafine grinding
6. Heap leaching
7. Barrick's thiosulfate technology

20.2 Ore Body Knowledge

20.2.1 Geometallurgy

Understanding of the geology and the uncertainties associated with the ore bodies is critical to the success of any mining operation. This is typically carried out using data generated through drill holes, geological mapping, geophysical surveys, and the geologist's interpretation. Ore body modeling and resource estimation is the foundation on which the business case for future mine development and operation is intricately dependent. One major input for resource estimation is quantitative mineralogy information, which is increasingly being recognized as more important than mere elemental assays. Interpretation of mineralogy through assays have been traditionally carried out but with increasing complexities and uncertainties associated with ore bodies, a direct way of determining mineralogy without the need to make certain assumptions is a positive trend. Having more mineralogy attributes for both valuables and gangue in the resource and reserve models allows for a better integration of geology and metallurgy, which is an emerging area widely known as "geometallurgy."

Recent advancements in process mineralogy have furthered the cause of integrating mine site geology and process plant to a great extent with a main focus on improving the value of the mine. Geometallurgy relates to the practice of combining geology and geo-statistics with extractive metallurgy to create a geologically based predictive model for mineral processing plants. It is used for risk management and mitigation during plant design and also to assist and enhance mine production planning (Kittler et al. 2011). McCullough et al. (2013) have also highlighted the challenges associated with recognizing and embedding the value of geometallurgy in the mining value chain. The main challenge is to permanently change the behavior of people and the processes they follow. This requires executive sponsorship, technology, data integration along with automated intelligent analysis to realize the full value of geometallurgy. Kittler et al. (2011) has emphasized the need for proper sample selection with a clear focus on spatial and grade distribution. It is best to avoid blending of samples for geometallurgy samples to capture ore variability and to retain their spatial provenance. The requirements for operational geometallurgy have been presented by David (2010).

Bye (2011) has documented various industrial case studies that demonstrate strategies for gaining value from geometallurgical studies. Values include both operational benefits such as proactive fragmentation control and better strategic planning process along with building of geometallurgical domain models.

20.2.2 Quantitative Gold Deportment for Complex Ore Bodies

With increasing complexities of gold deposits, it is becoming vitally important to have a deeper understanding of ore gold mineralogy and deportment behavior to identify the best processing route for maximizing the value from a deposit. Advances in mineralogy technologies have made it possible to accurately quantify gold deportment.

An important development in gold mineralogy is the ability to carry out quantitative gold deportment for refractory and double refractory ores. The characterization of these ores (e.g., Carlin types deposits) is challenging due to low gold ore grades, presence of variable proportion of preg-robbing total carbonaceous matters (TCM) with sulfide inclusions, visible and invisible gold in different sulfides phases (e.g., pyrite, chalcopyrite, chalcocite, and bornite) and also in iron oxide phases (e.g., magnetite, hematite, and goethite). This section will focus on Carlin type deposits to demonstrate the value of carrying out detailed gold deportment studies.

The gold ore body in north-eastern Nevada, the USA, is commonly known as Carlin type deposits. Mineralization covers an extensive area and numerous ore bodies have been localized by the complex structural and lithological controls. Most of the economic gold mineralization is hosted in limy to dolomitic mudstones. The characteristics of the host rocks that are believed to enhance their favorability to gold deposition are the presence of reactive carbonate, porosity, permeability, and the presence of iron, which can be sulfidized to form auriferous pyrite (Bettles 2002). Most of the gold at Goldstrike is located in arsenian pyrite overgrowths or pre-ore pyrite (Arehart et al. 2003).

Naturally occurring organic carbon or TCM is found in many of these ores within Carlin trend and has the ability to adsorb the gold cyanide from leach solution, commonly known as preg-robbing (Stenebraten et al. 2000; Helm et al. 2009; Hausen and Bucknam 1984). The proportion of TCM in these ores ranges from 0 to 8.5 % with varying preg-robbing activities from high to low which correspond to gold recoveries from ~20 to ~90 % (Stenebraten et al. 2000).

Quantitative gold deportment of these complex ores require an integrated approach to process mineralogy involving various tools such as optical microscopy, mineral analyzer (MLA, QEMSCAN), X-ray diffraction (XRD), Dynamic secondary ion mass spectrometry (D-SIMS), Time-of-flight—secondary ion mass spectrometry (TOF-SIMS), X-ray photoelectron spectroscopy (XPS), and Laser ablation microprobe-inductively coupled plasma mass spectrometry (LAM-ICPMS). Details of these quantitative gold deportment techniques are presented elsewhere (Chattopadhyay and Gorain 2012, 2014; Chryssoulis and McMullen 2005). These measurements are immensely useful but could be expensive and time consuming and therefore it is important to be prudent on representative sample selection with a clear purpose. Improper use of these tools could result in poor diagnosis and wrong interpretation.

Figure 20.2 shows morphology of different TCM in Carlin type deposits. These TCMs are typically fine grained irregular veins, interstitial stringers, and composite types, suggesting the complication of separating them from pyrite through flotation, for example.

Figure 20.3 shows different types of pyrite morphology for Carlin type ores.

Figure 20.4 shows quantitative gold deportment for one of the Carlin type deposits in Nevada. Information obtained in Fig. 20.4 is based on a novel gold deportment technique through extensive studies (Chattopadhyay and Gorain 2014). This provides an in-depth understanding of gold deportment in various ore types and process streams, which is vital for problem diagnosis and allows well-informed decision making such as defining strategies to develop innovative process solutions for metallurgical issues.

20.3 Comminution and Classification

Comminution costs are typically one of the largest cost items in a gold milling operation. Since the advent of Autogenous Grinding (AG) and Semi-Autogenous Grinding (SAG) milling technologies in the late 1950s, they have established themselves as the present standard and are com-

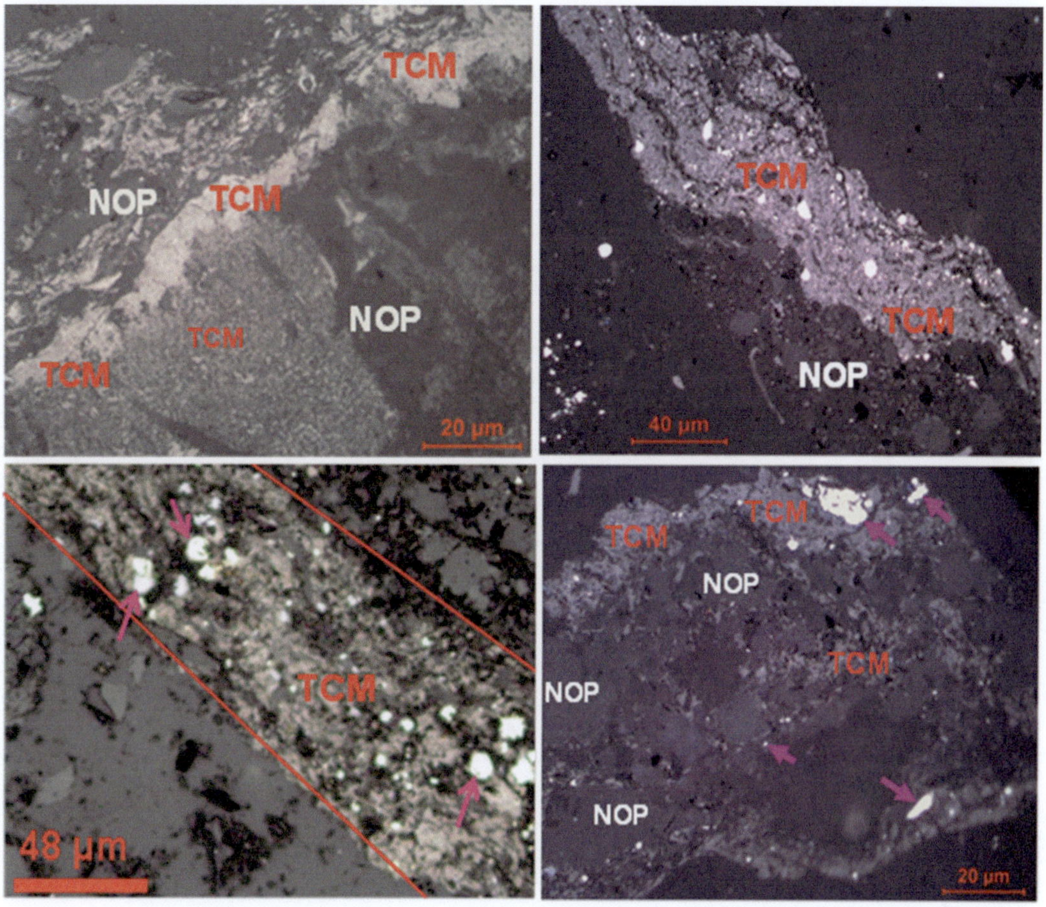

Fig. 20.2 Microphotographs of different types of total carbonaceous matter (*TCM*) in Carlin type deposits

monly used in the gold industry now. These technologies have replaced the previous "conventional" comminution circuits involving crushing-ball mill or rod mill–ball mill circuits. It is estimated that over 90 % of the gold ounces are produced by milling operations, and the majority of these ounces come from high aspect ratio SAG circuits with significant contributions from low aspect SAG circuit such as in South Africa (Mosher 2005). The main reasons for the attraction for SAG milling are circuit simplicity and typically lower capital and operating costs to meet the needs for operations with higher throughputs. SAG mills works well for handling clay rich sticky ores, which is a challenge for multistage crushing circuits requiring washing plants.

20.3.1 Crushing

Crushing is an important step in comminution that prepares the ROM ore for primary grinding and other downstream processes such as heap leach. The selection of the right crushers and crusher flowsheet is dependent on the upstream mining method and blasting characteristics and the downstream process requirements. As process plant throughputs are increasing, the need to reduce operating costs and capital cost per tonne of ore processed is critical. Both crusher and circuit design have evolved with larger crushers employing more horsepower and speed to treat higher throughputs at a reduced cost.

With the advent of SAG milling, cone crushers are now more or less eliminated from com-

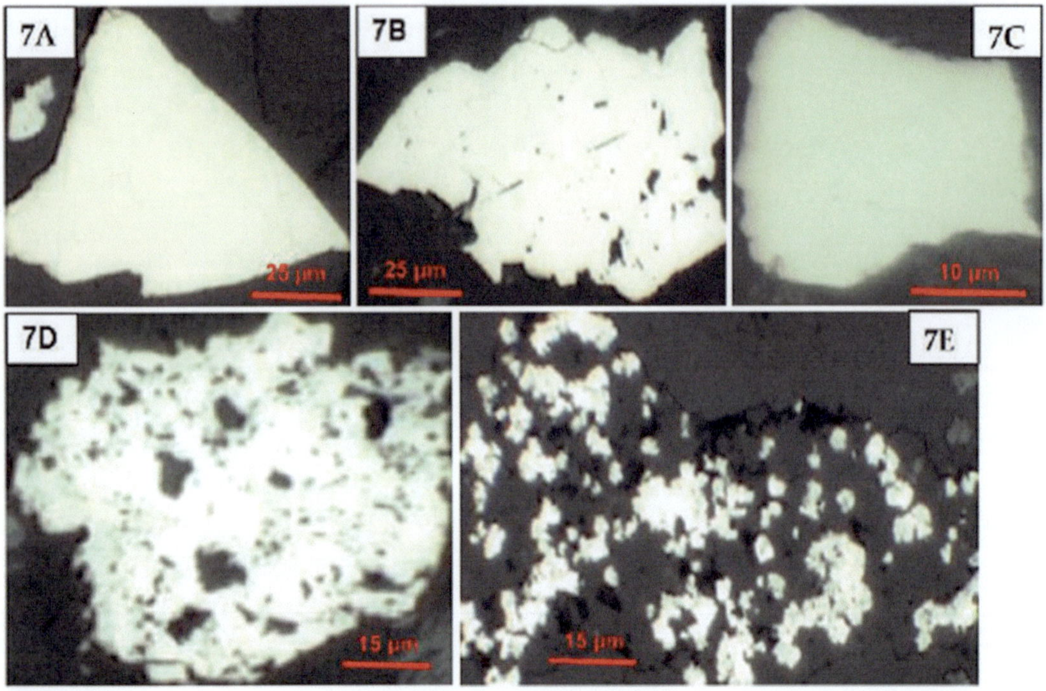

Fig. 20.3 Different types of pyrite identified in Carlin type deposits. *Plate 7A*: Coarse Pyrite, *Plate 7B*: Porous Pyrite, *Plate 7C*: Fine Pyrite, *Plate 7D*: Microcrystalline Pyrite, and *Plate 7E*: Disseminated Pyrite

minution flow sheets. Instead the cone crushers are employed in an SABC circuit to crush the recirculating pebbles. These pebbles tend to be more resistant to impact breakage in the SAG mill and the crusher creates ore surfaces more conducive to breakage in the grinding mills (Major 2002).

During the last 20 years, the size of gyratory crushers has not changed significantly, though their installed horsepower have increased allowing these crushers to treat higher throughputs. An example of the largest gyratory crusher is FLSmidth's 60′×113″ UD design with an installed power of 1 MW. Another feature of the new design of Gyratory crushers is the development of top service gyratory crusher which appears to enhance safety, reduce maintenance downtime, and lower cost of crusher installation (Erickson 2014).

20.3.2 HPGR

SAG mills, however, are less energy efficient in handling harder and abrasive ores. As the ore bodies are steadily becoming more competent along with increasing energy costs, the industry has been looking for alternatives to SAG. It is interesting to note that the traditional crusher and ball milling technology has returned with a more innovative approach to crush rocks using high pressure grinding rolls (HPGR).

Recent successes of HPGR in hard rock application such as at Freeport's Cerro Verde in Peru (Vanderbeek et al. 2006; Koski et al. 2011) and Newmont's Boddington in Western Australia (Dunne et al. 2007a, b; Hart et al. 2011) with a key target of significant grinding energy savings of 15–20 %. Both these operations have two stages of coarse crushing using gyratory and

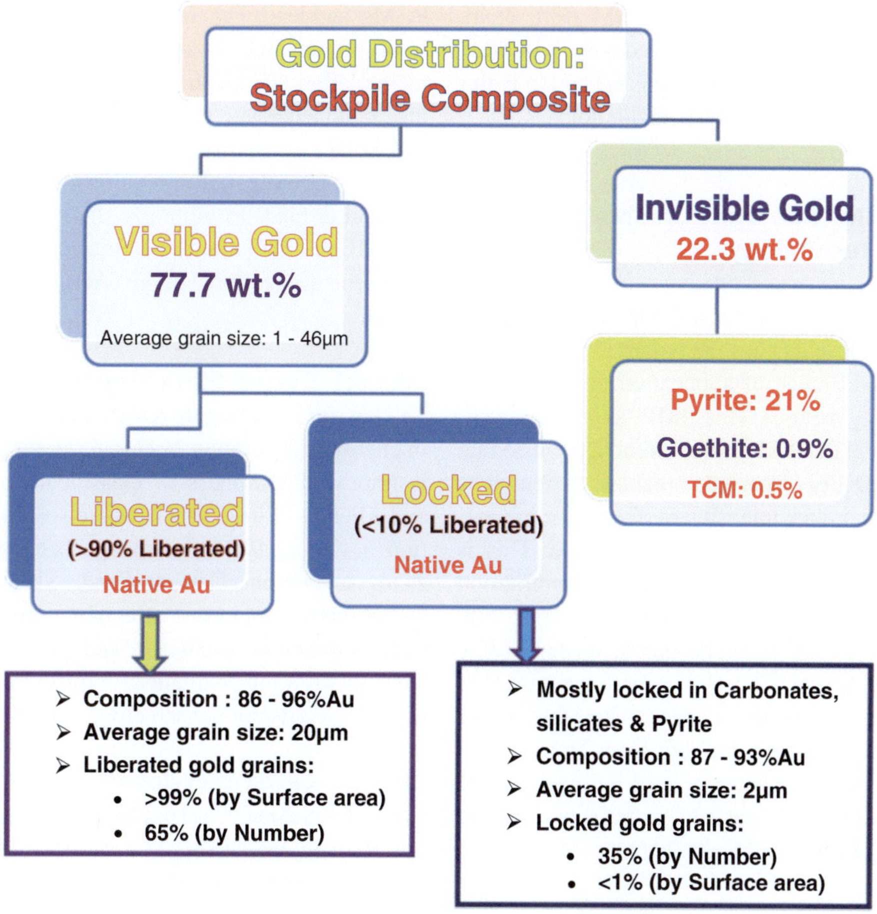

Fig. 20.4 Quantitative gold deportment for one of the Carlin type deposits in Nevada

cone crushers followed by a third stage of crushing involving HPGR and a single stage ball milling circuit.

20.3.3 Mine to Mill

Mine to Mill optimization is various operations over the years have shown significant benefits such as high mill throughput rates from reduced top size from mining through increased powder factor or blast energies (Gillot 2006; Kanchibotla 2014). New blasting technologies are emerging that have the potential to significantly increase powder factor through innovative blasting practices such as dual blast layers within a single blast event (Brent et al. 2013).

20.3.4 Selfrag

Selfrag is an emerging technology showing some promise in pre-weakening of mineral ores by high voltage pulses. This was pioneered in the 1950s but it is attracting some serious attention during the last few years. The key benefits of this technology compared to conventional comminution machines is that this is highly selective and its ability to weaken the ore with a potential for significant energy savings in further comminution (Shi et al. 2013). Studies on a Newcrest's Cadia east Cu-Au ore samples by JKMRC showed an energy reduction of 5 kWh/t, along with a potential to remove two MP1000 pebble crushers and one 10 MW ball mill. Simulation studies showed that Selfrag has the potential to reduce operating cost by A$ 19M (Shi, et al. 2014). Though this study showed some potential, a detailed capital cost estimate is needed to better evaluate the potential of this technology.

20.3.5 IMP Superfine Crusher Technology

The "IMP" superfine crusher technology is presently evolving and has the potential for developing a broad range of application options as the technology matures. This "IMP" superfine crushing concept evolved from the simple premise that the probability of particle breakage increases as the number and intensity of forces simultaneously contacting the particles increase. This is quite different from conventional comminution machines where breakage rate decreases with increased energy inefficiencies as feeds become finer and harder. The IMP superfine crusher is designed to provide a mechanism that could effectively deal with fine hard feeds by maintaining high breakage rates as hard particle became finer and single particle population rapidly expands (Kelsey and Kelly 2014).

20.3.6 Classification

Classification is an important aspect of any comminution circuits. It is well known that a closed circuit grinding circuit with a classification device results in significant benefits to the comminution process. In the early days, mechanical rake and spiral classifiers were used in closed circuit grinding circuits. Due to capacity limitations and high operational costs because of wear and tear, these technologies are not commonly used in the present comminution circuits. Hydrocyclones became popular since the early 1950s as they are high capacity devices that allowed the industry to meet the demands of increasing throughputs to treat low grade ores. Since hydrocyclone separation is a function of both size and density, a sharp size separation is not always possible for many ore containing high specific gravity minerals and metals such as gold, PGMs, and lead. This results in building of these small heavies, which are mostly liberated, in the recirculation stream in cyclone underflow leading to unnecessary grinding with poor gravity and flotation responses. The inefficiency of the hydrocyclone separation requires the use of high circulating loads to minimize the mill residence time. High circulating loads increase the power requirements per tonne of ore (Albuquerque et al. 2008).

20.3.6.1 Derrick Stack Sizer™

The key for major improvements in capacity and in energy consumption in closed circuit grinding is improved sharpness of classification (Hukki and Allenius 1968). This is where screen separation has a significant advantage because of its sharp separation. Derrick Stack Sizer™ is a recent innovation in fine screening which allows high separation efficiency and high tonnage capacity on a much smaller footprint that possible using conventional screens (Clark 2007).

20.3.6.2 Cavex Recyclone™

Cavex Recyclone™ is a double classification unit in one stage, which seems to increase the sharpness of separation and reduces the bypass of fines to the underflow. This technology uses wash water injection mainly to rupture the viscous layer to release the trapped fines for proper classification. Recent trials and installations in grinding circuits have shown to significantly reduce fines misreporting to mill from 30 to 15 % resulting in reduction in circulating load by around 50–60 % (Castro et al. 2009).

20.4 Pre-concentration and Ore Beneficiation

20.4.1 Pre-concentration

Removal of some waste as early as possible after blasting has the potential to reduce haulage costs to the mill in many situations. In addition, this provides an opportunity to upgrade the ore resulting in reduced energy and operating costs per unit of metal input to the plant. A higher head grade to the process plant typically results in better concentration ratio and better unit recoveries as well as the process now has to handle lower gangue content. This concept of pre-concentration is not new with application such as dense media separation and sorting on the surface since the 1930s (Munro et al. 1982), also applied underground (Lloyd 1979). It is worthwhile to note that not all ores are amenable to pre-concentration especially when the minerals are finely disseminated in the ore body, resulting in high losses of valuables in the waste product.

Ores types that are amenable to pre-concentration can add significant economic benefit, such as at the Kroondal platinum mines in South Africa where about 50 % of bulk mining feed of UG2 ores to the mill is rejected with a PGM recovery of 95 % (Holloway et al. 2009). The value of pre-concentration has now been demonstrated for a wide range of ore types at a coarse size range (Mohanty et al. 2000; McCullough et al. 1999; Schena et al. 1990). The value of pre-concentration should be looked in a holistic way involving integrated mining, processing, and waste disposal. Bamber (2008) has concluded based on his studies that the exploitation of a deposit with ore pre-concentration and waste disposal technologies integrated into the mining process prior to beneficiation on surface is superior to the conventional approach.

The following pre-concentration technologies appear to have significant potential and are slowly finding applications in different mining applications:

- Size classification
- Ore sorting
- Dense media separation
- Coarse particle flotation

20.4.1.1 Size Classification

Concentration by comminution and size classification alone has been found to be effective for some ore types (Burns and Grimes 1986; Sivamohan and Forssberg 1991; Logan and Krishnan 2012). Size classification presents a low cost option for rejection of waste, with maximum economic benefit at coarse particle sizes either at naturally arising ROM particle size distribution or at a coarse crush size, as close as possible to the mining face (Klein et al. 2002).

20.4.1.2 Ore Sorting

Ore sorting in the mining industry has evolved from manual sorting in the early days to basic optical sorting first used in mining in the 1970s using camera technology and digital image processing for industrial minerals. Arvidson and Wotruba (2014) have provided a review of the various applications of ore sorting technologies. The importance of ore sorting in improving economics of marginal deposits is increasing being realized by the mining industry (Lessard et al. 2014; Foggiatto et al. 2014).

At present, majority of automated ore sorters, outside of the diamond industry, are color or conductivity sorters (Bartram and Kowalczyk 2009). This is a significant innovation in mineral processing with an ability of pre-concentration with significantly improved economics. The optical sensors have quite a few applications but as the optical properties differ relatively little, laser based sensors could be useful. If the primary surface properties are distinct, NIR sensors are normally used. Novel sensor technologies are now necessary to extend the application to ore types beyond optical sensors that exploit material properties such as electrical conductivity, magnetization, molecular structure and thermal conductivity. Combination of sensors also have potential such as optical/NIR, optical/inductive or XRT/inductive (Arvidson and Wotruba 2014).

The recent development of LIBS (Laser Induced Breakdown Spectroscopy), focused on pattern recognition appears to have potential for mass ore sorting. SonicSampDrill and IHC Mining have been working on the development of the LIBS Ore Sorter (SonicSampDrill 2014). Though there is potential, much work is required to make this technology commercial for large throughputs.

20.4.1.3 Dense Media Separation

Dense media separation is widely used in coal, diamond, chromite, and iron ore industries since its introduction by Dutch Sate Mines (DSM) in 1947. Though used occasionally in other industries its application in precious and base metals viz. PGMs, gold, copper, and zinc are slowly gaining some momentum.

Napier-Munn et al. (2014) have provided a review of the various DMS technologies, with a focus on technologies that have been implemented by the industry. Tati nickel plant in Botswana uses DMS to reject 70 % of mass with recovery of ~2/3rd of the contained nickel and has allowed the mine to target a zero cutoff grade, thus allowing bulk mining and simplifying the mine plan with increased resources. Another impressive application of DMS has been in recovery of fine grained gold from waste rock dumps at the Witwatersrand gold fields in South Africa. Despite a low feed grade (0.2–0.3 g/t), DMS was successful in recovering 70 % of the gold into 30 % of the mass.

Typically DMS processes are compact processes with high throughput of 300–1000 t/h. Holloway et al. (2009) have concluded that DMS seems to be well positioned with tremendous growth potential to address the need for reducing energy costs, improving mine-mill integration and simplifying tailings handling through coarse ore pre-concentration. Bamber (2008) has carried out various studies and has highlighted the significant potential of DMS in underground applications.

20.4.1.4 Coarse Particle Flotation

Flotation of ore at a top size of 3 mm, followed by gravity concentration of the flotation tailings has been applied previously in South Africa to treat Witwatersrand gold ores resulting in production of a 40 % mass pull at an overall gold recovery of 98 % (Lloyd 1979). Coarse particle flotation of particle ranging from 3 to 5 mm is common in the potash industry. For many base metals and precious metals applications, coarse flotation is challenging in conventional and flash flotation machines for pre-concentration purposes.

Jameson (2014) has developed a fluidized bed flotation technology for coarse flotation focusing on pre-concentration at a coarse size (600–800 μm) with an estimated reduction in operating cost of 10–20 %. Though this is in experimental stage, nevertheless the concept has merit and deserves further attention. The Hydrofloat® technology from Eriez also uses fluidized bed for coarse particle flotation with commercial applications in grinding circuits similar to flash flotation, also with the possibility of producing a throwaway tails (Franco et al. 2015). These are interesting developments with implications for pre-concentration at a relatively finer size fraction.

Compared to other pre-concentration technologies, flotation offers the most potential especially for fine grained mineralogical complex ores as other technologies at a coarser grind will incur significant metal losses. A potential flow sheet involving HPGR product of 1–2 mm or SAG product of 800–1000 μm could be subjected to coarse particle flotation to reject a significant amount of feed and the coarse flotation concentration could then be ground to a finer size for conventional flotation.

20.4.2 Gravity Separation

Gravity separation has been used historically as the primary step to recover gold and silver in many gold plants. With the advent of improved gravity devices such as sluices in the early 1900s followed by introduction of jigs, spirals, strakes, Reichert cones, and shaking tables made a significant impact on recovering precious and other metals during the mid-1900s. Amalgamation to recover gold from these high grade gravity concentrate was common in the early days. With the advent of cyanidation, amalgamation has been phased out slowly due to health and environmental reasons.

With the rapid advances in carbon-in-pulp and carbon-in-leach circuits with possibility of obtaining high recoveries, the reliance on gravity as primary means of concentration was greatly reduced (Laplante and Gray 2005). This was further amplified by the move towards simplified and low capital and operating costs to treat low-grade oxide ore bodies with a head grade around 1 g/t. But as complications arose leading to poor recovery of coarse gold in cyanidation, there was a trend towards larger and more efficient gravity devices along with intensive cyanidation.

In most of the applications, gravity recovery targeted gravity recoverable gold (GRG) to supplement cyanidation or flotation, increasing overall gold recovery by 1–5 % and slightly reducing some of the costs associated with carbon stripping and regeneration. Economic justification of GRG recovery can be sometimes difficult as there is a view that these GRG can be recovered in an adequately designed flotation and cyanidation circuits. There is, however, some evidence that gravity circuits along with intensive leaching of these gravity concentrates can result in 0.5–1.5 % additional gold recoveries, at sites such as St. Ives and Kundana (Western Australia), Porgera (PNG) and Dome mine (Canada). Recovery improvement of 20 % has also been reported at the Penjom mine (Lewis 1990).

The recent resurgence of enhanced gravity separators such as Falcon, Knelson, Kelsey jig, and Multi-Gravity Separators (MGS) over the past three decades mainly to recover gold and silver from circulating loads in the grinding circuit is an important development. The ability to characterize GRG using an innovative approach by Laplante (2000) provided a rational basis for identifying opportunities with gravity concentration.

There are different enhanced gravity technologies commercially available. The Falcon concentrator utilizes a G force up to 300 g's for their continuous machines, which is a vertically aligned, open topped bowl shaped as a truncated cone mounted on a rotating shaft. The semicontinuous unit uses G forces ranging from 50 to 200 g's, with a fluidization based added to the upper zone of the bowl. Recent development is a unit with an ability to deliver a G force of 600 g's and targets particle sizes of 37–3 μm. The Knelson concentrators employ fluidized particle bed separation in a mechanically applied centrifugal force for both their continuous and semicontinuous machines. They operate by introducing water through a series of fluidization holes located in rings that circle the circumference of the bowl. The bowl is truncated cone shape and can deliver a G force of 200 g's. The semicontinuous machines are more popular in the industry. Kelsey jig is also an enhanced gravity device that can deliver a G force of up to 100. This technology has been successfully applied to many heavy metal applications with about 30–40 units installed globally. However, this is not as popular as Knelson and Falcon with a few thousand installations worldwide. Multi-Gravity Separator is another technology that seems to be the most efficient enhanced gravity separator, which uses riffling technology with three main component viz. cylindrical rotating drum, internal scraper network, and variable speed differential drive. Selective separation of fine particles are achieved along the internal surface of the rotating drum using the same basic principles employed by a conventional shaking table but replacing the table surface with a rotating drum resulting in many times the normal gravitational pull.

The parallel development of intensive leaching technology with enhanced gravity devices is a significant one for the gold and silver industry. During the last two decades, commercially available technologies such as Consep Acacia reactor and Gekko In-line Reactor (ILR) have been utilized in many gold operations with demonstrated benefits (Laplante and Gray 2005). The Acacia process uses high concentration of cyanide (15–25 g/L), 3–4 g/L caustic, and 2–10 g/L of a suitable oxidation along with a leaching temperature of 50–65 °C.

Flash flotation is an important pre-concentration process to recover free gold associated with base metal sulfides or arsenopyrite when ore GRG content is typically less than 25 % (Laplante and Dunne 2002).

20.4.3 Flotation

Flotation has been used as a means to pre-concentrate refractory ores even before the 1930s in Canada, Australia, and Korea using oils as flotation collectors to produce bulk low grade gold concentrates (Richart 1912; Taggart 1927). Until the late 1960s, most of the flotation activity centered in Canada came from flotation treatment of refractory, copper-gold and other complex ore types (Dunne 2005). During the late 1960s flotation of pyrite was a key focus in many South African gold operations to meet the needs of sulfuric acid market from the booming uranium industry. The gold boom in the 1980s and 1990s led to the processing of several refractory gold and copper-gold deposits in Australasia, Africa, and the Americas using flotation to produce concentrates with further treatment using bacterial and pressure leaching.

A review of the reagents used in gold flotation is presented elsewhere (Kappes et al. 2011). Flotation using xanthates as the primary collector is common for liberated, placer gold and even for many refractory ores. Dithiophosphates are widely used as secondary collectors along with xanthates in various operations. Monothiophosphates are normally more stable and selective for gold minerals with a high silver content. In acidic flotation circuits, mercaptobenzothiazole (MBT) is often preferred for gold bearing partially or sometimes fully oxidized pyrite. Modified thionocarbamates are used in copper-gold ores, which allows some selectively against iron sulfides even at a pH value less than 10. Many gold plants use a blended collector or separate collectors as needed based on mineralogy and ore chemistry.

Copper sulfate is commonly used as an activator of gold-bearing iron sulfides in flotation plants. The dosage, addition point, and the slurry pH range are critical parameters for maximizing the benefits from copper sulfate. Addition of copper sulfate before collector appears to be important to improve kinetics of iron sulfides. Addition of copper sulfate at pH values ranging from 7 to 10 may slow down floatability of iron sulfides. Excessive addition of copper sulfide may result in lower recovery due to oxidation of xanthate into dixanthogen in solution, thus reducing the effectiveness of collection. Copper sulfate addition also appears to impact froth stability. There is an optimum dosage below which results in high slimes recovery with a tenacious froth, whereas a higher dosage leads to a fragile froth behavior.

Sulfidization is also used for naturally oxidized or tarnished iron oxides due to unfavorable milling environment. Solution potential control at −450 mV using sodium sulfide or sodium hydrogen sulfide is typically used prior to flotation. Use of NaHS and silver ions has been reported to improve gold recovery at the Los Pelambres Mine in Chile (Chryssoulis 2001). Sulfide ions appear to act as flotation activators at low concentrations less than 10^{-5} M whereas at concentration above 10^{-5} M it acts as a depressant (Aksoy and Yarar 1989).

Flotation of Carlin type double refractory sulfide ores is considered as challenging (Bulatovic 1997; Kappes et al. 2010). Refractory gold ores commonly contain free gold, submicroscopic gold, carbonaceous material, base metals, pyrite, marcasite, arsenopyrite, and pyrrhotite, with clays and graphitic carbon are known to inhibit flotation behavior (Swash 1988). The Carlin refractory gold ores are mainly associated with arsenian pyrite, which is known to oxidize quickly resulting in reduced flotation kinetics. Newmont has developed the N_2TEC process to provide a reducing environment during grinding and flotation mainly to prevent oxidation of arsenian pyrite (Simmons 1997).

Mineralogy and gold deportment studies clearly suggest that an ultrafine grind (sub 10 μm) is necessary to maximize recovery of gold-bearing pyrite. A finer grind, however results in slower flotation kinetics and poor selectivity against carbonaceous matter resulting in a very high concentrate mass pull. Novel technologies are needed to address this challenge. Newmont has developed the N_2TEC process to float arsenian pyrite in a reducing environment to minimize oxidation of sulfides and this technology is presently used in Simmons (1997).

The deleterious preg-robbing organic carbon can also be removed by flotation and either discarded if the gold content is negligible (Taggart 1925) or separately treated with kerosene prior to cyanide leaching (Beer 1994).

With the trend of declining head grades requiring high throughputs to improve economics, the need for water has increased significantly over years. In temperate and equatorial latitudes, there is no shortage of water in general, although the disposal of used water is a significant issue. In other arid parts of the world such as in Chile, Western Australia, and South West USA, water is scarce or unavailable requiring the need to use sea, hypersaline or brackish water for mining.

Desalination of seawater is expensive as new desalination plants and associated pumping and pipeline systems can easily have a capital cost exceeding $500 million along with high operating costs (Blin and Dion-Ortega 2013). Barrick Gold has developed a new process, Air-Metabisulfite treatment (AMBS), which allows use of seawater or brackish water for pyrite depression in copper flotation with minimal metallurgical impact compared to that with the conventional lime based process (Gorain 2012). The AMBS treatment has also resolved the issue of molybdenum flotation in copper-molybdenum ores using seawater.

A holistic approach to reagent selection is becoming more important (Nagaraj and Farinato 2014) due to increased ore complexities and also the advent of quantitative mineralogy tools and better understanding of flotation cell hydrodynamics. In addition, due to stringent environmental regulations, there is a focus on developing greener flotation reagents to meet sustainability goals. More recent application of electrochemical and surface analytical techniques has allowed better understanding of the fundamentals of flotation mechanism and chemistry (Smart et al. 2007). This is an important development that is already helping in better understanding of flotation behavior of complex ores and in developing solutions to complex flotation problems (Ralston et al. 2007; Fuerstenau 2007; Gorain 2013).

20.4.4 Magnetic Separation

There are tens of thousands of low-intensity magnetic separation (LIMS) and thousands of high-intensity magnetic separations (HIMS) units used in the minerals industry today. A significant magnetic separation development that made a major impact in the minerals industry was the drum separator for magnetite ores, mainly to treat fine grained and low grade magnetite ores for producing high grade concentrates for sinter feed, pellets for blast furnace feed and also for direct reduction iron processes (Arvidson and Norrgran 2014). The Wetherill separator type is still in use today in various forms for separating precious metals, typically after calcination of ore feed. Use of magnetic separator to recover gold-bearing iron oxides from roaster leach tails is also practiced such as at Barrick's Goldstrike operation (Douglas and Semenyna 2013).

Due to declining grades, the need for sophisticated high capacity magnetic separators is becoming important. Drums for low-intensity magnetic separators in iron ore processing are becoming larger, with drum diameters around 1.5 m and widths exceeding 4 m. Other than the trend of increasing size, the use of cryogen-free superconducting coil system is also becoming prevalent, which is the most recent technology for superconducting magnetic separators using a niobium-titanium conductor to generate the magnetic field without the need for helium to cool the coil (Norrgran et al. 2009).

20.4.5 Electrostatic Separation

Electrostatic separation (ESS) is mainly used for mineral sands application, though earlier patents suggest applications in iron ore and phosphate separations. The relative electric conductivity property of the mineral surfaces is exploited to make the separation. For example, conductive minerals such as ilmenite, monazite, and rutile are separated from non-conductive silica and zircon present in mineral sands. ESS works best on clean surfaces and in many cases the feed needs to be washed and classified for efficient separation. ESS was also successfully implemented at several precious metals smelters recovering metal prills from slag, typically after the Merrill Crowe process. In addition, this technology has been successfully applied in gold mines to recover gold and silver from slag (Hearn 2014). Dry grinding is typically used to liberate metal prills, de-dusting and then ESS to recover these prills for direct smelting rather than the conventional wet milling, gravity tabling followed by drying before sending for smelting (Maki and Taylor 1987).

20.5 Cyanidation

20.5.1 Cyanide Leaching

The use of cyanide to leach gold has been known as early as 1783 and studied since then in the 1840s and 1850s by various researchers. Amalgamation was the primary recovery process, though chlorination has also been in use in various gold plants. Since the development and patenting of the cyanide leaching process by MacArthur and the Forest brothers during 1887–1888, the process was first commercialized at the Crown Mine (New Zealand) in 1889 (Rose and Merloc 2008; Marsden and Sass 2014). This technology spread globally very rapidly, with applications at Robinson Deep (Transvaal, South Africa) in 1890, Mercur (Utah, USA) and Calumet (California, USA) in 1891, El Oro (Mexico) in 1900, and La Belliere (France) in 1904. The cyanidation process was instrumental in rescuing a declining industry in South Africa allowing development of the famous gold discovery at Witwatersrand in Transvaal, South Africa. Gold production in South Africa increased almost ten times from less than 300,000 oz in 1888 to over three million ounces in 1898 (Stanley 1987). Various innovative developments came from South Africa that made a significant impact in the global gold industry. New process flowsheets involving cyanide leaching and zinc precipitation replaced amalgamation and gravity in most cases. Batch processing in vats of separate sand and slimes circuits was replaced by continuous process as we know today. Also milling in cyanide solution was tested and developed at Witwatersrand in 1892, though first commercially applied in New Zealand during 1897 (Marsden and Sass 2014).

The following innovations in the USA and elsewhere from 1904 to 1970s allowed a continuous leaching process with a shift to "all-slimes" treatment instead of separate sands and slimes treatment:

- Introduction of tube milling replacing stamp milling in 1904, allowing higher throughput along with a finer but uniform particle size distribution resulting in improved liberation and recovery (Rickard 1907)
- Mechanical advancements in solid–liquid separation during 1904–1908 including Dorr's thickener and the first CCD circuit implemented along with Butter's vacuum filter and Oliver's drum filter (Bain 1910)
- Development of Brown's air-agitated leach tanks (New Zealand) and "Pachuca" tanks (Mexico)

The recovery of gold and silver from cyanide solutions received significant attention resulting in efficiency of cementation through use of zinc dust instead of zinc shavings in zinc boxes, deaeration of pregnant solutions along with addition of a small amount of soluble lead salt to solutions prior to contact with zinc (Clennell 1915). This process is now referred to as "Merrill-Crowe" after the developers of this technology.

The next major development in gold and silver processing was the use of activated carbon and the carbon-in-pulp process, which was implemented first in the 1980s. The use of wood charcoal to recover gold from chlorination leach solution was patented in the late 1800s and was applied commercially at Mount Morgan and other operations in Australia, and was also studied by the US Bureau of Mines (USBM) along with many others during the 1920s and 1930s. Granular activated carbon was first used at the San Andreas de Copan plant in Honduras in

1949, where the loaded carbon was sold to smelter for gold and silver recovery. The Getchell mine in Nevada operated a 500 t/day CIP plant and a patent was granted to McQuiston and Chapman in 1951 for a CIP technique (Marsden and Sass 2014).

Activated carbon was used commercially in 1952 at the Carlton Mine in Cripple Creek, Colorado, which also had a reactivation kiln for carbon regeneration. These features allowed the carbon adsorption methods to be more economical than the Merrill-Crowe process and were an important development in acceptance of the CIP process worldwide (Marsden and Sass 2014). The Homestake mine in South Dakota was the first largest scale application of the CIP process in 1973. With further development in the carbon elution process at Anglo American Research Laboratory (AARL) along with research efforts at Mintek, CIP spread rapidly in South Africa at Modderfontein, President Brand, Randfontein Estates, Western Areas, Rand Mines, Beisa and 11 other installation during 1981 to 1984 (Boydell 1984). The first CIP plant in Australia was commissioned at Kambalda in 1981 and later at Norseman and Haveluck. The Jerritt Canyon Mine in the USA was commissioned in 1981.

The key enablers of the CIP circuits were development of the interstage screen, effective carbon elution techniques and the ability to regenerate the carbon using acid wash and thermal reactivation techniques. With better understanding of the processes involved, the separate leaching and CIP steps were combined mainly to reduce "preg-robbing" of gold and silver by naturally occurring carbonaceous matter in the ore, which is now referred to as Carbon-in-Leach (CIL) circuits. The CIP/CIL flowsheets became the new standard for gold and silver leaching by early 1980s. Despite lower silver loadings on carbon, it was also being applied to treat primary silver deposits as well as gold. In the early days, however, the carbon consumption was significant around 200–300 g/t but with pipe and agitator design modification and pre-conditioning of carbon along with rejection of fines and other modifications resulted in reduced consumption of 30–50 g/t. A typical CIL plant is shown in Fig. 20.5.

Fig. 20.5 Carbon-in-Leach plant

The Resin-in-Pulp technology was implemented in the former USSR during the late 1960s and 1970s, and later some plants were reportedly built in Russia, Uzbekistan, Kazakhstan, and Kyrgyzstan. The process of resin elution and regeneration is considered complex and therefore are not popular in the gold and silver industry. Commercial applications outside of former USSR are limited to Golden Jubilee in South Africa and Penjom in Malaysia.

20.5.2 Cyanide Detoxification

The accidental release of 50–100 t of cyanide in cyanide-bearing tailing solution from the Aural mining operation in January 2000 near Baia Mare in Romania sparked an immediate mission which eventuated the formation of a steering committee involving a wide range of stakeholders to establish a framework for the use of cyanide in the mining industry. As a result, the International Cyanide Management Code was developed in May 2002 and later in October 2002 the International Cyanide Management Institute (ICMI) was formed with the responsibility of administrating the code through a multi-stake team including representatives of mining companies. In February 2004, the Industry Advisory Group (IAG) was reconstituted by a group of gold producers, cyanide manufacturers, and transportation companies with an aim to coordinate and facilitate code implementation. More details on the history and its recent progress have been provided elsewhere (Gibbons 2005; Beazley and Woods 2013).

To meet the cyanide code guidelines (50 mg/L, weak acid dissociable cyanide, WAD) in tailings storage facilities, operations use different methods to suit their application. The preferred method is the natural decay of cyanide through the leach circuit allowing an acceptable discharge to the tailings facilities. Use of tailings thickener and sometimes tailings filtration also assist with recovering and recycling cyanide-bearing solution. Many operations use this method without applying any other form of cyanide destruction, but require extreme care in monitoring concentrations down the leach circuit. Use of dry stacked tailings such as at La Coipa in Chile allows cyanide recycling and could complement the detoxification efforts.

Cyanide destruction using biological oxidation has been successfully applied at the Homestake Mines in South Dakota since 1984 and was later implemented at other Homestake operations (Whitlock and Mudder 1986). Inco's method of using sulfur dioxide and air with copper as catalyst, developed in the early 1980s, has been the preferred and most cost-effective method of cyanide destruction where natural degradation is not possible.

Other method's involving hydrogen peroxide and Caro's acid have also been used successfully at different operations. Due to strong oxidizing conditions, both free and WAD cyanides are converted to cyanate very rapidly and is very useful where rapid detoxification is required such as in circuits with limited residence time availability.

Maelgwyn Mineral Services (MMS) have developed a cyanide destruction process called MMS CN-D Process™. The process utilizes MMS's proprietary Aachen Reactors to use oxygen, carbon, selected reagents, and catalysts to destroy cyanide to WAD levels well below 50 ppm (Adams and Glen 2011).

Barrick Gold has developed the LIXKill process involving Caro's acid, which is typically used to destroy free and WAD cyanide in tailings reclaimed solution before it is fed to the mill and autoclaves (Pekrul 2007). Use of alkaline chloride and hypochlorite oxidation of cyanide has been discontinued due to high consumption and costs along with the inclination to avoid any chloride into the cyanide leach system.

20.5.3 Cyanide Recovery

The AVR (Acidification, Volatization, Recovery) process for recovering cyanide involves acidifying cyanide-bearing solutions to generate hydrogen cyanide gas, volatilization with air to remove hydrogen cyanide (HCN) from solution followed by adsorption and recovery of HCN gas into a caustic scrubber system. This approach was used commercially at Flin Flon in Canada to treat cyanide-bearing process solution from about 1930 to 1995 (Marsden and Sass 2014). This process was modified to include high efficiency packed bed adsorption towers for improved stripping of HCN either from solution or directly from slurry thus avoiding solid–liquid separation (Goldstone and Mudder 1988). This modified AVR was used at Golden Cross in New Zealand, Delamar Silver Mines in Idaho during the 1990s and more recently at Rio Paracatu in Brazil and Cerro Vanguardia in Argentina.

AVR has not been used extensively in the industry mainly because most gold plants have provisions of cyanide recycling procedures such as through tailings thickening with reclaim solution collection systems along with effective management and control of cyanide concentration down the leach circuits. Another reason could be perceived safety issues associated with generation of HCN gas, despite decades of safe use of the AVR process.

20.5.4 Cyanidation Treatment of Copper-Gold Ores

Gold ores containing cyanide-soluble copper minerals such as chalcocite, covellite, bornite, cuprite, malachite, and azurite pose significant metallurgical challenges such as low gold recoveries in cyanidation as well as high cyanide consumption and detoxification costs. Moreover, the presence of copper in tailings supernatant tends to stabilize weak acid dissociable (WAD) cyanide in a form toxic to wildlife and less amenable to natural degradation processes.

The SART (Sulfidization, Acidification, Recycle and Thickening) process is a major development in treatment of copper-gold ores containing significant amount of cyanide-soluble copper. The SART process, as presently commercialized, was originally developed by SGS Lakefield Research and Teck for the Lobo-Marte project (MacPhail et al. 1998). This process involves copper sulfide precipitation with sodium hydrosulfide (NaHS) and sulfuric acid, resulting in release of cyanide as HCN. The solids are separated using thickening and filtration to produce a high grade copper sulfide solids. The thickener overflow solution is re-neutralized with lime and hydrogen cyanide is converted back into calcium cyanide, suitable for recycling to the cyanide leaching circuit.

The first commercial SART plant was commissioned in 2004 at Newcrest's Telfer Mine in Western Australia (Barter et al. 2001). Newmont has been operating a SART plant at their Yanacocha in Peru since December 2008 (Guzman and Acosta 2009). Other major SART operations are at Maricunga in Chile, Lluvia de Oro in Mexico, and Gedabek in Azerbaijan to treat pregnant gold-bearing solutions produced by heap leaching. BioteQ is the commercial supplier of the SART technology and have designed two commercial SART plants along with significant experience in design and operation of commercial sulfide precipitation plants (Lawrence and Lopez 2011; Littlejohn et al. 2013)

20.6 Oxidation Pretreatment

Refractory ores require some degree of pretreatment prior to cyanidation mainly to release gold locked as solid solution in the matrix of various gold-bearing iron sulfides, such as pyrite, arsenopyrite, and arsenian pyrite. Other minerals such as tellurides and stibnites also have varying degrees of refractory behavior. Refractory ores do not respond well to cyanidation with very low recoveries and are not economical with direct cyanidation processes. Various challenges with different types of refractory ores have been described elsewhere (Guay 1980; Han and Chen 1990; DeSomber et al. 1996; Gordon 1999; Lunt and Weeks 2005; Dunne et al. 2007a, b; Dunne 2012). The economics of refractory gold treatment dictate a higher gold content in the feed or higher gold price or both compared to the "break even" economics of "free" gold treatment. This is a result of the complex flowsheets required to recover and extract the gold. Consequently higher capital and operating costs are associated with refractory gold processing. Furthermore, from an operating perspective the cost and availability of utilities (electricity and water) along with water quality (especially chloride content) is important as refractory gold treatment has a greater demand for both. Reagent consumptions and unit costs for acid neutralization and cyanide usage are important components of operating costs as is the maintenance cost given the extreme operating conditions in refractory gold treatment. Downtime for major equipment repairs is also an important consideration.

There are three main commercially used pretreatment routes involving varying degrees of oxidation of sulfides present in an ore or in concentrate:

1. Roasting
2. Pressure oxidation
3. Bio-oxidation
4. Ultrafine grinding

20.6.1 Roasting

Roasting as a pretreatment process for refractory gold flotation concentrates has been practiced for more than 130 years. Up until 25 years ago, roasting was the standard process option for pretreating a refractory gold flotation concentrate. As environmental restrictions increased over the years and alternative processes became available, the application of roasting has decreased significantly.

During the early days (1860–1902), prior to the introduction of cyanidation and flotation processes in treatment of the refractory gold ores, gold recovery and extraction included combinations of crushing and grinding of the ore followed by roasting, fine grinding, chlorination, and precipitation of gold.

Thomas and Cole (2005) have provided details on the following four roaster development steps.

20.6.1.1 Rotary Kiln and Multiple Hearths

Rotary Kiln and multiple-hearth roasters were used in the early days before 1940s for refractory ores. The single Edward roaster was the most common in the gold industry. The Edwards roaster was developed in the 1890s in Australia, and became the preferred roaster around the world (Moss 1918; Kadenhe and Makende 1987; Loosley 1993). Wood, gas, and coal were used to heat the roasters. With the advent of flotation process in the early 1900s, upgrading of ores in concentrate allowed the autogenous roasting that did not require additional external fuel. Notwithstanding this advancement, good temperature control in the Edwards roaster was still an issue leading to sintering or under-roasting of the concentrate and thus poor gold extraction. Some successful applications of the Edwards roaster were Giant Yellowknife Mines Ltd., Ashanti Goldfields, and Government Roasting Plant in Zimbabwe.

20.6.1.2 Fluidized Bed

In 1947, the first Dorr-Oliver bubbling fluidized bed (BFB) roaster was built at the Campbell Red Lake Mine in Canada. Since that time all Edwards roasters have progressively been replaced by BFBs. For pyrite flotation concentrates with small amounts of arsenopyrite (<1 % As), a single stage BFB bed roaster is used. Roasting of a bulk arsenopyrite concentrate is done in two stages, the first stage with partial oxidizing conditions to volatilize the arsenic at 500 °C. The cleaned arsenic-trioxide fume is cooled and the dust is collected in baghouses at 105 °C. The second stage is a complete oxidizing roast step to oxidize the sulfides to provide a porous calcine amenable to cyanidation.

One major issue with the BFB roaster is the uneven distribution of heat between the top and the bottom (Hammerschmidt et al. 2005). This temperature differential is problematic as gold recovery is sensitive to temperature and roasting should be carried out over a small temperature range.

20.6.1.3 Circulating Fluidized Bed

The circulating fluidized bed (CFB) roasters were initially used in the alumina industry. During the 1950s and 1960s, Inco and Falconbridge commercialized the technology for the processing of pyrrhotite to recover nickel, copper, cobalt and iron. For every tonne of calcine production, almost 6 tonnes are recirculated to provide longer residence time to eliminate sulfur in the coarser size fractions. By the 1990s, this technology was utilized to treat certain refractory gold ores (Bunk and Bruyns 1993; Kosich 1992; Lahti 1996; Thomas and Cole 2005; Hammerschmidt et al. 2005; Hammerschmidt et al. 2011). With the introduction of a circulating fluidized bed (CFB), the temperature differential issue of BFB was resolved. The CFBs are higher capacity units. The first CFB roaster for refractory gold treatment was the Gidji roaster, Kalgoorlie, Western Australia, commissioned in 1989 (Biehl and Bunn 1990). When treating a flotation concentrate with a high content of arsenic, the preferred roasting option is still the two-stage BFB roaster (Goode 1993; Hammerschmidt et al. 2005).

20.6.1.4 Oxygenated Fluidized Bed

Freeport McMoRan developed a two-stage "whole" ore roasting process using oxygen as a fluidizing gas (Bunk and Bruyns 1993). The first whole-ore roaster using oxygen injection was built in 1989 at Big Springs, the USA. The process has significant advantages (Bunk and Bruyns 1993; Thomas and Cole 2005). The use of oxygen dramatically reduces the heat losses present in the exit gases. Also, expensive to maintain "intensive" heat and recovery systems are not required. The use of pure oxygen provides for good sulfur and preg-robbing carbon oxidization at very low temperatures that are not realized in traditional roasting with air only. Another improvement to the whole-ore roasting process is the use of dry grinding of the feed (Thomas and Cole 2005) to minimize the heat loss associated with water removal as steam. This also allows lower sulfide ores to be treated by roasting.

In 2000, Barrick Goldstrike commissioned an oxygen roasting circuit using dry grinding and Freeport McMoRan Inc. oxygenated roasting technology. Figure 20.6 shows a photograph of the Barrick Goldstrike roaster in Nevada, the USA.

20.6.2 Pressure Oxidation

The use of pressure oxidation in the aluminum industry (Bayer's process) dates back to the 1890s. Almost 60 years later, this was applied to the base metal sulfides, mainly due to the need to treat complex ore types to meet the increased demand for base metals. Also improvements in materials of construction of autoclaves for use in acid conditions at high temperatures and pressures along with availability of high-purity oxygen made it possible for use of pressure oxidation for base metals (Thomas and Cole 2005).

During 1946–1955, Chemical Construction Company (then a subsidiary of American Cyanamid Company) pioneered development of hydrometallurgical processes to treat nonferrous metal ores and concentrates. Their first client Sherritt-Gordon Mines Limited, a Canadian copper and nickel mining company, bought the rights to pressure oxidation based hydrometallurgical technologies in 1955. The first application was on nickel sulfide concentrates at Fort Saskatchewan, Alberta in Canada. Thereafter, two pressure-leaching plants were built to treat cobalt concentrates (Berezowsky et al. 1991).

It is remarkable as to how the technologies get transferred into other commodities such as treatment of zinc concentrates at Cominco zinc operation in Trail, British Columbia in Canada during the 1970s. During the 1980s several new acid-pressure-leaching plants were built to treat nickel-copper mattes for the platinum mining industry (Thomas 2005). This technology eventually got acceptance for treatment of nickel laterites such as at Moa Bay in Cuba (1959) and at Cawse and Murrin-Murrin in Western Australia during the late 1990s.

With successful application of pressure oxidation for different commodities, interest in the use of pressure oxidation to treat refractory gold and silver ores grew during 1970s and 1980s. Though some pilot installations for gold ores were carried out during this period, it was not until 1985 that the first gold application was at the Homestake-McLauglin mine to treat refractory pyrite ores. It appears that the first application of pressure

Fig. 20.6 Barrick's Goldstrike roaster in Nevada, the USA

oxidation to treat gold-bearing sulfide flotation concentrates was carried out at Sao Bento in Brazil during 1986. In 1988, an alkaline pressure oxidation facility was commissioned at the Barrick Mercur operation in Utah (Thomas 2005). These applications are important innovations for the gold industry that paved the way for many applications worldwide. A number of pressure oxidation plants were commissioned in Nevada including Goldstrike, Getchell, Lone Tree, and Tween Creeks during 1988–2000. Barrick's Goldstrike operation was the largest by tonnes of ore treated. During the same period, some operations were built with pressure oxidation viz., Lihir and Porgera in Papua New Guinea (McDonald 2003; Hille and Raudsepp 1995), Macraes in New Zealand along with Campbell Red Lake and Con in Canada. A summary of the gold pressure oxidation plants presently operating is shown in Table 20.1.

Table 20.1 indicates that the gold industry uses both whole ore and concentrate as feed for pressure oxidation. Concentrates are used as feed in cases where the ore is amenable to flotation. Many refractory ores such as Carlin type deposits have poor flotation response with high concentrate mass pulls and low flotation gold recovery. Ores such as at Porgera has a favorable flotation response with gold recoveries in the range from 92 % to 95 % with a concentrate mass pull and upgrade ratios (Au and S) of 14 % and 7 %, respectively.

Both acidic and alkaline pressure oxidation processes are used for whole-ore treatment depending on mineralogy and the carbonate content of the ores. The minimum sulfide sulfur target for autoclave feed ranges from 5 to 7 % for autogenous operation depending on ore mineralogy. Acidic treatment is used for treating concentrates due to high sulfide content and in some cases such as at Porgera where rejection of carbonates in flotation is important to reduce acid

Table 20.1 Gold pressure oxidation plants (based on Thomas 2005)

Operations	Company	Location	Start-up	Capacity (t/day)	Feed	Temperature (°C)
McLaughlin	Homestake	USA	1985	2730	Whole ore	180
Silo Bento	Gencor	Brazil	1986	240	Concentrate	190
Mercur	Barrick	USA	1988	680	Concentrate	225
Getchell	First Mississippi	USA	1989	2700	Whole ore	210
Goldstrike	Barrick	USA	1990	1825	Whole ore	225
Porgera	Barrick	PNG	1991	1215	Concentrate	195
Campbell	GoldCorp	Canada	1991	71	Concentrate	195
Goldstrike	Barrick	USA	1992	4700	Whole ore	225
Con	Royal Oak	Canada	1992	90	Concentrate	210
Goldstrike	Barrick	USA	1993	7065	Whole ore	225
Lone Tree	Newmont	USA	1994	2270	Whole ore	190
Twin Creeks	Newmont	USA	1996	7260	Whole ore	225
Lihir	Newcrest	PNG	1997	8100	Ore/concentrate	210
Hillgrove	Hillgrove	Australia	1999	24	Concentrate	220
Macraes	OceanaGold	New Zealand	1999	20	Concentrate	225
Pueblo Viejo	Barrick/Goldcorp	Dominican Republic	2012	24,000	Whole ore	230

consumption during acidulation of autoclave feed. Schematic of a typical pressure oxidation circuit treating pyrite concentrates is shown in Fig. 20.7.

Gold recoveries are typically higher for acidic pressure oxidation compared to that for alkaline pressure oxidation. The conversion of ferrous sulfate to ferric sulfate is important as ferrous sulfate increases cyanide consumption during leaching. In acid autoclaving, however, silver is precipitated as argentojarosite, which typically requires a lime-boil pretreatment stage for recovering silver. Temperatures in the range from 175 to 230 °C is optimum for acid autoclaving as this prevents production of elemental sulfur known to prevent complete oxidation and reduce gold recovery. Alkaline autoclave, though has significant advantages for treating high carbonate ores, but the alkaline chemistry results in production of oxidation products (iron oxides) and encapsulation of unoxidized sulfides thus impacting gold recovery (Thomas 2005). Alkaline autoclaving is normally preferred, if the carbonate content in ore is high, from a material construction point of view resulting in lower operating and capital costs.

The acidic pressure oxidation for concentrate treatment leverages high sulfide content to allow an autogenous reaction, hence there is no need for splash/flash tower on autoclave discharge for heat recovery. Autoclave discharge passes to a countercurrent decantation (CCD) wash circuit to allow recycling of thickener overflow to acidify the autoclave feed and to allow a partial recycle of thickener underflow to autoclave feed and prevent formation of elemental sulfur as an intermediate oxidation product (Berezowsky et al. 1991).

Pressure oxidation has a major advantage in treatment of arsenic-bearing gold ores and concentrates, and is an important technology for high arsenic gold deposits. Campbell Mine (now Goldcorp) has successfully treated high arsenic (~10 %) gold-bearing sulfide concentrates (S ~ 18 %) with studies showing economic and environmental advantages over roasting processes (Frostiak and Haugard 1992). Previous work has shown that arsenopyrite is oxidized producing sulfuric acid and a ferric arsenate compound is produced at a temperature above 150 °C, which is considered to be crystalline and stable at room temperature up to a pH of 8 (Neira Arenas and Monhemius 1998).

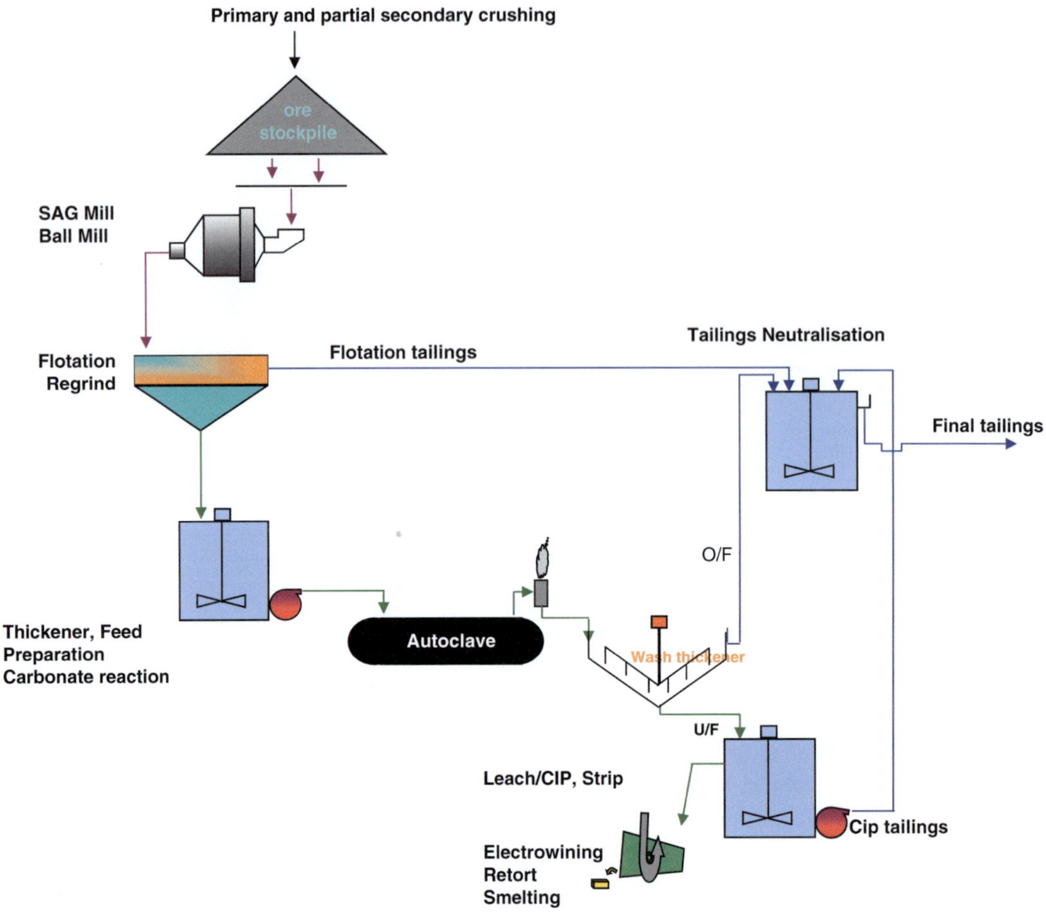

Fig. 20.7 Schematic of a typical pressure oxidation circuit treating pyrite concentrates

Newmont's high-temperature pressure oxidation process for treating double refractory gold ores is another innovation (Simmons 1994, 1996). In this process the "active carbon" is passivated through close control of temperature, oxidation potential and acidity of the finely ground ore or concentrate in autoclave to establish conditions under which the sulfides are oxidized. Newmont's Twin Creeks operation in Nevada uses this process to treat whole ore. Macraes Mine in New Zealand also uses a similar process to treat finely ground refractory flotation concentrates (Cadzow and Giraudo 2000).

Advantages of pressure oxidation over roasting:

- Higher gold recoveries with pressure oxidation for some ores with high arsenic content due to encapsulation of gold in the calcine and arsenic based products
- Autoclave vent gas handling requirements are significantly lower than roaster off gas handling and treatment
- Advantages for treating high carbonate ores in pressure oxidation as heat losses with roasting could be high
- Stable arsenic product (ferric arsenate) with pressure oxidation
- Dry grinding not necessary

Disadvantages with pressure oxidation over roasting

- High operating and capital cost with pressure oxidation
- Poor silver recovery due to silver jarosite formation resulting in additional pretreatment costs
- Treatment of high preg-robbing ores challenging unless non-cyanide routes are used such as the use of thiosulfate process at Goldstrike

20.6.3 Bio-oxidation

Interest in bio-oxidation treatment of refractory gold ores took prominence during the 1970s and 1980s, resulting in some successful pilot plant investigations such as at Fairview, South Africa, using BIOX® technology and at the Equity Silver Mines in British Columbia. In 1986, a 10 t/day plant was commissioned at Fairview to process flotation concentrates, which is still successfully operating. Sao Bento in Brazil during 1991 uses a bio-oxidation circuit to partially oxidized sulfide concentrates ahead of the pressure oxidation circuit to increase throughput. Ashanti-Obuasi in Ghana commissioned a 720 t/day agitated bio-oxidation process in 1994 to treat sulfide flotation concentrates, which was later expanded to 960 t/day and is being operated successfully.

The BIOX® process for the pretreatment of refractory sulfide ores was initially developed at the GENCOR Process Research in the 1970s and is now considered a mature and commercially viable process. Eleven new commercial BIOX® plants have been commissioned during this period with eight currently in operation. Harbour Lights in Western Australia was the first licensed BIOX® plant commissioned in 1992 with a 40 t/day capacity. The plant was successfully operated till 1994 when the ore and stockpile reserves were depleted and the mine was closed. Some recent applications include the Suzdal BIOX® plant in Kazakhstan, the Fosterville BIOX® plant in Australia (early 2005), the Bogoso BIOX® plant in Ghana, and the Jinfeng BIOX® plant in China (early 2007). The technology achieved another milestone with the successful commissioning of the 1500 m^3 live volume BIOX® reactors at Bogoso, the largest operating BIOX® reactors to date. The commissioning of Phase 1 of the Kokpatas plant in Uzbekistan was completed early in 2009 followed by the commissioning of Phase 2 in 2010. The Kokpatas BIOX® plant is currently the largest BIOX® plant in the world. About 21 million ounces of gold has been produced using BIOX® (Biomin 2015).

The first operation to use Mintek's BACOX process commercially was the Youanmi mine in Western Australia in 1994 (Miller and Brown 2005; Budden and Bunyard 1994). BACOX technology was then incorporated into the sulfide treatment plant at the Beaconsfield Gold Mine in Tasmania in 1998 (Holder 2007). The bacterial oxidation circuit installed at Beaconsfield uses the Mintek-BacTech technology. The bacteria are a combination of mesophilic iron and sulfur oxidizing cultures. The Laizhou Gol Metallurgy Plant (renamed to "Tarzan Biogold") of Shandong Province, China, also uses the BACOX technology (Miller et al. 2004). Interestingly the primary focus was to operate the plant as a toll treatment facility. The plant was commissioned in 2001 (Xie et al. 2008; Miller et al. 2004).

The preferred technology for the treatment of concentrates in China appears to be bio-oxidation followed by cyanide leaching and then zinc cementation. The Changchun Gold Research Institute (CCGRI) provides the process technology support for many bio-oxidation plants in China (Xie et al. 2008) using microorganism cultures isolated from hot water springs (moderate thermophiles) and mines in South China. The Olympiada Mining Combine, Russian Federation, developed and uses the BIONORD® technology (Sovmen et al. 2009). The process was developed to accommodate the harsh, extremely cold climatic conditions found at the Olympiada mine in the Krasnoyarsk Region of Russia.

High arsenic-bearing flotation concentrates with bio-oxidation results in conversion of iron, sulfur, and arsenic to ferric sulfate, sulfuric acid, and arsenic acid. The acidic solution exiting the bacterial reactors flows to a CCD circuit and the overflow solution is then sent to neutralization. In general, the acidic arsenic liquors arising from bio-oxidation have high ferric iron-to-arsenic ratios and this is ideal for the generation of ferric arsenate. A large quantity of air must be introduced into the biological reactor to ensure satisfactory oxidation rates. Many oxidation plants grind the flotation concentrate to enhance leach kinetics. The heat released during the oxidation process is significant. So it is important that this be dispersed efficiently since the bacteria do not function effectively outside their operating temperature range. The heat is collected by internal cooling coils inserted in the bio-oxidation reactors and the heat is then dispersed to the atmosphere through cooling towers (Ritchie and Barter 1997).

It is important to note that cyanide consumption is typically high in bio-oxidation process due to formation of polysulfides and sulfur followed by conversion to thiocyanate in the cyanide leach circuit (Ritchie and Barter 1997; Miller and Brown 2005). The bio-oxidation process is extremely sensitive to cyanide and thiocyanate, and at levels above 5 ppm it becomes extremely toxic to the bacteria. For this reason, process streams need to be separated to ensure thiocyanate does not find its way back into the bio-oxidation feed stream. Also the bio-oxidation process generates organic compounds that impact the activity of activated carbon and loading capacity of the carbon in the gold adsorption circuit (Ritchie and Barter 1997). Foaming in the cyanide leach section of a bio-oxidation plant is a problem that has impacted most bio-oxidation plants to some degree (Ritchie and Barter 1997). This has led to the prevalence of CIL circuits incorporating draft tube agitators. In addition anti-foaming and froth breaking agents are commonly added.

20.6.4 Ultrafine Grinding

Usually ultrafine grinding is not considered as a chemical oxidation process, but there is some evidence to suggest that for certain ore types, grinding ores to sub 10 μm enables sufficient liberation of gold-bearing iron sulfides. Further studies are required to understand the extent of oxidation of sulfides after ultrafine grinding. There are a number of ultrafine grinding technologies such as the Glencore technology's IsaMill and Metso's SMD Detritor. Fine grinding and ultrafine grinding have been found to be an option for treating telluride flotation concentrates before cyanide leaching (Liddell and Dunne 1988; Ellis and Gao 2002). The Albion process utilizes ultrafine grinding for treating refractory ores (Hourne and Turner 2012).

Ultrafine grinding is also employed ahead of pressure oxidation and bio-oxidation to enhance leaching kinetics. Fine grinding for roaster feed is detrimental due to low retention time in the roaster leading to inadequate oxidation of the sulfides and thus poor gold recovery.

Use of ultrafine grinding at Kalgoorlie Consolidated Gold Mines (KCGM) was the first commercial application for oxidative pretreatment before leaching with a gold recovery of over 90 %, thus successfully supplementing the Gidji roaster capacity (Ellis 2003). This paved the way for further applications such as at Cowal in Australia, Pogo in Alaska, and Kumtor in Kyrgyzstan for treating refractory and partly refractory gold ores.

20.7 Heap Leaching

Copper heap and dump leaches were common in southern Spain in the 1700s. Heap leaching of gold and silver only began in 1969 at Cortez (Kappes 2005), which is still in operation with major expansions over years and now a major producer of gold for Barrick. Heap leaching of gold and silver was developed at the US Bureau of Mines (USBM) in the late 1960s and early 1970s as a low cost option for low grade hydrothermal oxide ores. The first large-scale cyanide heap leach for gold and silver was commissioned at Newmont's Carlin Mine in 1971 processing low grade ores below the mill cutoff grade. The use of heap leach spread across the western USA in Nevada, New Mexico, California, Utah, Montana, Idaho, and South Dakota (Marsden and Sass 2014).

Typically, heap leaching is chosen for basic financial reasons—for a given situation it represents the best and safest return on investment (Kappes 2005). Heap leaching allows generation of revenue early in the life of a project to help delay large capital investments required for milling and additional mining operations. Gold recovery from heap leaches is typically ~70 %, although it can range from 50 to 90 %, whereas silver recovery is typically ~55 %. Heap leaching is highly dependent on the type of ore and mineralogy. Typical ore types amenable to heap-leach are Carlin-type sedimentary oxidized ores (e.g., Carlin, Goldstrike, twin Creeks), low sulfide acid volcanics or intrusives (Round Mountain in Nevada, Yanacocha in Peru), oxidized massive sulfides (Filon Sur in Spain, Hassai Mine in Sudan), Saprolites/Laterites (West Africa and Central America), Clay rich deposits (Buckhorn Mine in Nevada, Barney's Canyon Mines in Utah), and Silver-rich deposits (Coeur Rochester in Nevada and Comco in Bolivia). Figure 20.8 shows Barrick's Veladero operation in Argentina.

Some of the key enablers for heap leaching have been discussed in earlier publications (Marsden and Sass 2014; Kappes 2005). Effective agglomeration, liner systems, stacking, solution distribution, and irrigation systems are important considerations for any successful heap leach operations. Carbon-in-Column (CIC) systems were widely adopted since the early 1980s that had some advantages in handling large volumes of solution and also in avoiding the need for costly filtration or clarification systems. Centralized carbon handling facilities for CIL and CIC-generated carbon allowed significant cost savings for larger mining companies such as Barrick and Newmont with numerous operations in close proximity.

Heap leaching is now considered a proven technology beyond the conventional arid climatic regions to include extreme cold climates such as the Fort Knox operation in Alaska with temperatures down to −50 °F and Jinshan Gold Mine in China with temperatures reaching −32 °C. The common feature of these cold climate heap leach operations are the use of in-heap pregnant solution storage and the direct delivery of barren into

Fig. 20.8 Veladero heap leach operation in Argentina

Fig. 20.9 Copper and gold heap leach operations

the heap (Keane 2007). Figure 20.9 shows an overall view of copper and a gold heap leach operations.

20.8 Barrick's Thiosulfate Technology

Various reviews of non-cyanide based technologies for gold and silver leaching and their mechanisms have been carried out elsewhere (McMullen et al. 2005; Aylmore 2005; Muir and Aylmore 2005; Senanayake 2012; Senanayake and Zhang 2012). Some of these reviews suggest that thiosulfate and chloride leaching have the most potential to replace cyanide. Other reviews also suggest that this will, however, require further developments in reducing reagent consumption as well as better methods of recovering gold from solution.

Earlier efforts by gold mining companies such as Newmont in testing and demonstration of the concept of thiosulfate leaching of a bio-oxidized double refractory ore using a heap are noteworthy (Wan et al. 1993, 1994; Wan and Brierley 1997). During the late 1990s, Barrick has also studied the thiosulfate process to treat Goldstrike preg-robbing ores at a feasibility level based on success with many initial bench and pilot scale work using the ammonium thiosulfate-copper system for leaching and anionic exchange resin for extraction of the gold complex (Fleming et al. 2003). Despite reasonably high gold recoveries with thiosulfate in treatment of preg-robbing ores, the key challenge was the need for a cost-effective elution process to make it possible to strip gold from the resin. The environmental concerns with ground water contamination with the use of ammonia along with the need to overcome gold recovery challenges with recycled water containing residual ammonia were the main drivers in the search for a different source of thiosulfate to enable development of an industrial scale leaching process (Choi et al. 2013).

Extensive research and development efforts led by Barrick initiated, during 2008, focused on further development of the thiosulfate leaching technology using calcium thiosulfate (CaTS) as the lixiviant, as this resulted in not only high dissolution rates but also allowed mitigating environmental concerns from the use of ammonium salt. Work carried out on various ores from Barrick's Goldstrike and Cortez mines showed that gold recoveries for preg-robbing ores were not compromised using CaTS compared to that with ammonium salts. In addition, resins were shown to be the most effective means of gold extraction using the CaTS process instead of activated carbon used in CIL. The strong base type of resin was used in the development of the CaTS process by Barrick. Detailed investigations carried out in-house helped in better understanding of the Resin-in-Leach (RIL) systems compared to that with a conventional resin-free system such as CIL.

Prevention of gold surface passivation in thiosulfate solution in the presence of resin through preferential loading of polythionates (due to partial oxidation of thiosulfate) on resin surfaces has shown to enhance gold leaching and recovery in RIL. Optimization of retention time of RIL circuits is critical to achieve the highest gold recovery as excessive residence time can result in high concentration of polythionates on the resin surfaces which appears to drive gold complex back from resin into solution. In addition, the need for a lower pH range in CaTS is quite different to the high pH requirements from 9 to 10 with ammonium thiosulfate for higher gold recovery. It is important to note that the conventional CIL process involving countercurrent slurry and carbon flow configuration is not the best for thiosulfate based RIL circuits. A combination of cocurrent and countercurrent configuration has been shown to provide the best recovery in the CaTS process, where a portion of the resin is placed cocurrently with slurry flow in the first tank and the remaining resin is introduced in the last tank in a countercurrent manner to the slurry flow (Choi et al. 2013).

Based on a thorough understanding of the CaTS process through extensive bench studies, a demonstration plant was successfully operated at the Goldstrike's autoclave process facility, which operated for around 3 years since May 2010. Almost 1500 tonnes of different ore types were treated in this demonstration plant to ensure robustness of the process. This success is a testament to the collaborative efforts between various players within Barrick such as Strategic Technology Solutions (STS), regional metallurgical team, site personnel, Barrick Technology Centre (now AuTec) along with some external groups.

> Goldstrike achieved first gold production through its autoclaves in fourth quarter 2014, after being successfully retrofitted with Barrick's innovative and proprietary thiosulfate technology. The new thiosulfate circuit allows for continued production from the autoclaves and accelerates the cash flow from about four million stockpiled ounces. The expected average annual contribution is about 350 to 450 thousand ounces of production (including Cortez ore processed at Goldstrike) in the first full five years following implementation of this process. In 2015, Goldstrike's production is expected to exceed 1.0 million ounces with contributions from the thiosulfate process. (Barrick annual report 2014)

This plant is the first commercial plant of its kind. The success of this thiosulfate process is associated with the fact that this technology is a much more efficient gold lixiviant compared to cyanide in treatment of ores associated with high preg-robbing behavior. This success of CaTS at Goldstrike at this stage should not be construed as a technology that could be applied to all gold ores. Further work is needed to better understand the limitations and opportunities with this technology to treat various complex ores.

The technology has allowed the Goldstrike operation to play the role of a Barrick internal custom milling operation, accommodating the processing of ores and concentrates from various Barrick mines throughout Nevada. Figure 20.10 shows the thiosulfate leaching facility at Goldstrike, whereas Fig. 20.11 shows an aerial view of the overall Goldstrike complex including the thiosulfate leaching facility.

20.9 Concluding Remarks

It is becoming apparent that the future trend of gold mining and processing will involve ores with decreasing head grade with a lower gold-to-sulfur ratios. The proportion of gold production from highly refractory ore bodies involving various iron sulfides, carbonaceous matter, copper, arsenic, and mercury is expected to increase. Some of the key challenges with these deposits are suboptimal metallurgy, higher operating and capital costs, environmental issues associated with tailings and deleterious elements, increasing mining footprint, lack of resources including energy, quality water, skilled personnel, and ever-growing community issues. Major innovations with a holistic approach are essential to turn these marginal deposits into profitable mining operations in an environmentally friendly and sustainable manner. Some gold mining companies are already looking at breakthroughs and step-change

Fig. 20.10 The thiosulfate facility at Goldstrike in Nevada, the USA

Fig. 20.11 An aerial view of Barrick's Goldstrike operation (Nevada, USA) along with the thiosulfate processing facility

improvements with an integrated approach to problem solving (Bristow 2013; Dunne 2012; Kondos and Gorain 2012; Lakshmanan et al. 2012; Logan and Krishnan 2012; Phala 2012).

It is always best to address the root cause of a problem for enabling robust solutions to the present and future challenges of gold mining and processing. Our industry has been very innovative, no doubt, with a strong focus on making our existing mining paradigm safer, more efficient, and automated. We have been remarkably successful so far but to address the unique challenges we now face as an industry, there appears to be no choice but to look beyond our existing mining paradigm.

The root cause of most of the problems in the existing mining paradigm is the generation of significant amount of waste and the need for handling, processing, and storing this waste with valuables representing only a very small fraction. The ideal scenario will be the case for *zero waste mining*, in which mining will target only the valuables without the need to remove the host rock and also all mined material will be used to create value added products with no waste dumps or tailings disposal. This is a daunting task, nevertheless, but if realized the benefits are significant. The underlying premise is that zero waste mining is economically very attractive, environmentally friendly, and fully integrated with the needs of local communities, societies, and other stakeholders. Hence pursuing "zero waste mining" is worthwhile as our ultimate long term goal, but we must address this in small practical steps with a "horses for courses" approach.

Based on our experience and discussion with various players inside and outside the industry, some major step-change opportunities for gold and silver mining and processing being investigated or pursued by various players in the industry are presented as follows.

1. *Minimal removal of overburden to access ore body*:
   - *Bore hole mining*: Use of small diameter drill holes to access the ore body along with use of novel biotechnologies to mine and recover metals (Dunbar 2014; McMullen et al. 2005). This is conceptual at this stage and will need further investigations.
   - *In situ gold leaching*: In situ recovery is presently practiced successfully in the copper, uranium, and potash industry. Companies like Rio Tinto are looking at opportunities with in situ leaching as the "mine of the future" (Batterham 2008). In situ gold leaching has been experimented before by BHP in 1989 and different lixiviants such as chlorides, thiosulfate, humic acid, and biogenic reagents have been proposed previously (McMullen et al. 2005; Zammit et al. 2013). New technologies like Discrete Fracture Networking are evolving to make this practical in an environmentally friendly manner for most ore types (Dershowitz 2011).
2. *Selective liberation of ores early in the mining process*: It is becoming important to ask ourselves "Why create waste in the first place." Comminution begins with mining and a new generation of drilling and blasting for selective mining of ores is critical to avoid or reduce waste removal early-on in the mining process. This concept is referred to as "grade engineering" and is becoming a major focus of further development (CRC-ORE 2014).
3. *Minimal haulage of waste*:
   - *Pre-concentration*: This allows processing to be closer to the mine site. There is a growing interest in pre-concentration technologies such as sensor based mass sorting (ROM shovels/trucks), classification using screens, stream based ore-sorting, gravity and dense media separation, which is definitely a positive trend and important implications for the gold industry. Studies have suggested that integrated mining and waste rejection processes have high potential for deep underground mining (Dammers et al. 2013; Bamber 2008; Batterham 2003).
4. *Efficient comminution*: Technologies such as *Chemical Comminution* applied on a smaller mass after pre-concentration are also being investigated (Muir 2014). Recent innovations

in high intensity selective blasting to reduce footprint of comminution circuits is being tested and pursued by some operations. Technologies such as SelFrag to pretreat low grade ores (promoting fractures and high rock permeability) to minimize comminution energy is being used. This technology also has important implications in maximizing heap leach recoveries.

5. *Refractory ore processing*:
   - *Microbial gold processing*: Focus on new generation of low cost bio-heap leaching, bio-oxidation and cyanide destruction processes using native microbial population along with phylogenetic fingerprinting is gaining momentum and has major potential (Brierley 2014; Zammit et al. 2013; Follink 2010).
   - *Breakthrough flotation process*: With the advent of new flotation technologies to treat ultrafine particles, and with better understanding of chemistry through state-of-the-art surface analysis capabilities, the possibility of a flotation breakthrough is likely. This has potential for a major breakthrough for processing Carlin type double refractory deposits.
   - *Cost-effective sulfide oxidation*: Other than innovative bio-oxidation processes, there are significant opportunities to integrate novel ultrafine grinding with new generation of pressure oxidation technologies involving lower capital and operating costs to process fine grained refractory ores that are uneconomical using existing technologies.
   - *Alternatives to cyanide leaching*: New generation of alternative technologies are needed such as Process Research Ortech's novel chloride leach technology for treating complex ores (Au, Cu, Ag, TCM, As) with an ability to maximize recovery of gold along with valuable byproducts (Lakshmanan et al. 2012).

It is important to note that parallel development of many of these technologies is already happening. Also, not all these innovations will be relevant to every ore body as technologies must be tailored to suit the individual needs of an ore body. A systematic approach with a multidisciplinary collaboration involving various stakeholders is a must to realize the full benefits from a breakthrough technology (Gorain and Kondos 2015).

Acknowledgments The authors would like to thank various mining companies, centers of excellence, commercial laboratories, engineering companies, equipment vendors, reagent suppliers and consultants for their support, discussion, partnerships and collaboration in various projects over many years. The authors would also like to thank colleagues at Barrick Gold Corporation in corporate, region and operations for discussion on various matters presented in this chapter.

References

Adams, M. D. (Ed.). (2005). *Advances in gold ore processing* (Developments in mineral processing). Amsterdam: Elsevier.

Adams, M. D., & Glen, S. (2011). *Pilot scale and continuous bench testing of the MMS CN-D™ cyanide destruction process on an Australian gold plant tailings effluent*. Presented at the *World Gold Conference, Montreal, Canada.*

Aksoy, B. S., & Yarar, B. (1989). Natural hydrophobicity of native gold flakes and their flotation under different conditions. In G. S. Dobby & S. R. Rao (Eds.), *Processing of complex ores* (pp. 19–27). New York: Pergamon.

Albuquerque, L., Wheeler, J., Valine, S., Ganahl, B., & Barrios, G. (2008, October 22–24). Application of high frequency screens in closing grinding circuits. In *Proceedings of the Vth International Mineral Processing Seminar, Procemin 2008, Santiago, Chile.*

Anderson, C. J., Dunne, R. C., & Uhrie, J. L. (Eds.). (2014). *Mineral processing and extractive metallurgy: 100 years of innovation*. Englewood, CO: SME.

Arehart, G. B., Chakurian, A. M., Tretbar, D. R., Christensen, J. N., McInnes, B. A., & Donelick, R. A. (2003). Evaluation of radioisotope dating of Carlin-type deposits in the great basin, Western North America, and implications for deposit genesis. *Economic Geology, 98*(2), 235–248.

Arvidson, B. R., & Norrgran, D. (2014). Magnetic separation. In C. G. Anderson, R. C. Dunne, & J. L. Uhrie (Eds.), *Mineral processing and extractive metallurgy—100 years of innovation* (pp. 223–233). Englewood, CO: SME.

Arvidson, B. R., & Wotruba, H. (2014). Ore sorting. In C. G. Anderson, R. C. Dunne, & J. L. Uhrie (Eds.), *Mineral processing and extractive metallurgy—100 years of innovation* (pp. 209–211). Englewood, CO: SME.

Aylmore, M. G. (2005). Alternative lixiviants to cyanide for leaching gold ores. In M. D. Adams (Ed.), *Advances in gold ore processing* (Vol. 15, pp. 501–539). Amsterdam: Elsevier.

Bain, H. F. (1910). *More recent cyanide practice*. San Francisco: Mining and Scientific.

Bamber, A. S. (2008, April). *Integrated mining, pre-concentration and waste disposal systems for the increased sustainability of hard rock metal mining*. Ph.D. thesis, University of British Columbia, Vancouver, Canada.

Barrick Gold Corporation. (2014). Annual report 2014.

Barter, J., Lane, G., Mitchell, D., Kelson, R., Dunne, R., Trang, C. et al. (2001, February). Cyanide management by SART. In C. Young, L. Tidwell, & C. Anderson (Eds.), *Cyanide: Social, Industrial, and Economic Aspects, Annual Meeting of the Minerals, Metals and Materials Society* (pp. 549–562).

Bartram, K., & Kowalczyk, M. (2009, August 23–26). New developments in sensor based ore sorting. In *Proceedings of the 48th Conference of Metallurgists, Sudbury, Canada* (pp. 421–432).

Batterham, R. B. (2003, September 28–October 3). The mine of the future, will it be visible? In *Proceedings of the XXII International Mineral Processing Congress, Cape Town, South Africa*.

Batterham, R. B. (2008, September 24–28). Current practices shaping the science and practice of mineral processing. In *Proceedings of the XXV International Mineral Processing Congress, Beijing, China* (Vol. 1, pp. 55–59).

Beazley, R. J., & Woods, M. C. (2013, September 26–29). Cyanide code recertification—The three year challenge. In *Proceedings World Gold 2013, The Australasian Institute of Mining and Metallurgy, Brisbane, Australia* (pp. 247–254).

Beer, G. (1994). Processing options for the treatment of preg-robbing ores at the Macraes Gold Mine. In *Proceedings 6th AusIMM Extractive Metallurgy Conference 1994, AusIMM, Brisbane, Australia* (pp. 203–208).

Berezowsky, R. M. G. S., Collins, M. J., Kerfoot, D. G. E., & Torres, N. (1991). The commercial status of pressure leaching technology. *Journal of Metals, 43*, 9–15.

Bettles, K. (2002). Exploration and geology, 1962 to 2002, at the goldstrike property, Carlin Trend, Nevada. *Society of Economic Geologists Special Publication, 9*, 275–298.

Biehl, R., & Bunn, S. (1990). Roasting at Kalgoorlie Consolidated Gold Mines Ltd. In *Proceedings Pyrometallurgical Seminar, Murdoch University, Perth, Australia* (pp. 185–199).

BIOMIN Website. (2015). Retrieved from www.biomin.co.za

Blin, P. & Dion-Ortega, A. (2013, June/July). High and dry: With water in short supply, some are looking to the ocean for answers, *CIM Bulletin*.

Boydell, D. W. (1984). *Developments in plant practice for the recovery of gold in South Africa, Perth and Kalgoorlie regional conference on gold mining, metallurgy and geology*. Brisbane, Australia: AusIMM.

Brent, G. F., Dare-Bryan, P., Hawke, S., & Rothery, M. (2013, September 26–29). Ultra-high-intensity blasting—a new paradigm in mining. In *Proceedings of World Gold Conference, Brisbane, Australia* (pp. 395–400).

Brierley, J. A. (2014). Biohydrometallurgy innovations—Discovery and advances. In C. G. Anderson, R. C. Dunne, & J. L. Uhrie (Eds.), *Mineral processing and extractive metallurgy—100 years of innovation* (pp. 447–455). Englewood, CO: SME.

Bristow, M. (2013, September 26–29). Building a successful African gold mining business—managing risk and delivering long term returns. Presented at the *World Gold 2013, Brisbane, Australia* (p. 3).

Budden, J., & Bunyard, M. (1994). Pilot Plant test work and engineering design for the Bac-Tech bacterial oxidation plant at the Youanmi Mine. In *Proceedings International Conference and Workshop: Applications of Biotechnology to the Minerals Industry* (pp. 4.1–4.7).

Bulatovic, S. M. (1997). Flotation behaviour of gold during processing of porphyry copper-gold ores and refractory gold bearing sulfides. *Minerals Engineering, 10*(9), 895–908.

Bunk, S. A., & Bruyns, D. J. (1993). Review of the state of commercial art roasting. In *Randol Gold Forum* (pp. 175–178).

Burns, R., & Grimes, A. (1986, June). The application of pre-concentration by screening at Bougainville Copper Limited. In *Proceedings of the AUSIMM Mineral Development Symposium, Madang, Papua New Guinea*.

Bye, A. R. (2011, September 5–7). Case studies demonstrating value from geometallurgical studies. In *Proceedings of the First AusIMM International Geometallurgy Conference, Brisbane, Australia* (pp. 9–30).

Cadzow, M. D., & Giraudo, T S. (2000). Macraes gold project: Value creation through applied technology-pressure oxidation. In *Proceedings 2000 New Zealand Minerals and Mining Conference* (pp. 275–281).

Castro, E., Lopez, J., & Switzer, D. (2009). Maximum classification and efficiency through classification using the recyclone. In D. Malhotra, P. R. Taylor, E. Spiller, & M. LeVier (Eds.), *Recent advances in mineral processing plant design* (pp. 444–454). Englewood, CO: SME.

Chattopadhyay, A., & Gorain, B. K. (2012, September 24–28). Gold deportment studies on a copper-gold ore—a systematic approach to quantitative mineralogy focusing on diagnostic metallurgy, paper #615. In *Proceedings of the XXVI International Mineral Processing Congress, New Delhi, India* (pp. 778–788).

Chattopadhyay, A., & Gorain, B. K. (2014). Gold deportment studies on a copper-gold ore—a novel approach to quantify gold deportment for double refractory gold ores. In *Proceedings of the XXVII International Mineral Processing Congress, Santiago, Chile (Chapter 14)* (pp. 160–180).

Choi, Y., Baron, J. Y., Wang, Q., Langhans, J., & Kondos, P. (2013, September 26–29). Thiosulphate processing—From lab curiosity to commercial application. In *Proceedings of the World Gold 2013 Congress, Brisbane, Australia* (pp. 45–50).

Chryssoulis, S. L. (2001). Using mineralogy to optimize gold recovery by flotation. *Journal of Metals, 53*(12), 48–50.

Chryssoulis, S. L., & McMullen, J. (2005). Mineralogical investigation of gold ores. In M. D. Adams (Ed.), *Developments in mineral processing* (pp. 21–72). Amsterdam: Elsevier.

Clark, B. H. (2007, January 23–25). The Derrick stack sizer: Revolutionary advancements in wet screening. In *Proceedings of the 39th Canadian Mineral Processors Conference* (pp. 413–428).

Clennell, J. E. (1915). *The cyanide handbook* (2nd ed.). New York: McGraw Hill.

CRC-ORE. (2014). *Transforming mining productivity with grade engineering*. Retrieved from www.crcore.org.au/

Dammers, M., Schropp, C., Martens, P. N., Rattmann, L., & Robben, C. (2013, June 30–July 3). In *sixth International Conference on Sustainable Development in the Minerals Industry, Milos Island, Greece* (pp. 1–6).

David, D., (2010). Operational geometallurgy. In *Flotation Plant Optimization*, pp 201–209, AusIMM, January, Chapter 12.

Dershowitz, B. (2011). Discrete fracturing network modeling in support of in-situ leach mining. *Mining Engineering, 63*(11), 46–50.

Deschenes, G., Hodouin, D., & Lorenzen, L. (Eds.). (2005). *Treatment of gold ores: COM 2005*. Calgary, Canada: Metsoc, CIMM.

DeSomber, R. K., Fernandez, R., McAnany, L., Stolarski, E., & Schmidt, G. (1996). Refractory ore treatment at Newmont Gold Company. In *Proceedings Randol Gold Forum'96, Randol International Ltd., Golden, CO* (pp 239–243).

Douglas, W., & Semenyna, L. (2013, September 26–29). Magnetic recovery of gold bearing iron oxides at Barrick Goldstrike's roaster. In *Proceedings World Gold 2013, Brisbane, Australia* (pp. 79–85).

Dunbar, W. S. (2014, August 18). Will the mine of the future be a mine at all? In *Globe and Mail*.

Dunne, R. (2005). Flotation of gold and gold-bearing ores. In M. D. Adams (Ed.), *Advances in gold ore processing* (pp. 309–343). Amsterdam: Elsevier.

Dunne, R. (2012). Refractory gold treatment—state of the art and future challenges. In *Proceedings of the XXVI International Mineral Processing Congress, New Delhi, India, Paper # 1100* (pp. 1258).

Dunne, R., Buda, K., Hill, M., Staunton, W., Tjandrawan, V., & Wardell-Johnson, G. (2007). Assessment of options for economic of preg-robbing gold ores. In *Proceedings World Gold 2007, AusIMM, Brisbane, Australia* (pp. 205–212).

Dunne, R., Hart, S., Parker, B., & Veilette, G. (2007). Boddington gold mine—an example of sustaining gold production for 30 years. In *World Gold 2007 Conference, AusIMM, Brisbane, Australia* (pp. 213–230).

Ellis, S. (2003). *Ultrafine grinding—a practical alternative to oxidative treatment of refractory gold ores*. Brisbane, Australia: AusIMM.

Ellis, S., & Gao, M. (2002). The development of ultra-fine grinding at KCGM. In *SME Annual Meeting, Preprint 02-072, SME, Englewood, CO* (pp. 1–6).

Erickson, M. T. (2014). Innovations in comminution equipment: Crushers, high pressure grinding rolls, semi-autogenous grinding, ball mills and regrind mills. In C. J. Anderson, R. C. Dunne, & J. L. Uhrie (Eds.), *Mineral processing and extractive metallurgy: 100 years of innovation* (pp. 65–75). Englewood, CO: SME.

Ericsson, M. (2012, March). Mining technology—trends and development, *Polinares Consortium, working paper # 29*.

Fleming, C. A., McMullen, J., Thomas, K. G., & Wells, J. A. (2003). Recent advances in the development of an alternative to the cyanidation process: Thiosulfate leaching and resin in pulp. *Minerals and Metallurgical Processing, 20*, 1–9.

Foggiatto, B., Bueno, M., Lane, G., McLean, E., & Chandromohan, R. (2014). The economics of the large scale sorting. In *Proceedings of the XXVII International Mineral Processing Congress, Santiago, Chile (Chapter 16)* (pp. 11–20).

Follink, B. (2010). Bio-mineral processing—its potential to revolutionize the industry. In *Proceedings of the XXV International Mineral Processing Congress, AusIMM, Brisbane, Australia* (p. 61).

Franco, J. J. C., Castillo, M. F., Concha, J., Christodoulou, L., & Wasmund, E. B. (2015). Coarse gold recovery using flotation in a fluidized bed. In *Proceedings of the 13th Canadian Mineral Processors Conference, Ottawa, Canada* (pp. 439–447).

Frostiak, J., & Haugard, B. (1992). Start-up and operation of placer Dome's Campbell mine gold pressure oxidation plant. *Mining Engineering, 44*(8), 991–993.

Fuerstenau, D. W. (2007). A century of developments in the chemistry of flotation processing. In M. C. Fuerstenau, G. J. Jameson, & R. H. Yoon (Eds.), *Froth flotation: A century of innovation* (pp. 3–64). Englewood, CO: SME.

Gibbons, T. (2005). International cyanide management code. In M. D. Adams (Ed.), *Advances in gold ore processing* (pp. 182–199). Englewood, CO: SME. Chapter 8.

Gillot, P. (2006). Pit-to-plant optimization at Morila Gold Mines. In *Proceedings of the SAG 2006 conference, Vancouver, Canada*.

Goldstone, A. J., & Mudder, T. I. (1988). *Cyanide regeneration process* (US Patent No. 4994243).

Goode, J. R. (1993). Refractory gold ore: Causes, processes, testing and plants. In *Proceedings Annual SME Conference, 93–82, SME, Englewood, CO* (pp. 1–21).

Gorain, B. K. (2012). *Separation of copper minerals from pyrite using air-metabisulfite treatment, patent application global.*

Gorain, B. K. (2013, January 22–24). Developing solutions to complex flotation problems. In *Proceedings of the 34th Canadian Mineral Processors Conference, Ottawa, Canada* (pp. 293–312).

Gorain, B. K., & Kondos, P. (2015, January 20–22). Developing breakthrough solutions to complex metallurgical problems. In *Proceedings of the 36th Canadian Mineral Processors Conference, Ottawa, Canada* (pp. 314–330).

Gordon, L. (1999). Increase recovery from preg-robbing gold ore at Penjom Gold Mine. In *Proceedings Randol Gold and Silver Forum'99* (pp. 105–108).

Guay, W. J. (1980). How Carlin treats gold ores by double oxidation. *World Mining, March* (pp. 47–49).

Guzman, G., & Acosta, E. (2009, September 14–18). Proceso SART y su Influencia en el Proceso CIC Minera Yanacocha SRL. In *PERUMIN, 2009 Convención Minera, Arequipa, Peru.*

Hammerschmidt, J., Günter, J., & Holmström, A. (2011). Fluidized bed technology for gold ore and gold concentrate. In *Proceedings World Gold 2011, CIMM, Calgary, Canada* (pp. 171–182).

Hammerschmidt, J., Günter, J., & Kersteins, B. (2005). Roasting of gold ore in the circulating fluidized-bed technology. In M. D. Adams (Ed.), *Developments in mineral processing* (pp. 433–452). Amsterdam: Elsevier.

Han, K. N., & Chen, Y. (1990). Separation of carbonaceous material from Carlin ore by flotation. In *Proceedings Advances in Gold Silver Processing, SME, Englewood, CO* (pp. 121–130).

Hart, S., Parker, B., Rees, T., Manesh, A., & Megaffin, I. (2011). Commissioning and ramp up of the HPGR circuit at Newmont Boddington Gold. In *Proceedings of 2011 International Autogenous Grinding Semiautogenous Grinding and High Pressure Grinding Roll Technology, Vancouver, BC, Paper #041.*

Hausen, D. M., & Bucknam, C. H. (1984). Study of preg-robbing in the cyanidation of carbonaceous gold ore from Carlin, Nevada. In W. C. Park, D. M. Hausen, & R. D. Hagni (Eds.), *Proceedings of International Congress on Applied Mineralogy in Mineral Industry, Metallurgical Society of AIME, Los Angeles, CA* (pp. 833–856).

Hearn, S. (2014). Innovations in electrostatic separation. In C. J. Anderson, R. C. Dunne, & J. L. Uhrie (Eds.), *Mineral processing and extractive metallurgy: 100 years of innovation* (pp. 235–242). Englewood, CO: SME.

Helm, M., Vaughan, J., Staunton, W. P., & Avraamides, J. (2009). An investigation of the carbonaceous component of preg-robbing gold ore. In *World Gold Conference 2009, The Southern African Institute of Mining and Metallurgy.*

Hille, S., & Raudsepp, R. (1995). Pressure oxidation performance at Porgera. In *Proceedings 6th Mill Operators Conference, AusIMM, Brisbane, Australia* (pp. 27–35).

Holder, R. (2007). Beaconsfield Gold Mine-Ironing out the bugs. In *Proceedings 9th Mill Operators Conference, AusIMM, Brisbane, Australia* (pp. 185–193).

Holloway, A. R., Jones, L., & Lawrence, S. R. (2009, August 23–26). Pre-concentration of platinum group and base metal ores by dense medium separation. In *Proceedings of 48th Conference of Metallurgists, Sudbury, Canada* (pp. 373–384).

Hourne, M., & Turner, D. (2012). Commercialisation of the Albion process. In *Proceedings of the ALTA 2012 Gold Conference.*

Hukki, R. T., & Allenius, H. (1968). *A quantitative investigation of closed grinding circuit (AIME transactions)* (Vol. 241). Englewood, CO: SME.

Jameson, G. (2014). Experiments on the flotation of coarse composite particles. In *Proceedings of the XXVII International Mineral Processing Congress, Santiago, Chile, (Chapter 5)* (pp. 102–111).

Kadenhe, R. M., & Makende, E. S. (1987). Review of the roasting plant operation and services available to small mines at Kwewe, Zimbabwe. In *Proceedings African Mining Congress, IMM* (pp. 191–204).

Kanchibotla, S. (2014). Mine to mill value chain optimization—role of blasting. In C. J. Anderson, R. C. Dunne, & J. L. Uhrie (Eds.), *Mineral processing and extractive metallurgy: 100 years of innovation* (pp. 51–64). Englewood, CO: SME.

Kappes, D. W. (2005). Heap leaching of gold and silver ores. In M. D. Adams (Ed.), *Advances in gold ore processing* (Vol. 13, pp. 456–478). Amsterdam: Elsevier.

Kappes, R., Brosnahan, D., & Gathje, J. (2010). The effect of mineral liberation on the floatabilities of pyrite, arsenopyrite and arsenian pyrite for Carlin Trend ores. In *Proceedings XXV International Mineral Processing Congress* (pp. 2913–2926).

Kappes, R., Fortin, C., & Dunne, R. (2011). The current status of the chemistry of flotation in industry. In *Proceedings World Gold 2011, CIMM* (pp. 385–395).

Keane, J. (2007, October 3–6). Engineering and construction of a 20,000 TPD gold recovery process plant in North China. In *Precious Metals Processing symposium, Littleton, Colorado, SME.*

Kelsey, C., & Kelly, J. (2014). Super-fine crushing to ultra-fine size the "IMP" super-fine crusher. In *Proceedings of the XXVII International Mineral Processing Congress, Santiago, Chile (Chapter 9)* (pp. 239–251).

Kittler, P., Liebezeit, V., Ehrig, K., Macmillan, E., & Lower, C. (2011, September 5–7). 'It seemed like a good idea at the time … .'common mistakes in geometallurgy. *Presented at the First International AusIMM Geometallurgical Conference, Brisbane, Australia.*

Klein, B., Dunbar, W. S., & Scoble, M. (2002). Integrating mining and mineral processing for advanced mining systems. *CIM Bulletin, 95*(1057), 63–68.

Kondos, P., & Gorain, B. K. (2012). Development of breakthrough technologies for complex gold and copper-gold ores. In *Proceedings of the XXVI International Mineral Processing Congress, New Delhi, India, Paper # 1098* (pp. 2524–2532).

Kosich, D. Y. (1992). Independence roasters surpass expectation. *Mining World News, July* (pp. 2–3).

Koski, S., Vanderbeek, J. L., & Enriquez, J. (2011). Cerro Verde concentrator-four years operating HPGR's. In *Fifth International Conference on Autogenous and Semiautogenous, Grinding Technology, Vancouver, Canada*.

Lahti, P. A. (1996). Refractory gold ore processing methods at Jerritt Canyon. In *Randol Gold Forum'96* (pp. 225–233).

Lakshmanan, V. I., Sridhar, R., Roy, R., & Gorain, B. K. (2012). Recovery of gold and titanium dioxide. In *Proceedings of the XXVI International Mineral Processing Congress, New Delhi, India, Paper # 1069* (pp. 2688–2696).

Laplante, A. R. (2000). Testing requirements and insight for gravity gold circuit design. In *Randol Gold and Silver forum* (pp. 73–83).

Laplante, A. R., & Dunne, R. C. (2002). The gravity recoverable gold test and flash information. In *34th CMP Annual meeting, Paper 32* (pp. 105–123).

Laplante, A. R., & Gray, S. (2005). Advances in gravity gold technology. In M. D. Adams (Ed.), *Advances in gold ore processing* (Vol. 13, pp. 280–307). Amsterdam: Elsevier.

Lawrence, R. W., & Lopez, O. (2011, October 3–4). Improving the economics of gold-copper ore projects using SART technology. In *Gold 2011, Montreal, Canada*.

Lessard, J., deBakker, J., & McHugh, K. (2014). Development of ore sorting and its impact on mineral processing economics. *Minerals Engineering, 65*, 88–97.

Lewis, P. J. (1990). The treatment of oxidized and primary copper—Gold ores at Red Dome, Queensland, Australia. In *Randol Gold Forum'90, Square Valley, Randol International, Golden, CO*.

Liddell, K. S., & Dunne, R. C. (1988). Ultra fine grinding effect on leach recovery from refractory gold ores. In *Proceedings Randol International Gold Conference* (pp. 120–122).

Littlejohn, P., Kratochvil, D., & Hall, A. (2013). Sulfidisation-Acidification-Recycling-Thickening (SART) for complex gold ores. In *Proceedings of World Gold 2013* (pp. 149–155).

Lloyd, P. J. D. (1979). An integrated mining and extraction system for use on Witwatersrand gold mines. *Journal of the South African Institute of Mining and Metallurgy, 79*(1), 135–148.

Logan, A., & Krishnan, N. (2012). Newcrest technology—step change. In *Proceedings of the XXVI International Mineral Processing Congress, New Delhi, India, Paper # 1085* (pp. 3025–3037).

Loosley, A. (1993). Gold ore treatment at Emperor Gold Mining Company Ltd, Vatukoula, Fiji. In *Australasian Mining and Metallurgy, The Sir Maurice Mawby Memorial* (Vol. 2, pp. 953–954).

Lunt, D., & Weeks, T. (2005). Refractory gold ores case studies. In M. D. Adams (Ed.), *Developments in mineral operating plants processing* (pp. 73–96). Amsterdam: Elsevier.

MacPhail, P. K., Fleming C. A., & Sarbutt K. W. (1998, April). Cyanide recovery by the SART process for the Lobo-Marte project, Chile. In *Randol Gold Forum 1998, Denver, CO* (pp. 26–29).

Major, K. (2002). Types and characteristics of crushing equipment and circuit flowsheets. In A. L. Mular, D. N. Halbe, & D. J. Barratt (Eds.), *Mineral processing plant design, practice and control* (pp. 566–583). Englewood, CO: SME.

Maki, T. D., & Taylor, J. B. (1987). Precious metal slag treatment using an electrostatic separator. In *Small Mines Development in Precious Metals* (pp. 245–252).

Marsden, J. O., & Sass, S. A. (2014). Innovation in gold and silver extraction and recovery. In C. J. Anderson, R. C. Dunne, & J. L. Uhrie (Eds.), *Mineral processing and extractive metallurgy: 100 years of innovation* (pp. 361–387). Englewood, CO: SME.

McCullough, W. E., Bhappu, R. B., & Hightower, J. D. (1999). Copper ore preconcentration by heavy media separation for reduced capital and operating costs. Presented in *Cobre 99*.

McCullough, G., Mutsaers, B., & Jessop, J. (2013, September 30–October 2). Obstacles to embedding geometallurgy decision-making into operations and how to increase maturity of this critical capability. In *second AusIMM International Geometallurgy Conference, Brisbane, Australia*.

McDonald, A. (2003). Lihir gold five years on—Maximizing throughput and capital utilization. In *Proceedings 7th Mill Operators Conference, AusIMM, Brisbane, Australia* (pp. 115–120).

McMullen, J., Goode, J. R., & Kondos, P. D. (2005). The recovery of gold from low grade and refractory ores What's next? In G. Deschenes, D. Hodouin, & L. Lorenzen (Eds.), *Treatment of Gold Ores: Com2005, Calgary, Canada* (pp. 353–371).

Miller, P., & Brown, A. (2005). Bacterial oxidation of refractory concentrates. In M. D. Adams (Ed.), *Developments in mineral processing* (pp. 371–402). Amsterdam: Elsevier, 403–433.

Miller, P., Jiao, F., & Wang, J. (2004). The bacterial oxidation (BACOX) plant at Laizhou Shandong Province China-the first three years of operation. In *Proceeding Bac-Min 2004, AusIMM, Brisbane, Australia* (pp. 167–172).

Mohanty, J. K., Rao, D. S., Das, B., & Rao, R. B. (2000). Mineralogy and pre-concentration of the chromite overburden of the Sukinda Ultramafic belt. *CIM Bulletin, 93*(1038), 37–43.

Mosher, J. B. (2005). Comminution circuits for gold processing. In M. D. Adams (Ed.), *Advances in gold ore processing* (Vol. 12, pp. 253–277). Amsterdam: Elsevier.

Moss, F. (1918). The treatment of the auriferous sulphide ores of Kalgoorlie. In *Transaction Australian Institute of Mining and Metallurgy, Melbourne, Australia* (Vol. 99, pp. 40–67.

Muir, A. (2014). *Evolution in comminution, Keynote address in Comminution 2014, organized by MEI, 7-1-April, Cape Town, South Africa.*

Muir, D. M., & Aylmore, M. G. (2005). Thiosulfate as an alternative lixiviant for gold ores. In M. D. Adams (Ed.), *Advances in gold ore processing* (Vol. 14, pp. 541–560). Amsterdam: Elsevier.

Mular, A. L., Halbe, D. N., & Barratt, D. J. (Eds.). (2002). *Mineral processing plant design, practice and control, Littleton, CO* (Vol 2, pp. 1493–1693).

Müller, J., & Frimmel, H. E. (2010). Numerical analysis of historic gold production cycles and implications for future sub-cycles. *The Open Geology Journal, 4,* 29–34.

Munro, P. D., Schache, I. S., Park, W. G., & Watsford, R. M. S. (1982). The design, construction and commissioning of a heavy media plant for silver-lead-zinc ore treatment—Mount Isa Limited. In *Proceedings of the XIV International Mineral Processing Congress, Toronto* (Vol. 6, pp. 1–20).

Nagaraj, D. R., & Farinato, R. S. (2014). Major innovations in the evolution of flotation reagents. In C. J. Anderson, R. C. Dunne, & J. L. Uhrie (Eds.), *Mineral processing and extractive metallurgy: 100 years of innovation* (pp. 159–175). Englewood, CO: SME.

Napier-Munn, T. J., Bosman, J., & Holtham, P. (2014). Innovations in dense media separation technology. In C. J. Anderson, R. C. Dunne, & J. L. Uhrie (Eds.), *Mineral processing and extractive metallurgy: 100 years of innovation* (pp. 265–275). Englewood, CO: SME.

Neira Arenas, G., & Monhemius, A. J. (1998). The kinetics of pressure oxidation of arsenopyrite and arsenopyrite/pyrite mixtures by dissolved oxygen. In M. A. Sanchez, F. Vergara, & S. H. Castro (Eds.), *Environment and innovation in mining and mineral technology* (pp. 835–849). Concepcion, Chile: University of Concepcion.

Norrgan, D., Ashton, D., & Fears, P. (2009). *Installation and operation of the Eriez cryogen-free superconducting magnet at WBB Minerals, UK.*

Pekrul, J. (2007, October 3–6). Update on LIXKillat the autoclave. In *Precious Metals Processing—Advances in Primary and Secondary operations Symposium, SME, Littleton, CO.*

Phala, N. (2012). Innovation drivers and opportunities across the gold metallurgy value stream. In *Proceedings of the XXVI International Mineral Processing Congress, New Delhi, India, Paper # 1099* (pp. 4217–4230).

Ralston, J., Fornasiero, D., & Grano, S. (2007). Pulp and solution chemistry. In M. C. Fuerstenau, G. J. Jameson, & R. H. Yoon (Eds.), *Froth flotation: A century of innovation* (pp. 227–281). Englewood, CO: SME.

Richart, T. A. (1912). *The flotation process* (1st ed., pp. 54–55). San-Francisco: Mining and Scientific press.

Rickard, T. A. (1907). *Recent cyanide practice.* San Francisco: Mining and Scientific Press.

Ritchie, I, & Barter, J. (1997). Process and engineering design aspects of biological oxidation plants. In *Proceedings, BIOMINE 97, AusIMM, Brisbane, Australia* (pp. 14.41–14.48).

Rose, T. K., & Merloc, W. T. (2008). *The metallurgy of gold* (6th ed.). Wexford, Ireland: Wexford Press.

Schena, G. D., Gochin, R. J., & Ferrara, G. (1990). Pre-concentration by dense media separation—an economic evaluation. *Transactions of the Institute of Mining and Metallurgy, 99,* C21–C31.

Senanayake, G. (2012). Gold leaching by copper (II) in ammoniacal thiosulfate solutions in the presence of additives, Part I: A review of the effect of hard-soft and Lewis acid-base properties and interactions of ions. *Hydrometallurgy, 115–116,* 1–20.

Senanayake, G., & Zhang, X. M. (2012). Gold leaching by copper (II) in ammoniacal thiosulfate solutions in the presence of additives, Part II: Effect of residual Cu(II), pH and redox potentials on reactivity of colloid gold. *Hydrometallurgy, 115–116,* 21–29.

Shi, F., Krishnan, N., von der Weld, F., can-der-Wielen, K., Zuo, W., & Manlapig, E. (2014). A potential application of high voltage pulse technology in a gold-copper ore grinding circuit. In *Proceedings of the XXVII International Mineral Processing Congress, Santiago, Chile (Chapter 9)* (pp. 68–77).

Shi, F., Zuo, W., & Manlapig, E. (2013). Characterization of pre-weakening effect of ores by high voltages electric pulses using single particle tests. *Minerals Engineering, 50–51,* 69–76.

Simmons, G. L. (1994). Development of low temperature pressure oxidation at Lone Tree. In *SME Annual Meeting, Preprint, # 93–86, SME, Englewood, CO.*

Simmons, G. L. (1996). Pressure oxidation process development for treating carbonaceous ores at Twin Creeks. In *Randol Gold Forum'96* (pp. 199–208).

Simmons, G. L. (1997). Flotation of auriferous pyrite using Sante Fe Pacific's gold N_2TEC flotation process. In *SME Annual Meeting, Preprint No. 97–27, Society for Mining, Metallurgy and Exploration Inc., Denver, CO.*

Sivamohan, R., & Forssberg, E. (1991). Electronic sorting and other Pre-concentration methods. *Minerals Engineering, 4*(7–11), 797–814.

Smart, R. St. C., Skinner, W.M., Gersen, A.R., Mielczarskim, J., Chryssoulis, S. & Pratt, A.R. (2007). Surface characterization and new tools for research. In M.C. Fuerstenau, G.J. Jameson and R.H. Yoon (Eds.), Froth Flotation: A century of innovation (pp 283–337), Englewood, CO; SME.

SonicSampDrill. (2014). Retrieved from http://www.sonicsampdrill.com/news/sorting-ore-with-the-speed-of-light-1000-tons-per-hour-htm

Sovmen, V. K., Belyi, A. V., Danneker, M. Yu., Gish, A. A., & Teleutov, A. N. (2009). Biooxidation of refractory gold sulfide concentrate of Olympiada deposit. *J Trans Tech Publications, Switzerland* (pp. 71–74).

Stanley, G. G. (1987). *The extractive metallurgy of gold in South Africa, Monograph M7*. Johannesburg, South Africa: SAIMM.

Stenebraten, J. F., Johnson, W. P., & McMullen, J. (2000). Characterisation of goldstrike ore carbonaceous material, Part 2: Physical characteristics. *Mineral and Metallurgical Processing, 17*(1), 7–15.

Swash, P. M. (1988). A mineralogical investigation of refractory gold ores and their beneficiation with special reference to arsenical ores. *Journal of the South African Institute of Mining and Metallurgy, 88*(5), 25–28.

Taggart, A. F. (1925). Flotation of gold and silver. In: *Handbook of Mineral Dressing* (p. 796). New York: Wiley.

Taggart, A. F. (1927). Flotation of gold and silver. In: *Handbook of Mineral Dressing* (pp. 866–868). New York: Wiley.

Thomas, K. G. (2005). Pressure oxidation overview. In M. D. Adams (Ed.), *Developments in mineral processing* (pp. 346–369). Amsterdam: Elsevier.

Thomas, K. G., & Cole, A. P. (2005). Roasting developments-especially oxygenated roasting. In M. D. Adams (Ed.), *Developments in mineral processing* (pp. 403–433). Amsterdam: Elsevier.

Vanderbeek, J. L., Linde, T. B., Brack, W. S., & Marsden, J. O. (2006). *HPGR implementation at Cerro Verde: SAG2006* (Vol. 4, pp. 45–61).

Wan, R. Y., & Brierley, J. A. (1997). Thiosulfate leaching following bio-oxidation pretreatment for gold recovery from refractory carbonaceous-sulfidic ore. *Mining Engineering, 49*(8), 76–80.

Wan, R. Y., LeVier, K. M., & Clayton R. B. (1994). *Hydrometallurgical process for the recovery of precious metal values from precious metal ores with thiosulfate lixiviant* (US Patent No. 5,354,359).

Wan, R. Y., LeVier, K. M., & Miller, J. D. (1993). Research and development activities for the recovery of gold from noncyanide solutions. In *Hydrometallurgy: Fundamentals, technology and innovations* (pp. 4225–4436). Englewood, CO: SME-TMS.

Whitlock, J. L., & Mudder, T. I. (1986). The homestake wastewater treatment process: Biological removal of toxic parameters from cyanidation wastewaters and bioassay effluent evaluation. In R. W. Lawrance (Ed.), *Fundamental and applied biohydrometallurgy*. Amsterdam: Elsevier.

Xie, J. Y., Gao, J. C., Jin, S. B., & Han, X. G. (2008). Bio-oxidation plants of refractory gold concentrate in China. In *Proceedings XXIV International Mineral Processing Congress* (pp. 2737–2740).

Zammit, C. M., Cook, N., Ciobanu, C. L., Brugger, J., & Reith, F. (2013 September 26–29). Possible future of applications of biotechnologies in the gold industry—A prospective approach. In *Proceedings of World Gold 2013, Brisbane, Australia* (pp. 233–244).

21 Innovative Processes for By-product Recovery and Its Applications

V.I. Lakshmanan and Ram Ramachandran

This chapter covers briefly innovative processes for the production of by-product metals, their compounds, and application in various industries. In view of the vastness of the subject, the scope of the chapter is broad in nature and has not been treated in depth. Only a few applications have been described in some detail. They are use of by-product metals and their compounds in (a) photovoltaic cells, (b) fuel cells, and (c) polymer industry. They have been chosen because:

1. Photovoltaic cells are somewhat mature in their development—although some innovation is in progress.
2. Fuel cells are still in an evolving mode.
3. Polymer industry illustrates the application of by-products in a nonmetallurgical industry.

Assembly and production of photovoltaic cells and fuel cells are somewhat proprietary and their production is rarely published in detail.

To be commercially attractive, an ore must have a concentration of the desired metal to make the extraction of it technically and economically feasible. This concentration varies greatly for different metals; in some cases, the chemical and physical properties of the metal are a determining factor and the intermediate compound which determines the extraction processes is also a determining factor.

The chief sources of by-product metals are residues of operations for the recovery of other more common metals, whereby small quantities of by-product metals are concentrated to values which permit their extraction. Some examples are:

1. Platinum group metals from residues of nickel extraction processes and from copper anode slimes.
2. Germanium from zinc operations, bismuth and thallium from lead operations, and cadmium and indium from both zinc and lead operations.
3. Gallium from aluminum operations.
4. Selenium, tellurium from copper anode slimes.
5. Rhenium from molybdenum operations and from copper raffinate solutions in copper SX-EW operations.
6. Hafnium from zirconium operations.

Metals and metal compounds such as cadmium, indium, selenium, tellurium, bismuth, nickel sulfate, antimony oxide, sodium antimonite, copper arsenate—to mention a few—that are routinely produced and considered as by-products for commercial use are not described here.

V.I. Lakshmanan (✉)
Process Research Ortech Inc., 2350 Sheridan Park Drive, Mississauga, ON, Canada, L5K 2T4
e-mail: llakshmanan@processortech.com

R. Ramachandran
Consulting Engineer,
9650 E. Peregrine Place, Scottsdale, AZ 85262, USA

V.I. Lakshmanan et al. (eds.), *Innovative Process Development in Metallurgical Industry*,
DOI 10.1007/978-3-319-21599-0_21

Two examples, one production of a by-product "metal," vanadium from fly ash (a waste residue) and one "non-metal"—ammonium sulfate from uranium processing—produced as commercial products are described below. This is followed by description of the production of by-product metals and their compounds and their application in chosen industries.

21.1 Vanadium from Fly Ash

Fly ash is the by-product of treated petroleum or similar carbonaceous fuel materials. It has very fine particle size (100 % passing 2 mm) and is often collected in electrostatic precipitators or air filters used in the fuel processing plants. Fly ash is invariably composed of various metals and silicates that are originally present in the carbonaceous fuel intermixed with fine carbon.

Vanadium has many useful applications such as in catalysts, high purity alloys, and rechargeable batteries. During conventional hydrometallurgical processing for the recovery of vanadium from fly ash, sulfuric acid is used to leach fly ash to obtain an impure vanadium pentoxide. The impure vanadium pentoxide is treated with a sodium carbonate solution under atmospheric conditions and then precipitated with ammonia as ammonium metavanadate. The disadvantage of the sulfuric acid leaching is that all other metals present are leached together with vanadium and this requires subsequent elaborate purification of the solution to obtain high purity vanadium.

An innovative hydrometallurgical process for recovering vanadium and zeolitic aluminosilicates from fly ash involves flotation and pressure leaching (Lakshmanan et al. 1990). In this process, ground fly ash is wetted with tap water for preconditioning. The slurry of wetted, ground fly ash is subsequently conditioned with Varsol and MIBC frother and then treated in four-stage flotation. The overflow of the flotation stages is treated for carbon recovery and the underflow consisting of carbon depleted vanadium containing fine fly ash is subjected to liquid–solid separation. The cake is repulped and fed to an autoclave for leaching with sodium hydroxide. The pregnant leach liquor is fed to a solvent extracting circuit after clarification. The vanadium is stripped with dilute sulfuric acid from the loaded organic solution. The purified vanadium-containing solution is treated with ammonium hydroxide to precipitate vanadium red cake and subsequently calcined to produce vanadium pentoxide. A flow sheet showing the production of vanadium pentoxide from fly ash is shown in Fig. 21.1.

21.2 Ammonium Sulfate By-product from Uranium Processing

Another example of innovative processing for by-product recovery is the production of ammonium sulfate by crystallization (Melis et al. 1983). This innovative process was developed at Ontario Research Foundation and described in a paper by Melis and others (Melis et al. 1983). The uniqueness of the process is that this was the first time that ammonium sulfate was produced as a by-product in a metallurgical industry. During the pilot plant run, blended ore was fed to the grinding circuit through a vibrating feeder. The ground feed was subjected to either pressure leaching with oxygen or, alternatively, atmospheric leaching with Caro's acid. Leached pulp was fed to a countercurrent decantation circuit, in which Percol 351 was used as a flocculant. The clarified pregnant leach solution was fed to the solvent extraction circuit. The pregnant strip solution was reacted with ammonia and air to precipitate yellowcake. Dual media sand filter was used to clarify barren strip liquor. The clarified barren strip liquor was evaporated to produce crystallized ammonium sulfate.

21.3 Production of By-product Metals

By-product metals can be divided into two groups, viz., easily recoverable and not easily recoverable. Easily recoverable metals include bismuth, cadmium, cobalt, selenium, and tellurium, whereas, not easily recoverable metals

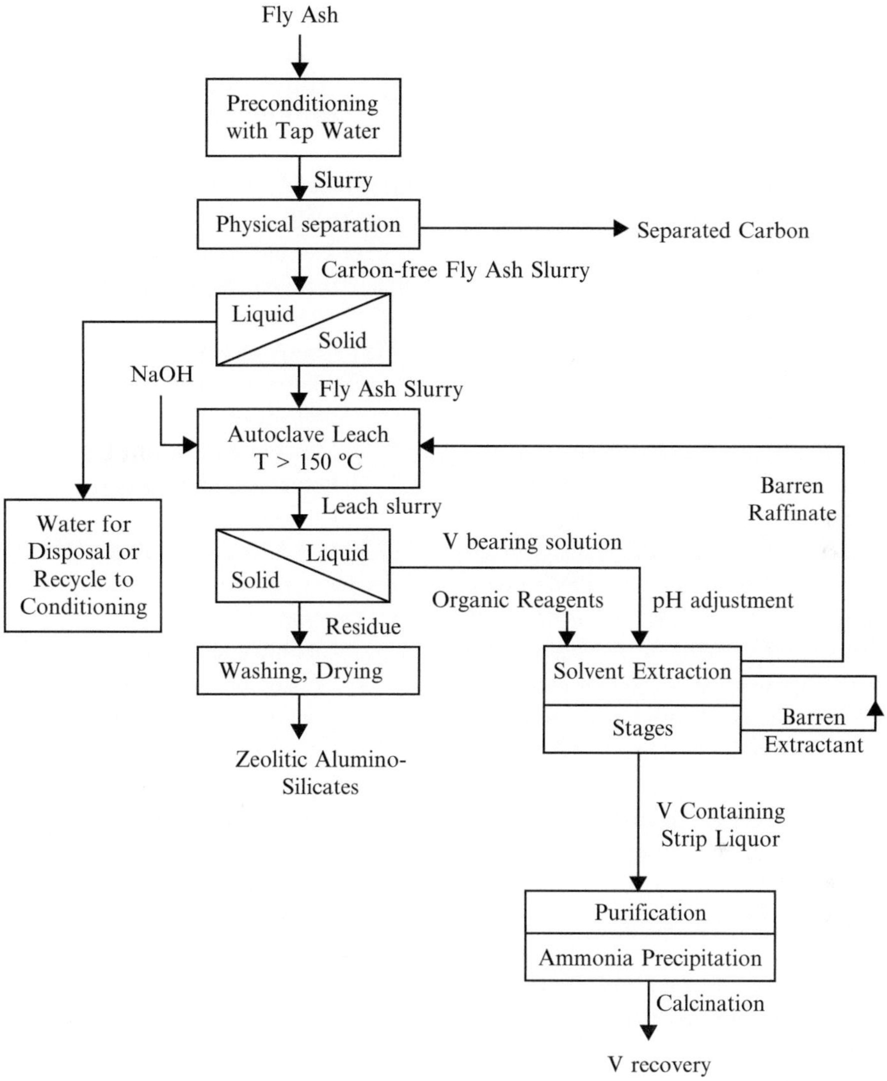

Fig. 21.1 Production of vanadium pentoxide from fly ash (Lakshmanan et al. 1990)

include: gallium, germanium, indium, precious metal group metals, rare earths, niobium, tantalum, and thallium. This characteristic of "not easily recoverable" requires recovery processes that are much different from the more classical ones used for the common metals.

The production of various by-product metals—easily and not easily recoverable—is well documented in literature (Hampel 1961) and will not be described in detail in this chapter. However, some general process steps used in the production of by-product metals are now mentioned. They are:

1. Electrolytic processes
   - Aqueous electrolysis (Cd, Cr, Co, Ga, In, Mn, and Tl)
   - Molten Salt Electrolysis (Al, Mg, and Na)
2. Thermite type reduction
   - Ti, Zr, Hf, Nb, Cb, Ta by Mg or Na reduction
   - Rare Earths by Ca reduction
3. Hydrogen reduction
   - Ge, Mo, W, Re, Co, PGM metals

4. Halide decomposition
    - Thermal decomposition of iodides of Ti, Hf, Cr, etc.
5. Inert Atmosphere Arc Melting
    - W, Mo, Ta, Cb by powder metallurgy
6. Electron Beam melting
    - Production of high melting point metals of high purity
7. Zone refining
    - Production of high purity metals by segregation of impurities

The last three process steps are not necessarily used in the production of by-product metals; they are used more so to produce metals of increasing purity—a very common requirement in the production of by-product metals. Most of the processes listed above are described in literature (Hampel 1961) and hence are not dealt with in this chapter. As an alternate, brief description of production of metals and metal compounds that are widely used in the photovoltaic cells will be briefly described in this chapter. Some other applications of by-product metals and their compounds, such as in the fuel cell industry and polymer industry will also be briefly mentioned here.

21.4 Metals and Materials Used in Photovoltaic Energy Industry

Since the discovery of the photoelectric effect in 1839 by Alexandre Becquerel, a French physicist, and the invention of the first practical photovoltaic cell in 1954 in Bell Laboratories, solar energy has come a long way in providing nonrenewable energy to society (Economist Technology Quarterly, March 10, 2007). In 2014, installed capacity of solar energy was 150 GW vs. 5 GW in 2005. Electricity production from solar energy grew 20 % per year for the last 15 years. There has been a continual effort to reduce the cost of electricity produced from solar cells. As of fourth quarter of 2012, thin-film Cd–Te solar panels produced electricity at $0.62/W (http://en.wikipedia.org/wiki/Solar_cell).

21.5 Materials Used in Solar Industry (http://en.wikipedia.org/wiki/Solar_cell)

Innovations in the production of various metals and materials have played a major role in the progress that has occurred in the solar industry. Some of them are listed below:

1. First generation cells.
    - Crystalline silicon.
        - Polysilicon.
        - Monocrystalline silicon.
        - Polycrystalline silicon.
        - Ribbon silicon.
        - Mono-like multisilicon.
2. Second generation cells.
    - Thin film cells.
        - Cadmium telluride.
        - Copper indium gallium arsenide.
        - Cadmium indium diselenide.
        - Silicon thin film—amorphous Si, protocrystalline Si, and nanocrystalline or microcrystalline silicon.
        - Gallium arsenide.
        - Germanium.
        - Other materials of interest to solar cells, viz., cuprous oxide, zinc phosphide, and tungsten diselenide.
3. Multijunction cells.
    - Multiple thin films—each photovoltaic cell grown on top of each other using metal organic vapor phase epitaxy. A typical triple-junction cell consists of semiconductors GaAs, Ge, and $GaInP_2$. GaAs-based multijunction devices are the most efficient solar cells to date.
    - Recently, Soitec Phoenix Labs, Phoenix, Arizona claims to have developed "multi-junction" photovoltaic cells that are certified to have a conversion efficiency of 46 %.

Considerable amount of innovative processes have been employed to develop and fabricate the above metals and intermetallic compounds that may be required as the actual active material.

Since very high purity is a requirement for these materials, the same innovative approach has been used to prepare very high purity elements such as gallium (4, 6, and 7 N), arsenic (4, 6, and 7 N), and germanium. Boron nitride and boron oxide—two intermetallic compounds—were developed to be used as containers for growing above multi-junction cells.

Continued research in the area of solar cells, specifically to improve conversion efficiency has resulted in very innovative approaches to produce various metals and materials of interest. To name a few that involves metals and metal compounds and/or complexes:

1. Quantum dot solar cells use CdS, CdSe, Sb_2S_3, PbS, PbSe, etc.
2. Incorporation of lanthanide-doped materials (Er^{+3}, Yb^{+3}, Ho^{+3}, or a combination); process known as upconversion.
3. Perovskite solar cells; has perovskite-structured material as active layer; it is a hybrid organic–inorganic tin or lead halide material.
4. Dye-sensitized solar cells (DSSC); Ruthenium metalorganic dye on a mesoporous layer of TiO_2.

As an example, some details of quantum dot cells are given here. In 2000, scientists at National Renewable Energy Laboratory (NREL) in Golden, Colorado predicted that Multiple Exciton Generation (MEG) using quantum dots could boost the efficiency of solar cells. In 2004, scientists at Los Alamos National Laboratory actually observed MEG in action in quantum confined materials. In 2011, Lusk and his colleagues at the Colorado School of Mines established that dot size determines what excitons are easily split up and small dots split up their favorite excitons the best. This idea led to the application of quantum dot ideas to various materials including silicon, with very encouraging results. One example was a solar cell made of lead selenide (PbSe) quantum dots (Lusk 2012).

Research work is in progress to develop materials for photovoltaic cells in the areas of nontraditional alloys, spontaneously ordered alloys, and interstitial semiconductors. The goal in the design and production of these materials is the ability to achieve band gap tuning and lattice matching simultaneously.

A growing demand for silicon from both the solar industry and the chip making industry—not to mention its cost—led to its shortage and the need to find new materials by the solar industry. One such innovation was the development of thin-film solar panels using (a) cadmium telluride and (b) copper indium gallium diselenide (CIGS).

Cadmium telluride is much cheaper than crystalline silicon, but its conversion efficiency is only 13.7 % nominal for CdTe and 18 % nominal for Si in 2014. In March 2014, First Solar has produced a thin-film PV module with 17 % efficiency, confirmed by NREL. Also, Sun Power reported the production of "Maxeon" cells (high energy density) that had a conversion efficiency of about 24 %. However, cadmium telluride's manufacturing cost is about $1.00/W cheaper than crystalline silicon. Copper indium gallium diselenide which has about 19 % efficiency promises to compete with silicon in cost and efficiency.

The exact process for the production and fabrication of materials used in the solar industry is somewhat proprietary due to the competitive nature of the business. As an example, a short summary of synthesis of a CdTe ingot is mentioned here. The processes are:

1. Liquid phase synthesis.
2. Vapor phase synthesis.
3. Solid phase synthesis.
4. Solution synthesis.
   - Vertical Bridgeman method.
   - Cold Travelling Heater Method.

The Cold travelling heater method can produce several kilograms of the CdTe compound in one run with very good stoichiometry. Details are outlined in a paper by R. Triboulet (Triboulet 2004). The process can also be extended to the production of CdZnTe. The main advantage of this process is that purity and homogeneity of the alloys can be maintained with a unique solidus composition.

As can be seen from the above descriptions, it is obvious that innovative processes have played

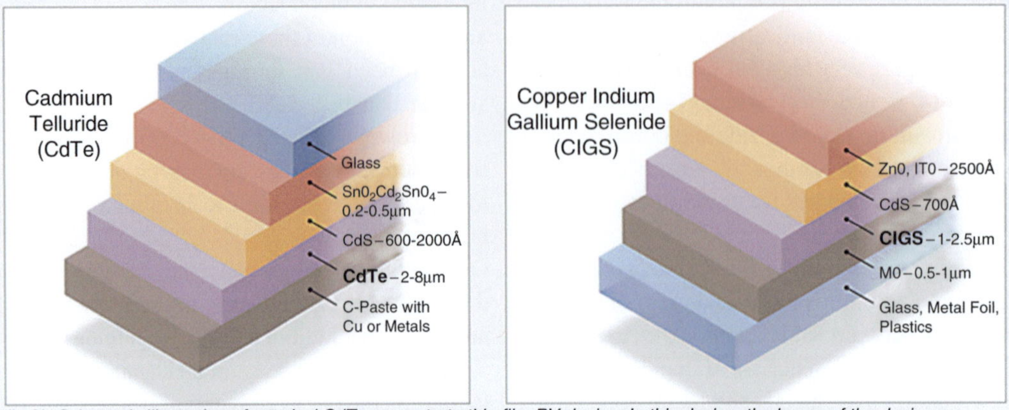

(Left). Schematic illustration of a typical CdTe superstrate thin-film PV device. In this design, the layers of the device are deposited onto a glass "superstrate" that allows sunlight to enter. The sunlight passes through the glass and priduces electrical current and voltage in the lower layers. The world-record NREL CdTe device is based on this structure and demonstrates a conversion efficiency of 17.0%.

(Right. Schematic illustration of a typical CIGS superstrate thin-film PV device. In this design, the layers of the device are deposited onto a glass, metal, or polymer substrate. Sunlight enters through the top layer of the device (hte transparent) conducting oxide) and produces electrical current and voltage in the lower layers. The world-record NREL CIGS device is based on this substrate structure and demonstrates a conversion efficiency of 20.0%.

Fig. 21.2 Typical thin-cell cadmium telluride and cadmium indium selenide photovoltaic cells (http://www.nrel.gov/pv/thinfilm.html)

a major role in the development of these intermetallic compounds and their application in the solar industry.

Typical thin-cell cadmium telluride and cadmium indium selenide photovoltaic cells are shown in Fig. 21.2. The schematic is self-explanatory and its working is well documented in literature. A cadmium chloride anneal of the CdTe/CdS solar cell changes the current transport mechanism favorably. However, use of $CdCl_2$ has adverse environmental issues. Recently, researchers at the University of Liverpool, England tested various other chlorides and established that use of $MgCl_2$ would yield comparable efficiency. This innovative approach of substituting $CdCl_2$ with $MgCl_2$ reduces health risks, production costs, and makes it easy for safe disposal (Bennington-Castro 2014).

21.6 Metals and Materials Used in Fuel Cell Industry

Fuel cells offer the possibility of zero-emissions electricity generation and increased energy security. In addition to high efficiency and low emissions, fuel cells are attractive for their modular and distributed nature and zero noise pollution. The fuel cell market is growing and is estimated that the stationary fuel cell market will reach about 50 GW by 2020.

Some of the most common fuel cells are listed below: (Haile 2003a)

1. Polymer Electrolyte Membrane—also known as proton exchange membrane (PEMFC).
2. Alkali Fuel Cell.
3. Phosphoric acid fuel cell.
4. Molten carbonate fuel cell.
5. Solid oxide fuel cell (SOFC).

A typical SOFC has (a) yttria-stabilized zirconia (YSZ) containing 8 mol% yttria as the electrolyte (YSZ), (b) a ceramic composite Ni+YSZ as anode, and (c) lanthanum strontium manganite (LSM—exact stoichiometry is not usually published). The PEMFC has (a) sulfonated polymers (Nafion™) as an electrolyte and (b) Pt nano particles are used as anodes and cathodes for hydrogen as a fuel. Solid electrolyte systems—PEMFC and SOFC—eliminate the need to contain corrosive liquids and are preferred over the other three fuel cells (Haile 2003b). Some details of the innovative approaches to achieve improvements in these cells are outlined here.

Efforts are in progress to reduce the operating temperatures of SOFC cells in the range 500–800 °C by using electrolytes with high ionic conductivity at around 600 °C. Three materials that have shown promise are doped ceria (CeO_2), doped lanthanum gallate ($LaGaO_3$), and barium zirconate ($BaZrO_3$). The first two are oxygen ion conductors and the third is a proton conductor. The ionic conductivity of CeO_2 is an order of magnitude greater than the conventional SOFC electrolyte, yttria-stabilized zirconia. Introduction of dopant ions, Sr and Mg onto the La and Ga sites, respectively, produces oxygen vacancy and hence the high oxygen ion conductivity. The proton transport in $BaZrO_3$ is achieved by doping it with yttrium on the Zr site to create oxygen vacancies (Haile 2003a).

Research is in progress to alleviate the problems associated with PEM fuel cells—problems not listed here—by developing "water-free" proton conducting membranes. Among the many materials under investigation, one inorganic compound being looked at California Institute of Technology (Haile 2003a) is cesium bi-sulfate ($CsHSO_4$). Other inorganic proton conductors are zirconyl phosphate and uranyl hydrogen phosphate.

Work is also in progress in the development of electrocatalysts with the objectives of increasing reaction kinetics, decreased Pt loadings, and decreasing anode susceptibility to CO poisoning. Much of the work in this area is focused on intermetallic alloys of Pt—specifically Pt–Ru alloy. Intermetallic compounds of Pt–Bi, viz., PtBi and $PtBi_2$ have also shown promise in tolerating CO poisoning of anodes (Haile 2003a).

In summary, innovations in the area of fuel cells run in parallel to the use of by-product metals and their compounds whose production involves innovative approaches. This is all the more critical when certain metal compounds with specific or unique properties need to be made and fabricated—again demanding innovative approaches. Additional breakthroughs are needed to transforming fuel cells from the technology of tomorrow to the technology of today (Haile 2003a).

21.7 Use of By-product Metal Compounds in Polymer Industry

There are a considerable number of by-product metals and/or their compounds used in nonmetallurgical industries. One such example is outlined below:

Germanium is a by-product from the processing of zinc and lead concentrates. Innovative processes were developed and are used for the production of germanium. Some details of the process steps are considered proprietary and not described in detail in literature. One germanium compound that is widely used is germanium dioxide (GeO_2)—either as a solid or as a solution. A typical application is in the production of polyethylene terephthalate (PET) catalysts. In this process, GeO_2 is dissolved in monoethylene glycol (MEG) and used as a PET polycondensation catalyst. It offers significant benefits in the quality of the final polymer, process benefits, and cost savings. The use of GeO_2 solution also results in higher clarity and more environmentally friendly polymer compared to an alternative catalyst, viz., Sb_2O_3 (Teck Resources Ltd, MSDS sheet).

Another compound $GeCl_4$—optic fiber grade—is used to produce optical fiber. Literature is filled with many such applications. The starting point of all such improvements is through innovative approaches to the production of such metal and metal compounds of by-product metals from the metallurgical industry (Teck Resources Ltd, MSDS sheet).

21.8 Summary

A few examples of innovative approaches to the use of by-product metals and their compounds are described in this subchapter.

Most recently, scientists at the University of Manchester, England are developing substances with electrical and optical properties that have been impossible to create before. It is called nano-scale Legos, building blocks on the atomic scale. This has been possible because of the innovative breakthrough creation of graphene (unfortunately not a metal!), a single sheet of carbon atoms.

Similar products that can be made as single sheet atoms are mica and hexagonal boron nitride. One such atomic scale Lego material that is being assembled is a five layer material, mostly graphene in combination with insulating boron nitride and semiconducting materials such as molybdenum sulfide and tungsten diselenide. As can be seen, this would not have been possible without the innovative approaches to the processing and production of single layer materials and metal compounds (Geim 2014).

In short, Nanoscale Legos represent something that has not been created before. Right now, the possibilities seem endless. There is no doubt these achievements would not have been possible but for the innovative processes that have been developed over a period of time by metallurgical and material engineers, from concept to commercialization.

References

Bennington-Castro, J. (2014). Salt-swap. *Scientific American, 311*(3), 20.

Geim, A. K. (2014). Atomic scale legos. *Scientific American, 311*(6), 50–51.

Haile, S. M. (2003a). Materials for fuel cells-review feature. *Materials Today*, 24–29.

Haile, S. M. (2003b). Fuel cell materials and components. *Acta Materialia, 51*(2003), 5981–6000.

Hampel, C. A. (Ed.). (1961). *Rare metals handbook* (2nd ed.). London: Chapman & Hall.

Lakshmanan, V. I., Melnbardis, D., Geisler, R. A., & McQueen, N. M. (1990, October 30). *Process for treatment of fly ash* (US Patent No. 4966761).

Lusk, M. (2012). Engineering a quantum leap in solar power. *Colorado School of Mines Magazine, 102*(2), 18–21.

Melis, L. A., Fraser, K. S., & Lakshmanan, V. I. (1983). The midwest uranium project—development of the milling process. *CIM Bulletin, 76*(852), 113–123.

Solar cell. Retrieved April 28, 2015, from http://en.wikipedia.org/wiki/Solar_cell

Teck Resources Ltd. Material Safety Data Sheet; (a) Germanium Dioxide Solution TGO, March 5, 2014; (b) Germanium Tetrachloride, 2012.

Triboulet, R. (2004). Fundamentals of CdTe synthesis. *Journal of Alloys and Compounds, 371*, 67–71.

General Reference

Bube, R. H. (2006). *Photovoltaic materials* (p. 281). London: Imperial College Press.

Conclusion

22

V.I. Lakshmanan, Raja Roy,
and Ram Ramachandran

In the last 100 years, the corporate Research and Development (R&D) has been the prime mover for generating new ideas. The active cooperation and alliance between scientists, engineers, and corporate management was an important factor in developing an innovation portfolio. However, execution of innovative ideas for a process is a very different problem. Successful implementation needs considerable financial support and cheer leading—in the form of a champion—from the corporate management.

Today, mining and metallurgical industry is facing many challenges in order to be sustainable. For industry to be sustainable, it needs to develop innovative technologies and improve process efficiency, energy, and water management. In addition, it needs to address societal needs.

The major seven issues facing the metallurgical industry are:

1. Lower ore grades with complex mineralogy.
2. Increasing cost of process water as a resource.
3. Increasing cost of energy as a resource.
4. Recycling of end-of-life products and recovery of value-added products.
5. Managing environmental issues to satisfy societal needs, viz., minimal environmental foot print.
6. Availability of finance for major mining and metallurgical projects.
7. Improvement in operational efficiencies and effectiveness.

Comments on these issues are outlined below:

In situ mining, underground concentration, and waste disposal should be evaluated to handle low-grade ores and mining in environmentally sensitive sites without contaminating ground water.

The importance of process water cannot be overemphasized; major mining companies in Chile and Australia are building reverse osmosis (RO) units on the coast to provide process water to their plants in the interior parts of the country. For example, BHP is building a 48 million imperial gallons per day (MIGD) RO unit—at a cost of US $103 million—in Chile to provide process water to Escondida mine—which is at a considerable distance from the sea and at a high altitude. Sea water is being evaluated as process water in mineral processing operations in South America. However, as a note of caution, corrosion and materials of construction would increase project costs.

V.I. Lakshmanan (✉) • R. Roy
Process Research Ortech Inc.,
2350 Sheridan Park Drive, Mississauga,
ON, Canada, L5K 2T4
e-mail: llakshmanan@processortech.com

R. Ramachandran
Consulting Engineer, 9650 E. Peregrine Place,
Scottsdale, AZ 85262, USA

V.I. Lakshmanan et al. (eds.), *Innovative Process Development in Metallurgical Industry*,
DOI 10.1007/978-3-319-21599-0_22

Continual efforts are in progress to lower energy usage and energy intensity in all mining and metallurgical processes. Innovative ideas will have to keep this objective always on the radar screen; in short, this has to be an ongoing process.

Recycling of intermediate products within a process eliminates or reduces waste; recycling of waste water and treated effluent decreases water usage; "Zero" liquid discharge should be the "mantra" for any metallurgical plant. Innovative processes for the treatment of end-of-life products to produce value-added metals should be developed—especially the metals that are in short supply.

Clean Air Act of 1963, Clean Water Act of 1977 and recent climate change initiatives have been primarily responsible for developing innovative technologies to meet the ever tightening environmental regulations. This is a monkey that will never get off the back of the metallurgical industry!! Constant vigil is needed by developing new innovative technologies to stay ahead of the environmental regulations.

There is competition for financial capital among many sectors of economy and to attract capital, mining and metallurgical industry needs to be competitive without compromising land, water, and air quality. This is a hard battle to fight but must be done.

Operational efficiencies/effectiveness need to be improved by state-of-the-art digital and real-time technology.

Since the mid-1980s, the demise of dynamic research organizations has been a major setback in the development of innovative processes and a cause for concern. Also, no innovation is possible without the availability of young researchers, scientists, and engineers trained in the field of chemical and process metallurgy. The metallurgical industry is facing a serious challenge in the dramatic decrease in the supply of young researchers, scientists, and engineers. Now is a good time to ask whether the industry can maintain its competitiveness and health with continued innovation based on R&D without the availability of young metallurgists (Sohn and Ramachandran 1998).

Skilled manpower from diverse fields such as research, engineering, finance, and resource management that includes all stakeholders across the value chain, be it in the industry, university, or government agency, need to come together to innovate in order to provide a good quality of life to the community and for the industry to sustain. There is a need for industry/academia consortia and working groups for innovative research and development in specific areas including technology, engineering and environment. A resurgence of interest in collaborative market-driven research between academia, industry, and government to spur further innovation in the industry is urgently needed.

It is the authors' fond hope that this book will be a wake-up call for the leaders of the mining and metallurgical industry to reactivate the concept of establishing a centrally located Corporate Research and Development group to develop innovative ideas. Some of these ideas would crystallize into processes that would benefit their company, the metallurgical industry, and society at large. The industry and government should also support universities financially to produce skilled manpower including researchers, scientists, and engineers needed for innovation to create wealth, jobs and good quality of life.

Reference

Sohn, H. Y., & Ramachandran, V. (1998, February 16–19). Advances in Sulfide Smelting Technology, R and D, Education. In J. A. Asteljoki, & R. L. Stephens (Eds.), *Proceedings of Sulfide Smelting-Current and Future Practices 1998, TMS Annual Meeting, San Antonio, TX* (pp. 3–37): Copyright 1998; quoted with permission from The Minerals, Metals and Materials Society.

Index

V.I. Lakshmanan et al. (eds.), *Innovative Process Development in Metallurgical Industry*,
DOI 10.1007/978-3-319-21599-0

MIX
Papier aus verantwortungsvollen Quellen
Paper from responsible sources
FSC® C105338

If you have any concerns about our products,
you can contact us on
ProductSafety@springernature.com

In case Publisher is established outside the EU,
the EU authorized representative is:
Springer Nature Customer Service Center GmbH
Europaplatz 3, 69115 Heidelberg, Germany

Printed by Libri Plureos GmbH
in Hamburg, Germany